THE CHRONOLOGY

OF

ANCIENT NATIONS

AN ENGLISH VERSION OF THE

ARABIC TEXT OF THE ATHÁR-UL-BÁKIYA OF ALBÍRÚNÍ,

OR

"VESTIGES OF THE PAST,"

COLLECTED AND REDUCED TO WRITING BY THE AUTHOR
IN A.H. 390—1, A.D. 1000.

TRANSLATED AND EDITED, WITH NOTES AND INDEX, BY

DR. C. EDWARD SACHAU
PROFESSOR IN THE ROYAL UNIVERSITY OF BERLIN.

LONDON:
PUBLISHED FOR THE ORIENTAL TRANSLATION FUND OF GREAT BRITAIN & IRELAND
BY WILLIAM H. ALLEN AND CO.
13 WATERLOO PLACE, PALL MALL.
PUBLISHERS TO THE INDIA OFFICE.

1879.

LONDON:
PRINTED BY W. H. ALLEN AND CO.

DEDICATED

TO THE MEMBERS

OF THE

COMMITTEE OF THE ORIENTAL TRANSLATION FUND (1878).

OSMOND DE BEAUVOIR PRIAULX.

EDWARD THOMAS, F.R.S.

JAMES FERGUSSON, F.R.S.

REINHOLD ROST, LL.D., SECRETARY.

AND TO THE MEMORY OF

THEODOR GOLDSTÜCKER, D.C.L.,

LATE PROFESSOR OF SANSKRIT IN THE UNIVERSITY OF LONDON.

PREFACE.

It was Sir Henry Rawlinson who first directed public attention to this work of Albîrûnî, in his celebrated article on Central Asia in the "Quarterly Review" for 1866, in which he gave some valuable information derived from his own manuscript copy, now the property of the British Museum. In offering the book, both in text and translation, to the learned world, I feel bound to premise that it is scarcely of a nature to attract the interest of the general reader. It appeals to minds trained in the schools of various sciences. Even competent scholars will find it no easy matter to follow our author through all the mazes of his elaborate scientific calculations. Containing, as it does, all the technical and historical details of the various systems for the computation of time, invented and used by the Persians, Sogdians, Chorasmians, Jews, Syrians, Ḥarránians, and Arabs, together with Greek traditions, it offers an equal interest to all those who study the antiquity and history of the Zoroastrian and Jewish, Christian and Muhammadan religions.*

The work of Albîrûnî has the character of a primary source. Oriental philologists are accustomed to see one book soon superseded by another, Barhebraeus by Ibn-al'athîr, Ibn-al'athîr by Al-Ṭabarî. Although it is likely enough

* By Christians, I understand the Melkite and Nestorian Churches, whilst the author does not seem to have known much more of the Jacobites than the name.

that on many subjects in this book we shall one day find better authenticated and more ancient information, I venture to say, that, as a whole, it will scarcely ever be superseded It is a standard work in Oriental literature, and has been recognised as such by the East itself, representing in its peculiar line the highest development of Oriental scholarship. Perhaps we shall one day find the literary sources themselves from which Albîrûnî derived his information, and shall be enabled to dispense with his extracts from them. But there are other chapters, *e.g.* those on the calendars of the ancient inhabitants of Central Asia, regarding which we shall, in all likelihood, never find any more ancient information, because the author had learned the subject from hearsay among a population which was then on the eve of dying out. As the first editor and translator of a book of this kind, I venture to claim the indulgence of the reader. Generations of scholars have toiled to carry the understanding of Herodotus to that point where it is now, and how much is wanting still! The work of generations will be required to do full justice to Albîrûnî. A classical philologist can edit a Greek text in a correct form, even though he may have no complete understanding of the subject-matter in all possible relations. Not so an Arabic philologist. The ambiguity of the Arabic writing—*proh dolor!*—is the reason why a manuscript expresses only three-quarters of the author's meaning, whilst the editor is compelled to supply the fourth quarter from his own knowledge and discernment. No number in any chronological table can be considered correct, as long as it is not proved by computation to be so, and even in the simplest historical narrative the editor and translator may most lamentably go astray in his interpretation, if there is something wrong with the method of his research.

I have boldly attacked the sometimes rather enigmatic style of the author, and if I have missed the mark, if the bewildering variety and multiplicity of the subject-matter have prevented me reaching the very bottom of every question, I must do what more or less every Oriental author does at the end of his work,—humbly ask the gentle reader to pardon my error and to correct it.

I. *The Author.*

The full name of the author is *Abû-Raiḥân Muḥammad b. 'Ahmad Albîrûnî.* He quotes himself as *Abû-Raiḥân* (*vide* p. 134, l. 29), and so he is generally called in Eastern literature, more rarely *Albîrûnî.*

The latter name means, literally, *extraneous*, being a derivative from the Persian بیرون which means *the outside* as a noun, and *outside* as a preposition. In our time the word is pronounced *Bîrûn* (or *Beeroon*), *e.g.* in Teheran, but the vowel of the first syllable is a *yâi-majhûl*, which means that in more ancient times it was pronounced *Bêrûn* (or *Bayroon*). This statement rests on the authority of the Persian lexicographers. That the name was pronounced in this way in Central Asia about the author's time, we learn from an indisputable statement regarding our author from the pen of Alsam'ânî, a philologist and biographer of high repute, who wrote only one hundred years after the author's death (*vide* Introduction to my edition of the text, p. xviii.).

He was a native of Khwârizm, or Chorasmia, the modern Khiva; to speak more accurately, a native either of a *suburb* (*Bêrûn*) of the capital of the country, both of which bore the same name *Khwârizm*, or of the *country-district* (also called *Bêrûn*) belonging to the capital.

Albîrûnî was born A.H. 362, 3. Dhû-alḥijja (A.D. 973,

4th September), and died A.H. 440, 2. Rajab (A.D. 1048, 11th December), aged seventy-five years.

The first part of his life he seems to have spent in Khwârizm, where he enjoyed the protection of the *House of Ma'mûn*, the rulers of the country. Originally vassals of the kings of Central Asia of the *House of Sâmân*, they became independent when the star of their masters began to sink, *i.e.* between A.H. 384–390. They were, however, not to play a great part in the history of the East, for so early as A.H. 407 their power was crushed by the great Maḥmûd of Ghazna, and their dominions annexed to his empire. Like Albîrûnî, other scholars also of high standing received protection and favours at the court of the Ma'mûnî princes.

The author is known to have lived some years also in Jurjân, or Hyrcania, on the southern shores of the Caspian sea, under the protection, and perhaps at the court, of Kâbûs ben Washmgîr Shams-alma'âlî, who ruled over Hyrcania and the adjoining countries at two different periods, A.H. 366–371 and 388–403. To this prince he has dedicated the present book, apparently about A.H. 390–391, (A.D. 1000).

During the years A.H. 400–407 he stayed again in his native country at the court of Ma'mûn b. Ma'mûn, as his friend and counsellor. He was a witness of the rebellion that broke out A.H. 407, of the murder of Ma'mûn, and of the conquest of the country by Maḥmûd of Ghazna, who, on returning, carried off him and other scholars to Afghanistan in the spring of A.H. 408.

Among his numerous works, we find mentioned a "Chronicle of Khwârizm," in which he probably had recorded all the traditions relating to the antiquity of his native country, and more especially the history of those events of which he had himself been a witness. This work seems to be lost. However, an extract of it has come down to us as the last part of the great chronicle of the royal house of Maḥmûd, composed by Albaihaḳî, the editior of which we owe to the industry and learning of the late W. H. Morley ("Bibliotheca Indica," Calcutta, 1862, pp. 834, &c.).

With A.H. 408 begins a new period in the author's life, when he enlarged the circle of his researches on mathematics, astronomy, geography, chronology, and natural sciences by his study of India, its geography and history, of the language and literature, manners and customs, of the Hindûs. It was the period when he gathered all those materials which he deposited towards the end of his life in his famous "Memoir on India."

After Albîrûnî had settled in Ghazna, he paid at least one more visit to his native country. He died, probably, at Ghazna. Whether he travelled much in other countries besides India, I have no means of proving. From the present book we can only infer that, besides his native country and Hyrcania, he also knew parts of Media, *e.g.* Rai (Rhagæ).

II. *His Work.*

Albîrûnî calls his work *Alâthâr Albâḳiya ʿan-il-Kûrûn Alkhâliya*, *i.e.* monuments or vestiges of generations of the past that have been preserved up to the author's time, meaning by *monuments* or *vestiges* the religious institutes of various nations and sects, founded in more ancient times, and, more or less, still practised and adhered to by the Oriental world about A.D. 1000.

With admirable industry the author gathers whatever traditions he can find on every single fact, he confronts them with each other, and inquires with critical acumen into the special merits or demerits of each single tradition. Mathematical accuracy is his last gauge, and wherever the nature of a tradition admits of such a gauge, he is sure to verify it by the help of careful mathematical calculation. To speak in general, there is much of the modern spirit and method of critical research in our author, and in this respect he is a phenomenon in the history of Eastern learning and literature. Authors of the first centuries of the Hijra sometimes betray a great deal of common sense and good method, sometimes also unmistakable traces of a marked individuality, whilst the later centuries are characterised by the very opposite. Then the author entirely disappears behind his book; all literary work sinks down to the level of imbecile compilation from good and bad sources; the understanding of the life and literature of the preceding centuries becomes rare and distorted. Common sense has gone never to return, and very seldom do we meet with a trace of scientific method or of the individuality of the author.

The fourth century is the turning-point in the history of the spirit of Islâm, and the establishment of the orthodox faith about 500 sealed the fate of independent research for ever. But for Alash'arî and Alghazzâlî the Arabs might have been a nation of Galileos, Keplers, and Newtons.

Originally I intended to give a complete *exposé* of the sources whence Albîrûnî has drawn his manifold information, but the material hitherto available for researches on the literary history of the east is still so scanty that I had to desist from my plan. This applies in particular to the east of the Khalifate, to Khurâsân. We are comparatively well informed regarding the literature of Mesopotamia, Syria, Egypt, and the farther west of Islâm, whilst we have very little information regarding the scientific and literary life east of Bagdad as it developed itself during the first three centuries of Abbaside rule, under the protection of the imperial governors and the later independent princes, *e.g.* the House of Sâmân.

It is to be hoped that Central Asia and Afghanistan, when once in the grasp of European influence, will yield us rich collections of valuable literary monuments. Hitherto manuscripts coming from those countries are seldom met with in the great libraries of Europe.

As for the *written* material which the author had at his disposal, he quotes many a book of which I elsewhere can scarcely find a trace. All the books, *e.g.* on Persian and Zoroastrian history and traditions, composed in early times, not only by Zoroastrians but also by Muslims, converts from the Zoroastrian creed, are altogether unknown in Europe; and it seems very probable that the bigoted people of later times have spared very little of this kind of literature, which to them had the intolerable smell of filthy idolatry.

As regards Persian history, Albîrûnî had an excellent predecessor in Alisfahânî, whom he follows frequently, and whom he was not able to surpass in many points.

From oral information Albirûnî seems to have learned all he knows of the chronology and calendar of the Zoroastrian populations of Persia, of his native country, and of Sogdiana (or Bukhârâ). In his time the majority of the country-people still adhered to Ahuramazda, and in most towns there must still have been Zoroastrian communities, so that Albîrûnî did not lack the opportunity for studying the manners and institutes of the then existing followers of Zoroaster. Unfortunately, the Zoroastrian creed had lost its clerical and political unity and constitution. The people practised their customs as they had seen their parents do, but they had no longer a correct understanding of their origin and meaning. Certainly a Mobedan-Mobed of the time of Ardashîr Bâbekân would have been able to give a more accurate and complete account of Zoroastrian life and religion; but still we must be thankful to Albîrûnî for his having preserved to posterity the festal calendars as used by Zoroastrians of his time when their religion was on the eve of dying out.

To *oral* information I ascribe also the author's admirable knowledge of the Jewish calendar. Jewish scholars will be able to say whether his informants were Ananites (Karaites) or Rabbanites. My critics do not seem to have noticed that Albîrûnî, a *Muslim*, is the first of all the scholars we know who has composed a scientific system of the Jewish chronology. He is much anterior to Moses Maimonides, also to Abraham bar Chiyya, being a contemporary of R. Sherîrâ and Hâî Gâôn, who seem to play a prominent part in the history of Jewish chronology.

With Nestorian Christians he must have been acquainted, as he speaks of the Nestorian communities of his native country. His report of the Melkite feasts, &c., may have been communicated to him by Nestorian priests from Syriac sources.

Albîrûnî wrote both in Arabic and Persian, as he has edited his "Kitâb-altafhîm" in both languages. There is a possibility of his having had a smattering of Hebrew and Syriac (*vide* pp. 18, 19), but of Greek he seems to have been ignorant, and whatever he relates on the authority of Greek authors—Ptolemy, Galen, Eusebius, &c.—must have been communicated to him by the ordinary channel of Syriac-Arabic translation. His study of Sanskrit falls into the latter half of his life.

From occasional notes in the book a description of the

author's character may be gleaned. He seems to have been a truth-loving man, attacking all kinds of shams with bitter sarcasms. He was not without a humoristic vein, and his occasional ironical remarks offer a curious contrast to the pervading earnestness of the tenor of his speech. As a Muslim he inclined towards the Shî'a, but he was not a bigoted Muslim. He betrays a strong aversion to the Arabs, the destroyers of Sasanian glory, and a marked predilection for all that is of Persian or Eranian nationality. Muslim orthodoxy had not yet become so powerful as to imperil the life of a man, be he Muslim or not, who would study other religions and publicly declare in favour of them. Daḳîḳî, a poet not long anterior to Albîrûnî, a favourite of the Muslim house of Sâmân, was allowed to sing—

> "Of all that is good and bad in the world,
> Daḳîḳî has chosen four things to himself:
> A woman's lips as red as rubies, the melody of the lute,
> The blood-coloured wine, and *the religion of Zoroaster.*"

Not long afterwards, at the court of the great Maḥmûd of Ghazna, these verses would probably have proved fatal to their author.

Habent sua fata libelli, and I may add, the fate of this book, from the time of its composition till the time when I began to study it, has not been a fortunate one. Only a few were able to understand it, few had an interest in having it copied.

In the form in which I offer the book to the reader it is not complete. Many most essential parts, both large and small, are missing, *e.g.* the chapter on Zoroaster, a most deplorable loss, arising probably from Muslim bigotry. However, I should think it does not require an apology from me to have edited the book in this mutilated form in which I have found it in the manuscripts. Should the favour of time bring to light one day a complete copy, I shall be happy if circumstances will allow me at once to edit the hitherto missing parts in text and translation.

The basis of my edition consists of two manuscripts of the seventeenth and one of the nineteenth century, all full of faults, and—what is worse!—agreeing with each other almost in every particular. In fact, all three copies represent one and the same original. Fortunately a chronological work offers this advantage, that in many cases mathematical examination enables the editor to correct the blunders of the tradition, *e.g.* in the numerous tables.

My notes are in the first place intended to give the calculations on which the tables rest. Besides, they contain contributions to the explications of certain difficult passages, short information on points of literary history, and, lastly, a few remarks on the text and corrections.

For all other introductory questions I refer the reader to the German preface to my edition of the text.

In offering my translation to the English reader, I desire to thank my friend, the Rev. Robert Gwynne, Vicar of St. Mary's, Soho, London, who not only corrected the whole manuscript, but also read the proof-sheets of the entire book.

EDWARD SACHAU.

Berlin, 24th May, 1879.

CONTENTS.

ERRATA.

p. 383, *delete the first two lines and insert*—

Sawîr. Perhaps identical with the Σαβειροί of Byzantine authors, from whom *Siberia* derived its name.

p. 451, col. 2, last line, *delete* China, 266, 10.

p. 452, col. 1, line 1, *after* Chinese *insert* 266, 10.

p. 460, after line 42 *insert*—

Poison-mountain (rarefied atmosphere), 263, 17.

p. 462, col. 2, after line 18 *insert*—

Tibetans, 263, 17.

DIRECTION TO THE BINDER.

Table of Kebi'ôth . . . . to face p. 154.

PREFACE.

IN THE NAME OF GOD, THE COMPASSIONATE, THE MERCIFUL.

Praise be to God who is high above all things (*lit.* those which are unlike, and those which are like to each other), and blessing be on Muḥammad, the elected, the best of all created beings, and on his family, the guides of righteousness and truth.

One of the exquisite plans in God's management of the affairs of his creation, one of the glorious benefits which he has bestowed upon the entirety of his creatures, is that categorical decree of his, not to leave in his world any period without a just guide, whom he constitutes as a protector for his creatures, with whom to take refuge in unfortunate and sorrowful cases and accidents, and upon whom to devolve their affairs, when they seem indissolubly perplexed, so that the order of the world should rest upon—and its existence be supported by—his genius. And this decree (that the affairs of mankind should be governed by a prophet) has been settled upon them as a religious duty, and has been linked together with the obedience towards God, and the obedience towards his Prophet, through which alone a reward in future life may be obtained—in accordance with the word of him, who is the truth and justice—and his word is judgment and decree, "O ye believers, obey God, and obey the prophets, and those among yourselves who are invested with the command." (Sûra iv. 62.)

Therefore, thanks be to God for those blessings, which he has bestowed upon his servants, by exalting our master, the commander, the prince, the glorious and victorious, the benefactor, Shams-alma'âlî, may God give him a long life, and give duration to his power and majesty, preserve through the course of time his excellence and his splendour, protect his whole house (*lit.* the areas inside and outside his house), prostrate all those who envy him, and all his enemies, (by exalting him) as a guide, who justly rules over his creatures, who furthers religion and truth, who fights for the altar and the hearth of the Muslims, and who protects their country against the mischief of evil-doers. And God has supported him by giving him a character similar to that with which he has blessed his Prophet, the bearer of his revelation; for he, whose name be praised, has said: "To thee has been given a high character." (Sûra lxviii. 4.) How wonderfully has he, whose name is to be exalted and extolled, combined with the glory of his noble extraction the graces of his generous character, with his valiant soul all laudable qualities, such as piety and righteousness, carefulness in defending and observing the rites of religion, justice and equity, humility and beneficence, firmness and determination, liberality and gentleness, the talent for ruling and governing, for managing and deciding, and other qualities, which no fancy could comprehend, and no human being enumerate! And how should a man wonder at this, it being undeniable that God has the power to combine the whole world in one individual (*i.e.* to create a microcosmos)! Therefore, may God permit the Muslims still for a long period to enjoy the kindness of his intentions, the ingenuity of his plans, and his evidently merciful and pitiful mind, with which he cares for them! May they from day to day derive more benefits from the perpetual shade of his majesty, to which they are accustomed! And may God assist by his kindness and mercy, high and low, to fulfil the works of obedience towards God, which are imposed upon them!

p. 4. **Dedication.—The Author's Method.**—A learned man once asked me regarding the eras used by different nations, and regarding the difference of their roots, *i.e.* the epochs where they begin, and of their branches, *i.e.* the months and years, on which they are based; further regarding the causes which led to such difference, and the famous festivals and commemoration-days for certain times and events, and regarding whatever else one nation practises differently from another. He urged me to give an explanation, the clearest possible, of all this, so as to be easily intelligible to the mind of the reader, and to free him from the necessity of wading through widely scattered books, and of consulting their authors. Now I was quite aware that this was a task difficult to handle, an object not easily to be attained or managed by anyone, who wants to treat it as a matter of logical sequence, regarding which the mind of the student is not agitated by doubt. However, from the majesty of our master, the prince, the glorious and victorious, the benefactor, Shams-alma'âlî—may God make his power to endure!—I derived strength in exerting my capabilities, and trying to do my utmost in order to explain the whole subject on the basis of that information which I have gathered either as an ear- or eye-witness, or by cogitation and study. Besides, I was encouraged by that robe of blessed service, in which I have dressed myself, to compose such an explanation for him, who occupies a high throne, that he may see herein a new sign of my service, and that thereby I may obtain the garments of such a glory, the memory and splendour of which will last as my heirloom in posterity through the flood of ages and generations.⌋ If, therefore, he—whose noble mind may God preserve!—will favour his servant by overlooking his audacity, and accepting his excuses, he follows the right idea, if it pleases God. And now I commence and say:

The best and nearest way leading to that, regarding which I have been asked for information, is the knowledge of the history and tradition of former nations and generations, because the greatest part of it consists of matters, which have come down from them, and of remains of their customs and institutes. And this object cannot be obtained by way of ratiocination with philosophical notions, or of inductions based upon the observations of our senses, but solely by adopting the information of those who have a written tradition, and of the members of the different religions, of the adherents of the different doctrines and religious sects, by whom the institutes in question are used, and by making their opinions a basis, on which afterwards to build up a system; besides, we must compare their traditions and opinions among themselves, when we try to establish our system. But ere that we must clear our mind from all those accidental circumstances which deprave most men, from all causes which are liable to make people blind against the truth, *e.g.* inveterate custom, party-spirit, rivalry, being addicted to one's passions, the desire to gain influence, etc. For that which I have mentioned, is the nearest way you could take, that leads to the true end, and the most efficient help towards removing all the clouds of uncertainty and doubt, which beset the subject. It is impossible in any other way to reach the same purpose, notwithstanding the greatest care and exertion. On the other hand, we confess that it is by no means easy to act upon that principle and that p. 5.
method, which we have laid down, that on the contrary from its recondite nature, and its difficulty, it might seem to be almost unattainable—on account of the numerous lies which are mixed up with all historical records and traditions. And those lies do not all on the face of it appear to be impossibilities, so that they might be easily distinguished and eliminated. However, that which is within the limits of possibility, has been treated as true, as long as other evidence did not prove it to be false. For we witness sometimes, and others have witnessed before us, physical appearances, which we should simply declare to be impossible, if something similar were related from a far remote time. Now the life of man is not sufficient to learn thoroughly the traditions of one of the many nations. How, therefore, could he learn the traditions of all of them? That is impossible.

The matter standing thus, it is our duty to proceed from what is near to the more distant, from what is known to that which is less known, to gather the traditions from those who have reported them, to correct them as much as possible, and to leave the rest as it is, in order to make our work help him, who seeks truth and loves wisdom, in making independent researches on other subjects, and guide him to find out that which was denied to us, whilst we were working at this subject, by the will of God, and with his help.

In conformity with our plan, we must proceed to explain the nature of day and night, of their totality, *i.e.* the astronomical day, and assumed beginning. For day and night are to the months, years, and eras, what one is for the numerals, of which they are composed, and into which they are resolved. By an accurate knowledge of day and night, the progress towards learning that which is composed of them and built upon them, becomes easy.

CHAPTER I.

ON THE NATURE OF DAY AND NIGHT, OF THEIR TOTALITY AND OF THEIR BEGINNINGS.

I SAY: Day and night (*i.e.* *νυχθήμερον*) are one revolution of the sun in the rotation of the universe, starting from and returning to a circle, which has been assumed as the beginning of this same Nychthemeron, whichsoever circle it may be, it being determined by general consent. This *circle* is a "great" circle; for each great circle is dynamically an horizon. By "*dynamically*" (*τῇ δυνάμει*), I mean that it (this circle) may be the horizon of any place on the earth. By the "*rotation of the universe*," I mean the motion of the celestial sphere, and of all that is in it, which we observe going round on its two poles from east to west.

The Setting of the Sun as the beginning of the Day.—Now, the Arabs assumed as the beginning of their Nychthemeron the point where the setting sun intersects the circle of the horizon. Therefore their Nychthemeron extends from the moment when the sun disappears from the horizon till his disappearance on the following day. They were induced to adopt this system by the fact that their months are based upon the course of the moon, derived from her various motions, and that the beginnings of the months were fixed, not by calculation, but by the appearance of the new moons. Now, full moon, the appearance of which is, with them, the beginning of the month, becomes visible towards sunset. Therefore their night preceded their day; and, therefore, it is their custom to let the nights precede the days, when they p. 6.
mention them in connection with the names of the seven days of the week.

Those who herein agree with them plead for this system, saying that darkness in the order (of the creation) precedes light, and that light suddenly came forth when darkness existed already; that, therefore, that which was anterior in existence is the most suitable to be adopted as the beginning. And, therefore, they considered absence of motion as superior to motion, comparing rest and tranquillity with darkness, and because of the fact that motion is always produced by some want and necessity; that weariness follows upon the necessity; that, therefore, weariness is the consequence of motion. Lastly, because rest (the absence of motion), when remaining in the elements for a time, does not produce decay; whilst motion, when remaining in the elements and taking hold of them, produces corruption. As instances of this they adduce earthquakes, storms, waves, &c.

The Rising of the Sun as the beginning of the Day.—As to the other nations, the Greeks and Romans, and those who follow with them the like theory, they have agreed among themselves that the Nychthemeron should be reckoned from the moment when the sun rises above the eastern horizon till the same moment of the following day, as their months are derived by calculation, and do not depend upon the phases of the moon or any other star, and as the months begin with the beginning of the day. Therefore, with them, the day precedes the night; and, in favour of this view, they argue that light is an *Ens*, whilst darkness is a *Non-ens*. Those who think that light was anterior in existence to darkness consider motion as superior to rest (the absence of motion), because motion is an *Ens*, not a *Non-ens*—is life, not death. They meet the arguments of their opponents with similar ones, saying, *e.g.* that heaven is something more excellent than the earth; that a working man and a young man are the healthiest; that running water does not, like standing water, become putrid.

Noon or Midnight as the beginning of the Day.—The greater part and the most eminent of the learned men among astronomers reckon the Nychthemeron from the moment when the sun arrives on the plane of the meridian till the same moment of the following day. This is an intermediate view. Therefore their Nychthemera begin from the visible half of the plane of the meridian. Upon this system they have built their calculation in the astronomical tables (the Canons), and have thereby derived the places of the stars, along with their equal motions and their corrected places, in the almanacks (*lit.* year-books). Other astronomers prefer the invisible half of the plane of the meridian, and begin, therefore, their day at midnight, as *e.g.* the author of the Canon (Zîj) of Shahriyârân Shâh. This does not alter the case, as both methods are based upon the same principle.

People were induced to prefer the meridian to the horizon by many circumstances. One was, that they had discovered that the Nychthemera vary, and are not always of the same length; a variation which, during the eclipses, is clearly apparent even to the senses.

The reason of this variation is the fact that the course of the sun in the ecliptic varies, it being accelerated one time and retarded another; and that the single sections of the ecliptic cross the circles (the horizons)

at a different rate of velocity. Therefore, in order to remove that variation which attaches to the Nychthemera, they wanted some kind of equation; and the equation of the Nychthemera by means of the rising of the ecliptic above the meridian is constant and regular everywhere on the earth, because this circle is one of the horizons of the p. 7.
globe which form a right angle (with the meridian); and because its conditions and qualities remain the same in every part of the earth. This quality they did not find in the horizontal circles, for they vary for each place; and every latitude has a particular horizon of its own, different from that of any other place, and because the single sections of the ecliptic cross the horizons at a different rate of velocity. To use the horizons (for the equation of the Nychthemera) is a proceeding both imperfect and intricate.

Another reason why they preferred the meridian to the horizon is this, that the distances between the meridians of different places correspond to the distances of their meridians on the equator and the parallel circles; whilst the distances between the horizontal circles are the same with the addition of their northern and southern declination. An accurate description of everything connected with stars and their places is not possible, except by means of that direction which depends upon the meridian. This direction is called "longitude," which has nothing in common with the other direction, which depends upon the horizon, and is called "latitude."

Therefore they have chosen that circle which might serve as a regular and constant basis of their calculations, and have not used others; although, if they had wished to use the horizons, it would have been possible, and would have led them to the same results as the meridian, but only after a long and roundabout process. And it is the greatest mistake possible purposely to deviate from the direct route in order to go by a long roundabout.

Day, Night, and the Duration of the Day of Fast.—This is the general definition of the day which we give, the night being included. Now, if we proceed to divide and to distinguish, we have to state that the words "*Yaum*" (day) in its restricted signification, and "*Nahâr*" (day), mean the same, viz., the time from the rising of the body of the sun till its setting. On the other hand, *night* means the time from the setting of the body of the sun till its rising. Thus these two terms are used among all nations by general consent, nobody disputing their meanings, except one Muslim lawyer, who has defined the beginning of the day to be the rise of dawn, and its end to be the setting of the sun, because he presumed that the day and the duration of fasting were identical. For this view of his he argues from the following word of God (Sûra ii. 183): "Eat and drink till you can distinguish a white thread from a black thread at the light of *dawn*. Thereupon fast the entire day till the *night*." Now, he has maintained that these two terms (*dawn* and *night*) are the two limits of the day (*beginning* and *end*). Between this view, however, and this verse of the Coran there is not the slightest connection whatsoever. For if the beginning of fasting was identical with the beginning of the day, his (God's) definition of something that is quite evident and well known to everybody, in such terms, would be like a pains-taking attempt to explain something void of sense. Likewise he has not defined the end of day and the beginning of night in similar terms, because this is generally known among all mankind. God orders that fasting should commence *at the rise of dawn*; but the end of fasting he does not describe in a similar way, but simply says that it should end at "*night*," because everybody knows that this means the time when the globe of the sun disappears. Hence it is evident that God, by the words of the first sentence (*i.e.* eat and drink till you can distinguish a white thread from a black thread at the light of dawn), does not mean the beginning of day.

A further proof of the correctness of our interpretation is the word of God (Sûra ii. 183): "It has been declared as lawful to you during the
p. 8. night of fasting to have intercourse (*lit.* to speak obscene things) with your wives," &c., to the passage, "Thereupon fast the entire day till the night." Thereby he extends the right of having intercourse with one's wife, and of eating and drinking, over a certain limited time, not over the entire night. Likewise it had been forbidden to Muslims, before this verse had been revealed, to eat and drink after night-prayer (the time when the darkness of night commences). And still people did not reckon their fasting by days and parts of the night, but simply by days (although the time of fasting was much longer than the day).

Now, if people say that God, in this verse (Sûra ii. 183), wanted to teach mankind the beginning of the day, it would necessarily follow that before that moment they were ignorant of the beginning of day and night, which is simply absurd.

Now, if people say the *legal* day is different from the *natural* day, this is nothing but a difference in words, and the calling something by a name, which, according to the usage of the language, means something else. And, besides, it must be considered that there is not the slightest mention in the verse of the day and of its beginning. We keep, however, aloof from pertinacious disputation on this subject, and we are willing to agree with our opponents as to the expressions if they will agree with us regarding the subject-matter.

And how could we believe a thing the contrary of which is evident to our senses? For evening-twilight in the west corresponds to morning-dawn in the east; both arise from the same cause, and are of the same nature. If, therefore, the rise of morning-dawn were the beginning of the day, the disappearance of evening-twilight would be its end. And actually some Shiites have been compelled to adopt such a doctrine.

Let us take it for granted that those who do not agree with us regarding that which we have previously explained, agree with us as to the fact that twice a year night and day are equal—once in spring and once in autumn. Further, that he thinks, like us, that we have the longest day when the sun stands nearest to the north pole; the shortest day when the sun is at the greatest distance from the north pole; that the shortest summer night is equal to the shortest winter day; and that the same meaning is expressed by the two verses of the Coran: "God makes night enter into day, and he makes day enter into night" (Sûra xxxv. 14), and "He wraps night around day, and he wraps day around night" (Sûra xxxix. 7). Now, if they do not know this, or pretend not to know it, at all events they cannot help admitting that the first half of the day is six hours long, and likewise the latter half. Against this they cannot pretend to be blind, because of the well known and well authenticated tradition which relates to the prerogatives of those who hasten to the mosque on a Friday, and which shows that their wages are the highest, although their time of work in the six hours from the beginning of the day till the time of the decline of the sun is the shortest. This is to be understood of the *Horæ temporales obliquæ* (ὧραι καιρικαί), not of the *Horæ rectæ*, which are also called *æquinoctiales* (ὧραι ἰσημεριναί).

Now, if we should comply with their wish, and acknowledge their assertions as truth, we should have to believe that an equinox takes place when the sun moves on either side of the winter solstice (*i.e.* near to the point of the winter-solstice either arriving there or leaving it); that this takes place only in some parts of the earth to the exclusion of others; that the winter night is not equal to the summer day, and that noon is not then when the sun reaches the midst between his rising and p. 9.
setting points. Whilst just the contrary of these necessary inferences from their theory is the conclusion generally accepted even by those who have only a slight insight into the matter. That, however, similar absurdities must follow out of their reasoning he only will thoroughly comprehend who is to some degree acquainted with the motions of the (celestial) globes.

If somebody will stick to what people say at dawn-rise, "*morning has come, night has gone;*" what is he to think of what they say when the sun is near setting, and becomes yellow—"evening has come, day has gone, night has come?" Such expressions merely indicate the approaching, the advancing, and the receding of the precise time in which people just happen to be. These phrases are to be explained as metaphors and metonymics. They are allowed in the usage of the language, *cf. e.g.* the word of God (Sûra xvi. 1): "The order of God has come; therefore do not *hurry* it."

Another argument in favour of our view is the following saying, which is attributed to the Prophet, to whom and to whose family may God be merciful: "The prayer of the day is silent." And the fact that people call the noon-prayer the "first" prayer, because it is the first of the two daily prayers; whilst they call the afternoon-prayer the "middle" prayer, because it is in the midst, between the first of the two daily prayers and the first of the prayers of the night.

My only object in all I have discussed in this place is to refute the opinion of those who think that those things which are necessary for certain philosophical or physical causes prove the contrary of that which is indicated by the Coran, and who try to support their opinion by the doctrine of one of the lawyers and commentators of the Coran. God helps to the right insight!

CHAPTER II.

ON THE NATURE OF THAT WHICH IS COMPOSED OF DAYS, VIZ., MONTHS AND YEARS.

I SAY: Year means one revolution of the sun in the ecliptic, moving in a direction opposite to that of the universal motion, and returning to the same point which has been assumed as the starting-point of *his* motion, whichsoever point this may be. In this way the sun includes in his course the four seasons, spring, summer, autumn, winter, and their four different natures; and returns always to the point whence he commenced. According to Ptolemy these revolutions are equal, because he did not find that the apogee of the sun moves; whilst they are unequal according to the authors of Sindhind and the modern astronomers, because their observations led them to think that the apogee of the sun moves. In each case, however, whether they be equal or different, these revolutions include the four seasons and their natures.

As to the length of such a revolution in days and fractions of a day, the results of the astronomical observations do not agree, but differ considerably. According to some observations it is larger; according to others less. However, in a short space of time this difference scarcely becomes perceptible; but in the long run of time, when this difference is being redoubled and multiplied many times, and is then summed up into a whole, a very great error becomes clearly manifest, on account of which the sages have strongly recommended us to continue making observations, and to guard against errors which possibly might have entered into them.

The difference of the observations regarding the length of one annual p. 10.
revolution of the sun does not arise from this cause, that people do not know how properly to institute such observations, and to gain thereby an accurate knowledge of the real state of the thing; but from this cause, that it is impossible to fix the parts of the greatest circle by

means of the parts of the smallest circle. I refer to the smallness of the instruments of observations in comparison with the vastness of the bodies which are to be observed. On this subject I have enlarged in my book, called *Kitâb-alistishhâd bikhtilâf-al'arṣâd.*

During this time, *i.e.* during one revolution of the sun in the ecliptic, the moon completes a little less than $12\frac{3}{8}$ revolutions, and has 12 lunations. This space of time, *i.e.* the 12 revolutions of the moon in the ecliptic, is, technically, the lunar year, in which the fraction (beyond the 12 revolutions), which is nearly 11 days, is disregarded. The same fact, further, is the reason why the ecliptic was divided into 12 equal parts, as I have explained in my book on the investigation of rays and lights; the same which I had the honour to present to His Highness. May God increase his majesty!

In consequence, people distinguish two kinds of years—the Solar year and the Lunar year. They have not used other stars for the purpose of deriving years from them, because their motions are comparatively hidden, and can hardly ever be found out by eyesight; but only by astronomical observations and experiments. Further (they used only sun and moon for this purpose), because the changes of the particles of the elements and their mutual metamorphoses, as far as time and the state of the air, plants and animals, etc., are concerned, depend entirely upon the motions of these two celestial bodies, because they are the greatest of all, and because they excel the other stars by their light and appearance; and because they resemble each other. Afterwards people derived from these two kinds of years other years.

The Solar Year.—According to the statement of Theon, in his Canon, the people of Constantinople, and of Alexandria, and the other Greeks, the Syrians and Chaldæans, the Egyptians of our time, and those who have adopted the year of Almu'ta-ḍid-billâh, all use the solar year, which consists of nearly $365\frac{1}{4}$ days. They reckon their year as 365 days, and add the quarters of a day in every fourth year as one complete day, when it has summed up thereto. This year they call an intercalary year, because the quarters are intercalated therein. The ancient Egyptians followed the same practice, but with this difference, that they neglected the quarters of a day till they had summed up to the number of days of one complete year, which took place in 1,460 years; then they intercalated one year, and agreed with the people of Alexandria and Constantinople as to the beginning of the year. So Theon Alexandrinus relates.

The Persians followed the same rule as long as their empire lasted;
p. 11. but they treated it differently. For they reckoned their year as 365 days, and neglected the following fractions until the day-quarters had summed up in the course of 120 years to the number of days of one complete month, and until the fifth parts of an hour, which, according to their opinion, follow the fourth parts of a day (*i.e.* they give the

solar year the length of $365\frac{1}{4}$ days and $\frac{1}{8}$ hour), had summed up to one day; then they added the complete month to the year in each 116th year. This was done for a reason which I shall explain hereafter.

The example of the Persians was followed by the ancient inhabitants of Khwârizm and Sogdiana, and by all who had the same religion as the Persians, who were subject to them, and were considered as their kinsmen, during the time when their empire flourished.

I have heard that the Pêshdâdian kings of the Persians, those who ruled over the entire world, reckoned the year as 360 days, and each month as 30 days, without any addition and subtraction; that they intercalated one month in every sixth year, which they called "intercalary month," and two months in every 120th year; the one on account of the five days (the Epagomenæ), the other on account of the quarter of a day; that they held this year in high honour, and called it the "blessed year," and that in it they occupied themselves with the affairs of divine worship and matters of public interest.

The character of the system of the ancient Egyptians, according to what the Almagest relates regarding the years on which its own system of computation was based, and of the systems of the Persians in Islâm, and the people of Khwârizm and Sogdiana, is their aversion to the fractions, *i.e.* the $\frac{1}{4}$ day and what follows it, and their neglecting them altogether.

The Luni-Solar Year.—The Hebrews, Jews, and all the Israelites, the Sâbians, and Ḥarrânians, used an intermediate system. They derived their year from the revolution of the sun, and its months from the revolution of the moon—with this view, that their feast and fast days might be regulated by lunar computation, and at the same time keep their places within the year. Therefore they intercalated 7 months in 19 lunar years, as I shall explain hereafter in the derivation of their cycles and the different kinds of their years.

The Christians agreed with them in the mode of the computation of their fasting and of some of their festivals, the cardinal point in all this being the Passover of the Jews; but they differed from them in the use of the months, wherein they followed the system of the Greeks and Syrians.

In a similar way the heathen Arabs proceeded, observing the difference between their year and the solar year, which is 10 days $21\frac{1}{5}$ hours, to speak roughly, and adding it to the year as one month p. 12.
as soon as it completed the number of days of a month. They, however, reckoned this difference as 10 days and 20 hours. This business was administered by the *Nasa'a* (the intercalators) of the tribe of Kinâna, known as the *Ḳalâmis*, a plural form of *Ḳalammas*, which signifies *a full-flowing sea*. These were 'Abû Thumâma and his ancestors:

I. 'Abû Thumâma Junâda ben
'Auf ben
'Umayya ben
Kala' ben

V. ‘Abbâd ben
Kala‘ ben
VII. Ḥudhaifa.

They were all of them intercalators. The first of them who held this office was—

VII. Ḥudhaifa ben
‘Abd ben
Fuḳaim ben
X. ‘Adiyy ben
‘Âmir ben
Tha‘laba ben
Mâlik ben
XIV. Kinâna.

The last of them, who held it, was 'Abû-Thumâma. The poet, who celebrates them, describes him in the following terms:—

"There is Fuḳaim! He was called Alḳalammas,
And he was one of the founders of their religion,
His word being obeyed, he being recognised as chieftain."

And another poet says:

"(He was) famous among the forerunners of Kinâna,
A celebrated man, of exalted rank.
In this way he spent his time."

Another poet says:

"The difference between the revolution of the sun and new-moon
He adds together and sums it up,
Till it makes out a complete month."

He (*i.e.* Ḥudhaifa) had taken this system of intercalation from the Jews nearly 200 years before Islâm; the Jews, however, intercalated 9 months in 24 lunar years. In consequence their months were fixed, and came always in at their proper times, wandering in a uniform course through the year without retrograding and without advancing. This state of things remained till the Prophet made his Farewell pilgrimage, and the following verse was revealed to him: "Intercalation is only an increase of infidelity, by which the infidels lead astray (people), admitting it one year and prohibiting it in another." (Sûra ix. 37.) The Prophet delivered an address to the people, and said: "Time has come round as it was on the day of God's creating the heavens and the earth," and, continuing, he recited to them the (just mentioned) verse of the Coran on the prohibition of the *Nasî'*, *i.e.* intercalation. Ever since they have neglected intercalation, so that their months have receded from their original places, and the names of the months are no longer in conformity with their original meanings.

As to the other nations, their opinions on this subject are well known. They are likely to have no other systems besides those we have mentioned, and each nation seems to follow the example of the system of their neighbours.

Years of the Indians.—I have heard that the Indians use the appearance of new-moon in their months, that they intercalate one lunar month in every 976 days, and that they fix the beginning of their era to the moment when a conjunction takes place in the first minute of any p. 13. zodiacal sign. The chief object of their searching is that this conjunction should take place in one of the two equinoctial points. The leap-year they call *Adhimâsa*. It is very possible that this is really the case; because, of all stars, they use specially the moon, her mansions and their subdivisions, in their astrological determinations, and not the zodiacal signs. However, I have not met with anybody who had an accurate knowledge of this subject; therefore I turn away from what I cannot know for certain. And God is my help!

'Abû-Muḥammad Alnâ'ib Alâmulî relates in his *Kitâb-alghurra*, on the authority of Ya'ḳûb ben Ṭâriḳ, that the Indians use four different kinds of spaces of time:

I. One revolution of the sun, starting from a point of the ecliptic and returning to it. This is the solar year.

II. 360 risings of the sun. This is called the middle-year, because it is longer than the lunar year and shorter than the solar year.

III. 12 revolutions of the moon, starting from the star *Alsharaṭân* (*i.e.* the head of Aries) and returning to it. This is their lunar year, which consists of 327 days and nearly $7\frac{2}{3}$ hours.

IV. 12 lunations. This is the lunar year, which they use.

CHAPTER III.

ON THE NATURE OF THE ERAS, AND THE DIFFERENT OPINIONS OF THE NATIONS REGARDING THEM.

ERA means a definite space of time, reckoned from the beginning of some past year, in which either a prophet, with signs and wonders, and with a proof of his divine mission, was sent, or a great and powerful king rose, or in which a nation perished by a universal destructive deluge, or by a violent earthquake and sinking of the earth, or a sweeping pestilence, or by intense drought, or in which a change of dynasty or religion took place, or any grand event of the celestial and the famous tellurian miraculous occurrences, which do not happen save at long intervals and at times far distant from each other. By such events the fixed moments of time (the *epochs*) are recognised. Now, such an era cannot be dispensed with in all secular and religious affairs. Each of the nations scattered over the different parts of the world has a special era, which they count from the times of their kings or prophets, or dynasties, or of some of those events which we have just now mentioned. And thence they derive the dates, which they want in social intercourse, in chronology, and in every institute (*i.e.* festivals) which is exclusively peculiar to them.

Era of the Creation.—The first and most famous of the beginnings of antiquity is the fact of the creation of mankind. But among those who have a book of divine revelation, such as the Jews, Christians, Magians, and their various sects, there exists such a difference of opinion as to the nature of this fact, and as to the question how to date
p. 14. from it, the like of which is not allowable for eras. Everything, the knowledge of which is connected with the beginning of creation and with the history of bygone generations, is mixed up with falsifications and myths, because it belongs to a far remote age; because a long interval separates us therefrom, and because the student is incapable of

keeping it in memory, and of fixing it (so as to preserve it from confusion). God says: "Have they not got the stories about those who were before them? None but God knows them." (Sûra ix. 71.) Therefore it is becoming not to admit any account of a similar subject, if it is not attested by a book, the correctness of which is relied upon, or by a tradition, for which the conditions of authenticity, according to the prevalent opinion, furnish grounds of proof.

If we now first consider this era, we find a considerable divergence of opinion regarding it among these nations. For the Persians and Magians think that the duration of the world is 12,000 years, corresponding to the number of the signs of the zodiac and of the months; and that Zoroaster, the founder of their law, thought that of those there had passed, till the time of his appearance, 3,000 years, intercalated with the day-quarters; for he himself had made their computation, and had taken into account that defect, which had accrued to them on account of the day-quarters, till the time when they were intercalated and were made to agree with real time. From his appearance till the beginning of the Æra Alexandri, they count 258 years; therefore they count from the beginning of the world till Alexander 3,258 years. However, if we compute the years from the creation of Gayômarth, whom they hold to be the first man, and sum up the years of the reign of each of his successors—for the rule (of Iran) remained with his descendants without interruption—this number is, for the time till Alexander, the sum total of 3,354 years. So the specification of the single items of the addition does not agree with the sum total.

Further, the Persians and Greeks disagree as to the time after Alexander. For they count from Alexander till the beginning of the reign of Yazdajird 942 years 257 days. If we deduct therefrom the duration of the rule of the Sasanian kings as far as the beginning of the reign of Yazdajird, as they compute it, viz., nearly 415 years, we get a remainder of 528 years as the time during which Alexander and the *Mulûk-al-ṭawâ'if* reigned. But if we sum up the years of the reign of each of the Ashkanian kings, as they have settled it, we get only the sum of 280 years, or,—taking into regard their difference of opinion as to the length of the reign of each of them,—the sum of not more than 300 years. This difference I shall hereafter try to settle to some extent.

A section of the Persians is of opinion that those past 3,000 years which we have mentioned are to be counted from the creation of Gayômarth; because, before that, already 6,000 years had elapsed—a time during which the celestial globe stood motionless, the natures (of created beings) did not interchange, the elements did not mix—during p. 15.
which there was no growth, and no decay, and the earth was not cultivated. Thereupon, when the celestial globe was set a-going, the first man came into existence on the equator, so that part of him in longitudinal direction was on the north, and part south of the line. The animals

were produced, and mankind commenced to reproduce their own species and to multiply; the atoms of the elements mixed, so as to give rise to growth and decay; the earth was cultivated, and the world was arranged in conformity with fixed norms.

The Jews and Christians differ widely on this subject; for, according to the doctrine of the Jews, the time between Adam and Alexander is 3,448 years, whilst, according to the Christian doctrine, it is 5,180 years. The Christians reproach the Jews with having diminished the number of years with the view of making the appearance of Jesus fall into the fourth millennium in the middle of the seven millennia, which are, according to their view, the time of the duration of the world, so as not to coincide with that time at which, as the prophets after Moses had prophesied, the birth of Jesus from a pure virgin at the end of time, was to take place. Both parties depend, in their bringing forward of arguments, upon certain modes of interpretation derived from the *Ḥisâb-al-jummal*. So the Jews expect the coming of the Messiah who was promised to them at the end of 1,335 years after Alexander, expecting it like something which they know for certain. In consequence of which many of the pseudo-prophets of their sects, as *e.g.* Al-râ'î, 'Abû-Îsâ Al-isfahânî, and others, claimed to be his messengers to them. This expectation was based on the assumption that the beginning of this era (Æra Alexandri) coincided with the time when the sacrifices were abolished, when no more divine revelation was received, and no more prophets were sent. Then they referred to the Hebrew word of God in the 5th book of the Thora (Deut. xxxi. 18), אנכי הסתר אסתיר פני מהם ביום ההוא, which means: "I, God, shall conceal *my being* till that day." And they counted the letters of the words הסתר אסתיר, the word for "concealing," which gives the sum of 1,335. This they declared to be the time during which no inspiration from heaven was received and the sacrifices were abolished, which is meant by God's *concealing himself*. The word "*being*" (פני=ذاتى) is here synonymous with "affair" (or "order, command"). In order to support what they maintain, they quote two passages in the Book of Daniel (xii. 11): מעת הוסר התמיד ולתת שקוץ שמם ימים אלף ומאתים ותשעים, which means: "Since the time when the sacrifice was abolished until impurity comes to destruction it is 1,290." and the next following passage (Dan. xii. 12): אשרי המחכה ויגיע לימים אלף שלוש מאות ושלשים וחמשה which means: "Therefore happy he who hopes to reach to 1,335." Some people explain the difference of forty-five years in these two passages so as to refer
p. 16. the former date (1,290) to the beginning of the rebuilding of Jerusalem; and the latter (1,335) to the time when the rebuilding would be finished. According to others, the first number is the date of the birth of Messiah, whilst the latter is the date of his public appearance. Further, the Jews say, when Jacob bestowed his blessing upon Judah (Gen. xlix. 10), he

informed him that the rule should always remain with his sons till the time of the coming of him to whom the rule belongs. So in these words he told him that the rule should remain with his descendants until the appearance of the expected Messiah. And now the Jews add that this is really the case; that the rule has not been taken from them. For the ראש גלותא, *i.e.* "the head of the exiles" who had been banished from their homes in Jerusalem, is the master of every Jew in the world; the ruler whom they obey in all countries, whose order is carried out under most circumstances.

The Christians use certain Syriac words, viz., ܝܫܘܥ ܡܫܝܚܐ ܦܪܘܩܐ ܪܒܐ, which mean "Jesus, the Messiah, the greatest redeemer." Computing the value of the letters of these words, they get the sum of 1,335. Now, they think that it was these words which Daniel meant to indicate by those numbers, not the above-mentioned years; because in the text of his words they are nothing but numbers, without any indication whether they mean years, or days, or something else. It is a prophecy indicative of the name of the Messiah, not of the time of his coming. Further, they relate that Daniel once dreamt in Babylonia, some years after the accession of Cyrus to the throne, on the 24th of the first month, when he had prayed to God, and when the Israelites were the prisoners of the Persians. Then God revealed to him the following (Dan. ix. 24–26): "'Ûrishlîm, *i.e.* Jerusalem, will be rebuilt 70 *Sâbû'*, and will remain in the possession of thy people. Then the Messiah will come, but he will be killed. And in consequence of his coming 'Ûrishlîm will undergo its last destruction, and it will remain a ruin till the end of time." The word *Sâbû'* (Hebrew שָׁבוּעַ) means a *Septennium*. Now, of the whole time (indicated in this passage) seven Septennia refer to the rebuilding of Jerusalem, which time is also mentioned in the Book of Zekharyâ ben Berekhyâ ben 'Iddô' (Zechariah iv. 2): "I have beheld a candlestick with seven lamps thereon, and with seven pipes to each lamp." And before this he says (iv. 9): "The hands of Zerubbabel have laid the foundation of this house, his hands also shall finish it." The time from the beginning of his rebuilding of the house (*i.e.* Jerusalem) till its end is 49 years, or 7 Septennia. Then, after 62 Septennia, they think, Jesus the son of Mary came; and in the last Septennium the sacrifices and offerings were abolished, and Jerusalem underwent its above-mentioned destruction, insomuch that no more divine revelation nor prophets were sent, as the Israelites were scattered all over the world, utterly neglected, not practising their sacrifices, nor having a p. 17.
place where to practise them.

In respect of all we have mentioned, each of the two parties makes assertions which they cannot support by anything but interpretations derived from the *Ḥisâb-al-jummal*, and fallacious subtilties. If the student would try to establish something else by the same means, and refute what they (each of the two parties) maintain, by similar arguments, it would not be difficult for him to search for them. As to what the Jews think of the continuance of the rule in the family of Juda, and which they transfer to the leadership of the exiles, we must remark that, if it was correct to extend the word "*rule*" to a similar leadership by way of analogy, the Magians, the Sabians, and others would partake of this, and neither the other Israelites nor any other nation would be exempt therefrom. Because no class of men, not even the lowest, are without a sort of rule and leadership with relation to others who are still inferior to them.

If we referred the numerical value of the word "*concealing*" in the Thora to that period from the earliest date which the Israelites assign to their exodus from Egypt till Jesus the son of Mary, this interpretation would rest on a better foundation. For the time from their exodus from Egypt till the accession of Alexander is 1,000 years according to their own view; and Jesus the son of Mary was born Anno Alexandri 304, and God raised him to himself Anno Alexandri 336. So the sum of the years of this complete period is 1,335 as the time during which the law of Moses ben 'Imrân existed, till it was carried to perfection by Jesus the son of Mary.

As to that which they derive from the two passages of Daniel, we can only say that it would be possible to refer them to something different, and to explain them in a different way; and more than that—that neither of their modes of interpretation is correct, except we suppose that the beginning of that number *precedes* the time when they were pronounced (by Daniel). For if it is to be understood that the beginning of both numbers (1,290 and 1,335) is one and the same time, be it past, present, or future, you cannot reasonably explain why the two passages should have been pronounced at different times. And, not to speak of the difference between the two numbers (1,290 and 1,335) the matter can in no way be correct; because the second passage ("Happy he who hopes to reach 1,335") admits, first, that the beginning of the number precedes the time when the passage was pronounced; so that it (the number) may reach its end one year, or more or less, after the supposed time; secondly, that the beginning of that number may be the very identical time when the passage was pronounced; or, thirdly, that it may be *after* this moment by an indefinite time, which may be smaller or greater. Now, if a chronological statement may be referred to all three spheres of time (past, present, and future), it cannot be referred to any one of them except on the basis of a clear text or an indisputable argument.

The first passage ("Since the time when the sacrifice was abolished, until impurity comes to destruction, it is 1,290") admits likewise of being referred, first, to the first destruction of Jerusalem; and, secondly, to
p. 18. its second destruction, which happened, however, only 385 years after the accession of Alexander.

Therefore the Jews have not the slightest reason to commence (in their calculations as to the coming of the Messiah) with that date with which they have commenced (viz., the epoch of the Æra Alexandri).

These are doubts and difficulties which beset the assertions of the Jews. Those, however, which attach to the schemes of the Christians are even more numerous and conspicuous. For even if the Jews granted to them that the coming of Messiah was to take place 70 *Septennia* after the vision of Daniel, we must remark that the appearance of Jesus the son of Mary did not take place at that time. The reason is this:—The Jews have agreed to fix the interval between the exodus of the Israelites from Egypt and the Æra Alexandri at 1,000 complete years. From passages in the books of the Prophets they have inferred that the interval between the exodus of the Israelites from Egypt and the building of Jerusalem is 480 years; and the interval between the building and the destruction by Nebucadnezar 410 years; and that it remained in a ruined state 70 years. Now this gives the sum of 960 years (after the exodus from Egypt) as the date for the vision of Daniel, and as a remainder of the above-mentioned millennium (from the exodus till Æra Alexandri) 40 years. Further, Jews and Christians unanimously suppose that the birth of Jesus the son of Mary took place Anno Alexandri 304. Therefore, if we use their own chronology, the birth of Jesus the son of Mary took place 344 years after the vision of Daniel and the rebuilding of Jerusalem, *i.e* about 49 Septennia. From his birth till the time when he began preaching in public are 4½ Septennia more. Hence it is evident that the birth (of Jesus) precedes the date which they have assumed (as the time of the birth of the Messiah).

For the Jews there follow no such consequences from their chronological system; and if the Christians should accuse the Jews of telling lies regarding the length of the period between the rebuilding of Jerusalem and the epoch of the Æra Alexandri, the Jews would meet them with similar accusations, and more than that.

If we leave aside the arguments of the two parties, and consider the table of the Chaldæan kings, which we shall hereafter explain, we find the interval between the beginning of the reign of Cyrus and that of the reign of Alexander to be 222 years, and from the latter date till the birth of Jesus 304 years; so that the sum total is 526 years. If we now deduct therefrom 8 years, for the rebuilding (of Jerusalem) commenced in the third year of the reign of Cyrus, and if we reduce the remainder to Septennia, we get nearly 75 Septennia for the interval between the vision (of Daniel) and the birth of Messiah. Therefore the birth of Messiah is later than the date which they (the Christians) have assumed.

If the Christians compute the Syriac words (ܝܫܘܥ ܡܫܝܚܐ ܦܪܘܩܐ ܘܚܝܐ), and believe that because of the identity of their numerical value with the number (1,335, mentioned by Daniel), these words were meant

(by Daniel) and not a certain number of years, we can only say that we cannot accept such an opinion except it be confirmed by an argument as indubitable as ocular inspection. For if you computed the numerical value of the following words: نجاة الخلق من الكفر بمحمّد ("the deliverance of the creation from infidelity by Muḥammad"), you would get the sum of 1,335. Or if you computed the words بشّر موسى بن عمران
p. 19. بمحمّد والمسيح باحمد ("the prophecy of Moses ben 'Imrân regarding Muḥammad; the prophecy of the Messiah regarding 'Aḥmad"), you would get the same sum, *i.e.* 1,335. Likewise, if you counted these words: يشرق برّيّة فاران بمحمّد الامّي ("*The plain of Fârân shines with the illiterate Muḥammad*"), you would again get the same sum (1,335). If, now, a man asserts that these numbers are meant to indicate a prophecy on account of the identity of the numerical values of these phrases with that of the Syriac words (ܡܚܡܕ ܡܫܝܚܐ ܦܪܘܩܐ ܙܟܐ), the value of his argument would be exactly the same as that of the Christians regarding those passages (in Daniel), the one case closely resembling the other, even if he should produce as a testimony for Muḥammad and the truth of the prophecy regarding him a passage of the prophet Isaiah, of which the following is the meaning, or like it (Isaiah xxi. 6–9): "*God ordered him to set a watchman on the watchtower, that he might declare what he should see. Then he said: I see a man riding on an ass, and a man riding on a camel. And the one of them came forward crying and speaking: Babylon is fallen, and its graven images are broken.*" This is a prophecy regarding the Messiah, "*the man riding on an ass,*" and regarding Muḥammad, "*the man riding on a camel,*" because in consequence of his appearance Babylon has fallen, its idols have been broken, its castles have been shattered, and its empire has perished. There are many passages in the book of the prophet Isaiah, predicting Muḥammad, being rather hints (than clearly out-spoken words), but easily admitting of a clear interpretation. And with all this, their obstinacy in clinging to their error induces them to devise and to maintain things which are not acknowledged by men in general, viz.: that "*the man riding on the camel,*" is Moses, not Muḥammad. But what connection have Moses and his people with Babel? And did that happen to Moses and to his people after him, which happened to Muḥammad and his companions in Babel? By no means! If they (the Jews) had one after the other escaped from the Babylonians, they would have considered it a sufficient prize to carry off to return (to their country), even though in a desperate condition.

This testimony (Isaiah xxi. 6–9) is confirmed by the word of God to Moses in the fifth book of the Thora, called *Almathnâ* (Deuteronomy xviii. 18, 19): "*I will raise them up a prophet from among their brethren like unto thee, and will put my word into his mouth. And he shall speak unto them all that I shall command him. And whosoever will not hearken unto the word of him who speaks in my name, I shall take revenge on him.*" Now I should like to know whether there are other brethren of the sons of Isaac, except the sons of Ishmael. If they say, that the brethren of the sons of Israel are the children of Esau, we ask only: Has there then risen among them a man like Moses—in the times after Moses—of the same description and resembling him? Does not also the following passage of the same book, of which this is the translation (Deut. xxxiii. 2), bear testimony for Muḥammad: "*The Lord came from Mount Sinai, and rose up unto us from Seir, and he shined forth from Mount Paran, accompanied by ten thousand of saints at his right hand?*" The terms of this passage are hints for the establishing of the proof, that the (anthropomorphic) descriptions, which are inherent in them, cannot be referred to the essence of the Creator, nor to his qualities, he being high above such things. His coming from Mount Sinai means his secret conversation with Moses there; his rising up from Seir means the appearance of Messiah, and his shining forth from Paran, where Ishmael grew up and married, means the coming of Muḥammad from thence as the last of all the founders of religions, accompanied by legions of saints, who were sent down from heaven to help, being marked with certain badges. He who refuses to accept this interpretation, for which p. 20. all evidence has borne testimony, is required to prove what kinds of mistakes there are in it. "*But he whose companion is Satan, woe to him for such a companion!*" (Sûra iv. 42.)

Now, if the Christians do not allow us to use the numerical values of Arabic words, we cannot allow them to do the same with the Syriac words which they quote, because the Thora and the books of those prophets were revealed in the Hebrew language. All they have brought forward, and all we are going to propound, is a decisive proof, and a clear argument, showing that the words in the holy books have been altered from their proper meanings, and that the text has undergone modifications contrary to its original condition. Having recourse to this sort of computing, and of using false witnesses, shows and proves to evidence, that their authors purposely deviate from the path of truth and righteousness. If we could open them a door in heaven, and they ascended thereby, they would say: "*Our eyes are only drunken. Nay, we are fascinated people.*" (Sûra xv. 15.) But such is not the case. The fact is that they are blind to the truth. We pray to God, that he may help and strengthen us, that he may guard us against sin, and lead us by the right path.

As to the doctrine of abrogation (of one holy book by another), and as to their fanciful pretension of having passages of the Thora which order him who claims to be a prophet after Moses to be put to death, we must state, that the groundlessness of these opinions is rendered evident by other passages of the Thora. However, there are more suitable places to speak of these opinions than this, and so we return to our subject, as we have already become lengthy in our exposition, one matter drawing us to another.

Now I proceed to state that both Jews and Christians have a copy of the Thora, the contents of which agree with the doctrines of either sect. Of the Jewish copy people think that it is comparatively free from confusion. The Christian copy is called the "*Thora of the Seventy,*" for the following reason: After Nebukadnezar had conquered and destroyed Jerusalem, part of the Israelites emigrated from their country, took refuge with the king of Egypt, and lived there under his protection till the time when Ptolemæus Philadelphus ascended the throne. This king heard of the Thora, and of its divine origin. Therefore he gave orders to search for this community, and found them at last in a place numbering about 30,000 men. He afforded them protection, and took them into his favour, he treated them with kindness, and allowed them to return to Jerusalem, which in the meanwhile had been rebuilt by Cyrus, Bahman's governor of Babel, who had also revived the culture of Syria. They left Egypt, accompanied by a body of his (Ptolemæus Philadelphus') servants for their protection. The king said to them: "I want to ask you for something. If you grant me the favour, you acquit yourselves of all obligations towards me. Let me have a copy of your book, the Thora." This the Jews promised, and confirmed their promise by an oath. Having arrived at Jerusalem, they fulfilled their promise by sending him a copy of it, *but* in Hebrew. He, however, did not know Hebrew. Therefore he addressed himself again to them asking for people who knew both Hebrew and Greek, who might translate the book for him, promising them gifts and presents in reward. Now the Jews selected seventy-two
p. 21. men out of their twelve tribes, six men of each tribe from among the Rabbis and priests. Their names are known among the Christians. These men translated the Thora into Greek, after they had been housed separately, and each couple had got a servant to take care of them. This went on till they had finished the translation of the whole book. Now the king had in his hands thirty-six translations. These he compared with each other, and did not find any differences in them, except those which always occur in the rendering of the same ideas. Then the king gave them what he had promised, and provided them with everything of the best. The Jews asked him to make them a present of one of those copies, of which they wished to make a boast before their own people. And the king complied with their wish. Now this is the copy of the Christians, and people think, that in it no alteration or transposition has taken place. The Jews, however, give quite a different account, viz. that they made the translation under compulsion, and that they yielded to the king's demand only from fear of violence and maltreatment, and not before having agreed upon inverting and confounding the text of the book. There is nothing in the report of the Christians which, even if we should take it for granted—removes our doubts (as to the authenticity of their Bible); on the contrary, there is something in it which strengthens them greatly.

Besides these two copies of the Thora, there is a third one that exists among the Samaritans, also known by the name of *Al-lâmasâsiyya*. To them, as the substitutes for the Jews, Nebucadnezar had given the country of Syria, when he led the Jews into captivity, and cleared the country of them. The Samaritans had helped him (in the war against the Jews), and had pointed out to him the weak points of the Israelites. Therefore, he did not disturb them, nor kill them, nor make them prisoners, but he made them inhabit Palestine under his protection.

Their doctrines are a syncretism of Judaism and Zoroastrianism. The bulk of their community is living in a town of Palestine, called *Nâbulus*, where they have their churches. They have never entered the precincts of Jerusalem since the days of David the prophet, because they maintain that he committed wrong and injustice, and transferred the holy temple from Nâbulus to *Aelia*, *i.e.* Jerusalem. They do not touch other people; but if they happen to be touched by anyone, they wash themselves. They do not acknowledge any of the prophets of the Israelites after Moses.

Now as to the copy which the Jews have, and on which they rely, we find that according to its account of the lives of the immediate descendants of Adam, the interval between the expulsion of Adam from Paradise till the deluge in the time of Noah, is 1,656 years; according to the Christian copy the same interval is 2,242 years, and according to the Samaritan copy it is 1,307 years. According to one of the historians, Anianus, the interval between the creation of Adam and the night of the Friday when the deluge commenced, is 2,226 years 23 days and 4 hours. This statement of Anianus is reported by Ibn-Albâzyâr in his *Kitâb-alḳirânât* (Book of the Conjunctions); it comes very near that p. 22.
of the Christians. However, it makes me think that it is based upon the methods of the astrologers, because it betrays evidently an arbitrary and too subtle mode of research.

Now, if such is the diversity of opinions, as we have described, and if there is no possibility of distinguishing—by means of analogy—between truth and fiction, where is the student to search for exact information?

Not only does the Thora exist in several and different copies, but something similar is the case with the Gospel too. For the Christians have four copies of the Gospel, being collected into one code, the first by Matthew, the second by Mark, the third by Luke, and the fourth by John; each of these four disciples having composed the Gospel in conformity with what he (Christ) had preached in his country. The reports, contained in these four copies, such as the descriptions of Messiah, the relations of him at the time when he preached and when he was crucified, as they maintain, differ very widely the one from the other. To begin with his genealogy, which is the genealogy of Joseph, the bridegroom of Mary and step-father of Jesus. For according to Matthew (i. 2–16), his pedigree is this:—

I. Joseph.	Zorobabel.	Joram.	Salmon.
Jacob.	Salathiel.	Josaphat.	Naasson.
Matthan.	Jechonias.	Asa.	Aminadab.
Eleazar.	Josias.	Abia.	Aram.
V. Eliud.	XV. Amon.	XXV. Roboam.	XXXV. Esrom.
Achin.	Manasses.	Solomon.	Phares.
Zadok.	Ezekias.	David.	Judas.
Azor.	Ahaz.	Jesse.	Jacob.
Elyakim.	Joatham.	Obed.	Isaac.
X. Abiud.	XX. Ozias.	XXX. Booz.	XL. Abraham.

Matthew in stating this genealogy commences with Abraham, tracing it downward (as far as Joseph). According to Luke (iii. 23–31) the pedigree of Joseph is this:—

I. Joseph.	Esli.	Salathiel.	Matthat.
Heli.	Nagge.	Neri.	Levi.
Matthat.	Maath.	Melchi.	Simeon.
Levi.	Mattathias.	Addi.	Juda.
V. Melchi.	XV. Semei.	XXV. Cosam.	XXXV. Joseph.
Janna	Joseph.	Elmodam.	Jonam.
Joseph.	Judas.	Er.	Elyakim.
Mattathias.	Joanna.	Joseph.	Melea.
Amos.	Rhesa.	Elieser.	Menan.
X. Naum.	XX. Zorobabel.	XXX. Jorim.	XL. Matathâ.
			Nathan.
			XLII. David.

This difference the Christians try to excuse, and to account for it, saying, that there was one of the laws prescribed in the Thora which ordered that, if a man died, leaving behind a wife but no male children, the brother of the deceased was to marry her instead, in order to raise up a progeny to the deceased brother; that, in consequence, his children were *genealogically* referred to the deceased brother, whilst as *to real birth* they were the children of the living brother; that, therefore,
23. Joseph was referred to two different fathers, that Heli was his father *genealogically*, whilst Yakob was his father *in reality*. Further, they say, that when Matthew had stated the real pedigree of Joseph, the Jews blamed him for it, saying: "His pedigree is not correct, because it has been made without regard to his *genealogical* relation." In order to meet this reproach, Luke stated his pedigree in conformity with the genealogical ordinances of their code. Both pedigrees go back to David, and that was the object (in stating them), because it had been predicted of the Messiah, that he would be "the son of David."

Finally, the fact that only the pedigree of Joseph has been adduced for Messiah, and not that of Mary, is to be explained in this way, that according to the law of the Israelites, nobody was allowed to marry any but a wife of his own tribe and clan, whereby they wanted to prevent confusion of the pedigrees, and that it was the custom to mention only the pedigrees of the men, not those of the women. Now Joseph and Mary being both of the same tribe, their descent must of necessity go back to the same origin. And this was the object in their statement and account of the pedigree.

Everyone of the sects of Marcion, and of Bardesanes, has a special Gospel, which in some parts differs from the Gospels we have mentioned. Also the Manichæans have a Gospel of their own, the contents of which from the first to the last are opposed to the doctrines of the Christians; but the Manichæans consider them as their religious law, and believe that it is the correct Gospel, that its contents are really that which Messiah thought and taught, that every other Gospel is false, and its followers are liars against Messiah. Of this Gospel there is a copy, called, "The Gospel of the Seventy," which is attributed to one *Balâmis*, and in the beginning of which it is stated, that Sallâm ben 'Abdallâh ben Sallâm wrote it down as he heard it from Salmân Alfârisî. He, however, who looks into it, will see at once that it is a forgery; it is not acknowledged by Christians and others. Therefore, we come to the conclusion, that among the Gospels there are no books of the Prophets to be found, on which you may with good faith rely.

Era of the Deluge.—The next following era is the era of the great deluge, in which everything perished at the time of Noah. Here, too, there is such a difference of opinions, and such a confusion, that you have no chance of deciding as to the correctness of the matter, and do not even feel inclined to investigate thoroughly its historical truth. The reason is, in the first instance, the difference regarding the period between the Æra Adami and the Deluge, which we have mentioned already; and secondly, that difference, which we shall have to mention, regarding the period between the Deluge and the Æra Alexandri. For the Jews derive from the Thora, and the following books, for this latter period 1,792 years, whilst the Christians derive from *their* Thora for the same period 2,938 years.

The Persians, and the great mass of the Magians, deny the Deluge altogether; they believe that the rule (of the world) has remained with them without any interruption ever since Gayômarth Gilshâh, who was, p. 24.
according to them, the first man. In denying the Deluge, the Indians, Chinese, and the various nations of the east, concur with them. Some, however, of the Persians admit the fact of the Deluge, but they describe it in a different way from what it is described in the books of the prophets. They say, a partial deluge occurred in Syria and the west at the time of Tahmûrath, but it did not extend over the whole of the then civilized world, and only few nations were drowned in it; it did not

extend beyond the peak of Ḥulwân, and did not reach the empires of the east. Further, they relate, that the inhabitants of the west, when they were warned by their sages, constructed buildings of the kind of the two pyramids which have been built in Egypt, saying: "If the disaster comes from heaven, we shall go into them; if it comes from the earth, we shall ascend above them." People are of opinion, that the traces of the water of the Deluge, and the effects of the waves are still visible on these two pyramids half-way up, above which the water did not rise. Another report says, that Joseph had made them a magazine, where he deposited the bread and victuals for the years of drought.

It is related, that Tahmûrath on receiving the warning of the Deluge —231 years before the Deluge—ordered his people to select a place of good air and soil in his realm. Now they did not find a place that answered better to this description than Ispahân. Thereupon, he ordered all scientific books to be preserved for posterity, and to be buried in a part of that place, least exposed to obnoxious influences. In favour of this report we may state that in our time in Jay, the city of Ispahân, there have been discovered hills, which, on being excavated, disclosed houses, filled with many loads of that tree-bark, with which arrows and shields are covered, and which is called *Tûz*, bearing inscriptions, of which no one was able to say what they are, and what they mean.

These discrepancies in their reports, inspire doubts in the student, and make him inclined to believe what is related in some books, viz. that Gayômarth was not the first man, but that he was Gomer ben Yaphet ben Noah, that he was a prince to whom a long life was given, that he settled on the Mount Dunbâwand, where he founded an empire, and that finally his power became very great, whilst mankind was still living in (elementary) conditions, similar to those at the time of the creation, and of the first stage of the development of the world. Then he, and some of his children, took possession of the *κλίματα* of the world. Towards the end of his life, he became tyrannical, and called himself Âdam, saying: "If anybody calls me by another name than this, I shall cut off his head." Others are of opinion that Gayômarth was Emîm (אימים ?) ben Lûd ben 'Arâm ben Sem ben Noah.

The astrologers have tried to correct these years, beginning from the first of the conjunctions of Saturn and Jupiter, for which the sages among the inhabitants of Babel, and the Chaldæans have constructed astronomical tables, the Deluge having originated in their country. For people say, that Noah built the ark in Kûfa, and that it was there that "*the well poured forth its waters*" (Sûra xi. 42; xxiii. 27); that the ark rested upon the mountain of Aljûdî, which is not very far from those regions. Now this conjunction occurred 229 years 108 days before the Deluge. This date they studied carefully, and tried by that to correct
p. 25. the subsequent times. So they found as the interval between the Deluge and the beginning of the reign of the first Nebukadnezar (*Nabonassar*),

2,604 years, and as the interval between Nebukadnezar and Alexander 436 years, a result which comes pretty near to that one, which is derived from the Thora of the Christians.

This was the era which 'Abû-Ma'shar Albalkhî wanted, upon which to base his statements regarding the mean places of the stars in his Canon. Now he supposed that the Deluge had taken place at the conjunction of the stars in the last part of Pisces, and the first part of Aries, and he tried to compute their places for that time. Then he found, that they—all of them—stood in conjunction in the space between the twenty-seventh degree of Pisces, and the end of the first degree of Aries. Further, he supposed that between that time and the epoch of the Æra Alexandri, there is an interval of 2,790 intercalated years 7 months and 26 days. This computation comes near to that of the Christians, being 249 years and 3 months less than the estimate of the astronomers. Now, when he thought that he had well established the computation of this sum according to the method, which he has explained, and when he had arrived at the result, that the duration of those periods, which astronomers call "*star-cycles,*" was 360,000 years, the beginning of which was to precede the time of the Deluge by 180,000 years, he drew the inconsiderate conclusion, that the Deluge had occurred once in every 180,000 years, and that it would again occur in future at similar intervals.

This man, who is so proud of his ingenuity, had computed these star-cycles only from the motions of the stars, as they had been fixed by the observations of the Persians; but they (the cycles) differ from the cycles, which have been based upon the observations of the Indians, known as the "*cycles of Sindhind,*" and likewise they differ from the *days of Arjabhaz*, and *the days of Arkand.* If anybody would construct such cycles on the basis of the observations of Ptolemy, or of the modern astronomers, he might do so by the help of the well known methods of such a calculation, as in fact many people have done, *e.g.* Muḥammad ben 'Isḥâķ ben 'Ustâdh Bundâdh Alsarakhsî, 'Abû-al-wafâ Muḥammad ben Muḥammad Albûzajânî, and I myself in many of my books, particularly in the *Kitâb-al-istishhâd bikhtilâf al'arṣâd.*

In each of these cycles the stars come into conjunction with each other in the first part of Aries once, viz. when they start upon and return from their rotation, however, at different times. If he ('Abû-Ma'shar) now would maintain, that the stars were created standing at that time in the first part of Aries, or that the conjunction of the stars in that place is identical with the beginning of the world, or with the end of the world, such an assertion would be utterly void of proof, although the matter be within the limits of possibility. But such conclusions can never be admitted, except they rest on an evident argument, or on the report of some one who relates the *origines* of the world, whose word is relied upon, and regarding whom in the mind (of the reader or hearer) this persuasion is established, that he had received divine inspiration and help.

p. 26. For it is quite possible that these (celestial) bodies were scattered, not united at the time when the Creator designed and created them, they having these motions, by which—as calculation shows—they must meet each other in *one* point in such a time (as above mentioned). It would be the same, as if we, *e.g.* supposed a circle, in different separate places of which we put living beings, of whom some move fast, others slowly, each of them, however, being carried on in equal motions—of its peculiar sort of motion—in equal times; further, suppose that we knew their distances and places at a certain time, and the measure of the distance over which each of them travels in one Nychthemeron. If you then ask the mathematician as to the length of time, *after* which they would meet each other in a certain point, or *before* which they had met each other in that identical point, no blame attaches to him, if he speaks of billions of years. Nor does it follow from his account that those beings *existed* at that (past) time (when they met each other), or that they would *still exist* at that (future) time (when they are to meet again); but this only follows from his account, if it is properly explained, that, *if* these beings really existed (in the past), or would still exist (in future) in that same condition, the result (as to their conjunctions) could be no other but that one at which he had arrived by calculation. But then the verification of this subject is the task of a science which was not the science of 'Abû-Ma'shar.

If, now, the man who uses the cycles (the star-cycles), would conclude that they, viz. the stars, if they stood in conjunction in the first part of Aries, would again and again pass through the same cycles, because, according to his opinion, everything connected with the celestial globe is exempt from growth and decay, and that the condition of the stars in the past was exactly the same, his conclusion would be a mere assumption by which he quiets his mind, and which is not supported by any argument. For a proof does not equally apply to the two sides of a contradiction; it applies only to the one, and excludes the other. Besides it is well known among philosophers and others, that there is no such thing as an *infinite* evolution of power (δύναμις) into action (πρᾶξις), until the latter comes into real existence. The motions, the cycles, and the periods of the past were computed whilst they in reality existed; they have decreased, whilst at the same time increasing in number; therefore, they are not *infinite*.

This exposition will be sufficient for a veracious and fair-minded student. But if he remains obstinate, and inclines to the tricks of overbearing people, more explanations will be wanted, which exceed the compass of this book, in order to remove these ideas from his mind, to heal what is feeble in his thoughts, and to plant the truth in his soul. However, there are other chapters of this book where it will be more

suitable to speak of this subject than here. The discrepancy of the cycles, not the discrepancy of the observations, is a sufficient argument for—and a powerful help towards—repudiating the follies committed by 'Abû-Ma'shar, and relied upon by foolish people, who abuse all religions, who make the cycles of Sindhind, and others, the means by which to revile those who warn them that the hour of judgment is coming, and who tell them, that on the day of resurrection there will be reward and punishment in yonder world. It is the same set of people who excite suspicions against—and bring discredit upon—astronomers and mathematicians, by counting themselves among their ranks, and by representing themselves as professors of their art, although they cannot even impose p. 27.
upon anybody who has only the slightest degree of scientific training.

Era of Nabonassar.—The next following era is the Era of the first Nebukadnezar (Nabonassar). The Persian form of this word (*Bukhtanaṣṣar*) is *Bukht-narsi*, and people say that it means "one who weeps and laments much"; in Hebrew, "Nebukadnezar," which is said to mean "Mercury speaking," this being combined with the notion that he cherished science and favoured scholars. Then when the word was Arabized, and its form was simplified, people said "*Bukhtanaṣṣar.*" This is not the same king who devastated Jerusalem, for between these two there is an interval of about 143 years, as the following chronological tables will indicate.

The era of this king is based upon the Egyptian years. It is employed in the Almagest for the computation of the places of the planets, because Ptolemy preferred this era to others, and fixed thereby the mean places of the stars. Besides he uses the cycles of Callippus, the beginning of which is in the year 418 after Bukhtanaṣṣar, and each of which consists of seventy-six solar years. Those who do not know them (these cycles), try to prove by what they find mentioned in Almagest, that they are of Egyptian origin; for Hipparchus and Ptolemy fix the times of their observations by Egyptian days and months, and then refer them to the corresponding cycles of Callippus. Such, however, is not the case. The first cycle, employed by those who compute the months by the revolution of the moon and the years by the revolution of the sun, was the cycle of eight years, and the second that of nineteen years. Callippus was of the number of the mathematicians, and one who himself—or whose people—considered the use of this latter cycle as part of their laws. Thereupon, he computed this cycle (of seventy-six years), uniting for that purpose four cycles of nineteen years.

Some people think that in these cycles the beginning of the months was fixed by the appearance of new moon, not by calculation, as people at that (remote) age did not yet know the calculation of the eclipses, by which alone the length of the lunar month is to be determined, and these calculations are rendered perfect; and that the first who knew the theory of the eclipses was Thales of Miletus. For after having frequently attended the lectures of the mathematicians, and having learned from them the science of form and motions (astronomy), he proceeded to discover the calculation of the eclipses. Then he happened to come to Egypt, where he warned people of an impending eclipse. When, then, his prediction had been fulfilled, people honoured him highly.

The matter, as thus reported, does not belong to the impossible. For each a[illegible] goes back to certain original sources, and the nearer it is to its origin, the more simple it is, till you at last arrive at the very origin itself. However, this account, that eclipses were not known before Thales, must not be understood in this generality, but with certain local restrictions. For some people refer this scholar (Thales) to the time of Ardashîr ben Bâbak, others to that of Kaiḳubâdh. Now, if he lived at
28. the time of Ardashîr, he was preceded by Ptolemy and Hipparchus; and these two among the astronomers of that age knew the subject quite sufficiently. If, on the other hand, he lived at the time of Kaiḳubâdh, he stands near to Zoroaster, who belonged to the sect of the Ḥarrânians, and to those who already before him (Zoroaster) excelled in science, and had carried it to such a height as that they could not be ignorant of the theory of the eclipses. If, therefore, their report (regarding the discovery of the theory of the eclipses by Thales) be true, it is not to be understood in this generality, but with certain restrictions.

Era of Philippus Aridæus.—The era of Philip, the father of Alexander, is based upon Egyptian years. But this era is also frequently dated from the death of Alexander, the Macedonian, the Founder. In both cases the matter is the same, and there is only a difference in the expression. Because Alexander, the Founder, was succeeded by Philip, therefore, it is the same, whether you date from the death of the former, or the accession of the latter, the epoch being a connecting link common to both of them. Those who employ this era are called *Alexandrines*. On this era Theon Alexandrinus has based his so-called "Canon."

Era of Alexander.—Then follows the era of Alexander the Greek, to whom some people give the surname *Bicornutus*. On the difference of opinions regarding this personage, I shall enlarge in the next following chapter. This era is based upon Greek years. It is in use among most nations. When Alexander had left Greece at the age of twenty-six years, prepared to fight with Darius, the king of the Persians, and marching upon his capital, he went down to Jerusalem, which was inhabited by the Jews; then he ordered the Jews to give up the era of Moses and David, and to use his era instead, and to adopt that very year, the twenty-seventh of his life, as the epoch of this era. The Jews obeyed his command, and accepted what he ordered; for the Rabbis allowed them such a change at the end of each millennium after Moses. And at that time just a millennium had become complete, and their offerings and sacrifices had ceased to be practised, as they relate. So they adopted his era, and used it for fixing all the occurrences of their months and days, as they had already done in the twenty-sixth year of his life, when he first started from home, with the view of finishing the millennium (*i.e.* so as not to enter upon a new one). When, then, the first thousand years of the Æra Alexandri had passed, the end of which did not coincide with any striking event which people are accustomed to make the epoch of an era, they kept the Æra Alexandri, and continued to use it. The Greeks also use it. But according to the report of a book, which Ḥabîb ben Bihrîz, the metropolitan of Mosul, has translated, the Greeks used to date—before they adopted the Æra Alexandri—from the migration of Yûnân ben Paris from Babel towards the west.

Era of Augustus.—Next follows the era of the king Augustus, the p. 2 first of the Roman emperors (*Cæsares*). The word "*Cæsar*" means in Frankish (*i.e.* Latin) "he has been drawn forth, after a cutting has been made." The explanation is this, that his mother died in labour-pains, whilst she was pregnant with him; then her womb was opened by the "Cæsarean operation," and he was drawn forth, and got the surname "*Cæsar.*" He used to boast before the kings, that he had not come out of the *pudendum muliebre* of a woman, as also 'Aḥmad ben Sahl ben Hâshim ben Alwalîd ben Ḥamla ben Kâmkâr ben Yazdajird ben Shahryâr used to boast, that the same had happened to him. And he (Augustus) used to revile people calling them "*son of the pudendum muliebre.*"

The historians relate, that Jesus, the son of Mary, was born in the forty-third year of his reign. This, however, does not agree with the order of the years. The chronological tables, in which we shall give a corrected sequence of events, necessitate that his birth should have taken place in the seventeenth year of his reign.

It was Augustus who caused the people of Alexandria to give up their system of reckoning by non-intercalated Egyptian years, and to adopt the system of the Chaldæans, which in our time is used in Egypt. This he did in the sixth year of his reign; therefore, they took this year as the epoch of this era.

Era of Antoninus.—The era of Antoninus, one of the Roman kings, was based upon Greek years. Ptolemy corrected the places of the fixed stars, dating from the beginning of his reign, and noted them in the Almagest, directing that their positions should be advanced one degree every year.

Era of Diocletianus.—Then follows the era of Diocletian, the last of the Roman kings who worshipped the idols. After the sovereign power had been transferred to him, it remained among his descendants. After him reigned Constantine, who was the first Roman king who became a Christian. The years of this era are Greek. Several authors of Canons have used this era, and have fixed thereby the necessary paradigms of the prognostics, the *Tempora natalicia*, and the conjunctions.

Era of the Flight.—Then follows the era of the Flight of the

Prophet Muḥammad from Makka to Madîna. It is based upon Lunar years, in which the commencements of the months are determined by the appearance of New Moon, not by calculation. It is used by the whole Muhammadan world. The circumstances under which this very point was adopted as an epoch, and not the time when the Prophet was either born or entrusted with his divine mission or died, were the following:—Maimûn ben Mihrân relates, that Omar ben Alkhaṭṭâb, when people one day handed over to him a cheque payable in the month *Sha'bân*, said:—"Which Sha'bân is meant? that one in which we are or the next Sha'bân?" Thereupon he assembled the Companions of the Prophet, and asked their advice regarding the matter of chronology, which troubled his mind. They answered: "It is necessary to inform ourselves of the practice of the Persians in this respect." Then they fetc Hurmuzân, and asked him for information. He said: "We have a computation which we call *Mâh-rûz, i.e.* the computation of months and days." People arabized this word, and pronounced مورخ (*Mu'arrakh*), and coined as
p. 30. its infinitive the word "*Ta'rîkh*." Hurmuzân explained to them how they used this Mâh-rûz, and what the Greeks used of a similar kind. Then Omar spoke to the Companions of the Prophet: "Establish a mode of dating for the intercourse of people." Now some said: "Date according to the era of the Greeks, for they date according to the era of Alexander." Others objected that this mode of dating was too lengthy, and said: "Date according to the era of the Persians." But then it was objected, that as soon as a new king arises among the Persians he abolishes the era of his predecessor. So they could not come to an agreement.

Alsha'bî relates, that 'Abû-Mûsâ Al'ash'arî wrote to Omar ben Alkhaṭṭâb: "You send us letters without a date." Omar had already organized the registers, had established the taxes and regulations, and was in want of an era, not liking the old ones. On this occasion he assembled the Companions, and took their advice. Now the most authentic date, which involves no obscurities nor possible mishaps, seemed to be the date of the flight of the Prophet, and of his arrival at Madîna on Monday the 8th of the month Rabî' I., whilst the beginning of the year was a Thursday. Now he adopted this epoch, and fixed thereby the dates in all his affairs. This happened A.H. 17.

The reason why Omar selected this event as an epoch, and not the time of the birth of the Prophet, or the time when he was entrusted with his divine mission, is this, that regarding those two dates there existed such a divergency of opinion, as did not allow it to be made the basis of something which must be agreed upon universally.

Further he (Alsha'bî) says: People say that He was born in the night of Monday the 2nd, or the 8th, or the 13th of Rabî' I.; others say that he was born in the forty-sixth year of the reign of Kisrâ Anôshîrwân. In consequence there is also a difference of opinions regarding the length of his ife, corresponding to the different statements regarding his birth.

Besides, the single years were of different length, some having been intercalated, others not, about the time when intercalation was prohibited. Considering further that after the Flight, the affairs of Islâm were thoroughly established, while heathenism decreased, that the Prophet was saved from the calamities prepared for him by the infidels of Makka, and that after the Flight his conquests followed each other in rapid succession, we come to the conclusion that the Flight was to the Prophet, what to the kings is their accession, and their taking possession of the whole sovereign power.

As regards the well known date of his death, people do not like to date from the death of a prophet or a king, except the prophet be a liar, or the king an enemy, whose death people enjoy, and wish to make a festival of; or he be one of those with whom a dynasty is extinguished, so that his followers among themselves make this date a memorial of him, and a mourning feast. But this latter case has only happened very seldom. *E.g.* the era of Alexander the Founder is reckoned from the time of his death, he having been considered as one of those from whom the era of the kings of the Chaldæans and the western kings was transferred to the era of the Ptolemæan kings, of whom each is called *Ptolemy*, which means *warlike*. Therefore, those to whom the empire was transferred, dated from the time of his death, considering it as a joyful event. It is precisely the same in the case of the era of Yazdajird ben Shahryâr. For the Magians date from the time of his death, because when he perished, the dynasty was extinguished. Therefore they dated from his death, mourning over him, and lamenting for the downfal of their p. 31.
religion.

At the time of the Prophet, people had given to each of the years between the Flight and his death a special name, derived from some event, which had happened to him in that identical year.

The 1st year after the Flight is "the year of the permission."
The 2nd year " "the year of the order for fighting."
The 3rd year " "the year of the trial."
The 4th year " "the year of the congratulation on the occasion of marriage."
The 5th year " "the year of the earthquake."
The 6th year " "the year of inquiring."
The 7th year " "the year of gaining victory."
The 8th year " "the year of equality."
The 9th year " "the year of exemption."
The 10th year " "the year of farewell."

By these names it was rendered superfluous to denote the years by the numbers, the 1st, the 2nd, etc., after the Flight.

Era of Yazdajird.—Next follows the era of the reign of Yazdajird ben Shahryâr ben Kisrâ Parwîz, which is based upon Persian non-intercalated years. It has been employed in the Canons, because it is easy and simple to use. The reason why precisely the era of this king among all the kings of Persia has become so generally known, is this that he ascended the throne, when the empire had been shattered, when the women had got hold of it, and usurpers had seized all power. Besides, he was the last of their kings, and it was he with whom Omar ben Alkhaṭṭâb fought most of those famous wars and battles. Finally, the empire succumbed, and he was put to flight and was killed in the house of a miller at Marw-i-Shâhijân.

Reform of the Calendar by the Khalif Almu'taḍid.—Lastly, the era of 'Aḥmad ben Ṭalḥa Almu'taḍid-billâh the Khalif was based upon Greek years and Persian months; however, with this difference, that in every fourth year one day was intercalated. The following is the origin of this era, as reported by 'Abû-Bakr Alṣûlî in his *Kitâb-al'aurâḳ*, and by Ḥamza ben Alḥasan Alisfahânî in his book on famous poems, relating to Naurûz and Mihrjân. Almutawakkil, while wandering about over one of his hunting-grounds, observed corn that had not yet ripened, and not yet attained its proper time for being reaped. So he said: "Ubaid-allâh ben Yaḥyâ has asked my permission for levying the taxes, whilst I observe that the corn is still green. From what then are people to pay their taxes?" Thereupon he was informed, that this, in fact, had done a great deal of harm to the people, so that they were compelled to borrow and to incur debts, and even to emigrate from their homes; that they had many complaints and wrongs to recount. Then the Khalif said: "Has this arisen lately during my reign, or has it always been so?" And people answered: "No. This is going on according to the regulations established by the Persian kings for the levying of the taxes at the time of Naurûz. In this their example has been followed by the kings of the Arabs." Then the Khalif ordered the Maubadh to be brought before him, and said to him: "This has been the subject of much research on my part, and I cannot find that I violate the regulations of the Persians. How, then, did they levy the taxes from their subjects—considering the beneficence and good will which they observed towards them? And why did they allow the taxes to be levied at a time like
32. this, when the fruit and corn are not yet ripe?" To this the Maubadh replied: "Although they always levied the taxes at Naurûz, this never happened except at the time when the corn was ripe." The Khalif asked: "And how was that?" Now the Maubadh explained to him the nature of their years, their different lengths, and their need of intercalation. Then he proceeded to relate, that the Persians used to intercalate the years; but when Islâm had been established, intercalation was abolished; and that did much harm to the people. The landholders assembled at the time of Hishâm ben 'Abdalmalik and called on Khâlid Alḳasrî; they explained to him the subject, and asked him to postpone Naurûz by a month. Khâlid declined to do so, but reported on the

subject to Hishâm, who said: "I am afraid, that to this subject may be applied the word of God: "*Intercalation is only an increase of heathenism*" (Sûra ix. 37). Afterwards at the time of Alrashîd the landholders assembled again and called on Yaḥyâ ben Khâlid ben Barmak, asking him to postpone Naurûz by about two months. Now, Yaḥyâ had the intention to do so, but then his enemies began to speak of the subject, and said: "He is partial to Zoroastrianism." Therefore he dropped the subject, and the matter remained as it was before.

Now Almutawakkil ordered 'Ibrâhîm ben Al'abbâs Alṣûlî to be brought before him, and told him, that in accordance with what the Maubadh had related of Naurûz, he should compute the days, and compose a fixed Canon (Calendar); that he should compose a paper on the postponement of Naurûz, which was to be sent by order of the Khalif to all the provinces of the empire. It was determined to postpone Naurûz till the 17th of Ḥazîrân. Alṣûlî did as he was ordered, and the letters arrived in the provinces in Muḥarram A.H. 243. The poet, Albuḥturî has composed a Ḳaṣîda on the subject in praise of Almutawakkil, where he says:—

"The day of Naurûz has returned to that time, on which it was fixed by Ardashîr.
Thou hast transferred Naurûz to its original condition, whilst before thee it was wandering about, circulating.
Now thou hast levied the taxes at Naurûz, and that was a memorable benefit to the people.
They bring thee praise and thanks, and thou bringest them justice and a present, well deserving of thanks."

However, Almutawakkil was killed, and his plan was not carried out, until Almu'taḍid ascended the throne of the Khalifate, delivered the provinces of the empire from their usurpers, and gained sufficient leisure to study the affairs of his subjects. He attributed the greatest importance to intercalation and to the carrying out of this measure. He followed the method of Almutawakkil regarding the postponement of Naurûz; however he treated the subject differently, inasmuch as Almutawakkil had made the basis of his computation the interval between *his* year (*i.e.* that year, in which he then happened to live), and the beginning of the reign of Yazdajird, whilst Almu'taḍid took the interval between *his* year and that year in which the Persian empire perished by the death of Yazdajird, because he—or those who did the work for him—held this opinion, that *since that time* intercalation had been neglected. This interval he found to be 243 years and 60 days + a fraction, arising from the day-quarters (exceeding the 365 days of the Solar year). These 60 p. 33
days he added at Naurûz of his year, and put Naurûz at the end of them, which fell upon a Wednesday, the 1st Khurdâdh-Mâh of that year, coinciding with the 11th of Ḥazîrân. Thereupon he fixed Naurûz in the

Greek months for this purpose, that the months of *his* year should be intercalated at the same time when the Greeks intercalate their years. The man who was entrusted with carrying out his orders, was his Wazîr 'Abû-alḳâsim 'Ubaid-allâh ben Sulaimân ben Wahb. To this subject the following verses of the astronomer 'Alî ben Yaḥyâ refer:—

"O thou restorer of the untarnished glory, renovator of the shattered empire!
Who hast again established among us the pillar of religion, after it had been tottering!
Thou hast surpassed all the kings like the foremost horse in a race.
How blessed is that Naurûz, when thou hast earned thanks besides the reward (due to thee for it in heaven)!
By postponing Naurûz thou hast justly made precede, what they had postponed."

On the same subject 'Alî ben Yaḥyâ says:—

"The day of thy Naurûz is one and the same day, not liable to moving backward,
Always coinciding with the 11th of Ḥazîrân."

Now, although in bringing about this measure much ingenuity has been displayed, Naurûz has not thereby returned to that place which it occupied at the time when intercalation was still practised in the Persian empire. For the Persians had already begun to neglect their intercalation nearly seventy years before the death of Yazdajird. Because at the time of Yazdajird ben Shâpûr they had intercalated into their year two months, one of them as the necessary compensation for that space of time, by which the year had moved backward (it being too short). The five Epagomenæ they put as a mark at the end of this intercalary month, and the turn had just come to Âbân-Mâh, as we shall explain hereafter. The second month they intercalated with regard to the future, that no other intercalation might be needed for a long period.

Now, if you subtract from the sum of the years between Yazdajird ben Shâpûr and Yazdajird ben Shahryâr 120 years, you get a remainder of nearly—but not exactly—70 years; there is much uncertainty and confusion in the Persian chronology. The *Portio intercalanda* of these 70 years would amount to nearly 17 days. Therefore it would have been necessary, if we calculate without mathematical accuracy, to postpone Naurûz not 60, but 77 days, in order that it might coincide with the 28th of Ḥazîrân. The man who worked out this reform, was of opinion, that the Persian method of intercalation was similar to the Greek method. Therefore he computed the days since the extinction of their empire. Whilst in reality the matter is a different one, as we have already explained, and shall more fully explain hereafter.

This is the last of those eras that have become celebrated. But perhaps some other nations, whose countries are far distant from ours, have eras of their own, which have not been handed down to posterity, p. 34. or such eras as are now obsolete. For instance, the Persians in the time of Zoroastrianism used to date successively by the years of the reign of each of their kings. When a king died, they dropped his era, and adopted that of his successor. The duration of the reigns of their kings we have stated in the tables which will follow hereafter.

Epochs of the Ancient Arabs.—As a second instance we mention the Ishmaelite Arabs. For they used to date from the construction of the Ka'ba by Abraham and Ishmael till the time when they were dispersed and left Tihâma. Those who went away dated from the time of their exodus, whilst those who remained in the country dated from the time when the last party of the emigrants had left. But afterwards, after a long course of time, they dated from the year when the chieftainship devolved upon 'Amr ben Rabî'a, known by the name of 'Amr ben Yaḥyâ, who is said to have changed the religion of Abraham, to have brought from the city of Balkâ the idol Hubal, and to have himself made the idols 'Isâf and Nâ'ila. This is said to have happened at the time of Shâpûr Dhû-al'aktâf. This synchronism, however, is not borne out by the comparison of the chronological theories of both sides (Arabs and Persians).

Afterwards they dated from the death of Ka'b ben Lu'ayy—till the *Year of Treason,* in which the Banû-Yarbû' stole certain garments which some of the kings of Ḥimyar sent to the Ka'ba, and when a general fighting among the people occurred at the time of the holy pilgrimage. Thereupon they dated from the Year of Treason till the *Year of the Elephants,* in which the Lord, when the Ethiopians were coming on with the intention of destroying the Ka'ba, brought down the consequences of their cunning enterprise upon their own necks, and annihilated them. Thereupon they dated from the era of the Hijra.

Some Arabs used to date from famous accidents, and from celebrated days of battle, which they fought among themselves. As such epochs the Banû-Ḳuraish, *e.g.* had the following ones :—

1. The day of Alfijâr in the sacred month.

2. The day of the Confederacy of Alfuḍûl, in which the contracting parties bound themselves to assist all those to whom wrong was done. Because the Banû-Ḳuraish committed wrong and violence against each other within the holy precinct of Makka.

3. The year of the death of Hishâm ben Almughîra Almakhzûmî, for the celebration of his memory.

(4) The year of the reconstruction of the Ka'ba, by order of the Prophet Muḥammad.

The tribes 'Aus and Khazraj used the following days as epochs :—

1. The day of Alfaḍâ.

2. The day of Alrabî'.
3. The day of Alruḥâba.
4. The day of Alsarâra.
5. The day of Dâḥis and Ghabrâ.
6. The day of Bughâth.
7. The day of Ḥâṭib.
8. The day of Maḍris and Mu'abbis.

Among the tribes Bakr and Taghlib, the two sons of Wâ'il, the following epochs were used:—

1. The day of 'Unaiza.
2. The day of Alḥinw.
3. The day of Taḥlâḳ-allimam.
4. The day of Alḳuṣaibât.
5. The day of Alfaṣîl.

These and other "*war-days*" were used as epochs among the different tribes and clans of the Arabs. Their names refer to the places where they were fought, and to their causes.

If, now, these eras were kept in the proper order in which chronological subjects are to be treated, we should do with them the same that we intend to do with all the other subjects connected with eras. However, people say that between the year of the death of Ka'b ben Lu'ayy and the year of Treason there is an interval of 520 years, and between the year of Treason and the year of the Elephants an interval of 110 years. The Prophet was born 50 years after the invasion of the Ethiopians, and between his birth and the year of Alfijâr there were 20 years. At
p. 35. this battle the Prophet was present, as he has said himself: "I was present on the day of Alfijâr. Then I shot at my uncles." Between the day of Alfijâr and the reconstruction of the Ka'ba there are 15 years, and 5 years between the reconstruction of the Ka'ba and the time when Muḥammad was entrusted with his divine mission.

Likewise the Ḥimyarites and the Banû Ḳaḥṭân used to date by the reigns of their Tubba's, as the Persians by the reigns of their Kisrâs, and the Greeks by the reigns of their Caesars. However, the rule of the Ḥimyarites did not always proceed in complete order, and in their chronology there is much confusion. Notwithstanding, we have stated the duration of the reigns of their kings in our tables, as also those of the kings of the Banû-Lakhm, who inhabited Ḥîra, and were settled there, and had made it their home.

Chorasmian Antiquities.—In a similar way the people of Khwârizm proceeded. For they dated from the beginning of the colonization of their country, A. 980 before Alexander. Afterwards they adopted as the epoch of an era the event of the coming of Siyâwush ben Kaikâ'ûs down to Khwarizm, and the rule of Kaikhusrû, and of his descendants over the country, dating from the time when he immigrated and extended his sway over the empire of the Turks. This happened 92 years after the colonization of the country.

At a later time they imitated the example of the Persians in dating by the years of the reign of each king of the line of Kaikhusrû, who ruled over the country, and who was called by the title of *Shâhiya*. This went on down to the reign of Âfrîgh, one of the kings of that family. His name was considered a bad omen like that of Yazdajird the Wicked, with the Persians. His son succeeded him in the rule of the country. He (Âfrîgh) built his castle behind Alfîr, A. Alexandri 616. Now people began to date from him and his children (*i.e.* by the years of his reign and that of his descendants).

This Alfîr was a fortress on the outskirts of the city of Khwârizm, built of clay and tiles, consisting of three forts, one being built within the other, and all three being of equal height; and rising above the whole of it were the royal palaces, very much like Ghumdân in Yaman at the time when it was the residence of the Tubba's. For this Ghumdân was a castle in Ṣan'â, opposite the great mosque, founded upon a rock, of which people say that it was built by Sem ben Noah after the Deluge. In the castle there is a cistern, which he (Sem) had digged. Others think that it was a temple built by Alḍaḥḥâk for Venus. This Alfîr was to be seen from the distance of 10 miles and more. It was broken and shattered by the Oxus, and was swept away piece by piece every year, till the last remains of it had disappeared A. Alexandri 1305.

Of this dynasty was reigning at the time when the Prophet was entrusted with his divine mission—

10. Arthamûkh ben
9. Bûzkâr ben
8. Khâmgrî ben
7. Shâwush ben
6. Sakhr ben
5. Azkâjawâr ben
4. Askajamûk ben
3. Sakhassak ben
2. Baghra ben
1. Âfrîgh.

When Ḳutaiba ben Muslim had conquered Khwârizm the second time, after the inhabitants had rebelled, he constituted as their king—

14. Askajamûk ben
13. Azkâjawâr ben
12. Sabrî ben
11. Sakhr ben
10. Arthamûkh,

p. 36. and appointed him as their *Shâh*. The descendants of the Kisrâs lost the office of the "*Wali*" (the governorship), but they retained the office of the *Shâh*, it being hereditary among them. And they accommodated themselves to dating from the Hijra according to the use of the Muslims.

Ḳutaiba ben Muslim had extinguished and ruined in every possible way all those who knew how to write and to read the Khwârizmî writing, who knew the history of the country and who studied their sciences. In consequence these things are involved in so much obscurity, that it is impossible to obtain an accurate knowledge of the history of the country since the time of Islâm (not to speak of pre-Muhammadan times).

The *Wilâya* (governorship) remained afterwards alternately in the hands of this family and of others, till the time when they lost both *Wilâya* (governorship) and Shâhiyya (Shâhdom), after the death of the martyr

22. 'Abû 'Abdallâh Muḥammad ben
21. 'Aḥmad ben
20. Muḥammad ben
19. 'Irâḳ ben
18. Manṣûr ben
17. 'Abdallâh ben
16. Turkasbâtha ben
15. Shâwushfar ben
14. Askajamûk ben
13. Azkâjawâr ben
12. Sabrî ben
11. Sakhr ben
10. Arthamûkh, in whose time, as I have said, the Prophet was entrusted with his divine mission.

This is all I could ascertain regarding the celebrated eras; to know them all is impossible for a human being. God helps to the right insight.

CHAPTER IV.

THE DIFFERENT OPINIONS OF VARIOUS NATIONS REGARDING THE KING CALLED DHÛ-ALḲARNAINI OR BICORNUTUS.

We must explain in a separate chapter what people think of the bearer of this name, of Dhû-alḳarnaini, for the subject interrupts, in this part of the course of our exposition, the order in which our chronology would have to proceed.

Now it has been said, that the story about him as contained in the Koran, is well-known and intelligible to everybody who reads the verses specially devoted to his history. The pith and marrow of it is this, that he was a good and powerful man, whom God had gifted with extraordinary authority and power, and whose plans he had crowned with success both in east and west; he conquered cities, subdued countries, reduced his subjects to submission, and united the whole empire under his single sway. He is generally assumed to have entered the darkness in the north, to have seen the remotest frontiers of the inhabitable world, to have fought both against men and demons, to have passed between Gog and Magog, so as to cut off their communication, to have marched out towards the countries adjoining their territory in the east and north, to have restrained and repelled their mischievous inroads by means of a wall, constructed in a mountain-pass, whence they used to pour forth. It was built of iron-blocks joined by molten brass, as is still now the practice of artisans.

When Alexander, the son of Philip, Alyûnânî (*i.e.* the Ionian, meaning the Greek) had united under his sway the Greek empire (*lit.* the empire of the Romans), which had previously consisted of single prin- p. 37
cipalities, he marched against the princes of the west, overpowering and subduing them, going as far as the *Green Sea*. Thereupon he returned to Egypt, where he founded Alexandria, giving it his own name. Then he marched towards Syria and the Israelites of the country, went down

to Jerusalem, sacrificed in its temple and made offerings. Thence he turned to Armenia and Bâb-al'abwâb, and passed even beyond it. The Copts, Berbers, and Hebrews obeyed him. Then he marched against Dârâ, the son of Dârâ, in order to take revenge for all the wrongs which Syria had suffered at the hands of Bukhtanaṣṣar (Nebukadnezzar) and the Babylonians. He fought with him and put him to flight several times, and in one of those battles Dârâ was killed by the chief of his body-guard, called Naujushanas ben Âdharbakht, whereupon Alexander took possession of the Persian empire. Then he went to India and China, making war upon the most distant nations, and subduing all the tracts of country through which he passed. Thence he returned to Khurâsân, conquered it, and built several towns. On returning to 'Irâḳ he became ill in Shahrazûr, and died. In all his enterprises he acted under the guidance of philosophical principles, and in all his plans he took the advice of his teacher, Aristotle. Now, on account of all this he has been thought to be Dhû-alḳarnaini, or Bicornutus.

As to the interpretation of this surname, people say he was called so because he reached the two "*horns*" of the sun, *i.e.* his rising and setting places, just as Ardashîr Bahman was called Longimanus, because his command was omnipotent, wherever he liked, as if he had only to stretch out his hand in order to set things right.

According to others he was called so because he descended from two different "generations" (*lit.* horns) *i.e.* the Greeks and Persians. And on this subject they have adopted the vague opinions which the Persians have devised in a hostile spirit, *viz.* that Dârâ the Great had married his mother, a daughter of King Philip, but she had an offensive odour, which he could not endure, and so he sent her back to her father, she being pregnant; that he was called a son of Philip, simply because the latter had educated him. This story of theirs they try to prove by the fact, that Alexander, when he reached Dârâ, who was expiring, put his head on his lap and spoke to him: "O my brother, tell me, who did this to you, that I may take revenge for you?" But Alexander so addressed him only because he wanted to be kind towards him, and to represent him (Dârâ) and himself as brethren, it being impossible to address him as king, or to call him by his name, both of which would have betrayed a high degree of rudeness unbecoming a king.

On Real and Forged Pedigrees.—However, enemies are always eager to revile the parentage of people, to detract from their reputation, and to attack their deeds and merits, in the same way as friends and partisans are eager to embellish that which is ugly, to cover up the weak parts, to proclaim publicly that which is noble, and to refer everything to great virtues, as the poet describes them in these words:—

"The eye of benevolence is blind to every fault,
But the eye of hatred discovers every vice."

Obstinacy in this direction frequently leads people to invent laudatory stories, and to forge genealogies which go back to glorious ancestors, as p. 38
has been done, *e.g.* for Ibr 'Abdalrazzâḳ Alṭûsî, when he got made for himself a genealogy out of the Shâhnâma, which makes him descend from Minôsheihr, and also for the house of Buwaihi. For 'Abû-'Isḥâḳ 'Ibrâhîm ben Hilâl Alsâbî, in his book called *Altâj* (the crown), makes Buwaihi descend from Bahrâm Gûr by the following line of ancestors:—

I. Buwaihi.
Fanâkhusrû.
Thamân.
Kûhî.
V. Shîrzîl junior.
Shîrkadha.
Shîrzîl senior.
Shîrânshâh.
Shirfana.
X. Sasanânshâh.
Sasankhurra.
Shûzîl.
Sasanâdhar.
XIV. Bahrâm Gûr the king.

'Abû-Muḥammad Alḥasan ben 'Alî ben Nânâ in his epitome of the history of the Buwaihides, says that—

I. Buwaihi was the son of
Fanâkhusra, the son of
Thamân.

Then some people continue—

Thamân, the son of
Kûhî, the son of
V. Shîrzîl junior;

whilst others drop Kûhî.

Then they continue—

Shîrzîl senior, the son of
Shîrânshâh, the son of
Shîrfana, the son of
Sasanânshâh, the son of
X. Sasankhurra, the son of
Shûzîl, the son of
Sasanâdhar, the son of
XIII. Bahrâm.

Further, people disagree regarding this Bahrâm. Those who give the Buwaihides a Persian origin, contend that he was Bahrâm Gûr, and continue the enumeration of his ancestors (down to the origin of the family Sâsân), whilst others who give them an Arabic origin, say that he was—

Bahrâm ben
Alḍaḥḥâk ben
Al'abyaḍ ben
Mu'âwiya ben
Aldailam ben
Bâsil ben
Ḍabba ben
'Udd.

Others, again, mention among the series of ancestors—

Lâhû ben
Aldailam ben
Bâsil,

and maintain that from this name his son Layâhaj derived his name.

He, however, who considers what I have laid down at the beginning of this book, as the *conditio sine quâ non* for the knowledge of the proper mean between disparagement and exaggeration, and the necessity of the greatest carefulness for everybody who wants to give a fair judgment, will be aware of the fact, that the first member of this family who became celebrated was Buwaihi ben Fanâkhusra. And it is not at all known that those tribes were particularly careful in preserving and continuing their genealogical traditions, nor that they knew anything like this of the family Buwaihi, before they came into power. It very rarely happens that genealogies are preserved without any interruption during a long period of time. In such cases the only possible way of distinguishing a just claim to some noble descent from a false one is the agreement of all, and the assent of the whole generation in question regarding that subject. An instance of this is the lord of mankind,—

I. Muḥammad, for he is the son of
'Abd-allâh ben
'Abd-almuṭṭalib ben
Hâshim ben
V. 'Abd-Manâf ben
Ḳusayy ben
Kilâb ben
Murra ben
Ka'b ben
X. Lu'ayy ben
Ghâ'ib ben
Fihr ben
Mâlik ben
Alnaḍr ben

XV. Kinâna ben
Khuzaima ben
Mudrika ben
'Ilyâs ben
Muḍar ben
XX. Nizâr ben
Ma'add ben
XXII. 'Adnân.

Nobody in the world doubts this lineage of ancestors, as they do not doubt either, that he descends from Ishmael, the son of Abraham. The ancestry beyond Abraham is to be found in the Thora. However, regarding the link of parentage between 'Adnân and Ishmael there is a considerable divergence of opinions, inasmuch as some people consider as the father the person whom others take for the son, and *vice versâ*, and as they add considerably in some places, and leave out in others. p. 39.

Further as to our master, the commander, the prince, the glorious and victorious, the benefactor, Shams-alma'âlî, may God give him a long life, not one of his friends, whom may God help, nor any of his opponents, whom may God desert, denies his noble and ancient descent, well established on both sides, although his pedigree back to the origin of his princely family has not been preserved without any interruption. On the one side he descends from Wardânshâh, whose nobility is well-known throughout Ghîlân; and this prince had a son, besides the prince, the martyr Mardâwîj. People say, that the son of Wardânshâh obeyed the orders of 'Asfâr ben Shîrawaihi, and that it was he, who suggested to him (his brother Mardâwîj) the idea of delivering the people from the tyranny and oppression of 'Asfâr. On the other side he descends from the kings of Media, called the Ispahbads of Khurâsân and the Farkhwârjarshâhîs. And it has never been denied that those among them, who belonged to the royal house of Persia, claimed to have a pedigree which unites them and the Kisrâs into one family. For his uncle is the Ispahbad—

I. Rustam ben
هروين ben
Rustam ben
Ḳârin ben
V. Shahryâr ben
هروين ben
Surkhâb ben
باو ben
Shâpûr ben
X. Kayûs ben
XI. Ḳubâdh, who was the father of Anôshîrwân.

May God give to our master the empire from east and west over all the parts of the world, as he has assigned him a noble origin on both sides. God's is the power to do it, and all good comes from him.

The same applies to the kings of Khurâsân. For nobody contests the fact, that the first of this dynasty—

I. 'Ismâ'îl was the son of
'Aḥmad ben
'Asad ben
Sâmân-khudâh ben
V. جسيمان ben
طغمات ben
نوشرد ben
Bahrâm Shûbîn ben
IX. Bahrâm Jushanas, the commander of the marches of Adharbaijân.

The same applies further to the original Shâhs of Khwârizm, who belonged to the royal house (of Persia), and to the Shâhs of Shirwân, because it is believed by common consent, that they are descendants of the Kisrâs, although their pedigree has not been preserved uninterruptedly.

The fact that claims to some noble lineage, and also to other matters, are just and well founded, always becomes known somehow or other, even if people try to conceal it, being like musk, which spreads its odour, although it be hidden. Under such circumstances, therefore, if people want to settle their genealogy, it is not necessary to spend money and to make presents, as 'Ubaid-allâh ben Alḥasan ben 'Aḥmad ben 'Abdallâh ben Maimûn Alḳaddâḥ did to the genealogists among the party of the
p. 40. Alides, when they declared his claim of descent from them to be a lie, at the time when he came forward in Maghrib; finally he succeeded in contenting them and in making them silent. Notwithstanding the truth is well known to the student, although the fabricated tale has been far spread, and although his descendants are powerful enough to suppress any contradiction. That one of them, who reigns in our time, is 'Abû-'Alî ben Nizâr ben Ma'add ben 'Ismâ'îl ben Muḥammad ben 'Ubaid-allâh the usurper.

I have enlarged on this subject only in order to show how partial people are to those whom they like, and how hostile towards those whom they hate, so that frequently their exaggeration in either direction leads to the discovery of their infamous designs.

That Alexander was the son of Philip is a fact, too evident to be concealed. His pedigree is stated by the most celebrated genealogists in this way:—

I. فيلفس Philip,
مضريو
هرمس Hermes,
هردس

V. ميطون Meton.

رومى Rome.

ليطى

يونان Yûnân.

يافث Yâfeth.

X. سوخون

رومية Rûmiya.

بزنطا Byzantium.

توفيل Theophil.

رومى Rome.

XV. الاصفر Al'aṣfar.

اليفز Elifaz.

العيص Esau.

اسحق Isaak.

XIX. ابرهيم Abraham.

According to another tradition Dhû-alḳarnaini was a man, called اطركس, who marched against Sâmîrus, one of the kings of Babel, fought with him, made him a prisoner and killed him; then he stripped off the skin of his head together with his hair and his two curls, got it tanned, and used it as a crown. Therefore, he was called Dhû-alḳarnaini (Bicornutus). According to another version he is identical with Almundhir ben Mâ-alsamâ, *i.e.* Almundhir ben Imru'ulḳais.

Altogether the most curious opinions are afloat regarding the bearer of this name, that, *e.g.*, his mother was a demon, which is likewise believed of Bilḳîs, for people say that her mother was a demon, and of 'Abdallâh ben Hilâl the juggler, for he was thought to be the devil's son-in-law, being married to his daughter. Such and similar ridiculous stories people produce, and they are far known.

It is related, that 'Umar ben Alkhaṭṭâb, when he heard one day people entering into a profound discussion on Dhû-alḳarnaini, said, "Was it not enough for you, to plunge into the stories on human beings, that you must pass into another field and draw the angels into the discussion?"

Some say, as Ibn Duraid mentions in his *Kitâb-alwishâḥ*, that Dhû-alḳarnaini was Alsa'b ben Alhammâl Alḥimyarî, whilst others take him for 'Abû-karib Shammar Yur'ish ben 'Ifrîḳîs Alḥimyarî, and believe that he was called so on account of two curls which hung down upon his shoulders, that he reached the east and west of the earth, and traversed its north and south, that he subdued the countries, and reduced the people to complete subjection. It is this prince about whom one of the princes of Yaman, 'As'ad ben 'Amr ben Rabî'a ben Mâlik ben Ṣubaiḥ ben 'Abdallâh ben Zaid ben Yâsir ben Yun'im Alḥimyarî boasts in his poems, in which he says:—

"Dhû-alḳarnaini was before me, a true believer, an exalted king on p. 41.
the earth, never subject to anybody.

He went to the countries of the east and west, always seeking imperial power from a liberal and bountiful (Lord).
Then he saw the setting-place of the sun, at the time when he sets in the well of fever-water and of badly smelling mud.
Before him there was Bilḳîs, my aunt, until her empire came to an end by the hoopoo."

Now it seems to me that of all these versions the last is the true one, because the princes, whose names begin with the word Dhû, occur only in the history of Yaman and nowhere else. Their names are always a compound, the first part of which is the word Dhû, *e.g.*, Dhû-almanâr, Dhû-al'adh'âr, Dhû-alshanâtir, Dhû-Nuwâs, Dhû-Jadan, Dhû-Yazan, and others. Besides, the traditions regarding this Yaman prince, Dhû-alḳarnaini, resemble very much that which is related of him in the Koran. As to the rampart which he constructed between the two walls, it must be stated that the wording of the Koran does not indicate its geographical situation. We learn, however, from the geographical works, as *Jighrâfiya* and the *Itineraria* (the books called *Masâlik wa-mamâlik, i.e.* Itinera et regna), that this nation, *viz.* Yâjûj and Mâjûj are a tribe of the eastern Turks, who live in the most southern parts of the 5th and 6th *κλίματα*. Besides, Muḥammad ben Jarîr Alṭabarî relates in his chronicle, that the prince of Âdharbaijân, at the time when the country was conquered, had sent a man to find the rampart, from the direction of the country of the Khazars, that this man saw the rampart, and described it as a very lofty building of dark colour, situated behind a moat of solid structure and impregnable.

'Abdallâh ben 'Abdallâh ben Khurdâdhbih relates, on the authority of the dragoman at the court of the Khalif, that Almu'taṣim dreamt one night, that this rampart had been opened (rendered accessible). Therefore he sent out fifty men to inspect it. They set out from the road which leads to Bâb-al'abwâb, and to the countries of the Lân and Khazar; finally they arrived at the rampart, and found that it was constructed of iron tiles, joined together by molten brass, and with a bolted gate. Its garrison consisted of people of the neighbouring countries. Then they returned, and the guide led them out into the district opposite Samarḳand.

From these two reports, it is evident that the rampart must be situated in the north-west quarter of the inhabitable earth. However, especially in this latter report, there is something which renders its authenticity doubtful, *viz.* the description of the inhabitants of that country, that they are Muslims and speak Arabic, although they are without the slightest connection with the civilized world, from which they are separated by a black, badly smelling country of the extent of many days' travelling; further, that they were totally ignorant as to both Khalif and the Khalifate. Whilst we know of no other Muslim nation which is separated from the territory of Islâm, except the

Bulghâr and the Sawâr, who live towards the end of the civilized world, in the most northern part of the 7th *κλίμα*. And these people do not make the least mention of such a rampart, and they are well acquainted with the Khalifate and the Khalifs, in whose name they read even the p. 42. *Khuṭba;* they do not speak Arabic, but a language of their own, a mixture of Turkish and Khazarî. If, therefore, this report rests on testimonies of this sort, we do not wish to investigate thereby the truth of the subject.

This is what I wished to propound regarding Dhû-alḳarnaini. Allâh knows best!

CHAPTER V.

ON THE NATURE OF THE MONTHS WHICH ARE USED IN THE PRECEDING ERAS.

HERETOFORE I have mentioned already that every nation uses a special era of its own. And in the same degree as they differ in the use of the eras, they differ regarding the beginning of the months, regarding the number of days of each of them, and the reasons assigned therefor. Of this subject, I mention what I have learnt, and do not attempt to find out what I do not know for certain, and regarding which I have no information from a trustworthy person. And first we give the months of the Persians.

Months of the Persians.—The number of the months of one year is twelve, as God has said in his book (Sura ix. 36): "With God the number of the months was twelve months, in the book of God, on the day when God created the heavens and the earth." On this subject there is no difference of opinion between the nations, except in the leap-years. So the Persians have twelve months of the following names:—

Farwardîn Mâh.	Mihr Mâh.
Ardîbahisht Mâh.	Âbân Mâh.
Khurdâdh Mâh.	Âdhar Mâh.
Tîr Mâh.	Dai Mâh.
Murdâdh Mâh.	Bahman Mâh.
Shahrêwar Mâh.	Isfandârmadh Mâh.

I have heard the geometrician 'Abû Sa'îd 'Aḥmad ben Muḥammad ben 'Abd-aljalîl Alsijzî relating of the ancient inhabitants of Sijistân, that they called these months by other names and commenced likewise with Farwardîn Mâh. The names are these—

I. كواك	III. اوسال
رهو	تيركيالوا

V. سرذروا	IX. آركبازوا
مرذروا	كزيهشت
نوزر	كزهن
هرانوا	XII. سازوا

Every one of the Persian months has 30 days, and to each day of a month they give a special name in their language. These are the names— p. 43.

I. Hurmuz.	XI. Khûr.	XXI. Râm.
Bahman.	Mâh.	Bâdh.
Ardîbahisht.	Tîr.	Dai-ba-dîn.
Shahrêwar.	Gôsh.	Dîn.
Isfandârmadh.	Dai-ba-mihr.	Ard.
VI. Khurdâdh.	XVI. Mihr.	XXVI. Ashtâdh.
Murdâdh.	Srôsh.	Asmân.
Dai-ba-âdhar.	Rashn.	Zâmyâdh.
Âdhar.	Farwardîn.	Mârasfand.
Âbân.	Bahrâm.	Anîrân.

There is no difference among the Persians as to the names of these days; they are the same for every month, and they follow in the same order. Only the days Hurmuz and Anîrân are called by some, the former *Farrukh*, the latter *Bih-rôz*.

The sum total of the days is 360, whilst, as we have already observed heretofore, the real year (*i.e.* the mean solar or tropical year) has $365\frac{1}{4}$ days. Those additional five days they called *Fanjî* (*Panjî*) and *Andargâh*, arabized *Andarjáh;* they are also called *Almasrûḳa* and *Almustaraḳa* (*i.e.* ἡμέραι κλοπιμαίαι), on account of their not being reckoned as part of any one of the months. They added them between Âbân Mâh and Âdhar Mâh, and gave them names, which are different from those of the days of each month. These names I never read in two books, nor heard them from two men, in the same way; they are these—

I. اهندكاه II. اهتدكاه III. اسفندكاه IV. اسفندمذكاه V. وهشتشكاه

In another book I found them in the following form:

I. اهنود II. اهتود III. اسفندمذ IV. اخشتر V. وهستوهشت

The author of the *Kitâb-alghurra,* Alnâ'ib Alâmulî gives them these names—

I. خونود II. استود III. اسفندمذ IV. وهوخوشتر V. وهشتوهشت p. 44.

Zâdawaihi ben Shâhawaihi in his book on the causes of the festivals of the Persians, mentions them in this form—

I. فنجه انوفته II. فنجه الدرنده III. فنجه اهوسته
IV. فنجه اوروردیان V. فنجه الدركاهان

I myself heard 'Abû-alfaraj ben 'Aḥmad ben Khalaf Alzanjânî say that the Mobad in Shîrâz had dictated them to him in this form—

I. اهنودكاه II. اهتودكاه III. اسبتمذكاه IV. وهوخشتركاه V. وهشتويشتكاه

And lastly, I have heard them from the geometrician 'Abû-alḥasan Âdharkhurâ, the son of Yazdânkhasîs, in this form—

I. هنود (Ahunavaiti.) II. اهتود (Ustavaiti.) III. اسبتمن (Ṣpentâmainyu.) IV. وهخشتر (Vohukhshathra.)

V. وهستوهشت (Vahistôisti.)

The sum total of their days, therefore, was 365. The quarter of a day (beyond the 365 days) they neglected in their computation, till these quarters of a day had summed up to the days of one complete month, which happened in 120 years. Then they added this month to the other months of the year, so that the number of its months became thirteen. This month they called *Kabîsa* (intercalary month). And the days of this additional month they called by the same names as those of the other months.

In this mode the Persians proceeded till the time when both their empire and their religion perished. Afterwards the day-quarters were neglected, and the years were no longer intercalated with them, and, therefore, they did not return to their original condition, and remained considerably behind the fixed points of time (*i.e.* real time). The reason was this that intercalation was an affair settled under the special patronage of their kings at a meeting of the mathematicians, literary celebrities, historiographers, and chroniclers, priests, and judges,—on the basis of an agreement of all those regarding the correctness of the calculation, after all the persons I have mentioned had been summoned to the royal court from all parts of the empire, and after they had held councils in order to come to an agreement. On this occasion money was spent profusely to such an extent, that a man who made a low estimate said, the cost had sometimes amounted to one million of denars. This same day was observed as the most important and the most glorious of all festivals; it was called the *Feast of Intercalation*, and on that day the king used to remit the taxes to his subjects.

The reason why they did not add the quarter of a day every fourth year as one complete day to one of the months or to the Epagomenæ, was this, that according to their views, not the days, but only the months are liable to being intercalated, because they had an aversion to increasing the number of the days; this was impossible by reason of the prescription of the law regarding the days on which *zamzama* (whispering prayer) must be said, if it is to be valid. If the number of days be increased by an additional day (the order of the days of *zamzama* according to the law, is disturbed).

p. 45. It was a rule that on each day a special sort of odoriferous plants and flowers was put before the *Kisrâs*, and likewise a special drink, in a well regulated order, regarding which there was no difference of opinion.

The reason why they put the Epagomenæ at the end of Âbân Mâh, between this month and Âdhar Mâh (*lacuna*).

The Persians believe that the beginning of their year was fixed by the creation of the first man, and that this took place on the day Hurmuz of Farwardîn Mâh, whilst the sun stood in the point of the vernal equinox in the middle of heaven. This occurred at the beginning of the seventh millenium, according to their view of the millennia of the world.

The astrologers hold similar opinions, *viz.* that Cancer is the horoscope of the world. For in the first cycle of Sindhind the sun stands in the beginning of Aries above the middle between the two ends of the inhabitable world. In that case, Cancer is the horoscope, which sign according to their tenets, as we have mentioned, signifies the commencement of rotation and growth.

Others say, that Cancer was called the horoscope of the world, because of all the zodiacal signs, it stands nearest to the zenith of the inhabitable world, and because in the same sign is the ὕψωμα of Jupiter, which is a star of *moderate nature ;* and as no growth is possible, except when *moderate* heat acts upon moist substances, it (*i.e.* Cancer) is fit to be the horoscope of the growth of the world.

According to a third view, Cancer was called so, because by its creation the creation of the four elements became complete, and by their becoming complete all growth became complete.

And other comparisons besides of a similar kind are brought forward by the astrologers.

Further, people relate: When Zoroaster arose and intercalated the years with the months, which up to that time had summed up from the day-quarters, time returned to its original condition. Then he ordered people in all future times to do with the day-quarters the same as he had done, and they obeyed his command. They did not call the intercalary month by a special name, nor did they repeat the name of another month, but they kept it simply in memory from one turn to another. Being, however, afraid that there might arise uncertainty as to the place, where the intercalary month would have again to be inserted, they transferred the five Epagomenæ and put them at the end of that month, to which the turn of intercalation had proceeded on the last occasion of intercalating. And as this subject was of great importance and of general use to high and low, to the king and to the subjects, and as it is required to be treated with knowledge, and to be carried out in conformity with nature (*i.e.* with real time), they used to postpone intercalation, when its time happened to occur at a period when the condition of the empire was disturbed by calamities; then they neglected intercalation so long, until the day-quarters summed up to two months. Or, on the other hand, they anticipated intercalating the year at once by two months, when they expected that at the time of the next coming intercalation circumstances would distract their attention therefrom, as it has been done in the time of Yazdajird ben Sâbûr, for no other motive but that of precaution. That was the last intercalation which they carried out, under the superintendence of a Dastûr, called Yazdajird Alhizârî. Hizâr was an estate in the district of Iṣṭakhr in Fârs, from which he received his name. In that intercalation the turn had come to Âbân Mâh; therefore, the Epagomenæ were added at its end, and there they have remained ever since on account of their neglecting intercalation.

Months of the Sogdians.—Now I shall mention the months of the Magians of Transoxiana, the people of Khwârizm and of Sughd. Their months have the same number, and the same number of days as those of the Persians. Only between the beginning of the Persian and the Transoxanian months there is a difference, because the Transoxanians
p. 46. append the five Epagomenæ to the end of their year, and commence the year with the 6th day of the Persian month Farwardîn, Khurdâdhrôz. So the beginning of the months is different until Âdhar Mâh; afterwards they have the same beginnings.

These are the names of the months of the Sughdians.

I. نوسرد	of 30 days.	VII. فغاز	of 30 days.
جرجن	"	ابانج	"
نيسن	"	فوغ	"
بساك	"	مسافوغ	"
اشناخندا	"	ژيمدا	"
مريخشندا	"	خشوم	"

Some people add a Jîm (ج) at the end of نيسن and خشوم, and pronounce نيسنج and خوشومج; they add a Nûn and a Jîm (نج) at the end of بساك and ژيمدا and pronounce بساكنج and ژيمدنج. They call each day by a special name, as is the custom with the Persians. These are the names of the thirty days—

1. خرمزد	11. خوير	21. رامن
2. جهيمز	12. ماخ	22. واد
3. اردخوشت	13. تيش	23. دست
4. خستشور	14. غش	24. دين
5. سبندارمذ	15. دست	25. ارذخ
6. ردد	16. ميخش	26. اسغاد
7. مردد	17. سروش	27. سمن
8. دست	18. رشن	28. رام جيد
9. اتش	19. فروذ	29. نشيند
10. ابجن	20. وشغشغر	30. نغر

Some people give the day خوير the name ميرٓ. The names of the five Epagomenæ are the following :—

I. خاوشست   II. اخشدن   III. رخشن   IV. وفاذن   V. اردمبيس   p.

Regarding these names the same difference exists among the Sughdians as among the Persians. They are also called by the following names :—

I. ژبورد   II. ببورد   III. سردرد   IV. ماجرد   V. ميرزدة

These five days they add at the end of the last month خشوم. The Sughdian system of intercalation agreed with the practice of the Persians, as also did their neglecting intercalation. The reason why there arose a difference between the beginnings of the Sughdian and the Persian years I shall describe hereafter.

Months of the Chorasmians.—The Khwârizmians, although a branch of the great tree of the Persian nation, imitated the Sughdians as to the beginning of the year and the place where they add the Epagomenæ. These are the names of their months—

I. روچنافو ناو سارجى	VII. اومرى
اردوشت فوسيرج انكام	ياناخن فاخسرثان راچيبك
هروداد فوخيرى	اروفو فيمحكا بحرفين
جيرى فارازاك	وثمرفونالكانج انكام
همداد	اخمن فوبرد انكام
اخشريورى	اسبندارمجى فوخشوم

Others abbreviate these names and use them in this form—

I. ناوسارچي	VII. اومرى
اردوست	ياناخن
هروداد	ارو
جيرى	ريمژد
همداد	ارسمن
اخشريوري	اسبندارمجى

The thirty days they call by the following names :—

1. ريمژد	11. اخير	21. رام
2. ازميىن	12. ماه	22. واد
3. اردوشت	13. جيرى	23. ددو
4. اخشريورى	14. غوشت	24. دينى
5. اسبندارمجى	15. ددو	25. ارجوخى
6. هروداد	16. فيغ	26. اشتاد
7. همداد	17. اسروف	27. اسمان p. 48
8. ددو	18. رشن	28. راث
9. ارو	19. روجن	29. مرسبند
10. ياناخن	20. اريشن	30. اونرغ

I have found that they begin the Epagomenæ, which are appended at the end of the month Ispandârmajî, with the same name by which they begin the days of the month; the second day they call Azmîn, the third Ardawasht, and so on till the fifth day Ispandârmajî. Then they return and commence anew with the first day ريمژد, the 1st of the month Nâwasârjî. They do not use or even know special names for the Epagomenæ, but I believe that this fact simply arises from the same confusion, regarding these names, which prevails among the Persians and Sughdians. For after Ḳutaiba ben Muslim Albâhilî had killed their learned men and priests, and had burned their books and writings, they became entirely illiterate (forgot writing and reading), and relied in every knowledge or science which they required solely upon memory. In the long course of time they forgot that on which there had been a divergence of opinion, and kept by memory only that which had been generally agreed upon. But Allâh knows best!

As to the three identical names of days (the 8th, 15th, and 23rd,—*Dai* in Persian, *Dast* in Sughdian, *Dadhû* in Khwârizmian), the Persians refer them to the following, and compound them with these, saying *Dai-ba-Âdar*, and *Dai-ba-Mihr*, and *Dai-ba-Dîn*. Of the Sughdians and Khwârizmians some do the same, and others connect the words in their language for "the first, the second, the third," with each of them.

In the early times of their empire the Persians did not use the week. For, first, it was in use among the nations of the west, and more particularly among the people of Syria and the neighbouring countries, because there the prophets appeared and made people acquainted with the first week, and that in it the world had been created, in conformity with the beginning of the Thora. From these the use of the week spread to the other nations. The pure Arabians adopted the week in consequence of
49. the vicinity of their country to that of the Syrians.

We have not heard that anybody has imitated the example of the Persians, Sughdians, and Khwârizmians, and has adopted their usage (of giving special names to the thirty days of the month, instead of dividing them into weeks), except the Copts, *i.e.* the ancient inhabitants of Egypt. For they, as we have mentioned, used the names of the thirty days till the time when Augustus, the son of Gajus, ruled over them. He wanted to induce them to intercalate the years, that they might always agree with the Greeks and the people of Alexandria. Into this subject, however, it would be necessary to inquire more closely. At that time precisely five years were wanting till the end of the great intercalation period. Therefore, he waited till five years of his rule had elapsed, and then he ordered people to intercalate one day in the months in every fourth year, in the same way as the Greeks do. Thereupon they dropped the use of the names of the single days, because, as people say, those who used and knew them would have required to invent a name for the intercalary day. They (the names of the days of the month) have not been handed down to posterity.

Months of the Egyptians.—The following are the names of their months:—

I. Thot	30 days.		VII. Phamenoth	30 days.	
Paophi	30	„	Pharmuthi	30	„
Athyr	30	„	Pachon	30	„
Choiak	30	„	Payni	30	„
Tybi	30	„	Epiphi	30	„
Mechir	30	„	ابيقا	30	„

These are the ancient names of the months. In the following we give the names which were modernized by one of their princes, after intercalation had been adopted:—

I. توت	VII. برمهات
بابه	برمودة
هتور	بشنس
كيهك	بونة
طوبة	ابيب
امشير	مسرى

Some people call the months كيهك, برمهات, بشنس, and مسرى by the names كياك, برمهوط, بشانس and ماسورى. These are the forms on which people agree; in some books, however, these names are found in forms somewhat different from those we have mentioned.

The five additional days they call, Ἐπαγομέναι, which means "*the small month*;" they are appended at the end of Mesori, and at the same place the intercalary day is added, in which case the Epagomenæ are six days. The leap-year they call الثقا, which means "*the sign.*" p.

Months of the People of the West.—'Abû-al'abbâs Alâmulî relates in his *Kitâb-dalâ'il-alḳibla*, that the *Western people* (of Spain?) use months, the beginnings of which agree with those of the Coptic months. They call them by the following names:—

I. May	30 days.		VII. November	30 days.	
June	30	„	December	30	„
July	30	„	January	30	„
August	30	„	February	30	„
September	30	„	March	30	„
October	30	„	April	30	„

Then follow the five Epagomenæ at the end of the year.

Months of the Greeks.—The months of the Greeks are always twelve in number. Their names are these:—

I. Ἰανουάριος	31 days.		IV. Ἀπρῖλις	30 days.	
Φεβρουάριος	28	„	Μαῖος	31	„
Μάρτιος	31	„	Ἰούνιος	30	„

VII.	Ἰούλιος	31 days.	X.	Ὀκτώβριος	31 days.
	Αὔγουστος	31 „		Νοέμβριος	30 „
	Σεπτέμβριος	30 „		Δεκέμβριος	31 „

The sum of the days of their year is 365, and as in all four years the four quarters of a day are summed up, they append it as one complete day to the month February, so that this month has in every fourth year 29 days. He who first induced people to intercalate the years was Julius, called Dictator, who ruled over them in bygone times, long before Moses. He gave them the months with such a distribution (of the days), and with such names as we have mentioned. He induced them to intercalate the day-quarters into them (the months) in every 1461st year, when the day-quarters had summed up to one complete year. So that (this intercalation) preserved these (the months, keeping them in agreement with real time). This intercalation they called the "*great one*," after they had called the intercalation, which takes place every four years, the "*small one*." This "small" intercalation, however, they did not introduce until a long period had elapsed after the death of the king (Julius Cæsar). A characteristic of their system is the division of the days of the months into weeks, for reasons which we have mentioned before.

51. The author of the *Kitâb-ma'khadh-almawâḳît* (method for the deduction of certain times and dates) thinks that the Greeks and other nations, who are in the habit of intercalating the *day-quarter*, had fixed the sun's entering Aries upon the beginning of April, which corresponds to the Syrian Nîsân, as the beginning of their era. And we confess that in his account he comes pretty near the truth. For astronomical observation has taught that the fraction which follows the (365) days of the solar year, is less than one complete quarter of a day, and we ourselves have observed that the sun's entering the first part of Aries precedes the beginning of Nîsân. Therefore that which he mentions is possible, and even likely.

Further on he says, speaking of the Greeks, that "they, on perceiving that the beginning of their year had changed its place, had recourse to the years of the Indians; that they then intercalated into their year the difference between the two years (viz., the Greek year and the solar year), and that in consequence the sun's entering the first part of Aries again took place at the beginning of Nîsân. If we on our side do the same, Nîsân returns to its original place." He has tried to give an example, but has not finished it, being incapable of doing so. On this occasion he has shown his ignorance, as he, in his account of the Greeks, has also rendered it evident that he is inimical to the Greeks, and partial to others. The fact is, that according to the Indian system he has converted the difference between the Greek year and the solar year into fractions, putting it down as 729 seconds. Then he changes also the day into seconds, and divides them by that difference. So he gets 118 years 6 months and $6\frac{2}{3}$ days. This would be the space of time in which the calendar would necessitate the intercalation of one complete day, on account of this plus-difference. Further, he says, "Now, if we intercalate the past years of the Greek era," which were at his time 1,225 years, "the sun's entering the first part of Aries again takes place at the beginning of Nîsân." But he has dropped his example, and has not intercalated the years. If he had done so, his conclusions would have led to the contrary of what he says and maintains, and the beginning of Nîsân would come near the sun's entering the first part of Taurus. For that date, which he wanted to treat as an example, would necessitate the intercalation of $10\frac{1}{3}$ days. Now the Greek year being too short (according to him), the beginning of Nîsân precedes the sun's entering the first part of Aries, and the time which it would be necessary to intercalate (*portio intercalanda*), would have to be added to the first of Nîsân, so as to proceed as far as to the 10th of it.

Now I should like to know which equinox this man, who is so partial to the Indians, meant. For the vernal equinox took place according to their system at that time six or seven days before the first of Nîsân. I should further like to know at what time the Greeks did what he relates of them. For they are so deeply imbued with, and so clever in geometry and astronomy, and they adhere so strictly to logical arguments, that they are far from having recourse to the theories of those who derive the bases of their knowledge from divine inspiration, when their artifices desert them and they are required to come forward with an argument; not to mention the sciences of philosophy and theology, physics and p. 52.
arts, cultivated among the Greeks. "However, everybody acts according to his own mode, and each community enjoys what they have got of their own." (Sûra xvii. 86.) That man had not read the Almagest, and had not compared it with the most famous book of the Indians, called the *Canon Sindhind*. The difference between them must be evident to anybody in whom the slightest spark of sagacity is left.

To something similar Ḥamza ben Alḥasan Alisfahânî has applied himself in his treatise on the Naurôz, at the time when he was partial to the Persian mode of treating the solar year, because they reckoned it as 365 days and $6\frac{81}{400}$ hours, while the Greeks neglected in their intercalation the fraction following the six hours. As a proof he adduced that Muḥammad ben Mûsâ ben Shâkir, the astronomer, had explained this subject, and had enlarged on it in one of his books on the solar year, and that he had produced the arguments for it, and pointed out the errors of the ancients, who had held erroneous views in this respect.

Now, we have examined the astronomical observations of Muḥammad ben Mûsâ, and of his brother 'Aḥmad, and we have found that they prove only that these fractions are less than six hours. The book, to which Alisfahânî refers, is attributed to Thâbit ben Ḳurra, because he was a *protégé* of those people, entirely mixed up with them, and because it was he who polished for them their scientific work. He had collected the materials of this book with the object of explaining the fact of the solar years not being always equal to each other, on account of the motion of the apogee. With all this he was compelled to assume equal circles, and equal motions along with their times, in order to derive thereby the mean motion of the sun. But he did not find equal circles, except those which move in an excentric plane, described (*viz.* the circles) round a point within it, which point is assumed exclusively for these circles. And this circle, which was sought for, extends the six hours by additional fractions (*i.e.* its time of revolution is 365 days 6 hours + *a fraction*), as Ḥamza has related. However, such a circle is not called a solar year, for the solar year is, as we have defined already, that one, in which all natural occurrences which are liable to growth and to decay return to their original condition.

Jewish Months.—The Hebrews and all the Jews, who claim to be related to Moses, have the following twelve months:—

I. Nîsân	of 30 days.	VI. Tishrî	of 30 days.
Jyâr	of 29 „	Marḥeshwân	of 29 „
Sîwân	of 30 „	Kislew	of 30 „
Tammûz	of 29 „	Ṭêbeth	of 29 „
Âbh	of 30 „	Shefaṭ (Shebhaṭ)	of 30 „
Elûl	of 29 „	Adhâr	of 29 „

53. The sum total of their days is 354, being identical with the number of days of the lunar year. If they simply used the lunar year as it is, the sum of the days of their year and the number of their months would be identical. However, after having left Egypt for the desert Al-tîh, after having ceased to be the slaves of the Egyptians, having been delivered from their oppression, and altogether separated from them, the Israelites received the ordinances and the laws of God, described in the second book of the Thora. And this event took place in the night of the 15th Nîsân *at full moon* and *spring time*. They were ordered to observe this day, as it is said in the second book of the Thora (Exodus xii. 17, 18): "Ye shall observe this day as an ordinance to your generations for ever on the fourteenth of the first month." By the "*first month*" the Lord does not mean Tishrî, but Nîsân; because in the same book he commands Moses and Aaron, that the month of passover should be the first of their months, and the beginning of the year (Exodus xii. 2).

Further, Moses spake unto the people: "Remember the day when ye came out from bondage. Therefore ye shall not eat leavened bread on this day in that month when the trees blossom." In consequence, they were compelled to use the solar year and the lunar months; the solar year in order that the 14th Nîsân should fall in the beginning of spring, when the leaves of the trees and the blossoms of the fruit trees come forth; the lunar months in order that, on the same day, the body of the moon should be lit up completely, standing in the sign of Libra. And as the time in question would naturally advance for a certain number of days (the sum of the days of twelve lunar months not being a complete year), it was necessary for the same reason to append to the other months those days, as soon as they made up one complete month. They added these days as a complete month, which they called the *First Adhâr*, whilst they called the original month of this name the *Second Adhâr*, because of its following immediately behind its namesake. The leap-year they called *'Ibbûr* (עִבּוּר), which is to be derived from *Me'ubbereth* (מְעֻבֶּרֶת), meaning in Hebrew, "*a pregnant woman.*" For they compared the insertion of the supernumerary month into the year, to a woman's bearing in her womb a foreign organism.

According to another opinion, the First Adhâr is the original month, the name of which without any addition was used in the common year, and the Second Adhâr is to be the leap-month, in order that it should have its place at the end of the year, for this reason, that according to the command of the Thôrâ, Nîsân was to be the first of their months. This, however, is not the case. That the Second Adhâr is the original month, is evident from the fact, that its place and length, the number of its days, the feast- and fast-days which occur in it, are not liable to any changes. And of all these days nothing whatsoever occurs in the First Adhâr of a leap-year. Further, they make it a rule that, during the Second Adhâr, the sun should always stand in the sign of Pisces, whilst in the First Adhâr of a leap-year he must be in the sign of Amphora.

Five Cycles.—Now for the leap-years they wanted a certain principle p. 54.
of arrangement as a help to facilitate their practical use. Therefore they looked out for cycles which were based upon solar years, consisting of lunar months. Of those cycles they found the following five:—

I. The cycle of 8 years consisting of 99 months, of which there are 3 leap-months.

II. The cycle of 19 years, called the *Minor Cycle*, consisting of 235 months, of which there are 7 leap-months.

III. The cycle of 76 years, consisting of 940 months, of which there are 28 leap-months.

IV. The cycle of 95 years, called the *Middle Cycle*, consisting of 1,176 months, of which there are 35 leap-months.

V. The cycle of 532 years, called the *Major Cycle*, consisting of 6,580 months, of which there are 196 leap-months.

Of these cycles they choose that one, the observation of which would be the easiest and simplest. This quality is peculiar to the cycles of 8 and of 19 years, with this difference, however, that the latter one agrees more closely with solar years. For this cycle contains, according to them, 6,939 days $16\frac{595}{1080}$ hours. Those small particles of an hour they call *Ḥalaḳs* (חֲלָקִים), of which 1,080 make one hour. If, therefore, you have got minutes, *i.e.* the 60th parts of an hour, and you want to change them into Ḥalaḳs, you multiply them by 18, and you get the corresponding number of Ḥalaḳs. And if you want the converse operation, you multiply the number of Ḥalaḳs by 200, and you get a sum of thirds of an hour (*i.e.* the 60th parts of a second); these fractions you can then raise to wholes.

Now, if we reduce this cycle (of 19 years) to fractions and change it into Ḥalaḳs, we get the following sum of Ḥalaḳs:—

179,876,755, expressed in Indian ciphers.

The solar year is, according to them, 365 days $5\frac{3791}{4104}$ hours long; this latter fraction is nearly identical with 990 Ḥalaḳs. If we now also reduce the solar year into Ḥalaḳs, we get the sum of—

9,467,190 Ḥalaḳs.

If you finally divide by this number the number of the Ḥalaḳs of the cycle of 19 years, you get as the quotient, 19 solar years, with a remainder
p. 55. of 145 Ḥalaḳs, which is nearly the 7th part of an hour and a fraction.

If we perform the same operation with the cycle of 8 years containing 2,923 days 12 hours and 747 Ḥalaḳs, we get as the sum of its Ḥalaḳs the number—

75,777,867

If we divide this sum by the sum of the Ḥalaḳs of the solar year, we get 8 solar years, and a remainder of 1 day 13 hours and 387 Ḥalaḳs, which is nearly $\frac{1}{5}+\frac{1}{6}$ (*i.e.* $\frac{11}{30}$) hour.

Hence it is evident that the cycle of 19 years comes nearest to real time, and is the best of all cycles which have been used. The other cycles are simply composed of duplications of the cycle of 19 years. Therefore the Jews preferred this cycle, and regulated thereby intercalation.

The three Ordines Intercalationis.—Now, although they agreed on the quality of the year as to the order of intercalation in the *Maḥzôr* (מחזור cycle), when it has to take place, and when not, they differed among each other regarding the nature of the beginning of the Maḥzôrs. And this has also produced a difference regarding the order of intercalation in the Maḥzôr. For some take the current year of the Æra Adami, of which you want to know whether it is a common year or a leap-year, and reduce the number of years to Maḥzôrs by dividing them by 19; then you get complete Maḥzôrs, and as a remainder, the years of the Maḥzôr not yet finished, including the current year. And then the order of the leap-years is fixed according to the formula בהזיגוח *i.e.* the 2nd, 5th, 7th, 10th, 13th, 16th, and 18th years.

Others take the years of the same Æra Adami, subtract one year, and fix the order of the leap-years in the remainder of the years of the incomplete Maḥzôr according to the formula אדוטבדהז, *i.e.* the 1st, 4th, 6th, 9th, 12th, 15th, 17th years. These two cycles are attributed to the Jews of Syria.

Others again subtract from the sum of years two years, and compute the order of the leap-years by the formula גבטבבג, *i.e.* the 3rd, 5th (5=3+2), 8th, 11th, 14th (5+3+3+3), 16th (16=14+2), and 19th (19=16+3) years.

This latter mode of arrangement is the most extensively diffused among the Jews; they prefer it to others, because they attribute its invention to the Babylonians. All three modes of computation are to be traced back to one and the same principle, on which there cannot be any difference of opinion, as is illustrated by the following circular figure:—

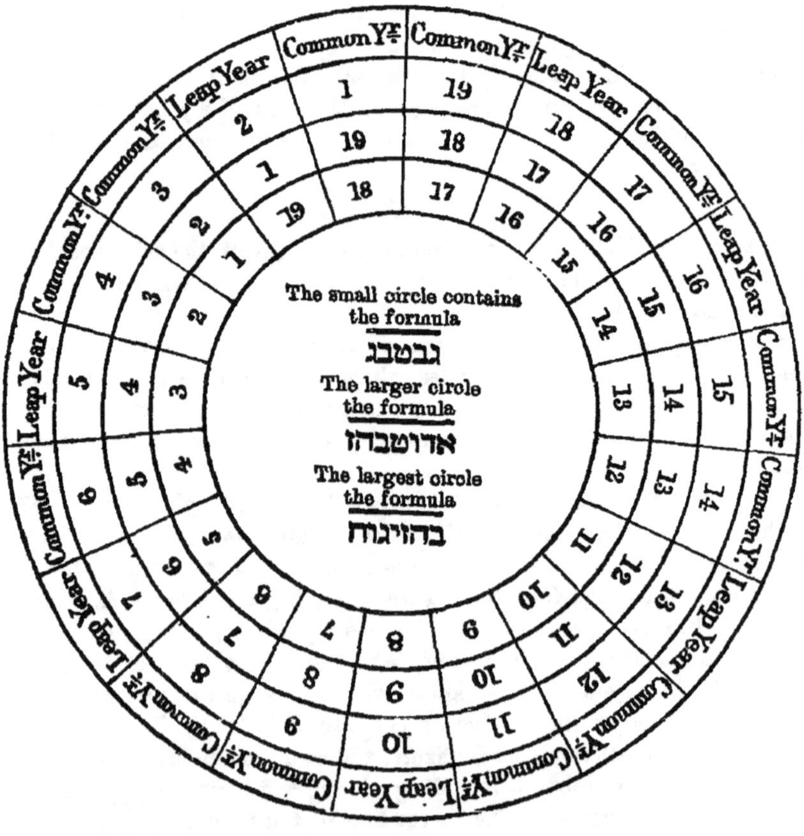
The small circle contains
the formula
גבטבבג
The larger circle
the formula
אדוטבדהז
The largest circle
the formula
בהזיגוה
Common Yr.
Leap Year

The first (outer) circle indicates the quality of the year, whether it is a common year or a leap-year. The three other circles contain the three *formulæ*, indicating the order of the leap-years in the Maḥzôrs; the
p. 56. second circle, the formula בהזיגוח; the third circle, the formula אדוטבהז; and the inner circle, the formula גבטבג.

The cycles which we have mentioned hitherto, are derived from the moon, though not exclusively. The solar cycle consists of 28 years, and serves to indicate on what days of the week the solar years commence. For if the Jewish year had simply 365 days without the quarter-day, the beginning of the year would in every seven years return to the same week-day. Since, however, they are intercalated once in four years, the beginning does not return to the same day, except in 28 years, *i.e.* 4 × 7 years. Likewise the other cycles, heretofore mentioned, do not, on being completed, return to the same week-day, except the largest cycle, on account of its arising from a duplication of the cycle of 19 years with the solar cycle.

The three kinds of the Jewish Year.—I say further: If the Jewish years had simply the first two qualities, *i.e.* were either common years or leap-years, it would be easy to learn their beginnings, and to distinguish between the two qualities which are proper to them, provided the above-mentioned formula of computation for the years of the Maḥzôr be known. The Jewish year, however, is a threefold one. For they have made an arrangement among themselves, that New Year shall not fall on a Sunday, Wednesday, or Friday, *i.e.* on the days of the sun and his two stars (Mercury and Venus); and that Passover, by which the beginning of Nîsân is regulated, shall not fall on the days of the inferior stars, *i.e.* on Monday, Wednesday, and Friday, for reasons on which we shall hereafter enlarge as much as possible. Thereby they were compelled either to postpone or to advance New Year and Passover, when they happened to fall on one of the days mentioned.

For this reason their year consists of the following three species:—

I. The year called حسارين, *i.e.* the imperfect one (חֲסֵרָה), in which the months Marḥeshwân and Kislêw have only 29 days.

II. The year called كسدران, *i.e.* the intermediate (כְּסִדְרָן), lit. *secundum ordinem suum*, in which Marḥeshwân has 29 days, and Kislêw 30 days.

III. The year called هلاميم, *i.e.* the perfect one (שְׁלֵימָא), in which both Marḥeshwân and Kislêw have 30 days.

Each of these three species of years may be either a common year or a leap-year. So we get a combination of six species of years, as we have here illustrated in the form of a genealogical diagram, and distributed in the following representation.

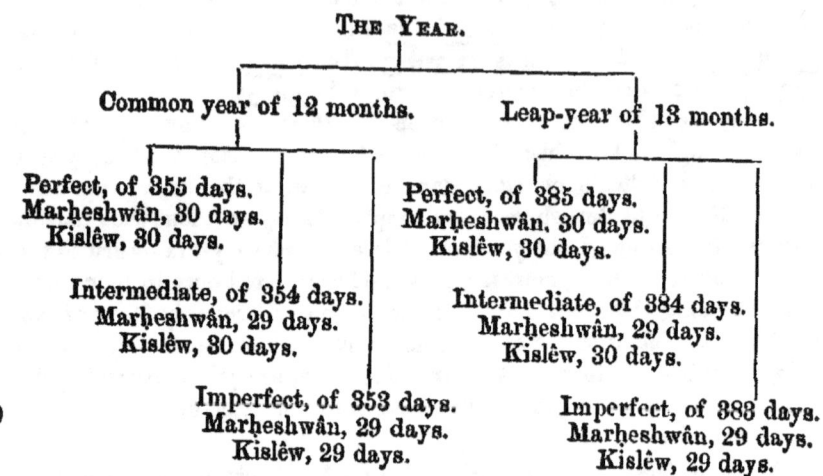

THE YEAR. p. 57

- Common year of 12 months.
  - Perfect, of 355 days. Marḥeshwân, 30 days. Kislêw, 30 days.
  - Intermediate, of 354 days. Marḥeshwân, 29 days. Kislêw, 30 days.
  - Imperfect, of 353 days. Marḥeshwân, 29 days. Kislêw, 29 days.
- Leap-year of 13 months.
  - Perfect, of 385 days. Marḥeshwân, 30 days. Kislêw, 30 days.
  - Intermediate, of 384 days. Marḥeshwân, 29 days. Kislêw, 30 days.
  - Imperfect, of 383 days. Marḥeshwân, 29 days. Kislêw, 29 days.

For the deduction of these differences they have many modes of computation as well as tables, which we shall not fail to explain hereafter.

Determination of New Moon.—Regarding their knowledge of the beginning of the month, and the mode in which it is computed and used, the Jews are divided into two sects, one of which are the Rabbanites. They derive the beginning of the month by means of calculation from the mean motions of the two luminaries (sun and moon), no regard being had as to whether new moon is visible already or not. For it was their object to have a conventional time, that was to begin from the conjunction of sun and moon. By the following accident they were, as they relate themselves, induced to adopt this system: at the time when they returned to Jerusalem, they posted guards upon the tops of the mountains to observe new moon, and they ordered them to light a fire and to make a smoke, which was to be a signal for them that new moon in fact had been seen. Now, on account of the enmity which existed between them and the Samaritans, these latter went and sent up the smoke from the mountain one day before new moon was seen. This practice they continued during several months, at the beginnings of which heaven always happened to be clouded. Finally, people in Jerusalem found out this, observing that new moon, on the 3rd and 4th of the month, rose above the horizon from the east. Hence it was evident that the Samaritans had deceived them. Therefore they had recourse to the scholars of their time, in order to be protected by a system of calculation against the deceitful practices of their enemies, to which they were exposed by their present method.

In order to prove that it was legally permitted to fix the beginning of the month by calculation instead of observation, they referred to the duration of the deluge. For they assert that Noah computed and fixed the beginnings of the months by calculation, because heaven was covered p. 58.

and clouded for so long as six months, during which time neither new moon nor any other phase of the moon could be observed.

The mathematicians, therefore, computed for them the cycles, and taught them how to find, by calculation, the conjunctions and the appearance of new moon, viz. that between new moon and the conjunction the time of 24 hours must elapse. And this comes near the truth. For if it was the corrected conjunction, not the mean one, the moon would in these hours move forward about 13 degrees, and her elongation from the sun would be about 12 degrees.

This reform was brought about nearly 200 years after Alexander. Before that time they used to observe the *Teḳûfôth* (תְּקוּפוֹת), *i.e.* the year-quarters, on the computation of which we shall enlarge hereafter, and to compare them with the conjunction of that month, to which the Teḳûfâ in question was to be referred. If they found that the conjunction preceded the Teḳûfâ by about 30 days, they intercalated a month in this year, *e.g.* if they found that the conjunction of Tammûz preceded the Teḳûfâ of Tammûz, *i.e.* the summer-solstice by about 30 days, they intercalated in that year a month Tammûz, so that it had one Tammûz and a second Tammûz (תַּמּוּז וְתַמּוּז). In the same way they acted with the other Teḳûfôth.

Some Rabbanites, however, deny that such guards were posted, and that they made a smoke as a signal. According to their opinion, the cause of the deduction of this system of calculation was the following: the scholars and the priests of the Israelites, feeling convinced that their people would be scattered and dispersed in consequence of the last destruction of Jerusalem, as they thought, were afraid that their compatriots, being scattered all over the world, and solely relying upon the appearance of new moon, which of course in different countries would be different for them, might, on account of this, fall into dissensions, and a schism in their doctrine might take place. Therefore they invented these calculations,—a work which was particularly attended to by Eliêzer ben Pârûaḥ, and ordered people to adhere to them, to use them, to return to them, wherever and under whatever circumstances they lived, so that a schism among them might be avoided.

The second sect are the *Milâdites*, who derive the beginning of the month from the conjunction; they are also called *Alḳurrâ* and *Al'ishma'iyya*, because they demand that people shall only follow the wording of the text, no regard being had to considerations and analogies, etc., even if it may be illogical and impracticable.

One party of them is called the *'Ânânites*, who derived their name from 'Ânân, the head of the emigration (ראש גלותא), who lived between 100 and 110 years ago. A head of the emigration must of necessity be one of the descendants of David; an offspring of another family would not be fit for this office. Their common people relate, that only he is qualified who, standing upright, can reach his knees with the tops of his fingers; just as people relate such things of the prince of the true believers, 'Alî ben 'Abî Ṭâlib, and of those of his descendants who are qualified for the Imâma and the rule of the community (the Muḥammadan world).

The genealogy of this 'Ânân is the following:—

I. ענן בן דניאל ב׳ שאול ב׳ ענן ב׳ דוד
VI. ב׳ חסדאי ב׳ כפנאי ב׳ בסתנאי ב׳ הנומר ב׳ זוטרא
XI. ב׳ רב הונא ב׳ שפטיה ב׳ הונא ב׳ נתן ב׳ אבא מר
XVI. ב׳ רבנא עקיבא ב׳ שבניא ב׳ זכאי ב׳ חזקיא ב׳ שמעיא
XXI. ב׳ שפתיא ב׳ יוחנן ב׳ רצוציתא ב׳ ענן ב׳ ישעיא
XXVI. ב׳ זכריא ב׳ ברכיא ב׳ עקוב ב׳ חנניא ב׳ בסודיא p. 59.
XXXI. ב׳ מעשיא ב׳ פדיה ב׳ זרבבל ב׳ שאלתיאל ב׳ יהויכין
XXXVI. ב׳ יהויקים ב׳ יהואחז ב׳ יאשיה ב׳ אחזיא ב׳ יהורם
XLI. ב׳ יהושפט ב׳ אסה ב׳ אביה ב׳ רחבעם ב׳ שלמה
XLVI. ב דוד

He opposed a community of Rabbanites in many of their observances. He fixed the beginning of the month by the appearance of the new moon in a similar way, as is prescribed in Islâm, not caring on what day of the week the beginning of the month happened to fall. He gave up the system of computation of the Rabbanites, and made the intercalation of a month depend upon the observation of barley-seed in 'Irâk and Syria between the 1st and the 14th Nîsân. If he found a first-fruit fit for friction and reaping, he left the year as a common year; if he did not find that, he intercalated the year. The mode of prognosticating the state of the corn was practically this, that one of his followers went out on the 23rd Shebâṭ, to examine—in Syria and the countries of a similar climate—the state of the barley-seed. If he found that the Safâ, *i.e.* the prickles of the beard of the ear of corn, had already come out, he counted from that day till Passover 50 days; if he found that it had not yet come out, he intercalated a month into the year. And some added the intercalary month to Shefaṭ, so that there was a *Shefaṭ* and an *U-Shefaṭ*; whilst others added it to Adhâr, so that there was an *Adhâr* and a *We-Adhâr*. The Ânânites mostly use Shefaṭ, not Adhâr, whilst the Rabbanites use exclusively Adhâr.

This system of prognosticating the state of the corn is a different one according to the difference of the air and the climate of the countries. Therefore it would be necessary to make a special rule for every place, and not to rely upon the rule made for one certain place, because this would not be applicable elsewhere.

Syrian Months.—The Christians in Syria, 'Irâk, and Khurâsân have combined Greek and Jewish months. For they use the months of the Greeks, but have adopted the 1st of the Greek October as the beginning of their year, that it might be nearer to the Jewish new-year, because Tishrî of the Jews always precedes that date a little. And they call their months by Syrian names, some of which agree with the Jewish names, whilst others differ. People have derived these names from the Syrians, *i.e.* the Nabatæans, the inhabitants of Sawâd; the Sawâd of 'Irâḳ being called Sûristân. But I do not see why they derive these months from them, because in Islâm they use the months of the Arabs, and at the time of heathenism they used the months of the Persians. Others say that Sûristân means Syria. If that be the case, the inhabitants of this country were Christians before the time of Islâm, and held a middle position between Jewish and Greek theories.

30. The names of their months are these :—

I.	Tishrîn ḳedîm	of 31 days.	VI.	Nîsân	of 30 days.
	Tishrîn ḥrâi	of 30 „		Iyâr	of 31 „
	Kânûn ḳedîm	of 31 „		Ḥazîrân	of 30 „
	Kânûn ḥrâi	of 31 „		Tammûz	of 31 „
	Shebâṭ	of 28 „		Âbh	of 31 „
	Adhâr	of 31 „		Îlûl	of 30 „

In the month Shebâṭ they intercalate one day every four years, so that it then has 29 days. Regarding the quality of their year they agree with the Greeks.

These months have become widely known, so that even the Muslims adopted them, and fixed thereby the dates of practical life. The words *Ḳedîm* (primus) and *Ḥrâi* (postremus) have been translated into Arabic, and in the word اير they have added an Elif, so as to make it ايار, because a single *yâ* (without *Tashdîd*) is disagreeable to the organ of the Arabs, if this Elif is not added.

Months of the Arabs.—The Arabs have the following twelve months :—

I.	Almuḥarram.	VII.	Rajab.
	Ṣafar.		Sha'bân.
	Rabî' I.		Ramaḍân.
	Rabî' II.		Sh wwâl.
	Jumâdâ I.		Dhû-alka'da.
	Jumâdâ II.		Dhû-alḥijja.

Regarding the etymology of these names various opinions have been advanced, *Almuḥarram*, *e.g.* was called so, because it was one of the *Ḥurum*, *i.e.* the four sacred months.

Ṣafar was called so, because in it people used to procure their provisions, going out in a company of men which was called *Ṣafariyya*.

The two months *Rabî'* were called so on account of the coming forth of the flowers and blossoms and of the continual fall of dew and rain.

All of which refers to the nature of that season which we call "autumn," but which the Arabs called "spring" (*Rabî‘*).

The two months *Jumâdâ* were called so, because in them the water *froze* (جمد).

Rajab was called so, because in it people formed the intention of travelling, there being no fear of the evils of war. For "*rujba*" means *sustentaculum* (a thing by means of which a tree is propped up), and hence people say, "*a propped up (murajjab) palm-tree which bears a heavy load of fruit.*"

Sha‘bân was called so, because in it the tribes were *dispersed.*

Ramaḍân was called so, because of the stones being *roasted* by the intense heat.

Shawwâl was called so, because of the increasing and the decreasing of the heat.

Dhû-alḳâ‘da was called so, because in it people *stayed* in their homes.

Dhû-alḥijja was called so, because in it people performed the *Ḥajj, i.e.* the pilgrimage.

We found, however, also other names of the months of the Arabs, which were given to them by their ancestors. They are the following:—

I. Almu'tamir.	VII. Al'aṣamm. p. 61.
Nâjir.	‘Âdil.
Khawwân.	Nâfiḳ.
Ṣuwân.	Wâghil.
Ḥantam.	Huwâ‘.
Zabbâ.	Burak.

The forms as well as the order of these names sometimes differ from what we have given. One of the poets, *e.g.* has comprised them in the following verses :—

"We have commenced with *Mu'tamir*, *Nâjira*, and *Khawwân*, to which follows *Ṣuwân*.
And with *Zabbâ* comes *Bâ'ida*, its next follower. Then comes the turn of '*Aṣamm*, in which hatred was deaf.
And *Wâghila*, *Nâṭila*, and *‘Âdila*, all three are noble and beautiful.
Then comes *Ranna*, and after it *Burak*. Now are complete the months of the year, as you may count with your fingers."

In the following we shall explain the meanings of these names according to the statements of the dictionaries :—

Almu'tamir means that it "*obeys*" all the decrees of fortune, which the year is going to bring.

Nâjir is derived from *najr*, which means "intense heat," as it is used in the following verse:—

"A stinking water, on account of which a man turns his face aside,
Even he who is tortured by thirst, if he tasted it in a '*boiling hot*' month."

Khawwân is the form فَعَّال of the verb "*to deceive*," and *Ṣuwân* is the form فُعَّال of the verb "*to preserve, to take care*." And these significations agreed with the natures of the months at the time when they were first employed as names for them.

Zabbâ means a "*great and frequently occurring calamity*." The month was called so, because in it there was much and frequent fighting.

Bâ'id, too, received its name from the fighting in it, for many people used to "*perish*" in it. This circumstance is expressed in the following proverb: "*All that is portentous happens between Jumâdâ and Rajab*." For in this month people were in great haste and eagerness to carry out whatever blood revenge or warlike expeditions they were upon, before the month Rajab came in.

'Aṣamm was called so, because in it people abstained from fighting, so that the clash of weapons was not heard.

Wâghil means "*one who comes to a drinking-party without having been invited*." This month was called so, because it suddenly comes in after Ramaḍân, and because in Ramaḍân there was much wine-drinking, on account of the next following months being the months of pilgrimage.

Nâṭil means "*a measure, a pot of wine*." The month was called so, because in it people indulged in drinking debauches, and frequently used that pot.

'Âdil is derived from "*'adl*" (which means either "to be just" or "to turn aside"). The month was called so, because it was one of the months of pilgrimage, when they used to abstain from the use of the Nâṭil, *i.e.* the wine-pot.

Ranna was called so, because the sheep were "*crying*" on account of the drawing near of the time when they were to be killed.

Burak was called so, because of the kneeling down of the camels on being led to the slaughtering-place.

A better versification of these names than the above-mentioned one is that by the Wazîr 'Ismâ'îl ben 'Abbâd:—

"You wanted to know the months of the pagan Arabs. Take them according to the order of Muḥarram (Ṣafar, etc.), of which they partake.
p. 62. First comes *Mu'tamir*, then *Nâjir*; and *Khawwân* and *Suwân* are connected by one tie.
Ḥanîn, *Zabbâ*, *'Aṣamm*, *Âdil*, *Nâfiḳ* with *Waghl*, and *Ranna* with *Burak*."

If the etymologies of these two classes of names of the months are such as we have related, we must suppose that between the two periods of giving the names there was a great interval of time. Or else our explanations and etymologies would not be correct. For in one class of the months the highest pitch of the heat is Ṣafar, whilst in the other it is Ramaḍân; and this (that the greatest heat should be either in Ṣafar or in Ramaḍân) is not possible at one and the same period, or at two periods which are not very far distant from each other.

Intercalation of the Ancient Arabs.—At the time of paganism the Arabs used their months in a similar way to the Muslims; their pilgrimage went wandering around through the four seasons of the year. But then they desired to perform the pilgrimage at such time as their merchandise (hides, skins, fruit, etc.) was ready for the market, and to fix it according to an invariable rule, so that it should occur in the most agreeable and abundant season of the year. Therefore they learned the system of intercalation from the Jews of their neighbourhood, about 200 years before the Hijra. And they used intercalation in a similar way to the Jews, adding the difference between their year and the solar year, when it had summed up to one complete month, to the months of their year. Then their intercalators themselves, the so-called *Ḳalâmis* of the tribe Kinâna, rose, after pilgrimage had been finished, delivered a speech to the people at the fair, and intercalated the month, calling the next following month by the name of that month in which they were. The Arabs consented to this arrangement and adopted the decision of the *Ḳalammas*. This proceeding they called "*Nasî*," *i.e. postponement*, because in every second or third year they *postponed* the beginning of the year for a month, as it was required by the progression of the year. One of their poets has said:—

"We have an intercalator, under whose banner we march;
He declares the months profane or sacred, as he likes."

The first intercalation applied to Muḥarram; in consequence Ṣafar was called Muḥarram, Rabî' I. was called Ṣafar, and so on; and in this way all the names of all the months were changed. The second intercalation applied to Ṣafar; in consequence the next following month (Rabî I.) was called Ṣafar. And this went on till intercalation had passed through all twelve months of the year and returned to Muḥarram. Then they commenced anew what they had done the first time.

The Arabs counted the cycles of intercalation and fixed thereby their dates. They said for instance: "From the time x till the time y the years have turned round one cycle."

But now, if notwithstanding intercalation it became evident that a month progressed beyond its proper place in the four seasons of the year, in consequence of the accumulation of the fractions of the solar year, and of the remainder of the *plus-difference* between the solar year and the lunar year, to which latter they had added this plus-difference, they made a second intercalation. Such a progression they were able to recognize from the rising and setting of the Lunar Mansions. This went on till the time when the Prophet fled from Makka to Madîna, when the turn of intercalation, as we have mentioned, had come to Sha'bân.

68. Now, this month was called Muḥarram, and Ramaḍân was called Ṣafar. Then the Prophet waited till the "*farewell pilgrimage*," on which occasion he addressed the people, and said: "The season, the time has gone round as it was on the day of God's creating the heavens and the earth." (Sûra ix. 38.) By which he meant that the months had returned to their original places, and that they had been freed from what the Arabs used to do with them. Therefore, the "*farewell pilgrimage*," was also called "*the correct pilgrimage*." Thereupon intercalation was prohibited and altogether neglected.

Months of the Themudeni.—'Abû-Bakr Muḥammad ben Duraid Al'azdî relates in his *Kitâb-alwishâḥ*, that the people Thamûd called the months by the following names:—

I.	Mûjib *i.e.* Muḥarram.	VII.	Haubal.
	Mûjir.		Mauhâ.
	Mûrid.		Daimur.
	Mulzim.		Dâbir.
	Muṣdir.		Ḥaifal.
	Haubar.		Musbil.

He says that they commenced their year with the month Daimur, *i.e.* Ramaḍân. The following is a versification of these names by 'Abû-Sahl 'Îsâ ben Yaḥyâ Almasîḥî:—

"The months of Thamûd are *Mûjib, Mûjir, Mûrid;* then follow *Mulzim* and *Muṣdir.*
Then come *Haubar* and *Haubal,* followed by *Mauhâ* and *Daimur.*
Then come *Dâbir,* and *Ḥaifal,* and *Musbil,* till it is finished, the most celebrated among them."

Arabic Names of Days.—The Arabs did not, like the Persians, give special names to the single days of the month, but they had special names for each three nights of every month, which were derived from the state of the moon and her light during them. Beginning with the first of the month, they called—

The first three nights (1st–3rd) *ghurar*, which is the plural of *ghurra*, and means *the first of everything*. According to others they were called so, because during them the new moon appeared like *a blaze on the forehead of a horse.*

The second three nights (4th–6th) *nufal*, from *tanaffala*, which means, "*beginning to make a present without any necessity.*" Others call them *shuhb*, *i.e.* the white nights.

The third three nights (7th–9th) *tusa'*, because the *ninth* night is the last of them. Others call them *buhr*, because in them the darkness of the night is particularly *thick*.

The fourth three nights (10th–12th) *'ushar*, because the *tenth* night is the first of them.

The fifth three nights (13th–15th) *bîḍ*, because they are *white* by the shining of the moon from the beginning of the night till the end.

The sixth three nights (16th–18th) *dura'*, because they are black at the beginning like the sheep *with a black head and a white body*. p. 64.
Originally the comparison was taken from a coat of mail in which people are clad, because the colour of the head of him who is dressed in it, differs from the colour of the rest of his body.

The seventh three nights (19th–21st) *ẓulam*, because in most cases they were *dark*.

The eighth three nights (22nd–24th) *ḥanâdis* (from *ḥindis*=extremely dark). Others call them *duhm*, on account of their being dark.

The ninth three nights (25th–27th) *da'âdî'*, because they are remainders (or last parts). Others derive it from the mode of walking of the camels, *viz.*, *stretching forth the one foot, to which the other is quickly following*.

The tenth three nights (28th–30th) *miḥâḳ*, on account of the *waning* of the moon and the month.

Besides, they distinguished certain nights of the month by special names, *e.g.* the last night of the month was called *sirâr*, because in it the moon *hides herself;* it was also called *faḥama* on account of there being no light in it, and *barâ'*, because the sun has nothing to do with it. Likewise the last day of the month was called *naḥîr*, because it is in the *naḥr* (throat) of the month. The 13th night is called *sawî'*, the 14th the night of "*badr*," because in it the moon is *full*, and her light complete. For of everything that has become complete you say *badara; e.g.* 10,000 dirhams are called one *badra*, because that is supposed to be the most complete and the last number, although it is not so in reality.

The Arabs used in their months also the seven days of the week, the ancient names of which are the following:—

1. 'Awwal, *i.e.* Sunday.
2. 'Ahwan.
3. Jubâr.
4. Dubâr.
5. Mu'nis.
6. 'Arûba.
7. Shiyâr.

They are mentioned by one of their poets in the following verse:—

"I strongly hope that I shall remain alive, and that my day (of death) will be either *'Awwal*, or *'Ahwan*, or *Jubâr*,

Or the following day, *Dubâr*, or if I get beyond that, either *Mu'nis* or *'Arûba* or *Shiyâr*."

Afterwards the Arabs gave them the following new names :—

Al-'aḥad, *i.e.* one.
Al-ithnân, „ two.
Al-thulathâ, „ three.
Al-'arbi'â „ four.
Al-khamîs „ five.
Al-jum'â, „ gathering.
Al-sabt, „ sabbath.

The Arabs fixed the beginning of the month by the appearance of new moon, and the same has been established as a law in Islâm, as the Lord has said (Sûra ii. 185): "They will ask thee regarding the new moons. Speak: they are certain moments of time for the use of mankind (in general) and for pilgrimage."

Determination of the length of Ramadan, the Month of Fasting.—Some years ago, however, a pagan sect started into existence somehow or other. They considered how best to employ the interpretation (of the Koran), and to attach themselves to the system of the exoteric school of interpreters who, as they maintain, are the Jews and Christians. For these latter have astronomical tables and calculations, by means of which they compute their months, and derive the knowledge of their fast days, whilst Muslims are compelled to observe new moon, and to inquire into the different phases of the light of the moon, and into that which is common to both her visible and invisible halves. But then they found that Jews and Christians have no certainty on this subject, that they differ, and that one of them blindly follows the other, although they had done their utmost in the study of the places of the moon, and in the researches regarding her motions (*lit.* expeditions) and stations.

Thereupon they had recourse to the astronomers, and composed their *Canons* and books, beginning them with dissertations on the elements of the knowledge of the Arabian months, adding various kinds of compu-
65. tations and chronological tables. Now, people, thinking that these calculations were based upon the observation of the new-moons, adopted some of them, attributed their authorships to Ja'far Al-ṣâdiḳ, and believed that they were one of the mysteries of prophecy. However, these calculations are based not upon the apparent, but upon the mean, *i.e.* the corrected, motions of sun and moon, upon a lunar year of 354⅓ days, and upon the supposition that six months of the year are complete, six incomplete, and that each complete month is followed by an incomplete one. So we judge from the nature of their *Canons*, and from the books which are intended to establish the bases on which the *Canons* rest.

But, when they tried to fix thereby the beginning and end of fasting, their calculation, in most cases, preceded the legitimate time by one day. Whereupon they set about eliciting curious things from the following word of the Prophet: " Fast, when she (new-moon) appears, and cease fasting when she re-appears." For they asserted, that the words " fast, *when* she appears " (صوموا لرؤيته), mean the fasting of *that* day, in the afternoon of which new-moon becomes visible, as people say, " *prepare yourselves to meet him* " (تهيّئوا لاستقباله), in which case the act of preparing precedes that of meeting.

Besides, they assert that the month of Ramaḍân has never less than thirty days. However, astronomers and all those who consider the subject attentively, are well aware that the appearance of new-moon does not proceed regularly according to one and the same rule for several reasons: the motion of the moon varies, being sometimes slower, sometimes faster; she is sometimes near the earth, sometimes far distant; she ascends in north and south, and descends in them; and each single one of these occurrences may take place on every point of the ecliptic. And besides, some sections of the ecliptic sink faster, others slower. All this varies according to the different latitudes of the countries, and according to the difference of the atmosphere. This refers either to different places where the air is either naturally clear or dark, being always mixed up with vapours, and mostly dusty, or it refers to different times, the air being dense at one time, and clear at another. Besides, the power of the sight of the observers varies, some being sharp-sighted, others dim-sighted. And all these circumstances, however different they are, are liable to various kinds of coincidences, which may happen at each beginning of the two months of Ramaḍân and Shawwâl under innumerable forms and varieties. For these reasons the month Ramaḍân is sometimes incomplete, sometimes complete, and all this varies according to the greater or less latitude of the countries, so that, *e.g.* in northern countries the month may be complete, whilst the same month is incomplete in southern countries, and *vice versâ*. Further, also, these differences in the various countries do not follow one and the same rule; on the contrary, one identical circumstance may happen to one month several consecutive times or with interruptions.

But even supposing that the use which they make of those tables and calculations were correct, and their computation agreed with the appearance of new-moon, or preceded it by one day, which they have made a fundamental principle, they would require special computations for each degree of longitude, because the variation in the appearance of new-moon does not depend alone upon the latitudes, but to a great extent p. 66
also upon the longitudes of the countries. For, frequently, new-moon is not seen in some place, whilst she *is* seen in another place not far to the west; and frequently she is seen in both places at once. This is one of the reasons for which it would be necessary to have special calculations

and tables for every single degree of longitude. Therefore, now, their theory is quite utopian, *viz.* that the month of Ramaḍân should always be complete, and that both its beginning and end should be identical in the whole of the inhabited world, as would follow from that table which they use.

If they contend that from the (above-mentioned) tradition, which is traced back to Muḥammad himself, the obligation of making the beginning and end of fasting precede the appearance of new-moon, follows, we must say that such an interpretation is unfounded. For the particle *Lâm* (لام التوقيت) relates to future time, as they have mentioned, and relates to past time, as you say, *e.g.*:—كُتب لكذا مضى من الشهر (“*dated from this or that day of the month*”), *i.e.* from that moment when *x* days of the month were past already, in which case the writing does not precede the past part of the month. And this, not the first mentioned, is the meaning of that tradition. Compare with this the following saying of the Prophet: “We are illiterate people, we do not write nor do we reckon the month thus and thus and thus,” each time showing his ten fingers, meaning a complete month or thirty days. Then he repeated his words, saying, “and thus and thus and thus,” and at the third time he held back one thumb, meaning an incomplete month or twenty-nine days. By this generally known sentence, the Prophet ordained that the month should be one time complete, and incomplete another time, and that this is to be regulated by the appearance of new-moon, not by calculation, as he says, “*we do not write, nor do we reckon (calculate).*”

But if they say that the Prophet meant that each complete month should be followed by an incomplete one, as the chronologists reckon, they are refuted by the plain facts, if they will not disregard them, and their trickery in both small and great things, in all they have committed, is exposed. For the conclusion of the first-mentioned tradition proves the impossibility of their assertion, *viz.* “*Fast when she (new-moon) appears, and cease fasting when she re-appears, but if heaven be clouded so as to prevent your observation, reckon the month Sha‘bân as thirty days.*” And in another tradition, the Prophet says, “*If a cloud or black dust should prevent you from observing the new moon, make the number thirty complete.*” For if the appearance of new-moon be known either from their tables and calculations, or from the statements of the authors of the canons, and if the beginning and end of fasting is to precede the appearance of new-moon, it would not be necessary to give full thirty days to the month Sha‘bân, or to count the month Ramaḍân as full thirty days, in case the horizon should be covered by a cloud or by dust. And this (*i.e.* to give full thirty days to Ramaḍân) is not possible, except by performing the fasting of the day in the evening of which the new-moon is first seen.

If, further, the month Ramaḍân were always complete, and its beginning were known, people might do without the observation of new-moon for the month Shawwâl. In the same way, the word of the Prophet: "*and cease fasting when she (new-moon) re-appears,*" is to be interpreted.

However, party spirit makes clear-seeing eyes blind, and makes sharp-hearing ears deaf, and instigates people to engage in things which no mind is inclined to adopt. But for this reason, such ideas would not p. 67.
have entered their heads, if you consider the traditions which occur in the books of the *Shî'a Zaidiyya,*—may God preserve their community!—and which have been corrected by their authorities,—may God bless them!—as for instance, the following: In the time of the Prince of the Believers ('Alî) people had been fasting twenty-eight days in the month of Ramaḍân. Then he ordered them still to perform the fasting of *one* day, which they did. The fact was that both consecutive months, Sha'bân and Ramaḍân, were imperfect, and there had been some obstacle which had prevented them from observing new-moon at the beginning of Ramaḍân; they gave the month the full number of thirty days, and at the end of the month the reality of the case became evident. Then there is the following saying, related to have been pronounced by 'Abû 'Abd-Allâh Alṣâdiḳ: "*The month of Ramaḍân is liable to the same increase and decrease as the other months.*" Also the following is reported of the same: "*If you observe the month Sha'bân without being able to see the new-moon, count thirty and then fast.*" The same 'Abû-'Abd-Allâh Alṣâdiḳ, on being asked regarding the new-moon, said: "*If you see the new moon, fast, and if you see her again, cease fasting.*" All these traditions in the code of the Shî'a refer only to the fasting.

It is astonishing that our masters, the family of the Prophet, listened to such doctrines, and that they adopted them as a uniting link for the minds of the community of the believers who profess to follow them, instead of imitating the example of their ancestor, the Prince of the Believers ('Alî), in his aversion to conciliating the obstinate sinners, when he spoke: "*I did not hold out an arm to those who lead astray*" (*i.e.* I did not lend support to them).

As regards the following saying, ascribed by tradition to Alṣâdik: "When you observe the new-moon of Rajab, count fifty-nine days, and then begin fasting;" and the following saying ascribed to the same: "If you see the new-moon of the month of Ramaḍân at the time when she appears, count 354 days, and then begin fasting in the next following year. For the Lord has created the year as consisting of 360 days. But from these he has excepted six days, in which he created the heavens and the earth; therefore they (these six days) are not comprehended in the number (of the days of the year)"—regarding these traditions we say, that, if they were correct, his (Alṣâdiḳ's) statement on this subject would rest on the supposition, that it (the month Ramaḍân) was really greater in one place, and did not follow the same rule everywhere, as we have heretofore mentioned. Such a method of accounting for the six days is something so subtle, that it proves the tradition to be false, and renders it void of authenticity.

In a chronicle I have read the following: 'Abû-Ja'far Muḥammad ben Sulaimân, Governor of Kûfa, under the Khalîf Manṣûr, had imprisoned 'Abd-alkarîm ben 'Abî-al'aujâ, who was the uncle of Ma'n ben Zâ'ida, one of the Manichæans. This man, however, had many protectors in Baghdâd, and these urged Manṣûr in his favour, till at last he wrote to Muḥammad ordering him not to put 'Abd-alkarîm to death. Meanwhile, 'Abd-alkarîm was expecting the arrival of the letter in his cause. He said to 'Abû-aljabbâr confidentially: "If the 'Amîr gives me respite for three days, I shall give him 100,000 dirhams." 'Abû-aljabbâr told this to Muḥammad, who replied: "You have reminded me of him, whilst I
p. 68. had forgotten him. Remind me of him when I return from the mosque." Then, when he returned, 'Abû-aljabbâr reminded him of the prisoner, whereupon he (Muḥammad) ordered him to be brought and to be beheaded. And now, knowing for certain that he was to be killed, he said, "By God, now that you are going to kill me, I tell you that I have put down 4,000 traditions (in my books), in which I forbid that which is allowed, and allow that which is forbidden. And verily, I have made you break your fast when you ought to have fasted, and I have made you fast when you ought not to have fasted." Thereupon he was beheaded, and afterwards the letter in his cause arrived.

How thoroughly did this heretic deserve to be the author of this subtle interpretation which they have adopted, and of its original (*i.e.* the text to which the interpretation refers)!

I myself have had a discussion with the originator of this sect, regarding the Musnad-tradition (*i.e.* such a tradition as is carried back by an uninterrupted chain of witnesses to Muḥammad himself). On which occasion I compelled him to admit that consequences, similar to those here mentioned, follow from his theories. But then in the end he declared, that the subject was one that of necessity resulted from the language (*i.e.* from the interpretation of the *Lâm-altauḳît*), and that the language has nothing whatever to do with the law and its corollaries. Thereupon, I answered: "May God have mercy upon you! Have not God and his Prophet addressed us in the language generally known among the Arabs? But the thing is this, that *you* have nothing whatever to do with the Arabic language; and also in the science of the law you are utterly ignorant. Leave the law aside and address yourself to the astronomers. None of them would agree with you regarding your theory of the perpetual completeness of the month of Ramaḍân; none of them thinks that the celestial globe and sun and moon distinguish the moon of Ramaḍân from among the others, so as to move faster or slower just in this particular month. The luminaries do not mark out this month in particular as do the Muslims, who distinguish it by performing their fasting in it.

However, arguing with people who are obstinate on purpose, and persevere in their obstinacy on account of their ignorance, is not productive of any good, either for the student or for the object of his researches. God speaks (Sûra lii. 44): "If they saw a piece of heaven falling down, they would say, 'It is only a conglomerated cloud.'" And further (Sûra vi. 7): "If we sent down to you a book (written) on paper, and they touched it with their hands, verily the unbelieving would say, 'This is nothing but evident witchcraft.'" God grant that we may always belong to those who follow and further the truth, who crush and expose that which is false and wrong!

Months of the Reformed Calendar of Almu'tadid.—The months of Almu'tadid are the Persian months, with the same names and the same order. But the Persian days are not used in these months, because to the Epagomenæ in every fourth year one day is added by way of intercalation; and so for that reason which we have mentioned, when speaking of the months of the Egyptians, the (Persian) names of the single days have been dropped. The order of intercalation used in these months agrees with that of the Greeks and Syrians.

As to the months of the other nations, Hindus, Chinese, Tibetans, Turks, Khazars, Ethiopians, and Negroes, we do not intend, although we have managed to learn the names of some of them, to mention them here, postponing it till a time when we shall know them all, as it does not agree with the method which we have followed hitherto, to connect that which is doubtful and unknown with that which is certain and known.

We have collected in the following table the names of the months which have been mentioned in the preceding part of this book, in order to facilitate the study of the various kinds of them. God leads to the truth!

P. 70. P. 69.

TABLE OF THE MONTHS.

…he beginning of their months is the second Naurôz.		The beginning of their months is the first Naurôz.		Beginning of the months is the appearance of New-Moon about the Vernal Equinox.	Their beginning is the appearance of New-Moon. Their number begins with Daimur which corresponds with Ramaḍân.	The beginning of the months is the appearance of New-Moon which is observed (not calculated).			
Khwârizmians.	Sughdians.	The ancient inhabitants of Sijistân.	Persians.	Jews.	Thamûd.	Pagan Arabs.	Muslim Arabs.	The inhabitants of جكارتك.	The inhabitants of Ḳubâ.
ناوسارجى	نوسرد	كواد	Farwardîn-Mâh	Tishrî	Mûjib	Almu'tamir	Almuḥarram	نوسرد	حلو
اردوست	جرجن	رهو	Ardîbahisht-Mâh	Marḥeshwân	Mûjir	Nâjir	Ṣafar	فدى نوسرد	اوين
هروداد	نيسنج	اوسال	Khurdâdh-Mâh	Kislêw	Mûrid	Khawwân	Rabî' the first.	سافول	حجش
جيرى	نساكنج	تيركيانوا	Tîr-Mâh	Ṭebeth	Mulzim	Buṣṣân	Rabî' the last.	سافت	لوليا
همداد	اهناخنداااكا	سريروا	Murdâdh-Mâh	Shebâṭ	Musdir	Ḥantam	Jumâdâ the first.	اوريس	لو
اخشريورى	مرخندا	مريروا	Shahrêwar-Mâh	Adhâr	Haubar	Zabbâ	Jumâdâ the last.	يسن	نر
اومرى	فغكان	مرور	Mihr-Mâh	Nîsân	Haubal	Al'aṣamm	Rajab.	سسك	مهر
تاقاخن	ابانج	هرانوا	Âbân-Mâh	Iyâr	Mauhâ	'Âdil	Sha'bân	جدل	الما
ارى	فوغ	اركيازوا	Âdhar-Mâh	Sîwân	Daimur	Nâfiḳ	Ramaḍân	هيات	نوا
ريمژد	مسانوغ	كريشت	Dai-Mâh	Tammûz	Dâbir	Waghl	Shawwâl	سيون	مفاة
اخمن	ژيمدنج	كرسن	Bahman-Mâh	Âbh	Ḥaiḳal	Huwâ'	Dhû-alka'da	مجسند	ىى
اسبندارمد	خشوم	ساروا	Isfandârmadh-Mâh	'Elûl	Musbil	Burak	Dhû-alḥijja	دريمنكان	اوفاة

I have not been able to learn how long these months are, nor what they mean, nor of what kind they are.	The beginning of these months is the conjunction taking place about the vernal equinox.	The beginning of the leap-year is the 29th of Âb, the beginning of the simple year is the 1st of Dai-Mâh.		The beginning of these months is the first day of the Syrian month Kânûn the last.		The beginning of these months is an assumed day which is not in relation to anything else.	
Turks.	Hindus.	People of the West.	Copts.	Ancient Greeks.	Greeks (Ρωμαῖοι).	Syrians.	Turks.
Ulugh Ây	بيشاك	May	Thôth	Audynæus	Yanuarius	Teshrîn the first	Sijkan
Küčük Ây	زيشت	June	Phaôphî	Peritius	Februarius	Teshrîn the last	Od
Birinj Ây	اسار	July	Athyr	Dystrus	Martius	Kânûn the first	Pârs
Ikinj Ây	سراوان	August	Choiak	Xanticus	Maïus	Kânûn the last	Tafshikhan
Altünj Ây	بهدربه	September	Tybi	Artemisius	Aprilius	Shebâṭ	Lû
Beshinj Ây	اسوج	October	Methîr	Daïsius	Yunius	Âdhâr	Yylan
Sekizinj Ây	كارث	November	Phamenôth	Panemus	Yulius	Nîsân	Yont
Tokuzunj Ây	منكس	December	Pharmûthî	Lôus	Augustus	Iyâr	Kuy
Onunj Ây	بوش	January	Pachôn	Gorpiæus	Septembrius	Ḥazîrân	Pîčîn
Türtünj Ây	ماك	February	Payni	Hyperberetæus.	Octombrius	Tammûz	Taghuk
Učünj Ây	باكر	March	Epiphi	Dius	Novembrius	Âbh	It
Yetinj Ây	جيتر	April	Mesori	Apellæus	Decembrius	Îlûl	Tunguz

P. 71.

p. 72.

CHAPTER VI.

ON THE DERIVATION OF THE ERAS FROM EACH OTHER, AND ON THE CHRONOLOGICAL DATES, RELATING TO THE COMMENCEMENTS AND THE DURATIONS OF THE REIGNS OF THE KINGS, ACCORDING TO THE VARIOUS TRADITIONS.

It is the special object at which I aim in this book, to fix the durations (of the reigns of the kings) by the most correct and perspicuous method. But, now, wishing to explain the derivation of the eras from each other in conformity with the usual mode of the *canons*, which specify the various kinds of calculation and of derivation (*e.g.* stating one era in the terms of another), and which contain rules and paradigms, I find this subject to be a very wide one, and the wish to embrace this whole science compels me to cause trouble both to myself and to the reader.

Agreeably to the method which I have adhered to from the beginning of this book, I shall explain the intervals between the epochs of the usual eras by a measure which is counted in the same way by all nations, *i.e.* by days; for, as we have already mentioned, both years and months are differently measured. Everything else is generally mentioned in years, but for the knowledge of the intervals between the epochs of the eras the statement in *days* is quite sufficient, since it has been impossible to obtain a knowledge of the real quality of the years of the various eras, and there has been but little need for the use of them.

Now, if we in some places wander about through various branches of science, and plunge into subjects which are not very closely connected with the order of our discussion, we must say that we do not do this because we seek to be lengthy and verbose, but as guided by the desire of preventing the reader from getting tired. For if the mind is continually occupied with the study of one single science, it gets easily tired and impatient; but if the mind wanders from one science to another, it is as if it were wandering about in gardens, where, when it is roving over one, another one already presents itself; in consequence of which, the mind has a longing for them, and enjoys the sight of them; as people say, "Everything that is new offers enjoyment."

Now let us begin with the traditions of those to whom a divine book was sent (Jews and Christians) regarding Adam, his children and their descendants. All this we shall fix in tables, in order to facilitate the pronunciation of their names, and the study of the different traditions regarding them. On this subject we combine the traditions of the Jews and Christians, placing them opposite to each other (in the same table). We commence by the help of God, under his guidance, and with his gracious support. p. 73.

The Names of the Descendants of Adam, who form the Chronological Chain of the Era, and the Chronological Differences between Christians and Jews regarding them.	How old they were when a son was born to them according to the Christians.	The sum of the years of the era according to the Christians.	How old they were when a son was born to them—according to the Jews.	How long each of them lived after a son had been born to him—according to the Jews.	How long altogether each of them lived—according to the Jews.	The sum of the years of the era—according to the Jews.
I.—Adam the father of mankind—till the birth of his son Seth . .	230	230	130	800	930	130
Seth ben Adam—till the birth of his son Enos	205	435	105	807	912	235
Enos ben Seth—till the birth of his son Cainan	190	625	90	815	905	325
Cainan ben Enos—till the birth of his son Mahalaleel . . .	170	795	70	840	910	395
V.—Mahalaleel ben Cainan—till the birth of his son Jared . .	165	960	65	830	895	460
Jared ben Mahalaleel—till the birth of his son Enoch . .	162	1122	162	800	962	622
Enoch ben Jared—till the birth of his son Methuselah . . .	165	1287	65	300	365	687
Methuselah ben Enoch—till the birth of his son Lamech . .	167	1454	187	782	969	874
Lamech ben Methuselah—till the birth of his son Noah . .	188	1642	182	595	777	1056
X.—Noah ben Lamech—till the birth of his son Shem . . .	500	2142	500	450	950	1556
Shem ben Noah—till the Deluge .	100	2242	100	500	600	1656
From the Deluge till the birth of Arphaxad ben Shem . .	2	2244	2	0	0	1658
Arphaxad ben Shem—till the birth of his son Salah . . .	135	2379	35	463	498	1693
Salah ben Arphaxad—till the birth of his son Eber . . .	130	2509	30	460	490	1723
XV.—Eber ben Salah—till the birth of his son Peleg	134	2643	34	396	430	1757
Peleg ben Eber—till the birth of his son Reu	130	2773	30	179	209	1787
Reu ben Peleg—till the birth of his son Serug	132	2905	32	175	207	1819
Serug ben Reu—till the birth of his son Nahor	130	3035	30	170	200	1849
Nahor ben Serug—till the birth of his son Terah	79	3114	29	119	148	1878
XX.—Terah ben Nahor till the birth of his son Abraham . . .	75	3189	70	135	205	1948

p. 74. Now, he who studies the numbers of years of this table, till the birth of Abraham, will become aware of the difference between the two systems (that of the Christians and that of the Jews).

The Jewish copy of the Thora, although stating the duration of the lives of Abraham, Isaac, Jacob, Levi, Kohath, and Moses, does not specify how old they were when a son was born to each of them, nor how long they lived after that; except in the case of Abraham, Isaac, and Jacob. For it is stated that Isaac was born unto Abraham when he was 100 years of age, and that he afterwards lived 75 years more; that Jacob was born unto Isaac when he was 60 years of age; that Jacob entered Egypt together with his sons, when he was 130 years of age, and that he after that lived 17 years more.

Now, the Israelites stayed in Egypt 210 years, according to the statement of the Jews, that between the birth of Abraham and that of Moses there was an interval of 420 years, and that Moses was 80 years of age, when he led the Israelites out of Egypt. From the second book of the Thora, however, we learn that the entire length of the sojourning of the Israelites in Egypt was 430 years. If, now, the Jews are asked to account for this difference, they maintain that that space of time is to be counted from the day when God made the treaty with Abraham, and promised him to make him the father of many nations, and to give to his descendants the country of Canaan as an inheritance. But we leave the matter to God, who knows best what they mean.

The chronological differences regarding the later periods of Biblical history, arising out of the three different copies of the Thora, are of the same kind as we have already explained.

How little care the Jews bestow upon their chronology is shown to evidence, by the fact, that they, all of them, believe in the first instance, that between their exodus from Egypt and Alexander there is an interval of 1,000 years, corrected (*i.e.* made to agree with the sun or real time) by intercalation, and that they rely on this number in their computation of the qualities of the years (whether they be perfect or imperfect or intermediate). But if we gather from their books which follow after the Thora, the years of every one of their rulers after Moses, the son of Amram, and add them together, we get a sum which already at the building of Jerusalem goes beyond the millennium by such a space of time as cannot be tolerated in chronological computations. If this sum were too small (less than a millennium), the difference might be accounted for by assuming that an interval between two persons might have been omitted. But a surplus in this case does not admit of any interpretation whatsoever.

Being unable to give a satisfactory answer to such a question, some of them assert that the accurate specification of these years was found in the records of the family of Juda, and that these records are no longer at their disposal, but have been carried off to the countries of the

Greeks. For after the death of Solomon, the Israelites were split into two parties. The tribes of Juda and Benjamin elected as their king the son of Solomon, whilst the ten tribes elected as their king Jeroboam, the client of Rehoboam, the son of Solomon. And thereupon he led them astray (to idolatry), as we shall mention hereafter in the chapter on the Jewish festivals. His children reigned after him, and both parties made war upon each other.

The following is a synopsis of the years of their rulers, who ruled over them after their exodus from Egypt, when they marched towards *Baḥr-al-ḳulzum* (the Red Sea) in order to pass it, and to march to *Altih*, a p. 75. desert in Alḥijâz, in the direction of Jerusalem; all of which rests on the authority of their chronicles. But they have another book which they call *Sêder-'ôlâm* (סדר עולם), *i.e.* the years of the world, which contains a less sum of years than that of *the books which follow after the Thora*, whilst in some respects it comes near to their original system. The statements of both these kinds of their historical records we have collected in the following synopsis.

The Names of the Rulers, Governors, Priests, and Judges of the Israelites till the Foundation of the Temple, which is a space of 480 years.	How long each of them ruled, according to the Biblical chronicle.	The sum of the years.	How long each of them ruled, according to Sêder-'Ôlâm.	The sum of the years.
The Israelites left Egypt and dwelt in the desert till the death of Moses .	40	40	40	40
Yehôshû'a ben Nûn, the successor of Moses	27	67	27	67
'Othnî'êl ben Ḳenaz	40	107	40	107
'Eglôn the king of Mô'âb and the Amalekites of the Banî-'Ammôn . .	18	125	—	—
'Êhûd ben Gêrâ, the left-handed, of the Ephraimites	80	205	80	187
Shamgar ben 'Anâth	20	225	—	—
Debôrâ the prophetess and her lieutenant Bârâḳ	40	265	40	227
The Midianites, the oppressors . .	7	272	7	234
Gid'ôn ben 'Ofrâ, of the tribe of Manasseh	40	312	} 43	277
'Abîmêlekh ben Gid'ôn	3	315		
Tôlâ' ben Pû'â, of the tribe of Ephraim	23	338	} 44	321
76. Yâ'îr from Gil'âd, of the tribe of Manasseh	22	360		
The sons of 'Ammôn the Philistine, *i.e.* the people of Palestine	18	378	18	339
Yiftaḥ from Gil'âd	6	384	6	345
'Ibṣân, also called Naḥshôn, from Bethlehem	7	391	7	352
'Êlôn	10	401	10	362
'Abdôn ben Hillêl	8	409	8	370
The Philistines	40	449	—	—
Shimshôn the giant of the tribe of Dân	20	469	20	390
The people without a ruler	10	479	—	—
'Êlî the priest	40	519	40	430
The ark in the hands of the enemies, until Samuel was sent	10	529	10	440
Samuel, till they asked him to give them a king, whereupon he made Ṭâlût their king	20	549	—	—
Saul, *i.e.* Ṭâlût	20	569	20	442!
David; he commenced building the Temple in the 11th year of his reign .	40	609	40	482
Solomon ben David—till he finished the Temple	3	612	3	485

The Names of the Kings and other Rulers of the Israelites from the Foundation of the Temple till its first Destruction, which is a space of 410 years,	How long each of them ruled, according to the Biblical chronicle.	The sum of the years.	How long each of them ruled, according to Sêder-'Ôlâm.	The sum of the years.	
					p. 77.
Solomon ben David—after the Temple was finished	37	649	37	522	
Reḥab'âm ben Solomon . . .	17	666	17	539	
'Abiyyâ ben Reḥab'âm	3	669	2	541	
'Âsâ ben 'Abiyyâ	41	710	41	582	
Yehôshâfâṭ ben 'Âsâ	25	735	23	605	
Yehôrâm ben Yehôshâfâṭ . . .	8	743	6	611	
'Aḥazyâ ben Yehôrâm	1	744	11	622	
'Athalyâ—till she was killed by Yô'âsh	6	750	6	628	
Yô'âsh ben 'Aḥazyâ—till he was killed by his people	40	790	40	668	
'Amazyâ ben Yô'âsh—till he was killed	29	819	29	697	
'Uzziyâ ben 'Amazyâ—till he died .	52	871	52	749	
Yôthâm ben 'Uzziyyâ—till he died .	16	887	16	765	
'Âḥâz ben Yôthâm—till he died . .	16	903	16	773 !	
Ḥizkiyyâ ben 'Âḥâz, the king of all the tribes	29	932	29	802	
Menashshê ben Ḥizkiyyâ . . .	55	987	55	857	
'Ammôn ben Menashshê	2	989	2	859	
Yôshiyyâ ben 'Ammôn—till he was killed by the king of Egypt . . .	31	1020	31	890	
Yehô'âḥâz ben Yô'shiyyâ—till he was made a prisoner by the king of Egypt	3	1023	—	—	
Yehôyâḳîm ben Yehô'âḥâz, set up by the king of Egypt	10	1033	11	901	
Yehôyâkhîn ben Yehôyâḳîm, till he was made a prisoner by Nebucadnezar .	3	1036	—	—	p. 78.
Ṣidḳiyyâ—till he rebelled against Nebucadnezar, when he was killed and the Temple destroyed	6	1042	11	912	
The Temple remained in ruins . . .	70	1112	70	982	
But according to another view between the time when they were led into captivity and Daniel there was an interval of	90	1202	90	1052 !	
From Daniel till the birth of the Messiah	483	1685	483	1535	
From the birth of the Messiah till the epoch of the flight of Muḥammad .	600	2285	600	2135	

It cannot be thought strange that you should find similar discrepancies with people who have several times suffered so much from captivity and war as the Jews. It is quite natural that they were distracted by other matters from preserving their historical traditions, more particularly at times of such distress, "when each woman who suckled a child forgot her child, and each pregnant woman gave birth to the burthen of her womb." (Sûra xxii. 2.)

Besides, the governorships and headships were not always held by one and the same tribe, but came to be divided (among several tribes) after the death of Solomon the son of David; then one part of them was held by the tribes of Juda and Benjamin, another part by the other tribes of the Israelites.

Further, their rule was not organized so well; nor their empire and government handed over from one to the other in such good order as to render it necessary for them both to preserve the dates when each of their rulers ascended the throne, and to record the duration of his reign, except by a rough method of computation. For some people maintain that, after the death of Joshua, Kûshân, the King of Mesopotamia, of the family of Lot, overpowered them, and held them under his sway during eight years; that then Othniel rose. And some people attribute to his rule more years, others less.

Frequently, one author thinks that some ruler reigned over them so-and-so many years, whilst another assigns to his rule a less number of years, and maintains that the former number represents the duration of his whole life (not that of his rule); or a third possibility is this, that by adding the two spaces of time, mentioned by the two authors, you get a common space of time for two rulers, during which they ruled simultaneously.

The chronological system of the Sêder-'ôlâm, although coming near to the sum (assumed by the generality of the Jews), differs considerably from the statements in detail; this applies specially to the time of the first building of the Temple, not to mention the uncertainty which hangs over those points of their history which we have spoken of before.

The length of the Human Life.—Some one among the inexperienced and foolish people of the *Ḥashwiyya* and *Dahriyya* sects, have rejected as incredible the long duration of life which has been ascribed to certain tribes in the past, specially to the patriarchs before the time of Abraham. Likewise they consider as monstrous what has been related of the huge size of their bodies. They maintain that all this lies altogether beyond the limit of possibility, drawing their conclusions from objects which they are able to observe in their own age. They have adopted the doctrine of astrologers, regarding the greatest possible gift (of years of life) which the stars are supposed to bestow upon mankind in the
p. 79. nativities, if the following constellation occurs: The sun must be at such a nativity both *mater familias* and *pater familias*, *i.e.* he must stand

in his *domus* (οἶκος), or in his *altitudo* (ὕψωμα), in a *cardo*, and in a *concordant* masculine quarter. In that case he bestows his *greatest years*, *i.e.* 120 years, to which the

Moon	- -	adds 25 years.
Venus	- -	„ 8 „
Jupiter	- -	„ 12 „

These are the *smallest years* of each of these three stars, for they are not able to add a greater number of years, if they have a *concordant aspect* (in relation to the horoscope). Further, the two unlucky among the stars (Saturn and Mars) must have no *aspect* to the horoscope, so as not to exercise any diminishing influence. The *Caput Draconis* must stand with the sun in the same sign of the ecliptic, but still sufficiently far from him, so as not to stand within the ὅροι ἐκλειπτικοί.

If this constellation occurs, it increases the gift (of years of life) of the sun by one fourth, *i.e.* 30 years. So the whole sum of years makes 215 years, which they maintain to be the longest duration of life which mortal man may reach, if it is not cut short by any accident. The natural duration of life is to be 120 years, because the existence of the world depends upon the sun; and this number of years represents the *greatest years* of the sun.

Those people have settled this question as it best pleases them. And if reality followed their desire, heaven and earth would be greatly the worse for it. They have built their theory on a basis, the contrary of which is approved of by astronomers, in so far as they ascribe "greatest years" to these planets. They say in their books that these planets used to bestow their "greatest years" in the *millennia* of the *fiery signs* of the zodiac, when in them the rule was exercised by the superior planets (Saturn, Jupiter, Mars), and when the years of the sun and of Venus were made to exceed by far the longest duration of life ascribed to any one of the patriarchs.

This man is their master in chronology; they trust in his word, and do not oppose his audacity. He actually maintains that man may live during the years of a "*middle conjunction*" (of Saturn and Jupiter), when the nativity coincides with the *transitus* of the conjunction from one *trigon* to another, whilst the *ascendens* is one of the two *houses* of either Saturn or Jupiter, when the sun is *materfamilias* in day-time, and the moon at night, exercises the greatest power; that the same is possible, if this same constellation occurs at the transitus of the conjunction to Aries and its *trigons*.

And the argument for the assertion, that the new-born human being may live during the years of the "greatest conjunction," *i.e.* about 960 years, until the conjunction returns to its original place, is of the same description.

He has explained and propounded this subject in the beginning of his book, " *De Nativitatibus.*"

This, now, is their belief in the gifts (of years of life) of the stars.

Regarding these years, which the single planets are supposed to bestow upon mankind, we have had a discussion with the astronomers who use them, in the *Kitâb altanbih 'alâ ṣinâ'at altamwih* (*i.e.* the book in which the swindling profession is exposed), and we have given a direction how to use the best method in all questions where these years occur in the book entitled, *Kitâb alshumûs alshâfiya lilnufûs.*

Now, personal observation alone, and conclusions inferred therefrom, do not prove a long duration of the human life, and the huge size of human bodies, and what else has been related to be beyond the limits of possibility. For similar matters appear in the course of time in manifold shapes. There are certain things which are bound to certain times, within which they turn round in a certain order, and which undergo transformations as long as there is a possibility of their existing. If they, now, are not observed as long as they *are* in existence, people think them to be improbable, and hasten to reject them as altogether 80. impossible.

This applies to all cyclical occurrences, such as the mutual impregnation of animals and trees, and the forthcoming of the seeds and their fruits. For, if it were possible that men did not know these occurrences, and then were led to a tree, stripped of its leaves, and were told what occurs to the tree of getting green, of producing blossoms and fruits, etc., they would certainly think it improbable, till they saw it with their own eyes. It is for the same reason that people, who come from northern countries, are filled with admiration when they see palm-trees, olive-trees, and myrtle-trees, and others standing in full-bloom at winter-time, since they never saw anything like it in their own country.

Further, there are other things occurring at times in which no cyclical order is apparent, and which seem to happen at random. If, then, the time in which the thing occurred has gone by, nothing remains of it except the report about it. And if you find in such a report all the conditions of authenticity, and if the thing might have already occurred before that time, you must accept it, though you have no idea of the nature nor of the cause of the matter in question.

Irregular Formations of Nature.—There are still other things which occur in like manner, but which are called "*faults of nature*" (*lusus naturæ*), on account of their transgressing that order which is characteristic of their species. I, however, do not call them "*faults of nature*," but rather a superfluity of material beyond the due proportions of the measure of everything. To this category belong, *e.g.* animals with supernumerary limbs, which occur sometimes, when nature, whose task it is to preserve the species as they are, finds some superfluous substance, which she forms into some shape instead of throwing it away;

likewise animals with imperfect limbs, when nature does not find the substance by which to complete the form of that animal in conformity with the structure of the species to which it belongs; in that case she forms the animal in such a shape, as that the defect is made to lose its obnoxious character, and she gives it vital power as much as possible.

This is illustrated by an example, which Thâbit ben Sinân ben Thâbit ben Ḳurra relates in his chronicle, *viz.* that he had seen near Surraman-ra'â an Indian chicken that had come out of the egg without a defect, and of complete structure; but its head had two beaks and three eyes. The same author reports, that to Tûzûn, in the days of his reign, people brought a dead kid with the round face, the jaws and teeth like those of man; but it had only one eye, and something like a tail on its forehead. Further, he relates that in the district Almukharrim, of Baghdâd, there was born a child, which died instantly; it was brought before Ghurûr-aldaula Bakhtiyâr at the time when his father Mu'izz-aldaula was still alive, and he examined it. It was one complete body without a defect, and without an addition, except that two protuberances rose from it, and upon these there were two complete heads, with complete lineaments, with eyes, ears, two nostrils, and two mouths; between the loins were genitals like those of a woman, out of which the orifice of the penis of a man was apparent.

Another report of his says, that one of the nobles of the Greeks sent to Nâṣir-aldaula, in the winter of A.H. 352, two men grown together by the stomach; they were Aramæans, and twenty-five years of age. He p. 81.
says, they were called *Multaḥiyâni* (*i.e.* two bearded men). They were accompanied by their father. They turned their faces towards each other, but the skin, which formed the common connecting link between them, was long, and besides susceptible of extending so far as to permit the one to rise from the side of the other. People describe them as having, each of them, separate and complete organs of generation; that they did their eating and drinking, and the *exoneratio alvi* at different times; that they used to ride on *one* animal, the one closely behind the other, but so as to turn their faces towards each other; that the one had an inclination for women, the other for boys.

There is no doubt that the *Vis Naturalis* (the creative power of nature), in all work it is inspired and commissioned to carry out, never drops any material unused, if it meets with such; and if there is abundance of material, the *Vis Naturalis* redoubles its creating work. Such a double-creation sometimes proceeds in this way, that one being comes into existence in close proximity to another, being at the same time something separate by itself, as, *e.g.* in the case of twins; sometimes a being comes into existence tied up to another being, as, *e.g.* in the case of the two Aramæans; at other times, again, a being comes into existence inserted into and mixed up with another one, as in that case which we mentioned before speaking of the two Aramæans.

The various kinds of double-creations of this and other descriptions are also found among the other animals (besides man). There are, *e.g.* said to be certain species of sea-fishes that are double ones. I mean to say, if you open such a fish, you find a similar one inside.

Frequently, too, the reduplication of formation may pass into a multiplication. All of which is also found among the plants. Look, for instance, at the double-fruits that are grown together, at the fruits with double kernels, which are included in one shell. An example of such a double-formation, of which the one thing is inserted into the other, is an orange, in the interior of which you find another orange of the same kind.

Frequently the *Vis Naturalis* has not succeeded in finishing the double-creation, and producing a complete whole. In which case, she increases the number of limbs, either in their proper places, as *e.g.* supernumerary fingers—for although they are more than usual and than is necessary, still they are found in that place which is appropriated to fingers,—or not in their proper places. And in this case it would be correct to call such a formation an *Error of Nature.* An instance of this is the cow that was in Jurjân at the time of the Ṣâḥib, and when the family of Buwaihi held the country under their sway. Everybody, both young and old, had seen it, and they related to me that it had on the bunch close to the neck a foreleg like its other two forelegs, quite complete, with its shoulder, its joints, and hoof; and that she moved it about as she liked, contracting and extending it.

This case may justly be considered an error (of nature), because that supernumerary limb was quite useless, and because it had neither its proper place nor direction.

Now, all these and similar classes (of uncommon creations), on which I have composed special books, would not be admitted as possible by anyone who did not witness them, because he would not find in them the conditions of authenticity.

Length of the Human Life.—The length of human life is taught by experience to be regulated by a genealogical ratio. For instance, with the Ḥimyarites and others, long life is a peculiarity. Besides long life occurs in one place to the exclusion of others, *e.g.* in Farghâna and Yamâma. For well-informed people relate that in those countries some people grow older than anywhere else. And in this respect they are still surpassed by the Arabians and Indians.

Of this same 'Abû-Ma'shar Albalkhî, the following story is related by 'Abû-Sa'îd Shâdhân in his *Kitâb-almudhâkara-bil'asrâr* (*i.e.* the book in which he brings mysterious subjects before the mind of the reader):—
p. 82. The nativity of a son of the King of Serendîb (Ceylon) was sent to him. His *Ascendens* was Gemini ♊, whilst Saturn stood in Cancer ♋, and the Sun in Capricorn ♑. Now, 'Abû-Ma'shar gave his judgment that he would live during the middle cycle of Saturn. Thereupon, I said to

him, "God forbid! The οἰκοδεσποτής moves backward in the crisis of *retrograde motion* in a *domus cadens* of the *cardines*, so as not to give more than its small cycle. You must subtract fifty years therefrom on account of the retrograde motion."

'Abû-Ma'shar: "Those people are the inhabitants of a κλίμα, of whom one knows beforehand that they live very long, so that they frequently live on in a decrepit state, whilst Saturn is their *companion*. I have been told that, if a man dies before reaching the middle cycle of Saturn, people wonder that he has died so soon. If, therefore, Saturn occupies the dignity of οἰκοδεσποτής in a κλίμα of his own, he does not, in most cases, give less than his great and middle cycles, except he be *in cadente domo*."

I: "But, surely he is *in cadente domo*."

'Abû-Ma'shar: "Quite so! He (Saturn) is *falling* out of the figure of the *Aspectus*, but he is not falling out of the *Directio*."

(Here is a lacuna.)

The mysteries of the second are numerous. It is likewise in a well beneath the earth. In this circumstance there is curious matter for astonishment. Now, in this place, they have admitted that in one κλίμα people live longer than in another.

In another place he ('Abû-Sa'îd Shâdhân) relates of the same 'Abû-Ma'shar, that he was in his company when he was asked by 'Abû-'Iṣma, the Wazîr of Ṣaffâr, regarding something in the signs of his nativity, which he ('Abû-'Iṣma) was alarmed about.

'Abû-Ma'shar: "Do you know of what age your father died?"

'Abû-'Iṣma: "Yes."

'Abû-Ma'shar: "Have you already reached the same age?"

'Abû-'Iṣma: "I have passed it already."

'Abû-Ma'shar: "Do you know at what age your mother died?"

'Abû-'Iṣma: "Yes. That age, too, I have passed already."

'Abû-Ma'shar: "Do you know how long your paternal grandfather lived?"

'Abû-'Iṣma: "Yes. But that I have not yet reached."

'Abû-Ma'shar: "Then consider whether that difference, which is indicated by your nativity, agrees with the life of your grandfather?"

'Abû-'Iṣma: "Yes, it does agree."

'Abû-Ma'shar: "In that case you are right to be alarmed." Then he proceeded to explain: "Nature is most powerful. For in any mishap that befals a man when he is as old as his father or mother or his paternal grandfather were at the time of their death, he is certain to perish, except there be strong evidence (to the contrary). This is clear, too, in plants and seeds. For there are certain species of them which are known to exist very long, whilst others soon meet with mishaps and exist only a short time."

Now, 'Abû-Ma'shar again admits in this place that the duration of life is regulated by a genealogical ratio. Therefore, that astrological theory, to which they cling, is devoid of sense, since they admit such a genealogical ratio as not impossible. On the contrary, it is necessary, as we have already mentioned.

If this sect will reject everything that does not occur in their time or place, so as to fall under their personal observation, if they do not themselves find this everlasting scepticism of theirs absurd, if they will not admit anything that has happened in their absence, we can only say that extraordinary occurrences do not happen at all times; and if they, indeed, happen in some one age, they have in the course of time and the passing of generations no other tie which connects them with posterity except the uninterrupted chain of tradition. Nay, if they would draw the last conclusions from their theory, they would be mere sophists, and would be compelled to disbelieve anybody who would tell them that there are still other countries in the world besides those in which they are living; and other absurdities of a similar kind would follow.

If you would listen to them on the subjects which they propound, you
p. 83. would find that they refer to the traditions of the Indians, and rely on various sorts of tricks which they attribute to them. By way of argument they always mention an Indian idol, cut out of stone, the neck of which is surrounded by numerous iron collars, which represent the Indian eras of 10,000 years, and, if counted, would amount to an enormous sum of years. But if you then tell them what they, *i.e.* the Indians, maintain, *viz.* that the King of Jamâlâbadhra, that town whence the *Myrobalana*, the *Phyllanthus emblica*, and the *Myrobalana bellerica* are exported, even at the age of 250 years, rode and hunted and married, and behaved altogether like a young man, and that all this was the consequence of a dietetic treatment, they will reject it, and declare that the Indians are evident liars, not really learned men, because they base their sciences upon inspiration, and that therefore their doctrines are not trustworthy. Besides, they will begin to speak of the subtlety of all the tenets of the Indians in all questions of law and religion, of reward and punishment (eschatology), and they will dwell on the various sorts of torture which they practise in castigating their own bodies.

It is this sect whom God means in the verse of the Koran (Sûra x. 40): "Nay, they have declared to be a lie something, the science of which they did not comprehend"; and in the other verse (Sûra xlvi. 10): "And as they would not be guided thereby, verily, they will say: That is an old lie." They admit only that which suits them, although it be feeble, and they avoid everything that differs from their dogma, although it be true.

I have read a book of 'Abû-'Abdallâh Alḥusain ben 'Ibrâhîm Alṭabarî Alnâtilî, a treatise on the duration of natural life, where he maintains that its greatest length is 140 solar years, beyond which no increase is possible. He, however, who denies this so categorically, is required to produce a proof, which the mind is obliged to accept, and in which it acquiesces. But he has not established the least proof for his assertion, except that in his premises he lays down the following theory :—

Three *Status Perfectionis* are peculiar to man—

I. His attaining to manhood (or womanhood), the time when he becomes able to propagate his own race. That is the beginning of the second *Seventh*.

II. When his thinking power ripens, and his intellect proceeds from δύναμις to ποίησις. That is the beginning of the sixth *Seventh*.

III. When he becomes able to govern himself, if he be unmarried; his family affairs, if he be married; his public affairs, if he exercise some public authority.

The sum of these three *Status Perfectionis* is to be 140 years.

We do not see by what proportion 'Abû-'Abdallâh has calculated these numbers. For there is no proportion nor progression apparent among them. Verily, if we conceded to him that there are three such *Status Perfectionis*, if we then counted them in the way he has done, and declared finally, pre-supposing we did not apprehend being required to establish a proof, that the sum of these *Status* is 100 or 1,000 or something like it, his method and ours would be quite the same. However, there is this difference, that *we* find, that in our time man attains those phases of development, which he represents as the characteristic signs of the *Status Perfectionis*, in quite other *Sevenths* and times than those which he mentions. God knows best his meaning!

As regards the (superhuman) size of the bodies (of former generations), we say, if it be not necessary to believe it for this reason, that we cannot observe it in our time, and that there is an enormous interval between us and that time, of which such things are related, it is therefore by no means impossible. It is the same, the like of which is related in the Thora of the bodies of the giants (Nephîlîm, Rephâ'îm, 'Enâḳîm), p. 84
and the belief in this has not been abandoned since the time when the Israelites saw them with their own eyes. Therefore everybody may attack and ridicule this subject, if he likes! If the Thora was read to them, and they read it themselves, though up to that moment they had not declared the readers of the Thora to be liars, yet even if the giants were something quite different from what they are described to be (*i.e.* less extraordinary), they would declare the reader of the Thora to be a liar, in case he related anything that is not borne out by their experience and observation. If, indeed, there had never been classes of men with bodies of an extraordinary vastness, God having given them an uncommon size (*vide* Koran, ii. 24), no recollection of them would have remained in the uninterrupted chain of human tradition, and people would not compare with them everybody who, in size exceeds their genus, as it is

known to us. For instance, the people of 'Âd have become proverbial in this sense. But how can I expect them to believe me regarding the people of 'Âd, since they reject even that which is much nearer to our time and much more apparent? They produce such arguments as do not counterbalance the very weakest of those arguments which are urged against them. They shun accepting the striking arguments, flying before them like fugitive asses that fly before a lion (Koran, lxxiv. 51). What would they say of the monuments of larger races of men which exist still at the present time, such as the houses which were cut into the solid rocks in the mountains of Midian, of the graves built in the rocks, and of bones buried in their interior, which are as large as camel-bones and even larger, of the bad smell of those localities, which is so strong that you cannot enter there without covering the nose with something? And it is the common consent of all who inhabit those places that they (the authors of those monuments) are "*the people of darkness.*" But, when they hear of "*the day of darkness,*" they only laugh in a mocking way, make grimaces in haughty disdain, turn up their noses in joy over their theories, and in the persuasion that they are infinitely superior to, and altogether distinguished from all common people. But God is sufficient for them; they will get the reward of their doings, and we that of ours!

Chronological Tables.—In some book I have found tables illustrative of the durations of the reigns of the kings of the Assyrians, *i.e.* the people of Moṣul, of the kings of the Copts, who reigned in Egypt, and of the Ptolemæan princes, each of whom was called Ptolemæus. For Alexander, when dying, ordered that every king of the Greeks after him should be called Ptolemæus, in order to frighten the enemies, because the word means "*the warlike.*" In the same book I have found the chronology of the later kings of the Greeks.

In this book, the interval between the birth of Abraham and Alexander was reckoned as 2,096 years, which is more than Jews, Christians, and astrologers (those who apply the conjunctions of Saturn and Jupiter to history) reckon.

Now I have transferred those identical tables into this place of my book. Time has not enabled me to correct the names of the kings on the basis of their true pronunciation. I hope, therefore, that everyone will endeavour to correct and amend them, who like myself wishes to facilitate the subject for the student, and to free him from fatigue of research. And nobody ought to transcribe these tables and the other ones except him who is well acquainted with the *Ḥurûf-al-jummal,* and honestly endeavours to preserve them correct. For they are corrupted by the tradition of the copyists, when they pass from hand to hand among them. Their emendation is a work of many years.

Names of the Kings of the Assyrians, *i.e.* the people of Mosul. They are 37 in number, and they reigned during 1305 years.	How long each of them reigned.	The sum of the years.	
			p. 85.
I. Bêlos	62	62	
Ninos. He built Ninive in Mosul. Abraham was born in the [43rd year] of his reign	52	114	
Semiramis the wife of Ninos. She founded the ancient Sâmarrâ west of *Surra-man-ra'â*	42	156	
Zamês the son of Ninos. Abraham was tried by him, and fled therefore to Palestine in the [23rd year] of his reign	38	194	
V. Areios	30	224	
Aralios	40	264	
Xerxes	30	294	
Armamithrês	38	332	
Bêlôchos	35	367	
X. Balaios	52	419	
Altadas	32	451	
Mamythos	30	481	
Manchaleus	30	511	
Sphairos	20	531	
XV. Mamylos	30	561	
Sparethus	40	601	
Askatades	40	641	
Amyntês	45	686	
Bêlochos	25	711	
XX. Balatorês	30	741	p. 86.
Lampridês	32	773	
Sôsarês	20	793	
Lampares	30	823	
Panyas	45	868	
XXV. Sôsamos	19	887	
Mithraios	37	924	
Tautanês. In his time Ilion was taken by the Greeks, who had made war upon it	31	955	
Teutaios	40	995	
Thinaios	30	1025	
XXX. Derkylos	40	1065	
Eupales. In his time David reigned over Israel	38	1103	
Laosthenês. In his time the Israelites were divided into two kingdoms	40	1143	
Piritiades	30	1173	
Ophrataios	20	1193	
XXXV. Ophratanês. On the 167th day of the 42nd year of his reign Homer was born, who is with the Greeks the first poet, as Imru'ul-Ḳais with the Arabs	50	1243	
Akraganês	42	1285	
XXXVII. Thônos Konkoleros	20	1305	

p. 87. Western authors relate that, during the reign of this last king (Thônos Konkoleros, *alias* Sardanapalus), the prophet Jonah was sent to Niniveh, and that a foreigner, called *Arbâk* (Arbaces) in Hebrew, *Dah-âk* in Persian, and *Ḍaḥḥâk* in Arabic, came forward against this king, made war upon him, put him to flight, killed him, and took possession of the empire, holding it till the time when the Kayânians, the kings of Babylonia, whom western authors are in the habit of calling Chaldæans, brought the empire under their sway. The reign of Arbaces lasted seventy-two years.

Here we must remark that the Chaldæans are not identical with the Kayânians, but were their governors of Babylonia. For the original residence of the Kayânians was Balkh, and when they came down to Mesopotamia, people took to calling them by the same name which they had formerly applied to their governors, *i.e.* Chaldæans.

According to some chronicler, Nimrôd ben Kûsh ben Ḥâm ben Noah, founded a kingdom in Babylonia twenty-three years after the Confusion of Languages. And that was the earliest kingdom established on earth. The Confusion of Languages happened contemporaneously with the birth of the patriarch Re'û. The same chronicler mentions other kings that rose after Nimrôd, until the empire passed into the hands of the Assyrian kings, the chronology of whom has been illustrated by the preceding table. The chronology of the kings that have been recorded, is represented by the following table:—

The Kings of Babylonia.	How long they reigned.	Sum of the years.
Nimrôd	69	69
قمنورس	85	154
Ṣâmîrus	72	226
Arpakhshadh	10	236
Babylonia ἀβασιλευτός, till it was occupied by the Assyrians	5	241

p. 88. For the kings of Babylonia, we have also found another chronological tradition, beginning with Nebukadnezar the First (*i.e.* Nabonassar), and ending with the time when in consequence of the death of Alexander ὁ Κτίστης, people began to date by the reigns of the Ptolemæan princes. This tradition, now, we have transferred into this book, having corrected the numbers for the durations of their reigns. As to the names, however, I have simply transcribed them letter by letter, since I have not had an opportunity to correct them according to their pronunciation. The following table contains this chronological tradition.

Table of the Kings of the Chaldæans.	How long each of them reigned.	The sum of the years.	
Bukhtanaṣṣar Primus. With him the era in the *Almagest* begins	14	14	
Nebucadnezar. Nadios	2	16	
Chinzêros	5	21	
Ilulaios	5	26	
Mardokempad	12	38	
Arceanus	5	43	
Ἀβασίλευτος	2	45	
Bilibes	3	48	
Aparanadios	6	54	
Erigebalos	1	55	
Mesesimordakos	4	59	
Ἀβασίλευτος δεύτερος	8	67	
Asaridinos	13	80	
Saosduchinos	20	100	
Nabopolassaros and Kiniladanos	22	122	
Nebucadnezar	21	143	
Bukhtanaṣṣar, who conquered Jerusalem	43	186	p. 89
ترحلالتعر	2	188	
Belṭeshaṣṣar	4	192	
Darius the Median, the First	17	209	
Cyrus, who rebuilt Jerusalem	9	218	
Cambyses	8	226	
Darius	36	262	
Xerxes	21	283	
Artaxerxes Primus	43	326	
Darius	19	345	
Artaxerxes Secundus	46	391	
Ochus	21	412	
فسرون	2	414	
Darius	6	420	
Alexander ben Macedo, ὁ κτίστης	8	428	
Henceforward people commenced to date from the reign of Philippus.			

p. 90.

Names of the Coptic Kings in Egypt. They are 34 in number, besides the Persians, and they reigned during 894 years.	How long each of them reigned.	The sum of the years.
I. Diospolitæ	178	178
Smendis	26	204
Susennês	101	305
Nephercherês	4	309
V. Amenôphthis	9	318
Osochôr	6	324
Psinachês	9	333
Psûsennês	35	368
Sesonchôsis	21	389
X. Osôrthôn	15	404
Takelôthis	*13	417
Petûbastis	25	442
Osôrthôn	9	451
Psammos	10	461
XV. اوفانيراس (Euphanias ?)	44	505
Sabakôn Æthiops	12	517
Sebichôs	12	529
Tarakos Æthiops	20	549
Ammeris Æthiops	12	561
p. 91. XX. Stephinathis	7	568
Nechepsôs	6	574
Nechaô	8	582
Psammêtichos	44	626
Nechepsô (?) Nechaô (?)	6	632
XXV. Psammûthis	17	649
Vaphris	25	674
Amasis	42	716
The Persians till Darius	114	830
Amyrtaios	6	836
XXX. Nepheritês	6	842
Achôris	12	854
Psammûthis and Muthâtos (!)	2	856
Nektanebês	13	869
Teôs	7	876
XXXV. Nektanebos	18	894
Henceforward people ceased to date by the reigns of these and the Chaldean kings and commenced to use the era of Alexander the Greek.		

* P. adds 58, L. adds 8, as the reading of another manuscript.

Here we add the chronological tables of the Ptolemæans and the Roman Emperors. Chronology since the time of Philippus (Aridæus) consists of three parts:—I. of *Anni Philippi*; II. of *Anni Augusti*; III. of *Anni Diocletiani*. The first are the non-intercalated years of the Alexandrians; the second are the intercalated years of the Greeks; and of the same kind as the second are the *Anni Diocletiani*. With this king a new era commences, because, when the empire had devolved upon him, it remained with his descendants, and because after his death the Christian faith was generally adopted. Another (later) era than the *Æra Diocletiani* has not been mentioned, although the rule several times slipped out of the hands of his family. God knows best! Here follow the tables:—

p. 92.

Names of the Kings of Macedonia, who are the Greeks (Ionians), also called Ptolemæans.	How long they reigned.	Sum of the years.
Philippus	7	7
Alexander II. filius Alexandri . . .	12	19
Ptolemæus filius Lagi ὁ λογικός. He conquered Palestine, went up to Jerusalem and led the Israelites into captivity. Afterwards he restored them to liberty and made them a present of the vases of their Temple	20	39
Ptolemæus Philadelphus. He caused the Thora to be translated into Greek . . .	38	77
Ptolemæus Euergetes Phuskon Primus . .	25	102
Ptolemæus Philometor	17	119
Ptolemæus Epiphanes Phuskon Secundus . .	24	143
Ptolemæus Philopator the Deliverer . .	35	178
Ptolemæus Euergetes Alexander Secundus .	29	207
Ptolemæus Soter the Iron-smith, *Artium Fautor* .	36	243
Ptolemæus Dionysius Optimus	29	272
Cleopatra, till the time when Gajus, in Latin Julius, became Dictator	—	275
Cleopatra, till the death of Gajus and the succession of his son Augustus.	4a. 6m.	279
Cleopatra, till the time when he (Augustus) killed her	14a. 6m.	294

The calling Cleopatra by the name of *Ptolemæus* is a point of discussion, on account of her being a woman. But as she resided in Alexandria, and was the queen of it, she was called by that name. Gajus, in Latin Julius, means "*king of the world.*"

	Names of the Roman Kings, *i.e.* the Cæsars who resided in Rome. They are the Banû-al'aṣfar, *i.e.* the descendants of Ṣephô ben 'Elîphâz ben Esau ben Isaak ben Abraham.	How long each of them reigned.	Sum of the years.
93.	Augustus Cæsar, after he had killed Cleopatra .	43	43
	Tiberius filius Augusti	22	65
	Gajus	4	69
	Claudius, who killed the Apostle Paul and Simeon Petrus	14	83
	Nero, who killed the believers . . .	14	97
	Vespasianus. One year after his accession to the throne he conquered Palestine, and having besieged the Jews in Jerusalem for three years, he destroyed it, killed many, scattered the rest over the empire, and abolished their religious rites .	10	107
	Titus	3	110
	Domitianus. In the 9th year of his reign Johannes the Evangelist was banished. Thereupon he hid himself on an island till the emperor's death. Then he left the island and dwelt in Ephesus .	15	125
	Nerva	1	126
	Trajanus	19	145
	Hadrianus. It was he who destroyed Jerusalem and forbade anyone entering it in the 18th year of his reign	21	166
	Antoninus. It was he who rebuilt Jerusalem. Galenus says that he composed a book on anatomy in the beginning of his reign . .	23	189
	Commodus	32	221
94.	Severus and Antoninus	25	246
	Antoninus alone. Towards the end of his reign Galenus died	4	250
	Alexander filius Mammææ. Mammæa means "*weak*"	13	263
	Maximinus	3	266
	Gordianus	6	272
	Philippus	6	278
	Decius, who occurs in the story of the Seven Sleepers	1	279
	Gallus	3	282
	Valerianus	15	287 !
	Claudius	1	288
	Aurelianus	6	294
	Probus	7	301
	Carus and Carinus	2	303

NAMES OF THE KINGS OF CHRISTENDOM.	How long each of them reigned.	Anni Diocletiani. p. 95.
Diocletianus	21	21
Constantinus. The first king who adopted Christianity. He built the walls of Constantinople. In the 1st year of his reign his mother, Helene, sought for the wood of the Cross, which she finally found. In the 19th year the bishops assembled in Nicæa and established the canons of Christianity	32	53
Constantinus (Constantius)	24	77
Julianus Apostata	2	79
Valentinianus	1	80
Valens. He was burned, in escaping, in a barn	14	94
Theodosius the Great	17	111
Arcadius, his son	13	124
Theodosius Minor. In his time Nestorius was excommunicated	42	166
Marcianus and his wife Pulcheria. In their time the Jacobites were excommunicated	6	172
Leo the Great. He belonged to the moderate party	18	190
Zeno Alarmînâkî. He was a Jacobite	17	207
Anastasius. He built Ammorium, and was a Jacobite	27	234
Justinus	9	243
Justinianus. He built the church in Ruhâ (Edessa)	37	280
Tiberius	14	294
Mauricius. He helped Kisrâ against Bahrâm Shûbîn	14!	298!
Phocas, who was besieged in Constantinople by Shahrbarâz, the general of Kisrâ	8	318! p. 96.
Heraclius the wise	31	349
Constantinus. He was murdered in the bath	1	350
Constantinus	27	377
Constantinus	16	393
Justinianus. The Greeks cut off his nose	10	403
Leontius. He was found to be a weak man, being decrepit. So he was dethroned	3	406
Tiberius. Apsimarus	7	413
Justinianus Rhinomêtos	6	419
Philippicus	3	422
Anastasius. Atlîmus (Artemius). He was dethroned, when he could not carry on the war	2	424
Theodosius. He was besieged by Maslama ben 'Abd-almalik	1	425
Leo the Great. He deceived Maslama and repulsed him from Constantinople	24	449
Constantinus, the son of Leo the Great	34	483
Leo Junior, the son of Constantinus Senior	4	487
Constantinus Junior, the son of Leo Junior	18	505
Augusta (Irene) ruled the Greek empire	5	510
Nicephorus and Stauracius the son of Nicephorus	18	528
Michael the son of Georgius	2	—
Leo, till he was murdered by Michael in the church	7	—
Michael Constantinopolitanus, the murderer of Leo ben Theophilus ben Michael Constantinopolitanus	7a. 5m.	—
Basilius the Slavonian, the last of their kings	3a. 5m.	—

p. 97.

The Kings of Constantinople, as Ḥamza Alisfahânî records them on the authority of the judge Alwaḳi', who took them from a book that belonged to the Greek Emperor.	How long each of them reigned.		Sum of the years.	
	Years.	Months.	Years.	Months.
Constantinus, the son of Helene, the Victorious	31	0	31	0
Constantinus, his son	24	0	55	0
Julianus his nephew	2	6	57	6
Theodosius	10	9	68	3
Gratianus. Valentinianus	6	0	74	3
Arcadius, the son of Theodosius	13	3	87	6
Theodosius, the son of Arcadius	42	0	129	6
Marcianus	29	0	158	6
Leo Senior	16	0	174	6
Leo Junior	1	0	175	6
Zeno	17	0	192	6
Anastasius	27	4	219	10
Anṭlis	11	9	231	7
Ḳastrôndas. During his reign the Prophet was born	38	3	269	10
Stephanus	4	3	273 !	1
Marcianus (Mauricius). During his reign the Prophet received his Divine mission	20	4	293	5
Phocas. During his reign the flight of the Prophet occurred	8	0	301	5
Heraclius and his son. During their reign the Prophet died	31	0	332	5
p. 98. Constantinus, the son of Heracles	25	0	367 ?	5
Constantinus, the son of Heracles' wife	17	0	384	5
Constantinus, the son of Heracles	10	0	394	5
Leo or Leon (Lâwî or Elyûn)	3	0	397	5
Tiberius	7	0	411 ?	5
Estînus (Justinianus)	6	0	417	5
Anastasius	6	0	423	5
Theodosius	2	0	425	5
Leo. During his reign the empire of the Banû-'Umayya was dismembered	25	3	450	8
Leo, the son of Constantinus. People think that he was a worthless character, notwithstanding the length of his reign	5	0	455	8
Constantinus, the son of Leo	9	10	465	6
Constantinus	6	5	471	11
Irene, who received the empire from her father	5	0	476	11
Nicephorus, at the time of Hârûn Alrashîd	8	11	485	10
Stauracius, his son	0	2	486	0
Michael, his son	7	5	[476 ?]	5
Theophilus, his son	22	3	498	8
Michael, the son of Theophilus. With this king the dynasty expires—at the time of the Khalif Almu'tazz	28	0	526	8
Basilius the Slavonian	20	0	546	8
Leo the son of Basilius. Anno Hijræ 278 at the time of Almu'taḍid	26	0	[572]	8
Alexander, the son of Basilius. He died from a tumour in the belly, A.H. 299	1	2	[573]	10
Constantinus, the son of Leo, A.H. 301.	—	—	—	—

Chronology of the Persians.—The Persians call the first man p. 99. *Gayômarth*, with the surname *Girshâh*, *i.e.* "*king of the mountain*," or, as others say, *Gilshâh*, *i.e.* "*king of the clay*," because at that time there was no other man in existence (but himself, there being nothing but clay). People say that his name (Gayômarth) means "*a living, rational, mortal being.*"

The chronology of the Persians beginning with Gayômarth is divided into three parts:—

A. Part I. From Gayômarth till the time when Alexander killed Darius, seized upon the provinces of the Persians, and transferred their scientific treasures to his own country.

B. Part II. From that time till the time when Ardashîr ben Bâbak came forward, and the Persian empire was re-established.

C. Part III. From that time till the time when Yazdajird ben Shahryâr was killed, when the empire of the Sasanian dynasty was dissolved and Islâm arose.

Regarding the beginning of the world, the Persians relate many curious traditions, how Ahriman, *i.e.* the devil, was born out of the thought of God and of his pride in the world. And also regarding Gayômarth: for God, being bewildered at the sight of Ahriman, was covered with sweat on the forehead; this he wiped off and threw away; and out of this sweat Gayômarth was born. Then God sent him to Ahriman, who overpowered him, and began to travel about in the world, always riding upon him. At last, Ahriman asked him what was the most odious and horrible thing to him. Whereupon he said, that on arriving at the gate of hell he would suffer a painful terror. On having arrived, then, at the gate of hell, he became refractory, and managed by various contrivances to throw off the rider. But now Ahriman remounted him, and asked him from what side he was to begin devouring him. Gayômarth answered: "From the side of the foot, that I may still for some time look at the beauty of the world," knowing quite well that Ahriman would do the contrary of what he told him. Then Ahriman commenced devouring him from the head, and when he had come as far as the testicles and the spermatic vessels in the loins, two drops of sperma fell down on the earth. And out of these drops grew two Rîbâs bushes (*Rheum ribes*), from among which Mêshâ and Mêshâna sprang up, *i.e.* the Persian Adam and Eve. They are also called Malhâ and Malhayâna, and the Zoroastrians of Khwârizm call them Mard and Mardâna.

This is what I have heard from the geometrician, 'Abû-alḥasan Âdharkhûr.

In a different form this tradition, regarding the origin of mankind, is related by 'Abû-'Alî Muḥammad ben 'Aḥmad Albalkhî, the poet, in the

Shâhnâma, who premises that he has corrected his report on the basis of the following sources :—

I. *Kitâb-siyar-almulûk* by 'Abdallâh ben Almukaffa'.
II. „ „ by Muḥammad ben Aljahm Albarmakî.
III. „ „ by Hishâm ben Alkâsim.
IV. „ „ by Bahrâm ben Mardânshâh, the Maubadh of the city of Sâbûr.
V. „ „ by Bahrâm ben Mihrân Alisbahânî.

Besides he has compared his account with that of the Zoroastrian Bahrâm of Herât. He says : Gayômarth stayed in Paradise 3,000 years, *i.e.* the millennia of Aries, Taurus, and Gemini. Then he fell down on the earth and lived there safely and quietly three other millennia, those of Cancer, Leo, and Virgo, till the time when all that is evil in the world was brought about by Ahriman. The story is as follows: that Gayômarth, who was called *Girshâh,* because *Gir* means in Pahlavî "*mountain,*" dwelt in the mountains (Aljibâl-Media), being endowed with so
100. much beauty that no living being could view him without becoming terrified and losing the control of its senses.

Now, Ahriman had a son called *Khrûra,* who one day met with Gayômarth, and was killed by him. Whereupon, Ahriman complained to God of Gayômarth ; and God resolved to punish him in order to keep those covenants that existed between him and Ahriman. So he showed him first the punishments of this world and of the day of resurrection and other things, so that Gayômarth at last desired to die, whereupon God killed him. At the same moment two drops of sperma fell down out of his loins on the mountain Dâmdâdh in Istakhr, and out of them grew two Rîbâs-bushes, on which at the beginning of the ninth month the limbs (of two human bodies) began to appear, which by the end of that month had become complete and assumed human shape. These two are Mêshâ and Mêshyâna. Fifty years they lived without any necessity for eating and drinking, joyfully and without any pain. But then Ahriman appeared to them in the shape of an old man, and induced them to take the fruit of the trees. He himself commenced eating them, whereupon he at once again became a young man. And now they (Mêshâ and Mêshyâna) began to eat. Then they were plunged into misfortunes and evils. Lust arose in them, in consequence of which they copulated. A child was born unto them, but they devoured it from sheer ravenousness. But then God inspired their hearts with mildness. Afterwards the wife gave birth to six other children, the names of whom are known in the Avastâ. The seventh birth produced Siyâmak and Frâvâk, who married and begot a son Hôshang.

Regarding the chronology of this first part, the lives of the kings and their famous deeds, they relate things which do not seem admissible to the mind of the reader. However, the aim of our undertaking being to collect and to communicate chronological material, not to criticize and correct historical accounts, we record that on which the scholars of the Persians, the Hêrbadhs, and Maubadhs of the Zoroastrians agree among themselves, and which is received on their authority. At the same time we collect the materials in tables, as we have done heretofore, in order that our work may proceed on the same plan which we have laid down for the chronologies of the other nations.

To the names of the kings we add their epithets, because they are distinguished by individual epithets, whilst as to the other kings, if they have any epithet at all, it is one common to their whole class, by which he as well as everybody else who reigns in his place is called. Those common epithets correspond to the Shâhânshâh of the Persians. A list of them we give in the following table:—

The Classes of Princes.	The Epithets that apply to the Princes of these Classes.
The Sâsânian kings of the Persians	Shâhânshâh and Kisrâ.
The Greek kings	Bâsilî, *i.e.* Cæsar.
The kings of Alexandria	Ptolemæus.
„ Yaman	Tubbaʻ.
„ the Turks, Chazar, and Tagharghuz	Khâkhân. p.101
„ the Ghuzz-Turks	Ḥanûta.
„ the Chinese	Baghbûr.
„ India	Balharâ.
„ Ḳannûj	Râbî.
„ the Ethiopians	Alnajâshî.
„ the Nubians	Kâbîl.
„ the islands in the eastern ocean	Maharâj.
„ the mountains of Ṭabaristân	Ispahbadh.
„ Dunbâwand	Masmaghân.
„ Gharjistân	Shâr.
„ Sarakhs	Zâdhawaihi.
„ Nasâ and Abîward	Bâhmana.
„ Kash	Nîdûn.
„ Farghâna	Ikhshîd.
„ Asrûshana	Afshîn.
„ Shâsh	Tudun.
„ Marw	Mâhawaihi.
„ Nîshâpûr	Kanbâr.
Samarkand	Tarkhûn.
„ Sarîr	Alḥajjâj.
„ Dahistân	Sûl.
„ Jurjân	Anâhpadh.

The Classes of Princes.	The Epithets that apply to the Princes of these Classes.
02. The kings of the Sclavonians	Ḳabbâr.
„ the Syrians	Nimrôdh.
„ the Egyptians	Pharaoh.
„ Bâmiyân	Shîr-i-Bâmiyân.
„ Egypt	Al'azîz.
„ Kâbul	Kâbul-Shâh.
„ Tirmidh	Tirmidh-Shâh.
„ Khwârizm	Khwârizm-Shâh.
„ Shirwân	Shirwân-Shâh.
„ Bukhârâ	Bukhârâ-Khudâh.
„ Gûzgânân	Gûzgân-Khudâh.

Individual epithets (of princes) were not in use before the reign of Islâm, except among the Persians.

Part I. is divided into three parts :—

1. *Pêshdâdhians*, those who ruled over the whole world, founded cities, discovered mines and produced the metals, and found out the elements of handicrafts and arts; who practised justice on earth, and worshipped God as is his due.

2. *Kings of Êlân*, which means "*people of the highlands.*" They did not rule over the whole earth. The first who divided the empires of the world was Frêdûn *the Pure*, for he divided them between his sons, as a poet, a descendant of the family of the Kisrâs, says—

" Then we have divided our empire in our time,
Just as people divide meat on a meat-board.
Syria and Greece as far as the setting-place of the sun
We have given to a champion, to *Salm*.
To *Tôz* the Turks were given, and so a cousin
Holds the country of the Turks.
And to *Êrân* Al'irâḳ was given by dint of force. He has
Obtained the rule, and we have obtained the benefits thereof."

3. *Kayânians*, the heroes. In their days the rule over the world became divided between the various nations.

Between those parts (of ancient Persian chronology) there are gaps, on account of which the order and progress of chronology are much troubled and obscured.

Here follow the kings of Part I., according to the opinion of the generality of the Persians.

The classes of the Kings.	THE NAMES OF THE PERSIAN KINGS of PART I.	Their Epithets.	How long each of them reigned.	Sum of the years.	
The first men.	*Gayômarth*	Girshâh .	30	30	p.103
	Till the birth of Mêshâ and Mêshâna, who is called "*Mater filiorum et filiarum.*" These two are the Persian Adam and Eve . .	—	40	70	
	Till Mêshâ and Mêshâna married . .	—	50	120	
	Till the birth of Hôshang . . .	—	93	213	
The Pêshdâdians, the Just.	*Hôshang* ben Afrâwâk ben Siyâmak ben Mêshâ	Pêshdâdh	40	253	
	Tahmûrath ben Wîjahân ben Înkahadh ben Hôshang—till the coming-forward of Bûdâsaf .	Zêbâwand	1	254	
	The same—after that event . . .	—	29	283	
	Jam ben Wîjahân. From the time when he ordered people to fabricate weapons till the time when he ordered them to spin and weave	Shêdh .	50	333	
	Till the time when he ordered people to divide themselves into four classes . . .	—	50	383	
	Till the time when he made war upon the demons and subdued them . . .	—	50	433	
	Till the time when he ordered the demons to break rocks out of the mountains and to carry them	—	100	533	
	Till the time when he ordered a wheeled-carriage to be constructed. It *was* constructed, and he rode upon it . . .	—	66	599	
	After that, people lived in health and happiness—till the time when he hid himself . .	—	300	899	
	He continued to be hidden—till he was seized by Aldaḥḥâk, who tore out his bowels and sawed him with a saw	—	100	999	
	Aldaḥḥâk ben 'Ulwân of the Amalekites or—with another name—Bêvarasp ben Arwandasp ben Zîngâs ben Barîshand ben Ghar, who was the father of the pure Arabians, ben Afrâwâk ben Siyâmak ben Mêshâ . .	Azhdahâk	1000	1999	
	Afrêdûn ben Athfiyân Gâo ben Athfiyân Nîgâo ben Athfiyân ben Shahrgâo ben Athfiyân Akhunbagâo ben Athfiyân Sipêdhgâo ben Athfiyân Dîzagâo ben Athfiyân Nîgâo ben Nêfurûsh ben Jam the King . . .	Almaubadh	200	2199	p.104
The kings of Êlân, the people of the highlands.	*Îraj.* He was killed by his brothers *Salm* and *Ṭôj*, who reigned after him. They were all three sons of Afrêdûn	Almusṭafâ	300	2499	
	Minôshjihr ben Gûzan, the daughter of Îraj—till the time when he killed Ṭôj and Salm, i.e. Sharm in Persian	Fêrôz .	20	2519	
	Till the time when the son of Ṭôj occupied Êrânshahr, and drove Minôshjihr out of the country	—	60	2579	
	Firâsiyâb ben Bashang ben Înat ben Rishman ben Turk ben Zabanasp ben Arshasp ben Ṭôj—till the time when Minûshjihr gained the victory over him and drove him away. Thereupon they made a treaty on the basis of the well-known arrow-shot . . .	—	12	2591	

The classes of the Kings.	The Names of the Persian Kings of Part I.	Their Epithets.	How long each of them reigned.	Sum of the years.
The kings of Êlân, the people of the highlands.	Minôshjihr—till his death	—	28	2619
	Tôzh the Turk occupying Al'irâk	Firâsiyâb	12	2631
	Zâb ben Tahmâsp ben Kamjahûbar ben Zû ben Hûshab ben Widinak ben Dûsar ben Minôshjihr together with— *Garshâsp, i.e.* Sâm ben Narîmân ben Tahmâsp ben Ashak ben Nôsh ben Dûsar ben Minôshjihr	The two companions.	5	2636
The Kayânians, the heroes.	*Kaikobôdh* ben Zagh ben Nûdhagâ ben Mâishû ben Nûdhar ben Minôshjihr	The First	100	2736
	Kaikâûs ben Kainiya ben Kaikobâdh—till he rebelled, whereupon he was taken prisoner by Shammar and afterwards delivered by Rustam ben Dastân ben Garshâsp the King	Nimrud	75	2811
	The same—from the latter event till his death	—	75	2886
	Kaikhusrû ben Siyâwush ben Kaikâûs—till the time when he went away as a holy pilgrim and hid himself	Humâyûn	60	2946
	Kailuhrâsp ben Kaiwajî ben Kaimanish ben Kaikubâdh—till he sent Bukhtanassar to Jerusalem, who destroyed it	The Bactrian	60	3006
	p.105. The same after that event		60	3066
	Kaiwishtâsp ben Luhrâsp—till the appearance of Zoroaster	Alherbadh	30	3096
	The same after that event	—	90	3186
	Kai Ardashîr—Bahman ben Isfandiyâr ben Wishtâsp	Tall in the body.	112	3298
	Khumânî the daughter of Ardashîr—Bahman	Cihrâzâd	30	3328
	Dârâ ben Ardashîr—Bahman	The great	12	3340
	Dârâ ben Dârâ—till he was killed by Alexander the Greek	The second	14	3354

The account of the chronology of this Part I., which we have given, is stated very differently in the *Kitâb-alsiyar*. Our account, however, comes nearest to that view regarding which people agree. The chronology of this same part, but in a different shape, I have also found in the book of Ḥamza ben Alḥusain Alisfahânî, which he calls "*Chronology of great nations of the past and present.*" He says that he has endeavoured to correct his account by means of the *Âbastâ*, which is the religious code (of the Zoroastrians). Therefore I have transferred it into this place of my book.

TABLE II. of PART I.

Names of the Pêshdâdhian Kings, taken from the Abastâ, beginning with Gayômarth.	How long each of them reigned.	Sum of the Years.	
			p.106.
Gayômarth the first man	40	40	
An interregnum of 170 years.	—	—	
Hôshang	40	80	
Ṭahmûrath	30	110	
Jam	616	726	
Bêwarasp	1000	1726	
Afrêdûn	500	2226	
Minôshcihr	120	2346	
Firâsyâb	12	2358	
An interregnum of unknown length.	—	—	
Zâb	9	2367	
Garshâsp together with Zâb . . .	3	2370	
An interregnum.	—	—	
Names of the Kayânian Kings.			p.107.
Kaikobâdh	126	2494	
Kaikâûs	150	2646	
Kaikhusrau	80	2726	
Kailuhrâsp	120	2846	
Kaibishtâsp	120	2966	
Kaiardashîr	112	3078	
Cihrâzâd	30	3108	
Dârâ ben Bahman	12	3120	
Dârâ ben Dârâ	14	3134	

p.108. Further, Ḥamza relates that he has found also this part of Persian chronology in the copy of the Maubadh, such as is exhibited in the following table:—

TABLE III. OF PART I.

NAMES OF THE PÊSHDÂDHIAN KINGS, taken from the Copy of the Maubadh.	How long each of them reigned.	Sum of the Years.
Gayômarth	30	30
Mêshâ and Mêshânâ—till they got children	50	80
Till their death	50	130
Interregnum	94	224
Hôshang	40	264
Ṭahmûrath	30	294
Jam—till he hid himself	616	910
He remained hidden	100	1010
Bêwarasp	1000	2010
Frêdûn	500	2510
Minôshcihr	120	2630
Zû and Garshâsp	4	2634
p.109. NAMES OF THE KAYÂNIAN KINGS.		
Kaikobâdh	100	2734
Kaikâûs	150	2884
Kaikhusrau	60	2944
Luhrâsp	120	3064
Bishtâsp	120	3184
Ardashîr	112	3296
Cihrâzâd	30	3326
Dârâ ben Bahman	12	3338
Dârâ ben Dârâ	14	3352

p.110. In the biographical and historical books that have been translated from the works of Western authors, you find an account of the kings of Persia and Babylonia, beginning with Frêdûn, whom they call, as people say, Yâfûl (Pûl ?), and ending with Dârâ, the last of the Persian kings. Now, we find that these records differ greatly (from Eastern records) as to the number of the kings and their names, as to the durations of their reigns, their history, and their description. I am inclined to think that they confounded the kings of Persia with their governors of Babylonia, and put both side by side. But if we altogether refrain from mentioning those records, we should deprive this book of something that forms a due part of it, and we should turn away the mind of the reader therefrom. We, now, exhibit this tradition in a special table of its own, in order to prevent confusion getting into the arrangement of the various systems and traditions of this book. Here it follows :—

The Kings of Persia, beginning with Frêdûn, according to Western authors.	How long each of them reigned.	Sum of the Years.	
Yâfûl, *i.e.* Frêdûn	35	35	
Tighlath Pilesar	35	70	
Salmanassar, *i.e.* Salm	14	84	
Sanherib ben Salmanassar, *i.e.* in Persian: Sanâraft	9	93	
Sardûm (Ezarhaddon), *i.e.* Zû ben Tûmâsp	8	96	
After him the following powerful kings reigned :—			
Kaikobâdh	49	145	p.111.
Sanherib II.	31	176	
Mâjam	33	209	
Bukhtanaṣṣar, *i.e.* Kaikâûs	57	266	
Evilad ben Bukhtanaṣṣar	1	267	
Belteshaṣṣar ben Evilad	2	269	
Dârâ Almâhî I., *i.e.* Darius	9	278	
Koresh, *i.e.* Kaikhusrau	8	286	
Cyrus, *i.e.* Luhrâsp	34	320	
Cambyses	80	400	
Dârâ II.	36	436	
Xerxes ben Dârâ, *i.e.* Khusrau I.	26	462	
Ardashîr ben Xerxes, called *μακροχείρ*, *i.e.* Longimanus	41	503	
Khusrau II.	30	533	
Sogdianus, Notos ben Khusrau	9	542	
Ardashîr ben Dârâ II.	41	583	
Ardashîr III.	27	610	
Arses ben Ochus	12	622	
Dârâ, the last king of Persia	16	638	

The Jews, Zoroastrians, Christians, and the various sects of them, p.112. relate the *origines mundi* and carry chronology down from them, having previously admitted the truth of such *origines*, and having gained certain views regarding them, on which people either agree or differ. He, however, who denies such *origines*, cannot adopt that which is built upon them, except after producing various sorts of interpretations which he adds of his own.

However, those *origines mundi*, *i.e.* Adam and Eve, have been used as the epoch of an era. {And some people maintain that *time* consists of cycles, at the end of which all created beings perish, whilst they grow at their beginning; that each such cycle has a special Adam and Eve of its own, and that the chronology of this cycle depends upon them.

Other people, again, maintain that in each cycle a special Adam and Eve exist for each country in particular, and that hence the difference of human structure, nature, and language is to be derived.

Other people, besides, hold this foolish persuasion, *viz.* that *time* has no *terminus a quo* at all; they take some dogmas from the founders of religions, in order to construct some system by means of them. Many philosophers of this class have built up such systems. You could hardly find a prettier tale of this kind than that one produced by Sa'îd ben Muḥammad Aldhuhlî in his book. For he says: "People lived in bitter enmity and strife with each other; the better among them were maltreated and oppressed by the worse. But then, at last, the just king, Pêshdâdh, transplanted them to a place, called *Firdaus* (Paradise), situated between Adan and Serendîb. It was a place where aloe, cloves, and various sorts of perfumes were growing, and all kinds of delicious things were to be found. There they dwelt, till one day a demon (*'Ifrît*) came upon them, the king of the wicked, and began quarrelling with them. In the same place Pêshdâdh found a boy and girl, the parents of whom were unknown. These he educated and called them *Mêshâ* and *Mêshâna*, and made them marry each other. Thereupon they committed sin, and so he drove them out of that country." The tale as it has been related, is extremely long. He says that the interval between the time of their settlement in Paradise, the beginning of all chronology, and their meeting the demon was one year; till the time when Mêsha and Mêshâna were found, two more years elapsed; till their marriage, forty-one years; till their death, thirty years; and till the death of Pêshdâdh, ninety-nine years elapsed. But then he ceases from going on with his chronological account and does not carry it on.

Chronology of the Ashkanians.—As to Part II. of Persian chronology from Alexander till the rise of Ardashîr ben Bâbak, it must be known that during this period the "*Petty Princes*" existed, *i.e.* those princes whom Alexander had installed as rulers over certain special districts, who were all totally independent of each other. To the same period belongs the empire of the Ashkânians, who held 'Irâḳ and the
p.113. country of Mâh, *i.e.* Aljibâl, under their sway. They were the most valiant among the "*Petty Princes*;" still the others did not obey them, but only honoured them for this reason, that they descended from the royal Persian house. For the first prince of the Ashkânians was *Ashk* ben Ashkân, called Afghûrshâh ben Balâsh ben Shâpûr ben Ashkân ben اس اشكار ben Siyâwush ben Kaikâûs.

Most Persian chroniclers have connected the reign of Alexander immediately with that of the first Ashkânian prince, by which that period was most improperly curtailed. Others say that the Ashkânians came into power some time after Alexander, whilst others go on blundering without any knowledge of the matter.

I shall relate in this place such of their traditions as I have learned, and shall endeavour, as much as is in my power, to amend that which is wrong, to refute that which is false, and to establish the truth, beginning with that which corresponds most nearly to the Table I. of Part I., I also call it Table I. (of Part II.) :—

Table of the Names of the Ashkânian Kings, corresponding to the Table I. of Part I.		How long each of them reigned.	Sum of the Years.
——	Their Surnames.		
Alexander the Greek	. . . . .	14	14
Ashk ben Ashkân	Khôshdih	13	27
Ashk ben Ashk ben Ashk	Ashkân	25	52
Shâpûr ben Ashk	Zarrîn	30	82
Bahrâm ben Shâpûr	Khûrûn	21	103
Narsî ben Bahrâm	Gîsûwar	25	128
Hurmuz ben Narsî	Sâlâr	40	168
Bahrâm ben Hurmuz	Rôshan	25	193
Fêrôz ben Bahrâm	Balâd	17	210
Kisrâ ben Fêrôz	Barâdih	20	230
Narsî ben Fêrôz	Shikârî	30	260
Ardawân ben Narsî	The last	20	280

Next follows what corresponds to the Table II. of the same Part I., p.114.
that which Ḥamza has taken from the Abastâ. This, again, I call the Tabula II., for the purpose of connecting those portions of the three parts of Persian chronology that bear the same name (as Table I., II., III. of Parts I., II., III.) with each other, and to bring the tables, thereby, into a good order. It will not be necessary to mention this another time :—

TABLE II. OF PART II. IN THE ARRANGEMENT OF THE TABLES.

Names of the Ashghânian Kings, according to Ḥamza.	How long each of them reigned.	Sum of the Years.
Alexander the Greek	14	14
Ashk ben Balâsh ben Shâpûr ben Ashkân ben Ash the hero	52	66
Shâpûr ben Ashk	24	90
Jûdhar ben Wîjan ben Shâpûr	50	140
Wîjan ben Balâsh ben Shâpûr, the nephew of the preceding	21	161
Jûdhar ben Wîjan ben Balâsh	19	180
Narsa ben Wîjan	30	210
Hurmuzân ben Balâsh, the uncle of the preceding	17	227
Fêrôzân ben Hurmuzân	12	239
Khusrau ben Fêrôzân	40	279
Balâsh ben Fêrôzân	24	303
Ardawân ben Balâsh ben Fêrôzân	55	358

To this I add that which in the order of the tables is the third one, which Ḥamza says he has taken from the copy of the Maubadh, in order that the subject may be carried on, as it has been done in the two preceding tables. Here follows the Table III. of Part II. :—

TABLE III. OF PART II.

p.115.

NAMES OF THE ASHKÂNIAN KINGS, taken by Ḥamza from the Copy of the Maubadh.	How long each of them reigned.	Sum of the Years.
Alexander the Greek	14	14
After him reigned a class of Greek princes, with their Persian vizirs, altogether 14 in number	68	82
Ashk ben Dârâ ben Dârâ ben Dârâ	10	92
Ashk ben Ashkân	20	112
Shâpûr ben Ashkân	60	172
Bahrâm ben Shâpûr	11	183
Balâsh ben Shâpûr	11	194
Hurmuz ben Balâsh	40	234
Fêrôz ben Hurmuz	17	251
Balâsh ben Fêrôz	12	263
Khusrau ben Malâdhân	40	303
Balâshân	24	327
Ardawân ben Balâshân	13	340
Ardawân the Great, ben Ashkânân	23	363
Khusrau ben Ashkânân	15	378
Bahâfirîd ben Ashkânân	15	393
Jûdhar ben Ashkânân	22	415
Balâsh ben Ashkânân	30	445
Narsî ben Ashkânân	20	465
Ardawân, the last	31	496

p.116. Next I shall produce what I found in the chronicle of 'Abû-alfaraj 'Ibrâhîm ben 'Aḥmad ben Khalaf Alzanjânî the mathematician. This man, on having taken pains to compare the discordant traditions with each other, gives the following account of the "*Petty Princes*," and the durations of their reigns, as is exhibited in the following table. He maintains that the Persians fixed only the historical tradition regarding the Ashkânian princes, not regarding the other "*Petty Princes*," and that the Ashkânians first brought 'Irâḳ and Jibâl under their sway *Anno Alexandri* 246.

The Ashkânians, according to the Chronicle of 'Abû-alfaraj.	How long each of them reigned.	Sum of the Years.
Alexander the Greek	14	14
The "*Petty Princes*"	246	260
Afghûrshâh	10	270
Shâpûr ben Ashkân	60	330
Jûdhar, Senior	10	340
Bîzan the Ashkânian	21	361
Jûdhar the Ashkânian	19	380
Narsî the Ashkânian	40	420
Hurmuz	17	437
Ardawân	12	449
Khusrau	40	489
Balâsh	24	513
Ardawân, Junior	13	526

We have also found a chronological synopsis of this same Part II. in the Shâhnâma by 'Abû-Manṣûr 'Abd-alrazzâḳ, such as we exhibit in the following table :—

The Ashkânians, according to the Shâhnâma.	How long each of them reigned.	Sum of the Years.
Ashk ben Dârâ, according to others a descendant of *Arish*	13	13
Ashk ben Ashk	25	38
Shâpûr ben Ashk	30	68
Bahrâm ben Shâpûr	51	119
Narsî ben Bahrâm	25	144
Hurmuz ben Narsî	40	184
Bahrâm ben Hurmuz	5	189
Hurmuz	7	196
Fêrôz ben Hurmuz	20	216
Narsî ben Fêrôz	30	246
Ardawân	20	266

The nature of this Part II. is brought to light by a comparative examination of these tables. It is a period that begins with Alexander's conquest of Persia, and ends with the rising of Ardashîr ben Bâbak and his seizing the empire out of the hands of the Ashkânians. Both these limits are well known, and generally agreed upon. How, then, can the interval between them be a matter of doubt to us? However, it must be kept in mind that we are not able to make out by a mere course of reasoning the duration of the rule of each of the Ashkânian princes, nor of the other "*Petty Princes*," nor the number of the persons who occupied the throne. For all this depends upon historical tradition, ank it is well-known to what mishap tradition has been subject. The least, now, we must try to do is to amend this Part II. as much as is in our power.

It is evident and not unknown to anybody, that the year in which Yazdajird came to the throne was *A. Alex.* 943. This undeniable date we shall keep in mind as a basis, and establish it as a gauge by which to measure all their records.

Let us first take the sum of years which we get from the Table I. of Part II., *i.e.* 280 years. Hereto we add that sum which we shall
p.118. exhibit in the Table I. of Part III. for the time from the beginning of the reign of Ardashîr till that of the reign of Yazdajird, in order to combine the like tables (*i.e.* Table I., II., III. of Part II. respectively, with Table I., II., III. of Part III.) with each other. This latter period is about 410 years. So we get a sum of

690 years,

which is less than our gauge by about 258 years. We shall drop this calculation and not take further notice of it.

Next we consider the sum of years contained in the Table II. of Part II., *i.e.* 358 years. Hereto we add the sum which will be exhibited by Table II. of Part III., corresponding to the sum that occurs in the first calculation, and we get the sum total of

818 years,

which is again less than our gauge by about 125 years.

We shall drop this calculation, too, and proceed to the Tables III. in Parts II. and III., and add them together in the same way as we have done with Table I. and II. Then we get the sum of

930 years,

which is again below our gauge by about thirteen years.

We drop this calculation, and do not further notice it. For chronology does not admit of this difference, although it may be so slight as nearly to approach the truth.

If we make the same calculation with the years exhibited in the book of 'Abû-alfaraj, combining the corresponding tables with each other, we get the sum of

949 years,

which exceeds our gauge by six years.

If we pass by this and add together the years as reported in the Shâhnâma for this Part II., with the result of any of the tables of Part III., this calculation would still less agree with our gauge (than the preceding ones).

Now we shall put aside all these calculations, and try to derive an emendation of them from the book of Mânî, called *Shâbûrkân*, since, of all Persian books, it is one that may be relied upon (as a witness) for the time immediately following the rise of Ardashîr (ben Bâbak). Besides, Mânî in his law has forbidden telling lies, and he had no need whatsoever for falsifying history.

Mânî, now, says in this book in the chapter of the coming of the prophet, that he was born in Babylonia *Anno Astronomorum Babyloniæ* 527, *i.e.* Anno Alex. 527, and four years after the beginning of the reign of the king *Âdharbân*, whom I believe to be Ardawân the Last. In the same chapter he says that he first received divine revelation when he was thirteen years of age, or *Anno Astronomorum Babyloniæ* 539, two years after the beginning of the reign of Ardashîr the king of kings.

Hereby Mânî states that the interval between Alexander and Ardashîr is 537 years, and that the interval between Ardashîr and the succession of Yazdajird is 406 years. And this result is correct, being based upon the testimony of a book, favoured by God with a long duration, which is used as a religious code.

Further, we are informed by traditions, the correctness of which is proved by their mutual agreement, that the last intercalation was carried out at the time of Yazdajird ben Shâpûr, and that the Epagomenæ were put at the end of that month, to which the turn of intercalation had p.119.
come, *viz.* the eighth month (Âbân-Mâh). If, now, we count the interval between Alexander and Ardashîr as 537 years, we find the interval between Zoroaster and Yazdajird ben Shâpûr to be nearly 970 years, in which eight leap months are due, since it was their custom to intercalate one month in every 120 years. But if we count that interval (between Alexander and Ardashîr) as 260–270 years, or something more, as 300 years, as most authors do, we get a sum of about 600 years, in which only five leap months would be due, whilst we have already mentioned their report stating that eight leap months are due in that period. The latter is therefore an irreconcileable supposition (*viz.* that the interval between Alexander and Ardashîr is not more than 260–300 years).

Likewise it is written in the books of astrologers, that the horoscope of the year in which Ardashîr (ben Bâbak) rose was about half of Gemini, and the horoscope of the year in which Yazdajird rose was the sixth degree of Cancer. If, now, we multiply $93\frac{1}{4}$ degrees, which is the surplus of the solar cycle over the whole days according to the Persians, by 407 years, we get the sum of $152\frac{3}{4}$ degrees. If we subtract this from the rising-place of the degree of the horoscope of that year, in which Yazdajird came to the throne, and take the arc of the remainder for the rising-place of the region of 'Irâk, which was the residence of the Kisrâs, the horoscope is half of Gemini close to the place, which the astrologers mention. If the years, however, are either more or less, the horoscope does not agree (with what it is reported to have been). So, of course, that which is confirmed by two witnesses is more trustworthy than that which is contradicted by many.

If we add to the 407 years, mentioned by the astrologers, the 537 years which are reported by the Shâbûrkân, we get the sum of 944 years. And that is the year of the Æra Alexandri for Yazdajird's accession to the throne. The surplus of one year is only possible in the reports of such authors as do not give detailed statements regarding the months and minor fractions of time, in consequence of the fact that the years of the Persians and Greeks commence at different times.

Ḥamza relates that Mûsâ ben 'Îsâ Alkisrawî, on having studied this subject, and perceived the confusion we have mentioned, said: "The interval between Alexander and Yazdajird's accession to the throne is 942 years. If we subtract therefrom 266 years for the period of the reign of the Ashkânians, we get for the rule of the Sasanians, from Ardashîr till the accession of Yazdajird, 676 years. In their own traditions the Persians have no such chronological system."

Further, he says: "Thereupon we studied and examined the number of their kings. And here be it noticed that they have forgotten the names of some of them, whom the chroniclers have not mentioned, blending together some of their names on account of their similarity. I shall enumerate them as they really are." Accordingly, he, *i.e.* Mûsa, has increased the durations of their reigns and their number, as we shall explain, when the order of our exposition comes to that subject, if God permits.

.120. **Chronology of the Sasanians.**—Now we proceed to treat of the third part of Persian chronology, the beginning of which is the rising of Ardashîr ben Bâbak of the family of Bahman ben Isfandiyâr. For he was the son of Bâbak Shâh ben Sâsân ben Bâbak ben Sâsân ben Bahâfirîd ben Mihrmish ben Sâsân senior ben Bahman ben Isfandiyâr. This part of chronology also is not free from the same defects that beset the former two parts, but still they are less considerable. I commence this part with the Table I., corresponding to the (first) tables of each of the two preceding parts, and I shall proceed hereafter with Table II. and III. If you gather the dates from the single tables of the three parts, you get the consecutive course of Persian chronology. Here follows Table I.

	Names of the Sâsânian Kings, corresponding to Tables Ia in Parts I. and II.	Their Surnames.	How long each of them reigned.			Sum of the Years.		
			Years.	Months.	Days.	Years.	Months.	Days.
p.121.	Ardashîr ben Bâbak	Bâbakân	14	10	0	14	10	0
	Shâpûr ben Ardashîr	Fîrdih	30	6	12	45	4	12
	Hurmuz ben Shâpûr	The Hero	1	10	0	47	2	12
	Bahrâm ben Hurmuz, who killed Mânî	Yazdajân	3	3	3	50	5	15
	5. Bahrâm ben Bahrâm	Shâhîdih	17	0	0	67	5	15
	Bahrâm ben Bahrâm ben Bahrâm	Sakânshâh	0	4	0	67	9	15
	Narsî ben Bahrâm ben Bahrâm	Nakhcîrkân	9	0	0	76	9	15
	Hurmuz ben Narsî	Kôhpad	7	5	0	84	2	15
	Shâpûr ben Hurmuz Dhû-al'aktâf	Hôba-sunbâ	72	0	0	156	2	15
	10. Ardashîr ben Hurmuz	The Beautiful	4	0	0	160	2	15
	Shâpûr ben Shâpûr	Sâbûr-aljunûd	5	4	0	165	6	15
	Bahrâm ben Shâpûr	Karmânshâh	11	0	0	176	6	15
	Yazdagird ben Shâpûr	Sceleratus	21	5	17	198	0	2
	Bahrâm ben Yazdagird	Gûr	18	10	0	216	10	2
	15. Yazdagird ben Bahrâm	Shâhdôst	18	8	28	235	2	0
	Frêdûn ben Yazdagird	Mardâna	27	0	0	262	2	0
p.122.	Balâsh ben Fêrôz	Karmân-mâna	4	0	0	266	2	0
	Kobâd ben Fêrôz	Nîk-râî	38	0	0	304	2	0
	Jâmâsp ben Fêrôz, the brother of the preceding	Nîkârêw	2	0	0	306	2	0
	20. Kobâd ben Fêrôz, the second time	Zindîk	4	0	0	310	2	0
	Kisrâ Anôshirwân—till the birth of the Prophet	The Just King	41	0	0	351	2	0
	The same afterwards	" "	7	7	0	358	9	0
	Hurmuz ben Kisrâ—till he was deposed and strangled	Turkzâd	9	7	10	368	4	10
	Kisrâ—till he carried the wood of the Cross away from Aelia	Parwîz the Glorious King	33	0	0	401	4	10
	The same afterwards—till the Flight of the Prophet	" "	0	1	8	401	5	18
	The same afterwards—till he was deposed, blinded, and killed	" "	4	10	22	406	4	10
	Kobâd ben Kisrâ—till he perished in the plague	Shîrawaihi	0	8	0	407	0	10
	25. Ardashîr ben Shîrawaihi, 7 years old	The Little One	1	6	0	408	6	10
	Shahrbarâz, whom Kisrâ had sent out to besiege Constantinople	Khurramân	0	1	8	408	7	18
	Pûrân, daughter of Kisrâ Parwîz. Her mother was Mary, the daughter of the Caesar	Fortunata	1	4	0	409	11	18
	Kisrâ ben Kobâd ben Hurmuz ben Kisrâ Parwîz	The Short One	0	10	0	410	9	18
	Fêrôz	Khôshdîd	0	1	20	410	11	8
	30. Azarmîdukht, daughter of Parwîz, till she was poisoned	The Just	0	6	0	411	5	8
	Farrukhzâd Khusrau, a child	"	0	1	0	411	6	8
	Yazdagird ben Shahryâr ben Kisrâ Parwîz, 15 years old	The Last King	20	0	0	431	6	8

After him the rule of the Arabians commenced.

p.123. The following Table II. rests on the authority of Ḥamza, who says that he has amended it by means of the Abastâ, and transcribed it from the *Kitâb-alsiyar-Alkabîr*.

TABLE II. of PART III.

Names of the Sâsânian Kings.	How long each of them reigned.			Sum of the Years.		
	Years.	Months.	Days.	Years.	Months.	Days.
Ardashîr Bâbak	14	6	0	14	6	0
Shâpûr ben Ardashîr	30	0	28	44	6	28
Hurmuz ben Shâpûr	1	10	0	46	4	28
Bahrâm ben Hurmuz	3	3	3	49	8	1
5. Bahrâm ben Bahrâm	17	0	0	66	8	1
Bahrâm ben Bahrâm ben Bahrâm	0	4	0	67	0	1
Narsî ben Bahrâm	9	0	0	76	0	1
Hurmuz ben Narsî	7	5	0	83	5	1
Shâpûr ben Hurmuz Dhû-al'aktâf	72	0	0	155	5	1
10. Ardashîr ben Hurmuz	4	0	0	159	5	1
Shâpûr ben Shâpûr	50	4	0	209	9	1
Bahrâm ben Shâpûr	11	0	0	220	9	1
Yazdagird ben Bahrâm Sceleratus	21	5	8	242	2	9
Bahrâm ben Yazdagird, Gûr	23	0	0	265	2	9
p.124. 15. Yazdagird ben Bahrâm	18	4	28	283	7	7
Fêrôz ben Yazdagird	27	0	1	310	7	8
Balash ben Fêrôz	4	0	0	314	7	8
Kobâd ben Fêrôz	43	0	0	357	7	8
Anôshirwân ben Kobâd	47	7	0	405	2	8
20. Hurmuz ben Anôshirwân	11	7	10	416	9	18
Parwîz ben Hurmuz	38	0	0	454	9	18
Shîrawaihi ben Parwîz	0	8	0	455	5	18
Ardashîr ben Shîrawaihi	1	6	0	456	11	18
Pûrândukht, daughter of Parwîz	1	4	0	458	3	18
25. Guskanasptadha	0	2	0	458	5	18
Azarmîdukht, daughter of Parwîz	1	4	0	459	9	18
Khurzâd Khusra	0	1	0	459	10	18
Yazdagird ben Shahryâr	20	0	0	479	10	18

The following Table III. in this Part is that one which Ḥamza says he transcribed from the copy of the Maubadh. p.125.

Names of the Sâsânian Kings, such as Ḥamza says he has taken from the Copy of the Maubadh.	How long each of them reigned.			Sum of the Years.		
	Years.	Months.	Days.	Years.	Months.	Days.
Ardashîr b. Bâbak (after having made war upon the "*Petty Princes*")	14	10	0	14	10	0
Shâpûr ben Ardashîr	30	0	15	44	10	15
Hurmuz ben Shâpûr	3	3	0	48	1	15
Bahrâm ben Hurmuz	17	0	0	65	1	15
5. Bahrâm Sakân-shâh	40	4	0	105	5	15
Narsa ben Bahrâm	9	0	0	114	5	15
Hurmuz ben Narsa	7	0	0	121	5	15
Shâpûr Dhû-al'aktâf	72	0	0	193	5	15
Ardashîr ben Hurmuz	4	0	0	197	5	15
10. Shâpûr ben Shâpûr	5	0	0	202	5	15
Bahrâm ben Shâpûr	11	0	0	213	5	15
Yazdagird Sceleratus	21	5	18	234	11	3
Bahrâm Gûr	19	11	0	254	10	3
Yazdagird ben Bahrâm	14	4	18	269	2	21
15. Fêrôz ben Yazdagird	17	0	0	286	2	21
Balâsh ben Fêrôz	4	0	0	290	2	21
Kobâd ben Fêrôz	41	0	0	331	2	21
Anôshirwân	48	0	0	379	2	21
Hurmuz ben Anôshirwân	12	0	0	391	2	21
20. Parwîz	38	0	0	429	2	21
Kobâd Shîrawaihi	0	8	0	429	10	21
Ardashîr ben Shîrawaihi	1	6	0	431	4	21
Pûrân, daughter of Parwîz	1	4	0	432	8	21
Fêrôz	0	1	0	432	9	21
25. Azarmîdukht	0	6	0	433	3	21
Khurradâdh Khusra	1	0	0	434	3	21
Yazdagird ben Shahryâr	20	0	0	454	3	21

p.126.

In the book of 'Abû-alfaraj Alzanjânî we have found the chronology of this Part differing from our accounts in the preceding three tables; we have added his account in this place, in conformity with what we have done in the preceding two Parts. And herewith the Chronological Table ends.

p.127.

Names of the Sâsânian Kings, according to the Tradition of 'Abû-alfaraj Alzanjânî.	How long each of them reigned.			Sum of the Years.		
	Years.	Months.	Days.	Years.	Months.	Days.
Ardashîr ben Bâbak	14	10	0	14	10	0
Shâpûr ben Ardashîr	31	6	18	46	4	18
Hurmuz ben Shâpûr	1	6	0	47	10	18
Bahrâm ben Hurmuz	3	3	3	51	1	21
5. Bahrâm ben Bahrâm	17	0	0	68	1	21
Bahrâm ben Bahrâm ben Bahrâm	4	4	0	72	5	21
Narsî ben Bahrâm	9	0	0	81	5	21
Hurmuz ben Narsî	9	0	0	90	5	21
Shâpûr ben Hurmuz Dhûal'aktâf	72	0	0	162	5	21
10. Ardashîr ben Hurmuz	4	0	0	166	5	21
Shâpûr ben Shâpûr	5	4	0	171	9	21
Bahrâm ben Shâpûr	11	0	0	182	9	21
Yazdagird Sceleratus	21	5	18	204	3	9
Bahrâm Gûr	18	11	3	223	2	12
15. Yazdagird ben Bahrâm	18	4	18	241	7	0
Hurmuz	7	0	0	248	7	0
Fêrôz ben Yazdagird	27	0	0	275	7	0
Balâsh ben Fêrôz	4	0	0	279	7	0
Kobâd and Tâmâsp, sons of Fêrôz	43	0	0	322	7	0
20. Anôshirwân ben Kobâd	47	7	5	370	2	5
p.128. Hurmuz ben Anôshirwân	11	7	15	381	9	20
Parwîz ben Hurmuz	38	0	0	419	9	20
Shîrawaihi ben Parwîz	0	7	0	420	4	20
Ardashîr ben Shîrawaihi	0	5	0	420	9	20
25. Khûhân, who besieged the Greeks	0	0	22	420	10	12
Kisrâ ben Kobâd	0	3	0	421	1	12
Pûrân, daughter of Parwîz	1	6	0	422	7	12
Gushanasptadha	0	2	0	422	9	12
Azarmîdukht, daughter of Parwîz	0	4	0	423	1	12
30. Farrukhzâd Khusrau	0	1	0	423	2	12
Yazdagird ben Shahryâr	20	0	0	443	2	12

Next we return to fulfil our promise of explaining the way in which p.129.
Alkisrawî works out the chronology of this Part III., having perceived the confusion of the former two parts, although we cannot help wondering very much at him and at his method. For, whilst trying and experimenting, he has subtracted from the period between Alexander and Yazdagird 266 years for the period of the Ashghânian rule. Ḥamza, however, records only that tradition, which he says he has taken from and amended by means of the *Abastâ*, and the other tradition which he says he has taken from the copy of the Maubadh. And according to both these traditions, this period is longer even than 350 years (Ḥamza-Abastâ, 358 years; Ḥamza-Maubadh, 496 years). Now it is necessary for us to use either of these two traditions, or to add to them that one which Alkisrawî holds to be correct (as a third tradition), in order not to use any other tradition but those which he himself mentions. Or did he possibly place his confidence in that one which we have mentioned, and derived from the Shahnâma (266 years)?

Further, now, as Alkisrawî has done this, and thinks that the existence of such confusion is an established fact, I should like to know why he refers it to the period of the Sâsânian, not to that of the Ashghânian rule. For there was much more opportunity for mistakes creeping into the chronology of the Ashghânians (than into that of the Sâsânians), because during their period the Persian empire was disorganized, everyone minded only his own affairs, and people were prevented by various circumstances from preserving their chronology. Such were, *e.g.* the calamities which Alexander and his Greek lieutenants brought upon them, further the conflagration of all the literature in which people delighted, the ruin of all fine arts which were the recreation and the desire of the people. And more than that. He (Alexander) burned the greatest part of their religious code, he destroyed the wonderful architectural monuments, *e.g.* those in the mountains of Istakhr, now-a-days known as the Mosque of Solomon ben David, and delivered them up to the flames. People even say that even at the present time the traces of the fire are visible in some places.

This is the reason why they have neglected a certain space of time in the first part of the period, between Alexander and Ardashîr, *viz.* when the Greeks reigned over them. And they did not begin to settle their chronology until their fright and terror had subsided in consequence of the establishment of the Ashkânian rule over them. Therefore the period preceding this event was much more liable to confusion (than the later period of the Sâsânians), because under the Sâsânians the empire was in good order, and the royal dignity was transmitted in their family in uninterrupted succession, whilst in the time of those (their predecessors) there was much confusion. This is proved by all the testimonies which we have produced in support of this our view.

Here follows the table containing the so-called emendation of Alkisrawî.

p.130.

Names of the Sâsânian Kings, as reported by Ḥamza, according to the Emendation of Alkisrawî.	How long each of them reigned.			Sum of the Years.		
	Years.	Months.	Days.	Years.	Months.	Days.
Ardashîr ben Bâbak .	19	10	0	19	10	0
Sâbûr-aljunûd . .	32	4	0	52	2	0
Hurmuz b. Sâbûr-aljunûd	1	10	0	54	0	0
Bahrâm ben Hurmuz .	9	3	0	63	3	0
5. Bahrâm ben Bahrâm .	23	0	0	86	3	0
Bahrâm ben Bahrâm ben Bahrâm . . .	13	4	0	99	7	0
Narsa ben Bahrâm .	9	0	0	108	7	0
Hurmuz ben Narsa .	13	0	0	121	7	0
Shâpûr Dhû-al'aktâf .	72	0	0	193	7	0
10. Ardashîr, brother of the preceding . .	4	0	0	197	7	0
Shâpûr ben Shâpûr Dhû-al'aktâf	82	0	0	279	7	0
Bahrâm, son of the preceding	12	0	0	291	7	0
Yazdagird ben Bahrâm, *Clemens*, Prince of Sharwîn . . .	82	0	0	373	7	0
Yazdagird ben Yazdagird, *Atrox*. . .	23	0	0	396	7	0
15. Bahrâm Gûr, son of the preceding . .	23	0	0	419	7	0
Yazdagird b. Bahrâm Gûr	18	5	0	437 !	0	0
Bahrâm ben Yazdagird .	26	1	0	463	1	0
Fêrôz ben Bahrâm .	29	0	1	492	1	1
Balâsh ben Fêrôz . .	3	0	0	495	1	1
20. Kubâd, brother of Balâsh	68	0	0	563	1	1
p.131. Anôshirwân ben Kubâd	47	7	0	610	8	1
Hurmuz ben Anôshirwân	23	0	0	633	8	1
Parwîz ben Hurmuz .	38	0	0	671	8	1
Shîrawaihi ben Hurmuz	0	8	0	672	4	1
25. Ardashîr ben Shîrawaihi	1	0	0	673	4	1
Shahrbarâz . . .	0	1	8	673	5	9
Bûrân, daughter of Kisrâ Parwîz . . .	1	0	0	674	5	9
Khushnushbanda (Gushanasptadha) . .	0	2	0	674	7	9
Khusrau ben Kubâd ben Hurmuz . . .	0	10	0	675	5	9
30. Fêrôz, a descendant of Ardashîr ben Bâbak .	0	2	0	675	7	9
Azarmîdukht, daughter of Parwîz . . .	0	4	0	675	11	9
Farrukhzâd ben Khusrau b. Parwîz. His mother was Girawaihi, sister of Bahrâm Shûbîn .	0	1	0	676	0	9
34. Yazdagird ben Shahryâr	20	0	0	696	0	9

On Titles in the Khalifate.—It is a theory of the astrologers that p.132.
none of the khalifs of Islâm and the other kings of the Muslims reigns longer than twenty-four years. As to the reign of Almuṭî' that extended to nearly thirty years, they account for it in this way, saying that already at the end of the reign of Almuttaḳî, and at the beginning of that of Almustakfî, the empire and the rule had been transferred from the hands of the family of 'Abbâs into those of the family of Buwaihi (Bûya, Bôya), and that the authority which remained with the Banî-'Abbâs was only a juridical and religious, not a political and secular affair, in fact something like the dignity of the *Rôsh-gâlûthâ* with the Jews, who exercises a sort of religious authority without any actual rule and empire. Therefore the 'Abbâside prince, who at present occupies the throne of the *Khilâfa*, is held by the astrologers to be only the (spiritual) head of Islâm, but not a king.

Already in ancient times astrologers used to prophesy this state of affairs. Such a prophecy you find, *e.g.* in the book of 'Aḥmad ben Alṭayyib Alsarakhsî, where he speaks of the conjunction of Saturn and Mars in the sign of Cancer. The same was distinctly declared by the Hindu Kanaka, the astrologer of Alrashîd, for he maintained that the reign of the Banî 'Abbâs would be transferred to a man who would come from Ispahân. He determined, also, the time when 'Alî ben Buwaihi, called 'Imâd-aldaula, should come forward in Ispahân (as a claimant to supreme power).

When the Banî-'Abbâs had decorated their assistants, friends and enemies indiscriminately, with vain titles, compounded with the word *Daula* (*i.e. empire*, such as Helper of the Empire, Sword of the Empire, etc.), their empire perished; for in this they went beyond all reasonable limits. This went on so long till those who were especially attached to their court claimed something new as a distinction between themselves and the others. Thereupon the khalifs bestowed double titles. But then also the others wanted the same titles, and knew how to carry their point by bribery. Now it became necessary a second time to create a distinction between this class and those who were directly attached to their court. So the khalifs bestowed triple titles, adding besides the title of Shâhinshâh. In this way the matter became utterly opposed to common sense, and clumsy to the highest degree, so that he who mentions them gets tired before he has scarcely commenced, that he who writes them loses his time and writing, and he who addresses them runs the risk of missing the time for prayer.

It will not do any harm, if we mention here the titles which, up to our time, have been bestowed by their majesties the khalifs. We shall comprise them in the following table.

p.133.

The Names of those on whom Titles were bestowed.	The Titles which were bestowed by Their Majesties the Khalifs.
Alḳâsim ben 'Ubaid-allâh.	Waliyy-al-daula.
His son.	'Amîd-al-daula.
'Abû-Muḥammad ben Ḥamdân	Nâṣir-al-daula.
His son	Sa'd-al-daula.
'Abû-alḥasan 'Alî ben Ḥamdân	Saif-al-daula.
'Alî ben Buwaihi	'Imâd-al-daula.
'Abû-alḥasan 'Aḥmad ben Buwaihi	Mu'izz-al-daula.
Alḥasan ben Buwaihi	Rukn-al-daula.
'Abû-Manṣûr Bakhtiyâr ben 'Abî-alḥasan.	'Izz-al-daula.
'Abû-'Isḥâḳ ben Alḥusain	'Umdat-al-daula.
'Abû-Ḥarb Alḥabashî ben 'Abî-alḥusain.	Sanad-al-daula.
'Abû-Manṣûr Bîsutûn ben Washm-gîr.	Ẓahîr-al-daula.
'Abû-Manṣûr Buwaihi ben Alḥasan	Mu'ayyid-al-daula.
Almarzubân ben Bakhtiyâr	'I'zâz-al-daula.
Kâbûs ben Washmgîr	Shams-al-ma'âlî.
'Abû-'Aḥmad Ḥârith ben 'Aḥmad	Waliyy-al-daula.
'Abû-Shujâ' Fanâkhusra ben Al-hasan.	'Aḍud-al-daula wa Tâj-al-milla.
'Abû-Kâlinjar ben Fanâkhusra	Fakhr-aldaula wa Falak-al-'umma.
'Abû-Kâlinjar Marzubân ben Fanâkhusra.	Ṣamṣâm-al-daula wa Shams-al-milla.
'Abû-alfawârîs ben Fanâkhusra	Sharaf-al-daula wa Zaman-al-milla.
'Abû-Ṭâlib Rustam ben 'Alî	Majd-al-milla wa kahf-al-'umma.
p.134. 'Abû-alḳâsim Maḥmûd ben Sabuk-tagîn.	Yamîn-al-daula wa 'Amîn-al-milla.
'Abû-Naṣr Khurra Fêrôz ben Fanâkhusra.	Bahâ-al-daula wa Ḍiyâ-al-milla wa Ghiyâth-al-'umma.
'Abû-alḥasan Muḥammad ben 'Ibrâhîm.	Nâṣir-al-daula.
'Abû-al'abbâs Tâsh Alḥâjib	Ḥusâm-al-daula.
'Abû-alḥasan Fâ'iḳ-alkhâṣṣa	'Amîd-al-daula.
'Abû-'Alî Muḥammad ben Muḥam-mad ben 'Ibrâhîm.	Nâṣir-al-daula.
Sabuktagîn, first	Mu'în-al-daula.
Afterwards he received the title of	Nâṣir-al-din wal-daula.
Maḥmûd ben Sabuktagîn	Saif-al-daula.
'Abû-alfawâris Bektûzûn Alḥâjib	Sinân-al-daula.
'Abû-alḳâsim Muḥammad ben 'Ibrâhîm.	Naṣîr-al-daula.
'Abû-Mansûr Alp Arslân Albâlawî	Mu'în-al-daula.

Also the Wazîrs of the Khalifs have received certain titles, compounded with the word *Dhû*, as *e.g. Dhû-al-yaminain, Dhû-al-ri'âsatain, Dhû-al-kifâyatain, Dhû-al-saifain, Dhû-al-ḳalamain, etc.*

The Buwaihi family, when, as we have mentioned, the power passed into their hands, imitated the example of the khalifs; nay, they made it still worse, and their title-giving was nothing but one great lie, when they called their Wazîrs, *e.g. Kâfi-al-kufât, Alkâfi Al'auḥad, 'Auḥad-alkufât.*

The family of Sâmân, the rulers of Khurâsân, had no desire for such titles, contenting themselves with their *kunyas* (such as *'Abû-Naṣr, 'Abû-al-ḥasan, 'Abû-Ṣâliḥ, 'Abû-al-ḳâsim, 'Abû-al-ḥârith*). In their lifetime they were called *Almalik, Almu'ayyad, Almuwaffaḳ, Almanṣûr, Almu'aẓẓam, Almuntaṣir*, and after their death, *Alḥamîd, Alshahîd, Alsa'îd, Alsadîd, Alraḍî, etc.* To their field-marshals, however, they gave the titles of *Nâṣir-aldaula, 'Imâd-aldaula, Ḥusâm-aldaula, 'Amîd-aldaula, Saif-aldaula, Sinân-aldaula, Mu'în-aldaula, Nâṣir-aldaula*, in imitation of the ways of the khalifs.

The same was done by Bughrâkhân, when he had come forward (to claim supreme power) A.H. 382, calling himself *Shihâb-aldaula.*

Some of them, however, have gone beyond this limit, calling themselves *'Amîr-al-'âlam* and *Sayyid-al-'umarâ.* May God inflict on them ignominy in this world, and show to them and to others their weakness!

As to the 'Amîr, the glorious Prince, may God give a long duration to his reign! (to whom this book is dedicated), His Majesty the Khalif addressed him in a letter, and offered to him titles, such as those compounded with the word *Daula* (e.g. *Saif-al-daula, Ḥusâm-al-daula*, etc.). But then he considered himself superior to them, and abhorred the idea of being compared with those who were called by such titles but only in a very metaphorical way. He, therefore, selected for himself a title the full meaning of which did not exceed his merits (*Shams-al-ma'âlî*, p.135.
i.e. *Sun of the Heights*). He has become—may God give a long duration to his power!—among the kings of the world like the *sun*, who illuminates the darkness, in which they live, by the rays of his *heights.* He has come into high favour with the khalifs as a prince of the Believers. They wanted to redouble and to increase his title, but his noble mind declined it. May God give him a long life; may he enlighten all the parts of the world by his justice, and bless them by his look; may He raise *his* affairs and those of the subjects who dwell in his shadow to perfection, increasing them everlastingly. God is almighty to do this, and sees and knows all the affairs of his slaves!

Intervals between the Eras.—After this digression we now return to the point whence we started, and proceed, after having finished the collection of chronological dates in the preceding tables. Next we must turn our attention towards fulfilling our promise of teaching the reader that knowledge by means of which he may compute the eras that are used in the Canons, for astronomical observations, and elsewhere, *e.g.* in commercial stipulations and contracts. To this we shall prefix a twofold *Ṭailasân*, which will indicate the intervals between the single eras in a constant measure, *i.e.* in days. In the lower half under the diagonal, you find the distances computed in days and written in Indian ciphers. In the upper half you find two kinds of numbers; the upper ones are these identical days written according to the sexagesimal system, whilst the lower ones are the same days in their various degrees (units, tenths, hundreds, etc.) transcribed from the Indian ciphers into the *Ḥurûf-aljummal.*

The following well-known calculation is an example of this system of notation. If we take $\{[(16^2)^2]^2\}^2$ or 16^{16}, and subtract 1 from the sum, we get the total sum of the reduplications of all the checks of the chessboard, if we commence with *one* for the first check. This sum, noted in Indian ciphers, is the following: 18,446,744,073,709,551,615; noted according to the sexagesimal system: 30. 30. 27. 9. 5. 3. 50. 40. 31. 0. 15.; and transcribed into the *Ḥurûf-aljummal:*

האואההטעגזמדזודדחא

If you transcribe these characters one after the other into Indian ciphers, you get the above-mentioned number.

Now, in the same way as this example, our *Ṭailasân* is to be understood. This threefold system of notation we use for no other purpose but this, that each mode should bear testimony to the other in case a doubt should arise regarding some of the characters and figures that denote the numbers.

We mention our method only in a summary way, and not at full
p.136. length, because the reader of this book must be more than a beginner in mathematics. We say, if a man wants to find an (unknown) era by the help of a known one, let him reduce the whole of the known era into days, and this sum is called "*The Basis.*" Then he must take the interval between the two eras, *viz.* the known and the unknown ones. This we call "*The Equation.*"

If, then, the known era (*i.e* its epoch) precedes the unknown one, he subtracts the equation from the basis. If, on the other hand, the known era (*i.e.* its epoch) follows the unknown one, he adds the equation to the basis. And the sum which he gets is the number of days of the unknown era.

Thereupon he must divide this sum of days by the number of days of that kind of year which is ascribed to the era in question. By this division he gets complete years. And the remainder of days is to be distributed over the months of the year according to the proper lengths which we have mentioned as peculiar to each of the different kinds of them.

Here are the days of the intervals between the epochs of the various eras represented in the twofold *Ṭailasân.* God is allwise!

TABLE SHOWING THE INTERVALS BETWEEN THE EPOCHS OF THE ERAS, CALCULATED IN DAYS.

p.137.	33.27.45. 6. جهوطهدا	38.46.18. 6. حطهجوجا	15.46.17. 6. هرططهجا	56. 1. 44. 5. واهجهجا	33. 34. 28. 5. جزجمحاا	28.14.12. 5. حومباا	54. 7. 43. 4. درطای	33. 55. 41. 4. ججطدای	13.56.58. 3. جزامج	**Æra Diluvii,** with Egyptian years and months.
	20.31.46. 2. فدطله	25.50.19. 2. هبدجن	2. 50. 18. 2. بفـططد	43. 5. 45. 1. جدجمرج	20. 38. 29. 1. دبج	15.18.13. 1. هطحجوب	41. 11. 44. ابطها	20. 59. 42. سردها	**Æra Nabonassari,** with Egyptian years and months.	860,173
	0. 32. 3. 2. كرددد	5. 51. 36. 1. هووحدج	42.50.35. 1. بدندج	23. 6. 2. 1. جهجهجب	0. 39. 46. سطروا	55. 18. 30. هجاطی	21. 12. 1. ادجد	**Æra Philippi,** with Egyptian years and months.	154,760	1,014,933
	39. 19. 2. 2. طرجمد	44.38.35. 1. دبجددج	21.38.34. 1. اعمج	2. 54. 0. 1. بدبطاب	39. 26. 45. ططهجوا	34. 6. 29. دطردی	**Æra Alexandri,** with Syrian years and months.	4,341	159,101	1,019,274
	5. 13. 33. 1. هعهجج	10. 32. 6. 1. لهطجب	47. 31. 5. 1. زمهجب	28. 47. 31. حددداا	5. 20. 16. هفجه	**Æra Augusti,** with Greek years and Egyptian months.	104,794.	109,135	263,895	1,124,068
	0. 53. 16. 1. فزوزب	5. 12. 50. هبرفا	42. 11. 49. بيرزا	23. 27. 15. جدوهه	**Æra Antonini,** with Greek years and Egyptian months.	58,805	163,599	167,940	322,700	1,182,87
	37. 25. 1. 1. زجاابب	42. 44. 34. بجبا	19. 44. 33. طهدابا	**Æra Diocletiani,** with Greek years and months.	55,643	114,448	219.242	223.583	378,343	1,238,516
	18. 41. 27. حزوطا	23. 0. 1. جبوج	**Æra Fugæ,** with lunar years and Arabic months.	121,459	177,102	235,907	340,701	345,042	499,802	1,359,975
	55. 40. 26. هوسط	**Æra Yazdagirdi,** with Persian years and months.	3623	125,082	180,725	239.530	344,324	348,665	503,425	1,363,598
	Æra Mu'tadidi, with Greek years and Persian months.	96,055	99,678	221,137	276,780	335.585	440,379	444,720	599,480	1,459,653

p.138. **The Chess Problem.**—For the solution of the chess problem (*lit.* for the reduplication of the chess and its calculation) there are two fundamental rules. The one of them is this :—

The square of the number of a check x of the 64 checks of the chessboard is equal to the number of that check the distance of which from the check x is equal to the distance of the check x from the 1st check.

For example: take the square of the number of the 5th check, *i.e.* the square of 16 $(16^2)=256$, which is the number belonging to the 9th check. Now, the distance of the 9th check from the 5th is equal to the distance of the 5th check from the first one.

The second rule is this :—

The number of a check x minus 1 is equal to the sum total of the numbers of all the preceding checks.

Example: The number of the 6th check is 32. And $32-1$ is 31, which is equal to the sum of the numbers of all the preceding checks, *i.e.* of—

$$1+2+4+8+16\ (=31).$$

If we take the square of the square of the square of 16, multiplied by itself (*i.e.* $\{[(16^2)^2]^2\}^2$ or 16^{16}), this is identical with taking the square of the number of the 33rd check, by which operation the number of the 65th check is to be found. If you diminish that number by 1, you get the sum of the numbers of all the checks of the chessboard. The number of the 33rd check is equal to the square of the number of the 17th check.

The number of the 17th check is equal to the square of the number of the 9th check.

The number of the 9th check is equal to the square of the number of the 5th check. And this (*i.e.* the number of the 5th check) is the abovementioned number 16.

'Abû-Raiḥân says in his *Kitâb-al-'arḳâm* (Book of the Ciphers): "I shall explain the method of the calculation of the chess problem, that the reader may get accustomed to apply it. But first we must premise that you should know, that in a progression of powers of 2 the single numbers are distant from each other according to a similar ratio. (*Lacuna?*) If the number of the reduplications, *i.e.* the number of the single members of a progression is an even one, it has two middle numbers. But if the number of the reduplications is an odd one, the progression has only one middle number.

The multiplication of the two ends by each other is equal to the multiplication of the two middle numbers. (In case there is only one middle number, its square is equal to the multiplication of the two end numbers.) This is one thing you must know beforehand. The other is this :—

If we want to know the sum total of any progression of powers of 2, we take the double of the largest, *i.e.* the last number, and subtract therefrom the smallest, *i.e.* the first number. The remainder is the sum total of these reduplications (*i.e.* of this progression).

Now, after having established this, if we add to the checks of the chessboard one check, a 65th one, then it is evident that the number which belongs to this 65th check, in consequence of the reduplications of powers of 2, beginning with 1, is equal to the sum of the numbers of all the checks of the chessboard minus the 1st check, which is the number 1, the first member of the progression. If, therefore, 1 is subtracted from this sum, the remainder is the sum of the numbers of all the checks of the chessboard.

If, now, we consider the 65th check and the 1st as the two ends of a progression, their medium is the 33rd check, the first medium.

Between the checks 33 and 1, the check 17 is the medium, the second medium.

Between the checks 17 and 1, the check 9 is the medium, the third medium. p.139.

Between the checks 9 and 1, the check 5 is the medium, the fourth medium.

Between the checks 5 and 1, the check 3 is the medium, the fifth medium.

Between the checks 3 and 1, the check 2 is the medium, the sixth medium, to which belongs the number 2.

Taking the square of 2 (2^2), we get a sum which is a product of the multiplication of the number of the 1st check by that of the 3rd check ($1 \times 4 = 2^2$). The number of the 1st check is 1. This product, then, is the fifth medium, the number of the 3rd check, *i.e.* 4.

The square of 4 is 16, which is the fourth medium in the 5th check.

The square of 16 is 256, which is the third medium in the 9th check.

The square of 256 is 65,536, which is the second medium in the 17th check.

The square of 65,536 is 4,294,967,296, which is the first medium in the 33rd check.

The square of 4,294,967,296 is 18,446,744,073,709,551,616.

If we subtract from this sum 1, *i.e.* the number of the first check, the remainder is the sum of the numbers of all the checks of the chessboard. I mean that number which at the beginning of this digression we have used as an example (of the threefold mode of numeral rotation).

The immensity of this number cannot be fixed except by dividing it by 10,000. Thereby it is changed into *Bidar* (sums of 10,000 dirhams).

The *Bidar* are divided by 8. Thereby they are changed into *'Aukár* (loads).

The *'Aukár* are divided by 10,000. Thereby the mules, that carry them, are formed into *Ḳuṭ'án* (herds), each of them consisting of 10,000.

The *Ḳuṭ'án* are divided by 1,000, that, as it were, they (the herds) might graze on the borders of *Wádis*, 1,000 kids on the border of each *Wádi*.

The *Wâdis* are divided by 10,000, that, as it were, 10,000 *mountains* should rise out of each *Wâdi*.

In this way, by dint of frequently dividing, you find the number of those *mountains* to be 2,305. But these are (numerical) notions that the earth does not contain.

God is allwise and almighty!

p.140. **Rules for the Reduction of the different Eras.**—Now we shall give a detailed exposition of the subject of this chapter (*i.e.* the derivation of the eras one from the other), which cannot be dispensed with. We must, however, postpone our exposition of the derivation of the *Æra Adami* and *Æra Diluvii* according to Jews and Christians, because they are connected with the years and months of the Jews. And these are very intricate and obscure, and offer many difficulties for calculation,—a chapter, part of which we have already mentioned before. For which reason we must direct our attention exclusively to this subject, and explain it in a special chapter. And now we commence with the detailed exposition of the eras, pre-supposing the number of days which form the intervals between the epochs of the eras and that day which is sought to be known. These days we call *Dies Paratæ*.

If we want to find the *Æra Diluvii*, according to 'Abû-Ma'shar, who uses it in his *Canon* (or calendar), we divide its *Dies Paratæ* by 365, whereby we get complete years. If there is a remainder of days, we change them into Egyptian months. The 1st of Tôt of this *Æra Diluvii* always coincides with the 18th of Bahman-Mâh in the non-intercalated *Æra Yazdagirdi*.

If we want to find the *Æra Nabonassari* or the *Æra Philippi*, we divide its *Dies Paratæ* by 365, whereby we get complete years. The remainder of days is distributed over the single months, to each month its proper portion. We begin with Tôt, the 1st of which always coincides with [the 1st of] Dai-Mâh in the non-intercalated *Æra Yazdagirdi*.

If we want to find the *Æra Alexandri*, we divide its *Dies Paratæ* by $365\frac{1}{4}$ days, *i.e.* we multiply the *Dies Paratæ* by 4, changing them into fourth-parts, and the sum total we divide by 1,461, *i.e.* by the days of the year reduced into fourth-parts. Thereby we get complete years. The remaining fourth-parts we raise again to whole days, dividing them by 4. Then we distribute them over the single months, to each month its proper portion, beginning with Tishrîn I. If there is a remainder of days that do not fill up one month, this remainder represents the date of that identical month. To the month Shubâṭ we must give twenty-nine days in a leap year, and twenty-eight days in a common year.

The leap year is recognized in this way, that we consider the remainder which we get after dividing the fourth-parts (of the *Dies Paratæ*) by 4. If the remainder is 2, the current year is a leap year. If the remainder is less or more (*i.e.* 1, or 3, 4), the year is a common year.

The reason of this is, that intercalation preceded the epoch of this era by two years, so that at the beginning of the era two fourth-parts of a day had already summed up. If, therefore, at the end of the era there is a remainder of two fourth-parts, these, together with the two fourth- p.141
parts at the beginning of the era, make up *one* complete day. In that case the year is a leap year.

(In this calculation the Syrian year and months are used.)

If we, however, compute this era according to the method of the Greeks, we subtract 92 from the number of its *Dies Paratæ*, because the beginning of the Greek year differs from that of the Syrian year. The remainder we compute in the same way as we have done according to the method of the Syrians. The remaining fourth-parts we raise to whole days, and distribute them over the single months, to each month its proper portion, commencing with *Januarius, i.e.* Kânûn the Last.

The leap year is ascertained in the same way that we have mentioned before.

If we want to find the *Æra Augusti*, we compute its *Dies Paratæ* in the same way as we have done with the *Æra Alexandri*, so as to get complete years and a remainder of fourth-parts of a day. These latter we change into days, and distribute them over the single months, to each month its proper portion, beginning with Tôt. If the year is a leap year, we count the *Epagomenæ, i.e.* the *small month*, as six days, whilst in a common year we count it as five days.

The leap year is recognized by there being no remainder of fourth-parts of a day after we have converted them into whole days. Of which the reason is this, that the leap year preceded the beginning of the era. On this subject (*the Epagomenæ*) there cannot be much uncertainty, since they are placed at the end of the year, and the 1st of Tôt always coincides with the 29th of the Syrian month Âbh.

Of the *Æra Antonini*, we compute the complete years in the same way that we have explained for the *Æra Augusti*. The remainder (of fourth-parts of a day) we divide by 4, and distribute the whole days over the single months, to each month its proper portion, beginning with Tôt. In a leap year we count the Epagomenæ as six days.

The leap year is recognized by there being one quarter of a day as a remainder of the fourth-parts (of a day).

Of the *Æra Diocletiani*, we compute the *Dies Paratæ* in the same way as we have done with the *Æra Augusti*, etc., so as to get complete years and to convert the fourth-parts again into complete days. Thereupon we distribute them over the single months, beginning with *Januarius, i.e.* Kânûn the Last. In a leap year we give to Februarius, *i.e.* Shubâṭ, twenty-nine days, in a common year twenty-eight days.

The leap year is recognized in the same way as for the *Æra Alexandri*, by there being two fourth-parts as the remainder of the fourth-parts of a day.

As regards the eras of the Arabs and their months, how they intercalated them, and in what order they arranged them in pagan times, this is a subject that has been utterly neglected. The Arabs were totally illiterate, and as the means for the perpetuation of their traditions they relied solely upon memory and poetry. But afterwards, when the generation of those who practised these things had died out, there was no further mention of them. There is no possibility of finding out such matters.

.142. If we want to find the *Era of the Hijra* as used in Islâm, we divide its *Dies Paratæ* by the mean length of the lunar year, *i.e.* $354\frac{1}{5}+\frac{1}{6}$ days ($354\frac{11}{30}$ days), which is effected by multiplying the number of days by 30, the smallest common denominator for both fractions, fifth and sixth parts. The sum we divide by 10,631, which is the product of 354 multiplied by 30, *plus* $\frac{11}{30}=\frac{1}{5}+\frac{1}{6}$.

The quotient represents complete lunar years, and the remainder consists of thirtieth-parts of a day. If we divide these by 30, we get again whole days, which we distribute over the single months, giving to one month thirty days, to the other twenty-nine alternately, beginning with Almuḥarram. The remainder of days that does not make up one complete month, represents the date of that identical month.

This is the method for the computation of the eras used in the *Canons*. But if there are still other methods which people adopt for this purpose, they all go back to one and the same principle.

As for the calculation which is based upon the appearance of new moon, it must be remarked that two imperfect months (of twenty-nine days) may follow each other as well as three perfect ones (of thirty days), that the lunar year may exceed the above-mentioned measure (of $354\frac{11}{30}$ days), whilst it may not attain this length at other times, the reason of which is the variation in the rotation of the moon.

Of the *Æra Yasdagirdi*, we divide the *Dies Paratæ* by 365, whereby we get complete years. The remainder we distribute over the single months, to each month its proper portion, beginning with Farwardîn-Mâh. In this way we come to know the era, the epoch of which is the beginning of his reign, that era which is used in the *Canons*.

If we, however, want to find the *Era of the Zoroastrians*, we subtract twenty years from the Æra Yazdagirdi. The remainder is the Era of the Zoroastrians. For they date from the year in which Yazdagird was killed and their national empire ceased to exist, not from the year in which he ascended the throne.

The *Era of Almu'taḍid-billâh* we compute in the same way as the *Æra Alexandri*. We give to each month its proper portion, as to the Persian months, beginning with Farwardîn-Mâh, and proceeding as far as the beginning of Âdhar-Mâh. If, then, the year is a leap year, which is recognized in the same way as in the Æra Alexandri, by there being a remainder of two fourth-parts of a day, (we count the *Andargâhs* or

Epagomenæ between Abân-Mâh and Âdharmâh) as six days, whilst in a common year we count them only as five days. New-Year (Naurôz) always coincides with the 11th of Ḥazîrân, for those reasons which we have already mentioned by the help and the support of God!

Now it would seem proper to add a chapter which is wanting in the *Canons*, and has not been treated by anybody except by 'Abû-al'abbâs Alfaḍl ben Ḥâtim Alnairîzî, in his commentary on Almagest. And still it is a subject of frequent occurrence, and those who have to employ it may not always know what to do with it. The thing is this, that you may be required to compute a date for a certain time, the known parts p.143. of which are various *species* that do not belong to one and the same *genus*. There is, *e.g.* a day the date of which within a Greek, Arabic, or Persian month is known; but the name of this month is unknown, whilst you know the name of another month that corresponds with it. Further, you know an era, to which, however, these two months do *not* belong, or such an era, of which the name of the month in question is not known. Example:—

(*a*) On the day Hurmuz
(*b*) in the month Tammûz
(*c*) in the year of the Hijra 391.

In this case the proper method is to compute the *Æra Alexandri* for the 1st of Muḥarram of A.H. 391. Thereby we learn with what month and day of the Arabian months the 1st of Tammûz coincides. Further, we compute the *Æra Yazdagirdi* for the 1st of Tammûz, whereby we learn on what day of Tammûz the day Hurmuz falls. In this way the three eras together with their *species* and *genera* are found out.

If besides these elements the name of the week-day is known, this is an aid and a help for obtaining a correct result. Example:—

(*a*) On Friday
(*b*) in the first third of Ramaḍân
(*c*) in the year of Yazdagird 370.

Here the right method would be first to compute the Arabic era for the Naurôz of this year of the *Æra Yazdagirdi*, and thereby to compute the first third of Ramaḍân. Then we consider the week-days, to find which of them are the beginnings of the months. Thereby we find what we wanted to find.

Likewise, if the week-day and its place within some month, together with some era, are known, and if also the name of the month is known, you can find this out in the same way as we have mentioned.

The student who thoroughly knows all these methods will be able to solve whatever question of this sort be put to him; he will find out everything, if he considers the subject as it ought to be considered. If those parts of such dates, the numerical values of which are known, should be composed of diverse elements, so that their *units* mean something different from what the *decades* (tenths) mean,—*e.g.* you say of a day: *the* 25*th*, referring the 5 to a Persian month and the 20 to a Greek month, of which either one is known, or of which the two are unknown; or if you say: *Anno* 345, referring the 5 to a Greek, the 40 to an Arabian, and the 300 to a Persian era,—in such cases the cleverness of the student will manage to solve the problem, although the calculations necessary for such a derivation may be very long.

God helps to find the truth!

CHAPTER VII.

ON THE CYCLES AND YEAR POINTS, ON THE MÔLÊDS OF THE YEARS AND MONTHS, ON THEIR VARIOUS QUALITIES, AND ON THE LEAP MONTHS BOTH IN JEWISH AND OTHER YEARS. p.14

Having in the preceding pages explained the derivation of the eras from each other, with the exception of the *Æra Adami* and *Æra Diluvii*, according to the systems of Jews and Christians, we shall now have to explain the method by which we may obtain a knowledge of these two eras. To this we shall prefix a treatise on the Jewish years and months, their cycles and the *Môlêds* of their years, followed by an investigation of the commencements of the years of other nations. And hereto we shall add such things as may prove a ready help towards obtaining the object in view.

Now we proceed to state that the *Æra Adami* is used by the Jews, the *Æra Diluvii* by the Christians. If the 1st of Tishrî coincided with the 1st of *Tishrîn Primus*, the *Æra Alexandri* would be equal to the *Æra Mundi, plus* 3,448 years, which is, according to Jewish doctrine, the interval between Adam and Alexander.

However, the 1st of Tishrî always falls between the 27th of Âbh and the 24th of Îlûl, *on an average*. Therefore, the *Æra Alexandri, minus* that time by which the beginning of the Jewish year precedes the beginning of the Christian one, is equal to the complete *Æra Adami, plus* the interval between Adam and Alexander.

The reason why the 1st of Tishrî always varies within those days (27th Âbh—24th Îlûl), is this, that on an average the Jewish passover always varies between the 18th of the Syrian month Adhâr and the 15th of Nîsân, which is the time of the sun's moving in the sign of Aries. For it is the opposition occurring within this time, on which all those circumstances depend which form the *conditio sine quâ non* for passover.

This, however, is only an approximate calculation. For if the solar year went on parallel with the days of the Greek year (*lacuna??*). But this is impossible, since we have found by astronomical observation that this fraction (beyond the 365 complete days of the year) is 5h. 46′ 20″ 56‴. Therefore the sun, rotating at the rate of velocity found by astronomical observation, reaches any place whatever of the ecliptic earlier than he would reach it by that rotation on which their method is based, in each 165 complete days (*sic !*)

145. We shall use, however, their own system, and shall now explain how we may find the beginning of their year, and how we may ascertain its nature, whether it be a *common* or a *leap year*, *imperfect*, *intermediate*, or *perfect*. Now, if we want to find this, we add to the date of the *Æra Alexandri*, for the 1st of the Syrian *Tishrîn Primus*, 3,448 years. Thereby we get the corresponding date of the *Æra Adami* for the 1st of (the Jewish) Tishrî, that falls either in the end of Âbh or in Îlûl, both of which months precede that *Tishrîn Primus* whence we started in this calculation.

If we, further, want to know whether the year of which we have found the beginning be a common year or a leap year, we subtract 2 from the number of years, and divide the remainder by 19; the quotient we get represents the number of complete *Minor Cycles*. The remainder we compare with Circle I. of the *Assaying Circle*. There we find in Circle II., opposite to the year of the cycle, an indication of its nature, whether it be a common or a leap year. Further, we find in Circle III. the *date* of the Syrian month on which the beginning of the year in question falls. And lastly, we find in Circle IV. the *name* of this Syrian month.

Here follows the diagram of the Assaying Circle :—

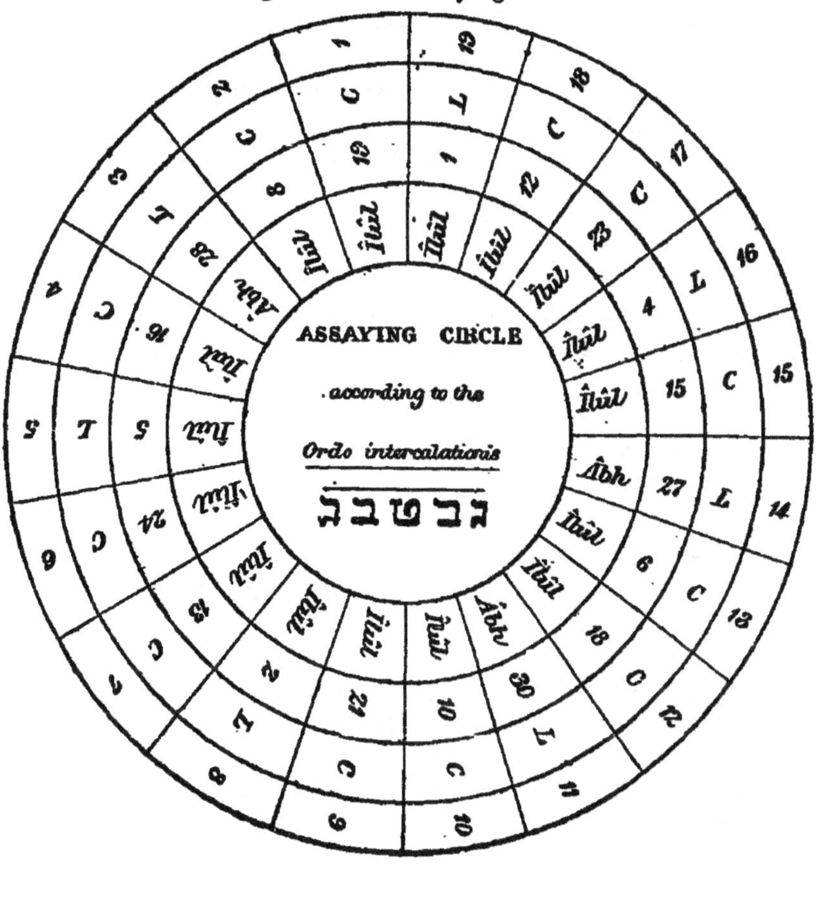

If the *Enneadecateris*, on being complete, returned to the same day of the week whence it started, which, as we have already mentioned, is not the case, we should have added in the *Assaying Circle* a circle V., for the indication of the days of the weeks on which the New-Year days of the single years of the Enneadecateris would fall. Under these circumstances, however, it is impracticable.

If we want to find the week-day with which the day indicated in Circle III. corresponds, we compute, by methods which will be hereafter explained, the commencements of either Âbh or Îlûl of the year in question, in whichsoever of these two months that day may fall. On having carried out this, we learn what we wanted to know.

This, our calculation regarding the 1st of Tishrî, is an average calculation, without any other correction being employed. But now the beginning of Tishrî frequently falls on such days which the Jews, as we have already mentioned, do not allow to be New-Year's day. Therefore it becomes necessary to fix it on a day earlier or later.

If we, now, want to acquaint ourselves with this correction (*lit.* equation), we must first know the conjunction of sun and moon at the beginning of Tishrî, according to the theory of the Jews themselves, not that of the astronomers. For between these two theories there are certain divergencies:—

I. They give to the lunar month, extending from conjunction to conjunction, the length of—

29d. 12h. 793 Ḥalaḳs,

which is equal to

(29d. 12h.) 44′ 3″ 20‴,

[whilst modern observers have found it to be

29d. 12h. 44′ 2″ 17‴ $21^{IV.}$] 12^{V}

Therefore the difference between the two computations is

1″ 2‴ $38^{IV.}$ 48^{V} p.146.

II. They give the solar year, if they reckon with mathematical accuracy, the length of

365d. $5\frac{3791}{4104}$h.

whilst modern astronomers have found it to be shorter.

III. Astronomers teach that that portion of the *Nychthemeron* which elapses between the time of conjunction and that moment when new moon becomes visible, varies according to the differences of both the longitudes and latitudes of the places, whilst the Jews compute it everywhere according to one and the same rule. We do not know for which particular place this mode of computation was originally calculated, but it seems rather likely that it was made for Jerusalem or its environs, for there was their central seat.

IV. They determine this space of time (between the conjunction and the appearance of new moon) by ὧραι καιρικαί. Whilst it is well known that it is not allowed to use them for the computation of conjunction, except on the equator.

V. They compute the conjunctions by the mean, not the apparent motion. Therefore passover frequently falls two complete days later than the real opposition—one day in consequence of the *Equations*, another day in consequence of their postponing passover from a *Dies illicita* to a *Dies licita*.

Computation of the Moled of a Year according to the Jewish System.—If we, now, want to find the *Môlêd* of a year, which term the Jews apply to the conjunction at the beginning of each month as well as the conjunction at the beginning of every cycle, we take the complete years of the *Æra Adami*, *i.e.* till the end of the year which is preceded by the month *Tishrî* in question. We convert the number of years into *Minor Cycles*, and multiply the number of cycles by 2d. 16h. 595^{H}, which you get as a remainder if you convert the days of the minor cycle into weeks. The product which arises we keep in mind.

Thereupon, we consider the remainder of years that do not fill up one complete minor cycle. How many of them are common years, how many leap years, we learn by the *Ordo intercalationis*,

בהזיגוח

(*i.e.* the 2nd, 5th, 7th, 10th, 13th, 16th, and 18th years of the cycle are leap years).

The number of common years we multiply by 4d. 8h. 876^{H}, the number of leap years by 5d. 21h. 589 . The product of these two multiplications we add to the sum we have kept in mind.

To the sum we always add

5d. 14h.,

which represents the interval between the time of the conjunction and the beginning of the night of Sunday that was the commencement of the first year of the *Æra Adami*.

Then we raise each 1,080 Ḥalaḳs to 1 hour, and add it to the other hours; each 24 hours we convert into 1 day, and add it to the other days. The sum of days that arises we convert into weeks, and the remainder of days that are less than a week is the distance of the *Môlêd* from the beginning of the night of Sunday. Now, that time to which in the last instance our calculation leads us, is the time of the conjunc-
p.147. tion at the beginning of *Tishrî*.

We have made such a computation for a year of the *Æra Alexandri*, in order to facilitate the process and to simplify the apparatus.

If you want to find the conjunction at the beginning of Tishrî, take the years of the *Æra Alexandri*, and subtract therefrom always 12 years, which are the remainder of the minor cycle at the epoch of the Æra

Alexandri, according to the *Ordo intercalationis* גבטבג. The remainder of years divide by 19; the quotient you get is the number of minor cycles.

Convert these minor cycles into great cycles, if they are of a sufficient number to give complete great cycles, and keep in mind what remainder of years you have got. They are the current years of the cycle in question, according to the *Ordo intercalationis* גבטבג.

The great cycles, if you get such, compare with the table of the great cycles, and take the number of days, hours, and Ḥalâḳîm which you find opposite them.

The small cycles compare with the table of the small cycles, and the number of days, hours, and Ḥalâḳîm which you find opposite them.

These two number add together, days to days, hours to hours, and Ḥalâḳîm to Ḥalâḳîm.

This sum add to the *Basis*, which is written in the table uppermost, and which is the Môlêd of the 12th year of the *Æra Alexandri*. Convert each 1,080 Ḥalâḳîm into an hour, each 24 hours into a day, and the days into weeks. The remainder of days you get is the distance between the beginning of the night of Sunday and the time of the conjunction. This is according to Jewish calculation.

We have used as the starting-point in this our calculation the beginning of the night for no other reason but this, that they commence the *Nychthemeron* with sunset, as we have mentioned in the first part of this book.

Here follows the table, computed by that method of calculation which we have explained in the preceding pages:— p.148.

The Numbers of the Small Cycles.	The Years of the Small Cycles.	Days.	Hours.	Ḥalâḳîm.
1	19	2	16	595
2	38	5	9	110
3	57	1	1	705
4	76	3	18	220
5	95	6	10	815
6	114	2	3	330
7	133	4	19	925
8	152	0	12	440
9	171	3	4	1,035
10	190	5	21	550
11	209	1	14	65
12	228	4	6	660
13	247	6	23	175
14	266	2	15	770
15	285	5	8	285

The Numbers of the Small Cycles.	The Years of the Small Cycles.	Days.	Hours.	Ḥalâḳîm.
16	304	1	0	880
17	323	3	17	395
18	342	6	9	990
19	361	2	2	505
20	380	4	19	20
21	399	0	11	615
22	418	3	4	130
23	437	5	20	725
24	456	1	13	240
25	475	4	5	835
26	494	6	22	350
27	513	2	14	945
28	532	5	7	460

149.

The Single Years of the Small Cycle.	Days.	Hours.	Ḥalâḳîm.	Leap Years.
1	5	21	589	—
2	3	6	385	—
3	0	15	181	L
4	6	12	770	—
5	3	21	566	L
6	2	19	75	—
7	0	3	951	—
8	4	12	747	L
9	3	10	256	—
10	0	19	52	—
11	5	3	928	L
12	4	1	437	—
13	1	10	233	—
14	5	19	29	L
15	4	16	618	—
16	2	1	414	L
17	0	22	1003	—
18	5	7	799	—
19	2	16	595	L

p.150.

The Numbers of the Great Cycles.	The Years of the Great Cycles.	Days.	Hours.	Ḥalâḳîm.
1	532	5	7	460
2	1064	3	14	920
3	1596	1	22	300
4	2128	0	5	760
5	2660	5	13	140
6	3192	3	20	600
7	3724	2	3	1060
8	4256	0	11	440
9	4788	5	18	900
10	5320	4	2	280
11	5852	2	9	740
12	6384	0	17	120
13	6916	6	0	580

Astronomical Computation of the Moled of a Year.—If a mathematician wants to know the time of conjunction as determined by astronomical observation, not that one which is found by the rules of the Jewish chronologers, he may use the (following) table, which we have tried to compute in the same way as the preceding ones, on the basis of the corrected observations that have been made not long before our time. For this purpose we have consulted the view of Ptolemy regarding the mean length of the month, the view of Khâlid ben 'Abdalmalik of Marwarûdh, according to his measurements made at Damascus, the view of the sons of Mûsâ ben Shâkir, and of others. Of all these, we found the most deserving to be adopted and followed that of the sons of Mûsâ ben Shâkir, because they spent their whole energy in endeavouring to find the truth; because they were unique in their age for their knowledge of, and their skill in, the methods of astronomical observations; because scholars bore witness of them to this effect, and warranted the correctness of their observations; and lastly, because there is a long interval between their observations and those of the ancients (Ptolemy, Hipparchus, etc.), whilst our time is not far distant from theirs (*i.e.* from the time when the sons of Mûsâ ben Shâkir made their observations). p.151.

Now we have computed the *Basis* according to their view, *viz.* the date of the conjunction at the beginning of the 13th year of the *Æra Alexandri.* It occurred at Baghdâd, $21^h.\ 20'\ 50''\ 14'''\ 29^{IV.}$ after noon on a Tuesday. And because the meridian of Jerusalem, on account of its more western longitude, is behind the meridian of Baghdâd by 14 *Times*, we have subtracted the corresponding space of time, *i.e.* 56 minutes from the date of the same conjunction at Baghdâd. So we get as a remainder the *Basis* for Jerusalem, *i.e.*—

$20^h.\ 24'\ 50''\ 14'''\ 29^{IV.}$ after noon.

He who calculates on this basis subtracts always 12 from the *incomplete* years of the *Æra Alexandri* (*i.e.* from the *Æra Alexandri*, including the current year), and converts the remainder into great and small cycles. He takes that portion of hours, minutes, seconds, etc. which corresponds in the tables to each of these numbers of great and small cycles. The remainder of single years he compares with the table of the consecutive years of the small cycle; he takes the values which he finds in the table opposite this number of years, and adds these three *Characters* (of the *Great Cycles*, the *Small Cycles*, and the *Consecutive Years* of the latter) together. This sum he adds to the *Basis*, and raises the hours and fractions of an hour to days and the corresponding wholes. Thereupon he converts the days into weeks, and the remainder which he gets is that time which has elapsed between the noon of Sunday at Jerusalem and the conjunction at the beginning of Tishrî.

Here follows the table *as based upon astronomical observations:*—

p.152.

The Numbers of the Small Cycles.	The Years of the Small Cycles.	Days.	Hours.	Minutes.	Seconds.	Thirds.	Fourths.
The *Basis.*	12	2	20	24	50	14	29
1	19	2	16	28	57	57	53
2	38	5	8	57	55	55	46
3	57	1	1	26	53	53	39
4	76	3	17	55	51	51	32
5	95	6	10	24	49	49	25
6	114	2	2	53	47	47	18
7	133	4	19	22	45	45	11
8	152	0	11	51	43	43	4
9	171	3	4	20	41	40	57
10	190	5	20	49	39	38	50
11	209	1	13	18	37	36	43
12	228	4	5	47	35	34	36
13	247	6	22	16	33	32	29
14	266	2	14	45	31	30	22
15	285	5	7	14	29	28	15
16	304	0	23	43	27	26	8
17	323	3	16	12	25	24	1
18	342	6	8	41	23	21	54
19	361	2	1	10	21	19	47
20	380	4	17	39	19	17	40
21	399	0	10	8	17	15	33
22	418	3	2	37	15	13	26
23	437	5	19	6	13	11	19
24	456	1	11	35	11	9	12
25	475	4	4	4	9	7	5
26	494	6	22	33	7	4	58
27	513	2	13	2	5	2	51
28	532	5	5	31	3	0	44

p.153.

The Single Years of the Small Cycles.	Days.	Hours.	Minutes.	Seconds.	Thirds.	Fourths.
1	5	21	32	29	45	35
2	3	6	20	57	13	49
3 L	0	15	9	24	42	3
4	6	12	41	54	27	38
5 L	3	21	30	21	55	52
6	2	19	2	51	41	27
7	0	3	51	19	9	41
8 L	4	12	39	46	37	55
9	3	10	12	16	23	30
10	0	19	0	43	51	44
11 L	5	3	49	11	19	58
12	4	1	21	41	5	33
13	1	10	10	8	33	47
14 L	5	18	58	36	2	1
15	4	16	31	5	47	36
16 L	2	1	19	33	15	50
17	0	22	52	3	1	25
18	5	7	40	30	29	39
19 L	2	16	28	57	57	53

p.154.

The Numbers of the Great Cycles.	Their Years.	Days.	Hours.	Minutes.	Seconds.	Thirds.	Fourths.
1	532	5	5	31	3	0	44
2	1064	3	11	2	6	1	28
3	1596	1	16	33	9	2	12
4	2128	6	22	4	12	2	56
5	2660	5	3	35	15	3	40
6	3192	3	9	6	18	4	24
7	3724	1	14	37	21	5	8
8	4256	6	20	8	24	5	52
9	4788	5	1	39	27	6	36
10	5320	3	7	10	30	7	20
11	5852	1	12	41	33	8	4
12	6384	6	18	12	36	8	48
13	6916	4	23	43	39	9	32

(In this our calculation of the conjunction) we have used *noon* as p.155.
terminus a quo for no other reason but this, that we may more easily find the equation for the môlêd by this method than by using the horizons (*i.e.* reckoning from sunset, as the Jews do).

The hours of the longest day for the latitude of Jerusalem are 13h. *plus* a fraction. Therefore the calculation of the Jews by ὧραι καιρικαί is incorrect, except in case the conjunction at the beginning of Tishrî should coincide with the autumnal equinox. This, however, never happens. On the contrary, the conjunction at the beginning of Tishrî always either precedes or follows the autumnal equinox by a considerable space of time, as we have explained heretofore.

Relation between the beginning of the Year and its Character. —If we, now, make out the time of the conjunction by the traditional calculation of the Jews, or by means of the table which we have constructed according to their theory, we arrive at the knowledge of the beginning of the year and of its character, whether it be imperfect, intermediate, or perfect, whilst we have already previously learnt how to know whether the year be a common or a leap year. Thereupon we look in the Table of Limits for a space of time in the week within the limits of which the conjunction as found by our calculation falls. If the year be a leap year, we look into the column of leap years; if it be a common year, we look into the column of common years. Having made out this, we find opposite the indication of the week-day on which the year commences, and of the quality of the year. Once knowing the beginning of the year (its precise date in the week) and its quality, and combining with it our knowledge as to whether the year is a common or a leap year, we come to know the beginning of the next following year.

Here follows the Table of the Limits :—

p.156.

The Limits of the Time Spheres as distributed over the Week, in Common Years.	New-Year's Day.	Character of the Year.
From noon of Saturday till 9h. 204H. in the night of Sunday	2	Imperfect.
From 9h. 204H. in the night of Sunday till 8h. 589H. in the day of Monday, if the preceding year is a leap year; till Noon of Monday, if the preceding year is a common year	2	Perfect.
From 8h. 589H. in the day of Monday, or From noon of Monday, till 9h. 204H. in the night of Tuesday	3	Intermediate.

The Limits of the Time-Spheres as distributed over the Week, in Common Years.	New-Year's Day.	Character of the Year.
From 9h. 204H. in the night of Tuesday till 9h. 204H. in the night of Thursday	5	Intermediate.
From 9h. 204H. in the night of Thursday till Noon of Thursday	5	Perfect.
From noon of Thursday till 0h. 208H. in the night of Friday, if the following year is a common year; till 9h. 204H. in the night of Friday, if the following year is a leap year	7	Imperfect.
From 0h. 208H. in the night of Friday, or From 9h. 204H. in the night of Friday, till Noon of Saturday.	7	Perfect.

p.15

The Limits of the Time Spheres as distributed over the Week, in Leap Years.	New-Year's Day.	Character of the Years.
From noon of Saturday till 8h. 491H. in the day of Sunday	2	Imperfect.
From 8h. 491H. in the day of Sunday till Noon of Monday	2	Perfect.
From noon of Monday till Noon of Tuesday	3	Intermediate.

The Limits of the Time Spheres as distributed over the Week, in Leap Years.	New-Year's Day.	Character of the Years.
From noon of Tuesday till 11h. 695H. in the night of Wednesday	5	Intermediate.
From 11h. 695H. in the night of Wednesday till Noon of Thursday	5	Perfect.
From noon of Thursday till 8h. 491H. in the day of Friday	7	Imperfect.
From 8h. 491H. in the day of Friday till Noon of Saturday.	7	Perfect.

158. Further, of these conditions and qualities there are certain ones which exclusively attach to the year in case its beginning falls on a certain day of the week, the other conditions being excluded. If you call this circumstance to help, it will prove an aid towards obtaining the object in view.

In the following figure we represent this subject by means of divisions and ramifications:—

The Year
is either

a common year — or — a leap year.

Thursday (*i.e.* if New-Year's day is a Thursday). The year cannot be Imperfect.

Thursday. It cannot be Intermediate.

In both Common and Leap Years.

Tuesday. It is always Intermediate.

Monday. It can never be Intermediate.

Saturday. It can never be Intermediate.

159. Further, of these conditions there are certain ones which may happen in two consecutive years, whilst others cannot. If we comprise them in a *Ṭailasân*, it will afford a help towards utilizing this circumstance, and will facilitate the method. We must look into the square which belongs

in common to the two qualities of the two years; in that square it is indicated whether the two years of two such qualities can follow each other or not.

		3 2 1 Imperfect.	Qualities of the years.
	5 4 Intermediate.	1 Cannot follow each other.	1 Imperfect.
6 Perfect.	4 Cannot follow each other.	2 Can follow each other.	4 2 Intermediate.
6 Can follow each other.	5 Can follow each other.	3 Can follow each other.	6 Perfect. 3

The reason why two intermediate years cannot follow each other is p.160
this, that their ends and beginnings cannot be brought into concord with each other, as the *Table of Equation* at the end of this book will show.

The reason why two imperfect years cannot follow each other is this, that the perfect months among the months of the cycle (Enneadecateris) preponderate over the imperfect ones. For the small cycle comprises 6,940 days, *i.e.* 125 perfect months and only 110 imperfect ones.

For the same reason, three months which are perfect *according to the appearance of new moon*, can follow each other, whilst of the imperfect months not more than two can follow each other. And their following each other is possible only in consequence of the variation of the motions of the two great luminaries (sun and moon), and of the variation of the setting of the zodiacal signs (*i.e.* the varying velocity with which the sun moves through the various signs of the Ecliptic).

In what Period the beginning of the Jewish Year returns to the same Date.—If the conjunctions at the beginnings of two consecutive great cycles (of 532 years) coincided with each other (*i.e.* if they were cyclical in such a way as to begin always at the same time of the week), we should be able to compute the qualities of the Jewish years by means of tables, comprising the years of a great cycle, similar to the *Chronicon* of the Christians. However, the môlêds of these cycles do not return to the same time of the week except in 689,472 years, for the following reason:

T. *Character* of the small cycle, *i.e.* the remainder which you get by dividing its number of days by 7, is 2d. 16h. 595Ḥ. This fraction is not raised to one whole, except in a number of cycles, which is equal to the number of Ḥalâḳîm of one *Nychthemeron*, *i.e.* 25,920. Because fractions are not raised to wholes, except when multiplied by a number which is equal to the complete number of the same kind of fractions of one whole (*i.e.* by the denominator).

But as both the number of the Ḥalâḳîm of the *Nychthemeron* (25,920) and the number of the remainder of the Ḥalâḳîm of the cycles (595) may be divided by 5, the fractions will be raised to wholes if multiplied by a number of cycles, which is equal to $\frac{1}{5}$ of the Ḥalâḳîm of the Nychthemeron, *i.e.* 5184.

Now, the conjunction (at the beginning of the year) does not return to the same time of the week except in a number of cycles which is the sevenfold of this number (5184), *i.e.* 36,288. And this is the number of cycles which represent the above-mentioned number of years (*viz.* 689,472).

In general, conjunction and opposition return to the same place (*i.e.* happen again at the same time of the week) in each 181,440 months, which is the product of the multiplication of the number of Ḥalâḳîm of one *Nychthemeron* (25,920) by 7.

Comparison between the Jewish Era and the Era of Alexander. —Since it is not possible to use this method for chronological purposes, we have not thought it proper to deviate from the traditional method, inasmuch as it tries to bring near that which is distant, and simplifies and facilitates that which is difficult and intricate. It is sufficient for us to know the beginnings and the qualities of the years, and the corresponding days of the Syrian months on which the days of New Year fall, for such a number of years as that the student will not require more in the majority of cases. This information we have recorded in three tables:—

I. The first represents the day of the week on which the year commences; the *Tabula Signorum.*

II. The second, or *Tabula Qualitatum*, shows the qualities of the years.

The letter ح (ח) designates an *Imperfect* year, because in their language it is called חסרין.

The letter ك (כ) means an *Intermediate* year, because they call it כסדרן.

p.161. The letter ش (ש) means a *Perfect* year, because they call it שלימים.

III. The *Tabula Integritatum et Quantitatum*, representing the days on which the Jewish New Year falls, the days of Âbh in red ink, the days of Îlûl in black ink.

Using these tables you take the *Æra Alexandri* for the current year, beginning with Tishrîn I., which falls always (a little) later than Tishrî. The whole number of years you compare with the vertical column of years; the single years (of the periods of nineteen years) you compare with the horizontal column of years. Then you find in the square which is common to both, that which you wanted, if God permits!

pp.162 –167. [Here follow the three tables, which I have united into one.]

TABLE OF KEBÎ'ÔTH.

Years of the Æra Alexandri.	0	1	2 L.	3	4 L.	5	6	7 L.	8	9	10 L.	11	12	13 L.	14	15 L.	16	17	18 L.	Years of the Jewish Æra Mundi.	Years of the Christian Æra.
1305	3 ב 19 I	7 ש 8 I	5 ש 29 A	5 ב 17 I	2 ח 6 I	7 ש 24 I	5 ב 14 I	2 ש 2 I	2 ח 22 I	5 ב 10 I	2 ש 30 A	2 ש 18 I	7 ח 8 I	3 ב 27 A	2 ש 15 I	7 ח 4 I	5 ב 22 I	2 ש 11 I	7 ש 1 I	4754	998
1324	7 ח 20 I	3 ב 8 I	7 ש 28 A	7 ש 17 I	5 ח 6 I	3 ב 24 I	7 ש 13 I	5 ש 3 I	5 ב 22 I	2 ח 11 I	5 ש 30 A	5 ב 19 I	2 ש 7 I	7 ח 28 A	5 ב 15 I	2 ש 4 I	2 ח 23 I	3 ב 11 I	2 ש 31 A	4773	1012
1343	2 ש 20 I	7 ח 9 I	3 ב 28 A	2 ש 16 I	7 ח 6 I	5 ש 23 I	3 ב 13 I	7 ש 2 I	7 ש 22 I	5 ב 11 I	2 ח 31 A	7 ש 18 I	5 ב 8 I	2 ח 27 A	7 ש 14 I	5 ש 4 I	5 ב 24 I	2 ח 12 I	5 ש 31 A	4792	1031
1362	5 ב 20 I	2 ש 9 I	7 ח 29 A	5 ב 16 I	2 ש 5 I	2 ח 25	5 ש 12 I	3 [illegible] 2 I	2 ש [illegible]1 I	7 ח 11 I	3 ב 29 A	2 ש 17 I	7 ש 7 I	5 ח 28 A	3 ב 14 I	7 ש 3 I	7 ש 23 I	5 ב 13 I	2 ח 1 I	4811	1050
1381	7 ש 19 I	5 ב 9 I	2 ח 29 A	7 ש 15 I	5 ש 5 I	5 ב 25 I	2 ח 14 I	5 ש 1 I	5 ב 21 I	2 ש 10 I	7 ח 31 A	5 ב 17 I	2 ש 6 I	7 ש 27 A	7 ח 16 I	3 ב 3 I	2 ש 22 I	7 ש 12 I	5 ח 2 I	4830	1069
1400	3 ב 19 I	7 ש 8 I	5 ח 29 A	3 ב 16 I	7 ש 4 I	7 ש 24 I	5 ב 14 I	2 ח 3 I	7 ש 20 I	5 ב 10 I	2 ש 30 A	2 ח 19 I	5 ב 6 I	2 ש 26 A	2 ש 15 I	7 ח 5 I	5 ב 22 I	2 ש 11 I	7 ח 1 I	4849	1088
1419	5 ש 19 I	3 ב 8 I	7 ש 28 I	7 ח 17 I	3 ב 5 I	2 ש 23 I	7 ש 13 I	5 ח 3 I	3 ב 21 I	7 ש 9 I	5 ש 30 A	5 ב 19 I	2 ח 8 I	5 ש 26 A	5 ב 15 I	2 ח 4 I	7 ש 22 I	5 ב 11 I	2 ש 31 A	4868	1107
1438	2 ח 20 I	5 ש 8 I	3 ב 28 A	2 ש 16 I	7 ח 6 I	5 ב 24 I	2 ש 12 I	7 ח 2 I	5 ש 20 I	3 ב 10 I	7 ש 29 A	7 ש 18 I	5 ב 8 I	2 ח 28 A	7 ש 14 I	5 ח 4 I	3 ב 22 I	7 ש 11 I	5 ש 31 A	4887	1126
1457	5 ב 20 I	2 ח 9 I	5 ש 28 A	5 ב 16 I	2 ח 5 I	7 ש 23 I	5 ב 13 I	2 ש 1 I	2 ח 21 I	5 ש 9 I	3 ב 30 A	2 ש 17 I	7 ח 7 I	3 ב 26 A	2 ש 14 I	7 ש 3 I	7 ח 23 I	3 ב 11 I	7 ש 31 A	4906	1145
1476	7 ש 19 I	5 ב 9 I	2 ח 29 A	7 ש 16 I	5 ח 5 I	3 ב 23 I	7 ש 12 I	5 ש 2 I	5 ב 21 I	2 ח 10 I	5 ש 29 A	5 ב 18 I	2 ש 6 I	7 ח 27 A	5 ב 14 I	2 ש 3 I	2 ש 22 I	7 ח 12 I	3 ב 31 A	4925	1164
1495	2 ש 19 I	7 ש 8 I	5 ח 29 A	3 ב 16 I	7 ש 5 I	7 ש 24 I	5 ב 14 I	2 ח 8 I	7 ש 21 I	5 ב 10 I	2 ח 30 A	7 ש 17 I	5 ב 7 I	2 ש 26 A	2 ח 15 I	5 ש 8 I	5 ב 23 I	2 ש 11 I	7 ח 1 I	4944	1183
1514	5 ב 19 I	2 ש 8 I	7 ח 28 A	5 ש 15 I	3 ב 5 I	2 ש 24 I	7 ח 13 I	3 ב 1 I	2 ש 20 I	7 ש 10 I	5 ח 30 A	3 ב 17 I	7 ש 6 I	5 ש 27 A	5 ב 15 I	2 ח 4 I	7 ש 22 I	5 ב 12 I	2 ח 31 A	4963	1202
1533	7 ש 18 I	5 ב 8 I	2 ש 28 A	2 ח 16 I	5 ש 4 I	5 ב 24 I	2 ש 13 I	7 ח 2 I	5 ב 20 I	2 ש 9 I	7 ח 30 A	5 ש 16 I	3 ב 6 I	7 ש 26 A	7 ח 15 I	3 ב 2 I	2 ש 21 I	7 ש 11 I	5 ח 1 I	4982	1221
1552	3 ב 18 I	7 ש 7 I	5 ש 28 A	5 ב 17 I	2 ח 5 I	7 ש 23 I	5 ב 13 I	2 ח 2 I	7 ש 19 I	5 ב 9 I	2 ש 29 A	2 ח 18 I	5 ש 5 I	3 ב 26 A	2 ש 14 I	7 ח 4 I	5 ב 21 I	2 ש 10 I	7 ש 31 A	5001	1240
1571	7 ח 20 I	3 ב 7 I	7 ש 27 A	7 ש 16 I	5 ח 6 I	3 ב 23 I	7 ש 12 I	5 ש 2 I	5 ב 22 I	2 ח 10 I	5 ש 29 A	5 ב 18 I	2 ש 7 I	7 ח 27 A	5 ב 14 I	2 ש 3 I	2 ח 23 I	5 ב 10 I	2 ש 30 A	5020	1259
1590	2 ש 19 I	7 ח 9 I	3 ב 27 A	2 ש 15 I	7 ח 5 I	5 ש 23 I	3 ב 12 I	7 ש 1 I	7 ש 21 I	5 ב 11 I	2 ח 30 A	7 ש 17 I	5 ב 7 I	2 ח 27 A	7 ש 13 I	5 ש 3 I	5 ב 23 I	2 ח 12 I	5 ש 30 A	5039	1278
1609	5 ב 19 I	2 ש 8 I	7 ח 29 A	5 ב 15 I	2 ש 4 I	2 ח 24 I	5 ש 12 I	3 ב 1 I	2 ש 20 I	7 ח 10 I	3 ב 29 A	2 ש 16 I	7 ש 6 I	5 ח 27 A	3 ב 14 I	7 ש 2 I	7 ש 22 I	5 ב 12 I	2 ח 1 I	5058	1297
1628	7 ש 18 I	5 ב 8 I	2 ח 28 A	7 ש 15 I	5 ש 4 I	5 ב 24 I	2 ח 13 I	5 ש 1 I	5 ב 20 I	2 ש 9 I	7 ח 30 A	5 ב 17 I	2 ש 5 I	7 ש 26 A	7 ח 15 I	3 ב 3 I	2 ש 21 I	7 ש 11 I	5 ח 1 I	5077	1316
1647	3 ב 19 I	7 ש 7 I	5 ח 28 I	3 ב 15 I	7 ש 4 I	7 ש 23 I	5 ב 13 I	2 ח 2 I	7 ש 20 I	5 ב 9 I	2 ש 29 A	2 ח 18 I	5 ב 6 I	2 ש 25 A	2 ש 14 I	7 ח 4 I	5 ב 22 I	2 ש 10 I	7 ח 31 A	5096	1335
1666	5 ש 18 I	3 ב 8 I	7 ש 27 A	7 ח 16 I	3 ב 4 I	2 ש 23 I	7 ש 12 I	5 ח 2 I	3 ב 20 I	7 ש 9 I	5 ש 29 A	5 ב 18 I	2 ח 7 I	5 ש 26 A	5 ב 14 I	2 ח 3 I	7 ש 21 I	5 ב 11 I	2 ש 30 A	5115	1354
1685	2 ח 19 I	5 ש 7 I	3 ב 28 A	2 ש 15 I	7 ח 5 I	5 ב 23 I	2 ש 12 I	7 ח 1 I	5 ש 19 I	3 ב 9 I	7 ש 29 A	7 ש 17 I	5 ב 7 I	2 ח 27 A	7 ש 14 I	5 ח 3 I	3 ב 21 I	7 ש 10 I	5 ש 31 A	5134	1373
1704	5 ב 19 I	2 ח 8 I	5 ש 27 A	5 ב 16 I	2 ח 4 I	7 ש 22 I	5 ב 12 I	2 ש 1 I	2 ח 20 I	5 ש 8 I	3 ב 29 A	2 ש 17 I	7 ח 6 I	3 ב 25 A	2 ש 13 I	7 ש 3 I	7 ח 22 I	3 ב 10 I	7 ש 30 A	5153	1392
1723	7 ש 19 I	5 ב 8 I	2 ח 28 A	7 ש 15 I	5 ח 5 I	3 ב 22 A	7 ש 11 I	5 ש 1 I	5 ב 21 I	2 ח 9 I	5 ש 28 A	5 ב 17 I	2 ש 6 I	7 ח 26 A	5 ב 13 I	2 ש 2 I	2 ש 22 I	7 ח 11 I	3 ב 30 A	5172	1411
1742	2 ש 18 I	7 ש 8 I	5 ח 28 A	3 ב 15 I	7 ש 4 I	7 ח 24 I	3 ב 11 I	7 ש 31 A	7 ש 20 I	5 ב 10 I	2 ח 29 A	7 ש 16 I	5 ב 6 I	2 ש 26 A	2 ח 14 I	5 ש 2 I	5 ב 22 I	2 ש 11 I	7 ח 31 A	5191	1430
1761	5 ב 18 I	2 ש 7 I	7 ח 28 A	5 ש 14 I	3 ב 4 I	2 ש 23 I	7 ח 13 I	3 ב 31 A	2 ש 19 I	7 ש 9 I	5 ח 30 A	3 ב 16 I	7 ש 5 I	5 ש 26 A	5 ב 15 I	2 ח 3 I	7 ש 21 I	5 ב 11 I	2 ח 31 A	5210	1449
1780	7 ש 17 I	3 ב 7 I	2 ש 27 A	2 ח 16 I	5 ש 3 I	5 ב 23 I	2 ש 12 I	7 ח 2 I	5 ב 19 I	2 ש 8 I	7 ח 29 A	5 ש 16 I	3 ב 5 I	7 ש 25 I	7 ח 14 I	3 ב 2 I	2 ש 20 I	7 ש 10 I	5 ח 31 I	5229	1468
1799	3 ב 18 I	7 ש 6 I	5 ש 27 A	5 ב 16 I	2 ח 5 I	7 ש 22 I	5 ב 12 I	2 ח 1 I	7 ש 19 I	5 ב 8 I	2 ש 28 A	2 ח 17 I	5 ש 5 I	3 ב 25 A	2 ש 13 I	7 ח 3 I	5 ב 21 I	2 ש 9 I	7 ש 30 A	5248	1487
1818	7 ח 19 I	3 ב 7 I	7 ש 26 A	7 ש 15 I	5 ח 5 I	3 ב 23 I	7 ש 11 I	5 ש 1 I	5 ב 21 I	2 ח 10 I	5 ש 28 A	5 ב 17 I	2 ש 6 I	7 ח 27 A	5 ב 13 I	2 ש 2 I	2 ח 22 I	5 ב 10 I	2 ש 29 A	5267	1506

The beginnings of the Jewish Months.—Let us suppose we did not know by means of the *Tabula Quantitatum* on what precise date in the months Âbh or Îlûl the Jewish New Year falls, but we knew from the *Tabula Signorum* on what day of the week it falls, and we had previously learnt from the *Assaying Circle* on what date of Âbh or Îlûl on an average it falls (no regard being had to the *Daḥiyyôth*). In this case we should be sufficiently informed to know in what way to advance or to postpone the date of the Syrian month if this day of the week should be incompatible with *Rôsh-hashshânâ*, so as to get at last the legitimate New-Year's day (*lacuna*) more particularly as the three festivals are to be found with perfect accuracy in the preceding three tables. p.168.

(In this way) we obtain a knowledge of the era of the Jews, of the beginning of their year, and of its complicated nature. Hence we proceed to learn the beginnings of the single months of their year, either by distributing over the months their proper portions of days in conformity with the two qualities of the year in question (whether it be ש, כ or ח or a common or leap year), or by means of the *Tabula Initiorum Mensium*. You compare the *Rôsh-hashshânâ* with the *Table of the Signum* (week-day) of Tishrî; in the table of common years, if the year be a common year; in the table of leap years, if the year be a leap year. At the side of this column you find another, which indicates whether the year be imperfect, intermediate, or perfect. After having made out this, you find in the corresponding squares the beginning of each complete month, and the two beginnings of each incomplete month. For the Jews assign to each month which is preceded by a complete month two beginnings (two first days), *viz.* one day which is in reality the beginning of the month, and the preceding day, or the 30th day of the preceding complete month. This you must keep in mind, for it is part of their bewildering terminology. God is allwise and almighty!

TABLE SHOWING ON WHAT DAYS OF THE WEEK THE BEGINNING OF THE MONTHS FALLS THROUGHOUT THE YEAR. p.169.

Table of Common Years.

'Elûl.	Âbh.	Tammuz.	Siwân.	'Iyâr.	Nîsân.	Adhâr.	Shebhâṭ.	Têbeth.	Kislêw.	Marḥeshwân.	Quality of the Year.	Signum initii mensis Tishrî.
4 III.	2	1 VI.	6	5 IV.	3	2 I.	7	6 V.	4 III.	2 I.	Perfect . .	7
2 I.	7	6 V.	4	3 II.	1	7 VI.	5	4	3	2 I.	Imperfect .	7
6 V.	4	3 II.	1	7 VI.	5	4 III.	2	1 VII.	6 V.	4 III.	Perfect . .	2
4 III.	2	1 VII.	6	5 IV.	3	2 I.	7	6	5	4 III.	Imperfect .	2
6 V.	4	3 II.	1	7 VI.	5	4 III.	3	1 VII.	6	5 IV.	Intermediate	3
2 I.	7	6 V.	4	3 II.	1	7 VI.	5	4 III.	2 I.	7 VI.	Perfect . .	5
1 VII.	6	5 IV.	3	2 I.	7	6 V.	4	3 II.	1	7 VI.	Intermediate	5

p.170.

Table of Leap Years.

'Elûl.	Âbh.	Tammuz.	Sîwân.	'Iyâr.	Nîsân.	'Adhâr Secundus.	'Adhâr Primus.	Shebhâṭ.	Ṭêbeth.	Kislêw.	Marḥeshwân.	Quality of the Year.	Signum initii mensis Tishrî.
6 V.	4	3 II.	1	7 VI.	5	4 III.	2 I.	7	6 V.	4 III.	2 I.	Perfect. .	7
4 III.	2	1 VII.	6	5 IV.	3	2 I.	7 VI.	5	4	3	2 I.	Imperfect .	7
1 VII.	6	5 IV.	3	2 I.	7	6 V.	4 III.	2	1 VII.	6 V.	4 III.	Perfect. .	2
6 V.	4	3 II.	1	7 VI.	5	4 III.	2 I.	7	6	5	4 III.	Imperfect .	2
1 VII.	6	5 IV.	3	2 I.	7	6 V.	4 III.	2	1 VII.	6	5 IV.	Intermediate	3
4 III.	2	1 VII.	6	5 IV.	3	2 I.	7 VI.	5	4 III.	2 I.	7 VI.	Perfect. .	5
2 I.	7	6 V.	4	3 II.	1	7 I.	5 IV.	3	2	1	7 VI.	Imperfect .	5

p.171. They were induced to assume two *Rôsh-Ḥôdesh*, as I am inclined to think, by the circumstance that originally they counted the complete month as 29 days *pure* (*i.e.* without any fraction), and that is in fact the correct time of the interval between two consecutive conjunctions. Into the 30th day, however, fall the fractions of the synodic month (*i.e.* the first 12 hours 793 Ḥalâḳîm of the 30th day belong to the preceding month, whilst the latter 11h. 287H. belong to the following month). Therefore they referred this 30th day to the month that had passed, so that thereby it became in reality complete, and to the incomplete month (just commencing), so that this latter one got two beginnings (*i.e.* the latter 11h. 287H. of the 30th day, and the first whole day of the new month). But God knows best what they intended!

Computation of the beginning and middle of the Months according to Jewish and Astronomical Systems.—If we now want to know the time of conjunction at the beginning of a month, or the time of opposition in the middle of the month, according to the system of the Jews, we derive them from the *Table of Môlêds and Fortnights*, where we find the Conjunction opposite the môlêd of each month, and the Opposition opposite its Fortnight; for the common year in the column of common years; for the leap years in the column of leap years. The number we find we add to the Môlêd Tishrî, *i.e.* to the conjunction at the beginning of Tishrî; the fractions we convert into wholes, the days into weeks. In this way we find what we wanted to know.

If we want to learn the same according to the doctrine of the astronomers, we make the same calculation with the *Table of Conjunctions and Oppositions*, using the table of common years if the year in question be a common year, and the table of leap years if the year in question be a leap year, *and* with the conjunction at the beginning of Tishrî as computed by the astronomers. In this way we arrive at the knowledge of both conjunctions and oppositions which we wanted.

Here follow the tables :—

Pp. 172, 173.

TABLE OF THE MÔLÊDS AND FORTNIGHTS.

The Môlêds and Fortnights of the Months.	Common Year.			The Môlêds and Fortnights of the Months.	Leap Year.		
	Days.	Hours.	Halâkim.		Days.	Hours.	Halâkim.
Môlêd Tishrî .	0	0	0	Môlêd Tishrî . .	0	0	0
Its fortnight .	0	18	396½	Its fortnight .	0	18	396½
Môlêd Marḥeshwân	1	12	793	Môlêd Marḥeshwân	1	12	793
Its fortnight .	2	7	109½	Its fortnight .	2	7	109½
Môlêd Kislêw . .	3	1	506	Môlêd Kislêw .	3	1	506
Its fortnight .	3	19	902½	Its fortnight .	3	19	902½
Môlêd Ṭêbeth .	4	14	219	Môlêd Ṭêbeth .	4	14	219
Its fortnight .	5	8	615½	Its fortnight .	5	8	615½
Môlêd Shebhâṭ .	6	2	1012	Môlêd Shebâth .	6	2	1012
Its fortnight .	6	21	328½	Its fortnight .	6	21	328½
Môlêd 'Adhâr .	0	15	725	Môlêd 'Adhâr I. .	0	15	725
Its fortnight .	1	10	41½	Its fortnight .	1	10	41½
Môlêd Nîsân . .	2	4	438	Môlêd 'Adhâr II. .	2	4	438
Its fortnight .	2	22	834½	Its fortnight .	2	22	834½
Môlêd 'Iyâr . .	3	17	151	Môlêd Nîsân . .	3	17	151
Its fortnight .	4	11	547½	Its fortnight . .	4	11	547½
Môlêd Sîwân . .	5	5	944	Môlêd 'Iyâr . .	5	5	944
Its fortnight .	6	0	260½	Its fortnight .	6	0	260½
Môlêd Tammuz .	6	18	657	Môlêd Sîwân . .	6	18	657
Its fortnight .	0	12	1053½	Its fortnight .	0	12	1053½
Môlêd Âbh . .	1	7	370	Môlêd Tammuz .	1	7	370
Its fortnight .	2	1	766½	Its fortnight .	2	1	766½
Môlêd 'Elûl . .	2	20	83	Môlêd Abh . .	2	20	83
Its fortnight .	3	14	479½	Its fortnight . .	3	14	479½
				Môlêd 'Elûl . .	4	8	876
				Its fortnight .	5	3	192½

pp.174, 175.

Table of Conjunctions and Oppositions.

The Conjunctions and Oppositions of the Months.	Common Year.						The Conjunctions and Oppositions of the Months.	Leap Year.					
	Days.	Hours.	Minutes.	Seconds.	Thirds.	Fourths.		Days.	Hours.	Minutes.	Seconds.	Thirds.	Fourths.
Conjunction of Tishri.	0	0	0	0	0	0	Conjunction of Tishri.	0	0	0	0	0	0
Its full moon .	0	18	22	1	8	$40\frac{4}{7}$	Its full moon .	0	18	22	1	8	$40\frac{4}{7}$
Conjunction of Marheshwân.	1	12	44	2	17	$21\frac{1}{7}$	Conjunction of Marheshwân.	1	12	44	2	17	$21\frac{1}{7}$
Its full moon .	2	7	6	3	26	$1\frac{5}{7}$	Its full moon .	2	7	6	3	26	$1\frac{5}{7}$
Conjunction of Kislêw.	3	1	28	4	34	$42\frac{2}{7}$	Conjunction of Kislêw.	3	1	28	4	34	$42\frac{2}{7}$
Its full moon .	3	19	50	5	43	23	Its full moon .	3	19	50	5	43	23
Conjunction of Ṭêbeth.	4	14	12	6	52	$3\frac{4}{7}$	Conjunction of Ṭêbeth.	4	14	12	6	52	$3\frac{4}{7}$
Its full moon .	5	8	34	8	0	$44\frac{1}{7}$	Its full moon .	5	8	34	8	0	$44\frac{1}{7}$
Conjunction of Shebhâṭ.	6	2	56	9	9	$24\frac{5}{7}$	Conjunction of Shebhâṭ.	6	2	56	9	9	$24\frac{5}{7}$
Its full moon .	6	21	18	10	18	$5\frac{2}{7}$	Its full moon .	6	21	18	10	18	$5\frac{2}{7}$
Conjunction of 'Adhâr.	0	15	40	11	26	46	Conjunction of 'Adhâr I.	0	15	40	11	26	46
Its full moon .	1	10	2	12	35	$26\frac{4}{7}$	Its full moon .	1	10	2	12	35	$26\frac{4}{7}$
Conjunction of Nîsân.	2	4	24	13	44	$7\frac{1}{7}$	Conjunction of 'Adhâr II.	2	4	24	13	44	$7\frac{1}{7}$
Its full moon .	2	22	46	14	52	$47\frac{5}{7}$	Its full moon .	2	22	46	14	52	$47\frac{5}{7}$
Conjunction of 'Iyâr.	3	17	8	16	1	$28\frac{2}{7}$	Conjunction of Nîsân.	3	17	8	16	1	$28\frac{2}{7}$
Its full moon .	4	11	30	17	10	9	Its full moon .	4	11	30	17	10	9
Conjunction of Siwân.	5	5	52	18	18	$49\frac{4}{7}$	Conjunction of 'Iyâr.	5	5	52	18	18	$49\frac{4}{7}$
Its full moon .	6	0	14	19	27	$30\frac{1}{7}$	Its full moon .	6	0	14	19	27	$30\frac{1}{7}$
Conjunction of Tammuz.	6	18	36	20	36	$10\frac{5}{7}$	Conjunction of Siwân.	6	18	36	20	36	$10\frac{5}{7}$
Its full moon .	0	12	58	21	44	$51\frac{2}{7}$	Its full moon .	0	12	58	21	44	$51\frac{2}{7}$
Conjunction of Âbh.	1	7	20	22	53	32	Conjunction of Tammuz.	1	7	20	22	53	32
Its full moon .	2	1	42	24	2	$12\frac{4}{7}$	Its full moon .	2	1	42	24	2	$12\frac{4}{7}$
Conjunction of 'Elûl.	2	20	4	25	10	$53\frac{1}{7}$	Conjunction of Âbh.	2	20	4	25	10	$53\frac{1}{7}$
Its full moon .	3	14	26	26	19	$33\frac{5}{7}$	Its full moon .	3	14	26	26	19	$33\frac{5}{7}$
							Conjunction of 'Elûl.	4	8	48	27	28	$14\frac{2}{7}$
							Its full moon .	5	3	10	28	36	55

We also find what we want to know regarding the Jewish years, by computing the next opposition (or full moon) after the vernal equinox, occurring in that space of time within the limits of which the Jewish passover varies; then we consider on what day within this time it falls, reckoning the day from one sunrise to the next one. If the opposition occurs on one of *Dies Licitæ*, that day is the day of passover; if, however, it occurs on one of the *Dies illicitæ*, *i.e.* the days of the three inferior planets, we postpone passover to the second (the next following) day. This postponement of passover they call in their language דְּחִי *Daḥî*. p.176.

Then you make the same computation in order to find the passover of the preceding year. To the *Signum* (*i.e.* week-day) of this latter passover you add two, whereby you get the day of the 1st of Tishrî that lies in the middle between the two passovers. Then you count the days intervening between the two passovers; if they exceed the number of days of a solar year, that year in which the latter passover lies is a leap year; if they are less, the year is a common year.

In this chapter you may learn the primary qualities of the year (its being common or intercalary), but not its secondary qualities (its being perfect, intermediate, or imperfect). For frequently passover has been postponed, when it ought to have been advanced according to the theory of the Jews, or it has been advanced when, according to them, it ought to have been postponed. Therefore, you get no exact information as to the quality of the year, whether it be perfect, intermediate, or imperfect. Frequently, even the opposition occurred near to one of the limits of that space of time, within which passover varies, whilst each of the places of sun and moon, as made out from appearance, was at variance with its mean place, on account of the alternate acceleration and retardation of the motion of sun and moon, in conformity with the total sum of their *Universal Equations*. Therefore, such an opposition not being fit to be employed, either the preceding or the following opposition was adopted.

For this reason there is a difference between the Jewish computation and this (astronomical) method, to such a degree that frequently according to the Jews the year was a leap year, whilst this astronomical calculation proves it to have been a common year, and *vice versâ*.

Likewise there is a difference between Jews and Christians regarding the leap year, as we shall explain in the chapter on the Christian Fast, if God permits. If, now, there is a difference between them, and they are willing to accept our decision, we shall consider the two oppositions of their two passovers, and shall say, that that opposition at which the moon moves in the middle part of Spica or of Cancer, or the sun is about to leave Aries, is to be rejected according to both systems, whilst the contrary is to be adopted. To the lover of truth, the correctness of these two assertions will be apparent, if the conditions we have mentioned are observed.

The Cycles of Yobel and Shabu'.—The Jews have still other cycles, *e.g.* the cycle of *Yôbêl* and the cycle of *Shâbû'*, *i.e.* of seven years. The first years of both cycles are called "*restitution years.*" For God says, regarding the cycle of seven years, in the third book of the Thora (Levit. xxv. 2–7): "When ye come into the land of Canaan, ye shall sow and reap and prune your vineyards six years. But in the seventh year ye shall not sow nor gather your grapes, but leave them to your servants and maids, and to those who sojourn with you, and to the cattle and the birds."

p.177. The same command God repeats in the second book of the Thora (Exodus xxiii. 10, 11): "And six years thou shalt sow thy land, and gather in the produce thereof. But the seventh year thou shalt let it rest, and shalt leave thy produce during that year to the poor and the cattle."

Likewise their religion and law allow a poor man to sell his child to a rich man, *i.e.* to give it in hire to him, to do service unto him; but not for sexual intercourse, for that requires a marriage-portion and a marriage-contract. The child does him service during the cycle of *Shâbû'*, and it is set free, unless it does not choose to be set free. For God says in the second book of the Thora (Exod. xxi. 2–6): "If anyone of you buy a servant from among the Israelites, six years he shall serve, but in the seventh year he will go out of his possession, and will be free to go where he pleases, he and his wife, if he have got one. But if the servant say, I love my master and will not leave his service, then his master shall bring him near the door-post, and shall bore his ears with an awl, and shall keep him as a servant as long as he pleases."

The cycle of Yôbêl was wanted on account of the following command of God in the third book of the Thora (Levit. xxv. 8–13): "You shall sow the land seven times seven, which is forty-nine years. Then you shall cause the trumpet to sound throughout all your land, and you shall hallow it for the fiftieth year. You shall not sow nor reap. And in the fiftieth year the restitution shall take place." "The land shall not be sold for ever, for the land is mine and you are its inhabitants and sojourners with me" (Levit. xxv. 23). "Everything that has been sold is to be restored in the fiftieth year. You shall sell according to the number of years," *i.e.* the remaining years of the cycle of Yôbêl (Levit. xxv. 13–15).

In the same book (Levit. xxv. 39, 40), God says: "If thy brother be waxen poor, and be sold unto thee, thou shalt not compel him to serve as a bond servant, but as a hired servant and as a sojourner until the year of restitution."

Because of the circumstances brought about by these regulations they required these two cycles, in order that in their sales the higher and lower prices should always correspond to the remaining number of years of the cycle. There are still other religious regulations of theirs which rendered them necessary. If, *e.g.* a servant does not wish to be set free, and remains in the condition of a servant during the whole cycle of Yôbêl, he cannot be retained after that period.

Now, if you want to know how many years have elapsed of each of the two cycles (at a certain time), take the years of the *Æra Adami*, including the current year, subtract therefrom 1,010, or add thereto 740; divide the sum by 350, and neglect the quotient. The remainder, however, compare with the column of numbers in the *Tabula Legum*, opposite which you find the statement of the number of years which have elapsed in each of the two cycles.

Here follows the *Tabula Legum*:

TABULA LEGUM. p.178, 179

Column of the Numbers.	Cycle of Yôbêl.	Cycle of Shâbû'.	Numbers.	Yôbêl.	Shâbû'.	Numbers.	Yôbêl.	Shâbû'.	Numbers.	Yôbêl.	Shâbû'.	Numbers.	Yôbêl.	Shâbû'.	Numbers.	Yôbêl.	Shâbû'.	Numbers.	Yôbêl.	Shâbû'.
1	1	1	26	26	5	51	1	2	76	26	6	101	1	3	126	26	7	151	1	4
2	2	2	27	27	6	52	2	3	77	27	7	102	2	4	127	27	1	152	2	5
3	3	3	28	28	7	53	3	4	78	28	1	103	3	5	128	28	2	153	3	6
4	4	4	29	29	1	54	4	5	79	29	2	104	4	6	129	29	3	154	4	7
5	5	5	30	30	2	55	5	6	80	30	3	105	5	7	130	30	4	155	5	1
6	6	6	31	31	3	56	6	7	81	31	4	106	6	1	131	31	5	156	6	2
7	7	7	32	32	4	57	7	1	82	32	5	107	7	2	132	32	6	157	7	3
8	8	1	33	33	5	58	8	2	83	33	6	108	8	3	133	33	7	158	8	4
9	9	2	34	34	6	59	9	3	84	34	7	109	9	4	134	34	1	159	9	5
10	10	3	35	35	7	60	10	4	85	35	1	110	10	5	135	35	2	160	10	6
11	11	4	36	36	1	61	11	5	86	36	2	111	11	6	136	36	3	161	11	7
12	12	5	37	37	2	62	12	6	87	37	3	112	12	7	137	37	4	162	12	1
13	13	6	38	38	3	63	13	7	88	38	4	113	13	1	138	38	5	163	13	2
14	14	7	39	39	4	64	14	1	89	39	5	114	14	2	139	39	6	164	14	3
15	15	1	40	40	5	65	15	2	90	40	6	115	15	3	140	40	7	165	15	4
16	16	2	41	41	6	66	16	3	91	41	7	116	16	4	141	41	1	166	16	5
17	17	3	42	42	7	67	17	4	92	42	1	117	17	5	142	42	2	167	17	6
18	18	4	43	43	1	68	18	5	93	43	2	118	18	6	143	43	3	168	18	7
19	19	5	44	44	2	69	19	6	94	44	3	119	19	7	144	44	4	169	19	1
20	20	6	45	45	3	70	20	7	95	45	4	120	20	1	145	45	5	170	20	2
21	21	7	46	46	4	71	21	1	96	46	5	121	21	2	146	46	6	171	21	3
22	22	1	47	47	5	72	22	2	97	47	6	122	22	3	147	47	7	172	22	4
23	23	2	48	48	6	73	23	3	98	48	7	123	23	4	148	48	1	173	23	5
24	24	3	49	49	7	74	24	4	99	49	1	124	24	5	149	49	2	174	24	6
25	25	4	50	50	1	75	25	5	100	50	2	125	25	6	150	50	3	175	25	7

p.180, 181

Numbers.	Yôbêl.	Shâbû'.	Numbers.	Yôbêl.	Shâbû'.	Numbers.	Yôbêl.	Shâbû'.	Numbers.	Yôbêl.	Shâbû'.	Numbers.	Yôbêl.	Shâbû'.	Numbers.	Yôbêl.	Shâbû'.	Numbers.	Yôbêl.	Shâbû'.
176	26	1	201	1	5	226	26	2	251	1	6	276	26	3	301	1	7	326	26	4
177	27	2	202	2	6	227	27	3	252	2	7	277	27	4	302	2	1	327	27	5
178	28	3	203	3	7	228	28	4	253	3	1	278	28	5	303	3	2	328	28	6
179	29	4	204	4	1	229	29	5	254	4	2	279	29	6	304	4	3	329	29	7
180	30	5	205	5	2	230	30	6	255	5	3	280	30	7	305	5	4	330	30	1
181	31	6	206	6	3	231	31	7	256	6	4	281	31	1	306	6	5	331	31	2
182	32	7	207	7	4	232	32	1	257	7	5	282	32	2	307	7	6	332	32	3
183	33	1	208	8	5	233	33	2	258	8	6	283	33	3	308	8	7	333	33	4
184	34	2	209	9	6	234	34	3	259	9	7	284	34	4	309	9	1	334	34	5
185	35	3	210	10	7	235	35	4	260	10	1	285	35	5	310	10	2	335	35	6
186	36	4	211	11	1	236	36	5	261	11	2	286	36	6	311	11	3	336	36	7
187	37	5	212	12	2	237	37	6	262	12	3	287	37	7	312	12	4	337	37	1
188	38	6	213	13	3	238	38	7	263	13	4	288	38	1	313	13	5	338	38	2
189	39	7	214	14	4	239	39	1	264	14	5	289	39	2	314	14	6	339	39	3
190	40	1	215	15	5	240	40	2	265	15	6	290	40	3	315	15	7	340	40	4
191	41	2	216	16	6	241	41	3	266	16	7	291	41	4	316	16	1	341	41	5
192	42	3	217	17	7	242	42	4	267	17	1	292	42	5	317	17	2	342	42	6
193	43	4	218	18	1	243	43	5	268	18	2	293	43	6	318	18	3	343	43	7
194	44	5	219	19	2	244	44	6	269	19	3	294	44	7	319	19	4	344	44	1
195	45	6	220	20	3	245	45	7	270	20	4	295	45	1	320	20	5	345	45	2
196	46	7	221	21	4	246	46	1	271	21	5	296	46	2	321	21	6	346	46	3
197	47	1	222	22	5	247	47	2	272	22	6	297	47	3	322	22	7	347	47	4
198	48	2	223	23	6	248	48	3	273	23	7	298	48	4	323	23	1	348	48	5
199	49	3	224	24	7	249	49	4	274	24	1	299	49	5	324	24	2	349	49	6
200	50	4	225	25	1	250	50	5	275	25	2	300	50	6	325	25	3	350	50	7

p. 182 **On the Tekufoth or Year-points.**—Besides the cycles we have mentioned, the have other cycles called *Teḳûfôth* תְּקוּפוֹת. *Teḳûfâ* means with them he commencement of each of the quarters of the year. Therefore

the *Teḳûfâ of Nîsân* is the vernal equinox,
the *Teḳûfâ of Tammuz*, the summer solstice,
the *Teḳûfâ of Tishrî*, the autumnal equinox,
and the *Teḳûfâ of Ṭêbeth*, the winter solstice.

The interval between two consecutive Teḳûfôth they determine equally at one-fourth of the days of the year, *i.e.* 91 d. $7\frac{1}{2}$ h. And on this rule they have based their calculations for the determination of the Teḳûfôth, (which were rendered necessary for this reason, that) the Jewish priests forbade the common people (the laity) to take any food at the hour of the Teḳûfâ, maintaining that this would prove injurious to the body. This, however, is nothing but one of the snares and nets which the Rabbis have laid for the people, and by which they have managed to catch them and to bring them under their sway. The thing has come to this, that people do not start on any undertaking unless they are guided by Rabbinical opinions and Rabbinical directions, without asking any other person's advice, as if the Rabbis were Lords beside the *Lord*. But God makes his account with them!

The Jews maintain, too, that at the hours of the *Môlêds* of the months the water becomes turbid; and one Jew, who is considered a wise and learned man, told me that he himself had witnessed it. If this be the truth, it must, of course, be explained by the results of astronomical observation, not by means of their traditional system of chronological computation. On the whole, we do not deny the abstract possibility of such a fact. For the students of physical sciences maintain that marrow and brain, eggs, and most moist substances increase and decrease with the increase and decrease of the moonlight; that the wine in casks and jugs begins to move so as to get turbid with sediment; and that the blood during the increase of moonlight runs from the interior of the body towards the outer parts, whilst during its decrease it sinks back into the interior of the body.

The nature of the *Lapis Lunæ* is still more strange than all this; for it is, as Aristotle says, a stone with a yellow dot on the surface. This dot increases together with the increasing moonlight, so as to extend over the whole surface of the stone when the moon has become full; afterwards it decreases again in the same proportion as the moonlight.

The Jew who told me this is a trustworthy authority, to whose account no suspicion attaches. Therefore these appearances, as related by the Jews, are not impossible in the abstract.

The intervals between the Teḳûfôth, as reckoned by the Jewish scholars, are identical with those of Ptolemy, *i.e.*,

From the Teḳûfâ	of	Tishrî	to the Teḳûfâ	of	Ṭêbeth = $88\frac{1}{8}$ d.
,,	,,	Ṭêbeth	,,	,,	Nîsân = $90\frac{1}{8}$ d.
,,	,,	Nîsân	,,	,,	Tammuz = $94\frac{1}{2}$ d.
,,	,,	Tammuz	,,	,,	Tishrî = $92\frac{1}{2}$ d.

This gives a sum of $365\frac{1}{4}$ days.

In the computation of the Teḳûfôth they do not reckon the year with mathematical accuracy. For, as we have already mentioned, if they reckon with mathematical accuracy, they fix he solar year at

$$365 \text{ d. } 5\frac{3791}{4104} \text{ h}$$

p. 183 **Computation of the Distance of the Apogee from the Vernal Point.**—If we, now, know the days of the year-quarters, we know also the place of the apogee of the solar sphere.

If we want to know the place of the apogee, such as it was at the time of their observations, we must find the mean motion of the sun for one day.

We multiply the fractions of one *Nychthemeron*,

i.e. 98,496,

which they call the *Solar Cycle*, by 360; and the product we divide by the length of the solar year, after it has been converted into the same kind of fractions,

i.e. 35,975,351,

which number they call the *Basis*.

By this method, as they have described it, you find the mean motion of the sun for one Nychthemeron to be about

$$0^\circ\ 59'\ 8''\ 17'''\ 7^{\text{IV}}\ 46^{\text{V}}.$$

For one day stands in the same proportion to all the days of the solar year as that portion of degrees of the sphere, which the sun traverses in one day, to the whole circle.

Now we draw the circle $\overline{a\,b\,c\,d}$, representing the solar sphere as homocentric with the Ecliptic, around the centre h. Then you make

$\bar{a}$ the beginning of Aries;
$\bar{b}$ the beginning of Cancer;
$\bar{c}$ the beginning of Libra;
$\bar{d}$ the beginning of Capricorn.

Further we draw the two diameters $\overline{a\,h\,c}$ and $\overline{b\,h\,d}$.

Already before, in recording their theory, we have mentioned that the sun requires more time to traverse the quarter $\overline{a\,b}$ than the other quarters. Therefore the centre of the *Excentric Sphere* must lie in this quarter.

Let $\bar{x}$ be the centre of the *Excentric Sphere*. Around it we draw the circle $\overline{s\,t\,f\,n}$, touching the homocentric sphere, as a representation of the *Excentric Sphere*. The point of contact is $\bar{t}$.

Then we draw the line $\overline{t\,x}$, the diameter $\overline{r\,x\,m\,k}$ parallel with the diameter $\overline{a\,h\,c}$ through the centre $\bar{x}$, and finally the radius $\overline{l\,x}$, which we prolong as a straight line as far as $\bar{s}$, parallel with the diameter $\overline{b\,h\,d}$.

Because, now, the sun in his mean motion traverses the half circumference $\overline{a\ b\ c}$, *i.e.* the sum of the vernal and summer quarters, in 187 days, the section $\overline{z\ f\ n}$ of the *Excentric Sphere* is equal to

$$184^\circ\ 18'\ 52''\ 43'''\ 12^{IV}.$$

If we subtract from this the half circle $\overline{r\ t\ f\ k}$, *i.e.* 180 degrees, we get as a remainder the sum of $\overline{z\ r}$ and $\overline{k\ n}$, *i.e.*

$$4^\circ\ 18'\ 52''\ 43'''\ 12^{IV}.$$

However, these two ($\overline{z\ r}$ and $\overline{k\ n}$) are equal, since the two diameters are parallel. Therefore each of them is

$$2^\circ\ 9'\ 26''\ 21'''\ 36^{IV}.$$

And the sine of each of them, *i.e.* the line $\overline{x\ s}$, is equal to

$$0^\circ\ 2'\ 15''\ 30'''\ 57^{IV},$$

if you take the radius $\overline{l\ x}$ as 1 degree.

Since the sun traverses the quarter $\overline{a\ b}$ in $94\frac{1}{2}$ days, the section $\overline{z\ t\ f}$ of the *Excentric Sphere* is equal to

$$93^\circ\ 8'\ 34''\ 38'''\ 44^{IV}.$$

And because $\overline{z\ l}$ is the sum of $\overline{z\ r}$, which is known, and of $\overline{r\ l}$, which is the quarter of a circle, we find, on subtracting $\overline{z\ l}$ from $\overline{z\ f}$, $\overline{l\ f}$ to be equal to the remainder, *i.e.*

$$0^\circ\ 59'\ 8''\ 17'''\ 8^{IV}.$$

The sine of $\overline{l\ f}$ according to the same measure is

$$0^\circ\ 1'\ 1''\ 55'''\ 35^{IV}.$$

This is the line $\overline{x\ m}$, which is equal to $\overline{s\ h}$.

Therefore, in the rectangular triangle $\overline{x\ s\ h}$, the two sides $\overline{x\ s}$ and $\overline{s\ h}$ are known, whilst the longest side is unknown. Now, we take the p. 184
squares of each of the two sides $\overline{x\ s}$ and $\overline{s\ h}$ and add them together. This gives

287, 704, 466, 674 eighths.

If we take the root of this number, we get

$$0^\circ\ 2'\ 28''\ 59'''\ 40^{IV},$$

which is the distance between the two centres, equal to the sine of the *Greatest Equation*.

If we look for the corresponding arc in the Sine Tables, we get

$$2^\circ\ 22'\ 19''\ 12'''\ 16^{IV},$$

which is the *Greatest Equation* (*lacuna*) one degree. For half (!) of $\overline{hx}$, measured by $\overline{xt}$ as 1 degree, stands in the same proportion to $\overline{xt}$ as (*lacuna*).

If we, now, want to know how long the line $\overline{xh}$ is, if measured by the line $\overline{hxt}$ as 1 degree, we multiply $\overline{xh}$ by 1 degree and divide the sum by $\overline{hx}$ plus 1 degree. Thereby we find $\overline{xh}$, as measured by the line $\overline{th}$, as 1 degree.

For $\overline{hx}$, if measured by $\overline{ht}$ as 1 degree, stands in the same proportion to $\overline{xt}$ as $\overline{xh}$, if measured by $\overline{xt}$ as 1 degree, to the sum of $\overline{hx}$ plus 1 degree, *i.e.* $\overline{xt}$.

In this way the distance between the two centres in its proportion to each of the two diameters, that of the homocentric and that of the excentric sphere, becomes known.

Further we draw the line $\overline{tu}$ at right angles to the diameter $\overline{ahc}$. Now the two triangles $\overline{tuh}$ and $\overline{xsh}$ are similar, and their corresponding sides are proportional to each other.

Now, everybody who knows trigonometry knows that in a triangle the side $\bar{\alpha}$ stands in the same proportion to the side $\bar{\beta}$ as the sine of the angle opposite the side $\bar{\alpha}$ to the sine of the angle opposite the side $\bar{\beta}$.

Therefore $\overline{hx}$, which is known, stands in the same proportion to $\overline{xs}$, which is also known, as the sine of the right angle $\overline{xsh}$, *i.e.* $\overline{ht}$ the *Sinus Totus*, to the sine of the angle $\overline{shx}$, *i.e.* the line $\overline{tu}$, which we wanted to find.

Finally we compute this line, as we compute the unknown number out of four numbers which stand in proportion to each other. So we get

$$0^\circ\ 54'\ 34''\ 19'''\ 48^{IV}\ 30^{V}.$$

The corresponding arc is

$$65^\circ\ 26'\ 29''\ 32''',$$

which is the line $\overline{at}$, or the distance of the apogee from the vernal equinox. And that is what we wanted to demonstrate.

Here follows the figure of the circle.

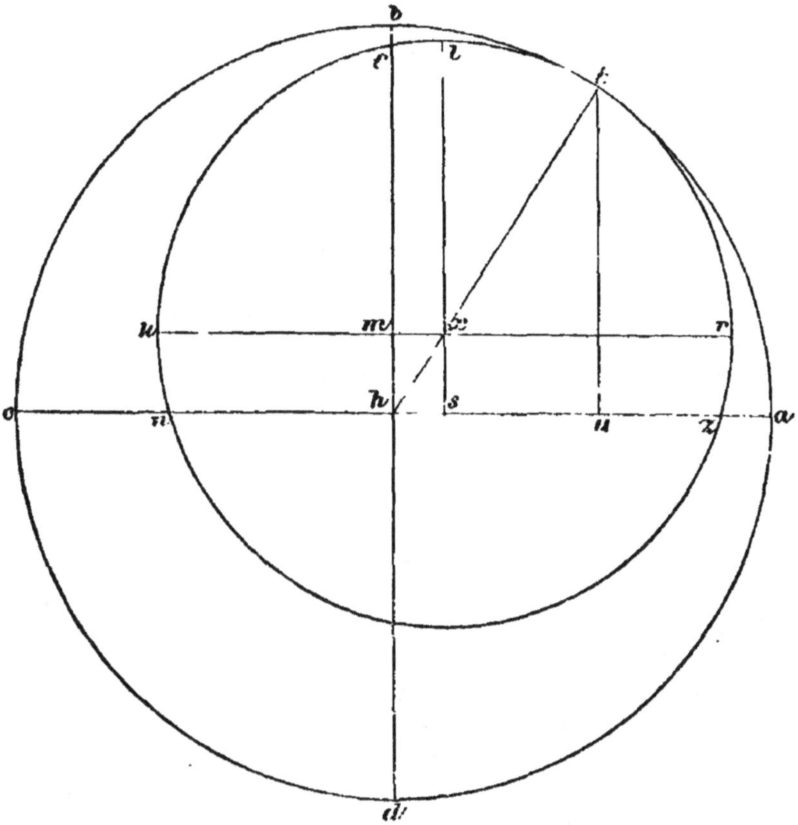

This is the method of the ancient astronomers for the calculation of the apogee. Modern astronomers, knowing that it is extremely difficult and next to impossible to determine the times of the two solstices, preferred in their observations of the four points $\overline{a\ b\ c\ d}$ the middle parts of the year-quarters, *i.e.*, the middle parts of the *Firm Signs* (*i.e.* of Taurus, Leo, Scorpio, Amphora). The method, however, which my master 'Abû-Naṣr Manṣûr b. 'Alî b. 'Irâḳ, a freedman of the 'Amîr-almu'minîn, has found out for the solution of the preceding problem, requires the determination of three points of the ecliptic, chosen *ad libitum*, and an accurate knowledge of the length of the solar year. In p. 185
my *Kitâb-alistishhâd bikhtilâf-ala'rṣâd* I have shown that this method is as much superior to that of modern astronomers as the method of the latter is superior to that of the ancient astronomers.

If I plunge into subjects foreign to the plan of this book, it is only for the purpose of leading the reader, as it were, about in the gardens of wisdom, so as to prevent his mind and eye becoming weary and getting a dislike (to continue the reading of this book). Let me hope that the reader will accept this apology of mine.

Computation of the Tekufoth according to the Jewish System. —Now we return to our subject and say: If the Jews want to find the year-quarters, *i.e.* the *Teḳûfôth* of some year, they take the years of the *Æra Adami*, the current year included, and convert them into Solar Cycles (dividing them by 28). As for the remaining years, they take for every single year 30 hours, *i.e.* $1\frac{1}{4}$ day. The number of weeks which are contained in this sum they disregard, so as to get finally a number of days less than seven. These days they count either from the beginning of the night of Wednesday, or they increase them by 3 and count the sum from the beginning of the night of Sunday. This brings them to the Teḳûfâ of Nîsân, *i.e.* the vernal equinox of the year in question.

In the preceding we have already explained the intervals between the single Teḳûfôth according to both views, the common and the learned one. If, therefore, one of the Teḳûfôth is known, thereby the other ones are known too.

Their counting the sum of days from the beginning of the night of Wednesday is for no other reason but this, that some of them maintain that the sun was created on Wednesday the 27th of Îlûl, and that the Teḳûfâ of Tishrî (autumnal equinox) took place at the end of the third hour of the day of Wednesday the 5th of Tishrî. Further, they make the sun traverse the two year-quarters of spring and summer in 182d. 15h., in case they do not reckon with mathematical accuracy, as we have before mentioned. Now, if we convert these 182d. 15h. into weeks, the days disappear, and we get only a remainder of 15h. If we, further, reckon from the Teḳûfâ of Tishrî backward, and we count these hours, we come as far as the beginning of the first hour of the night of Wednesday. And that is the moment whence the computation we have mentioned starts.

Others among the Jews maintain that the sun was created in the first part of Aries at this same moment whence the computation of the Teḳûfôth starts; that he was in conjunction with the moon, so as to form the Môlêd of Nîsân, 9h. 642H. after the creation. The solar year, if not computed with mathematical accuracy, is $365\frac{1}{4}$ days. If we convert it into weeks, we get as a remainder $1\frac{1}{4}$ day, which is the surplus of each Teḳûfâ over the corresponding one of the preceding year (the *Character* of the Teḳûfâ). Therefore we take this Character for each of the remaining years. If we begin (in the computation of the Teḳûfôth) from the beginning of the Solar Cycle either from the beginning of day or night, we come back at the end of the cycle to the same moment whence we started.

According to this mode of calculation we have computed the Teḳûfôth of a Solar Cycle. Now take the years of the *Æra Adami*, the current
p. 186 year included, convert them into Solar Cycles which you disregard; the remainder of years compare with the column of the Cycle till you find the corresponding number. Then you find opposite, the interval be-

tween the Teḳûfâ of Nîsân and the beginning of the night of Sunday in the current year in question; there you find, too, the next following three Teḳûfôth and the *Dominus Horæ, i.e.* the presiding planet of that hour in which the Teḳûfâ falls. For they mention these *Domini* together with the Teḳûfôth and call them "*Horoscopes of the Hours.*" If the hours you get are less than 12, they are hours of the night; if they are more, they are hours of the day. So you may subtract therefrom 12 hours, and the remainder represents the corresponding hour of the day.

TABLE OF TEḲÛFÔTH. p. 187 -191

Column of the Solar Cycle.	The Months of the Four Teḳûfôth.	The intervals between the Teḳûfôth and the beginning of the Night of Sunday.			*The Masters of the hours in which the Teḳûfôth occur.*
		d.	h.	H.	
1st year .	Nîsân . . .	4	18	0	Shabbethâî.
	Tammuz . .	5	1	540	„
	Tishrî . .	5	9	0	Ṣêdeḳ.
	Têbeth . .	5	16	540	„
2nd year .	Nîsân . . .	6	0	0	Ma'adhîm.
	Tammuz . .	6	7	540	„
	Tishrî . .	6	15	0	Ḥammâ.
	Têbeth . .	6	22	540	„
3rd year .	Nîsân . . .	0	6	0	Nôgah.
	Tammuz . .	0	13	540	„
	Tishrî . .	0	21	0	Kôkhabh Ḥammâ.
	Têbeth . .	1	4	540	„
4th year .	Nîsân . . .	1	12	0	Lebhênâ.
	Tammuz . .	1	19	540	„
	Tishrî . .	2	3	0	Shabbethâî.
	Têbeth . .	2	10	540	„
5th year .	Nîsân . . .	2	18	0	Ṣêdeḳ
	Tammuz . .	3	1	540	„
	Tishrî . .	3	9	0	Ma'adhîm.
	Têbeth . .	3	16	540	„
6th year .	Nîsân . . .	4	0	0	Ḥammâ.
	Tammuz . .	4	7	540	„
	Tishrî . .	4	15	0	Nôgah.
	Têbeth . .	4	22	540	„
7th year .	Nîsân . . .	5	6	0	Kôkhabh Ḥammâ.
	Tammuz . .	5	13	540	„
	Tishrî . .	5	21	0	Lebhênâ.
	Têbeth . .	6	4	540	„
8th year .	Nîsân . . .	6	12	0	Shabbethâî.
	Tammuz . .	6	19	540	„
	Tishrî . .	0	3	0	Ṣêdeḳ.
	Têbeth . .	0	10	540	„

Column of the Solar Cycle.	The Months of the Four Teḳûfôth.	The intervals between the Teḳûfôth and the beginning of the Night of Sunday. d.	h.	chl.	*The Masters of the hours* in which the Teḳûfôth occur.
9th year .	Nîsân . . .	0	18	0	Ma'adhîm.
	Tammuz . .	1	1	540	,,
	Tishrî . .	1	9	0	Ḥammâ.
	Têbeth . .	1	16	540	,,
10th year .	Nîsân . . .	2	0	0	Nôgah.
	Tammuz . .	2	7	540	,,
	Tishrî . .	2	15	0	Kôkhabh Ḥammâ.
	Têbeth . .	2	22	540	,,
11th year .	Nîsân . . .	3	6	0	Lebhênâ.
	Tammuz . .	3	13	540	,,
	Tishrî . .	3	21	0	Shabbethâî.
	Têbeth . .	4	4	540	,,
12th year .	Nîsân . . .	4	12	0	Ṣêdeḳ.
	Tammuz . .	4	19	540	,,
	Tishrî . .	5	3	0	Ma'adhîm.
	Têbeth . .	5	10	540	,,
13th year .	Nîsân . . .	5	18	0	Ḥammâ.
	Tammuz . .	6	1	540	,,
	Tishrî . .	6	9	0	Nôgah.
	Têbeth . .	6	16	540	,,
14th year .	Nîsân . . .	0	0	0	Kôkhabh Ḥammâ.
	Tammuz . .	0	7	540	,,
	Tishrî . .	0	15	0	Lebhênâ.
	Têbeth . .	0	22	540	,,
15th year .	Nîsân . . .	1	6	0	Shabbethâî.
	Tammuz . .	1	13	540	,,
	Tishrî . .	1	21	0	Ṣêdeḳ.
	Têbeth . .	2	4	540	,,
16th year .	Nîsân . . .	2	12	0	Ma'adhîm.
	Tammuz . .	2	19	540	,,
	Tishrî . .	3	3	0	Ḥammâ.
	Têbeth . .	3	10	540	,,
17th year .	Nîsân . . .	3	18	0	Nôgah.
	Tammuz . .	4	1	540	,,
	Tishrî . .	4	9	0	Kôkhabh Ḥammâ.
	Têbeth . .	4	16	540	,,
18th year .	Nîsân . . .	5	0	0	Lebhênâ.
	Tammuz . .	5	7	540	,,
	Tishrî . .	5	15	0	Shabbethâî.
	Têbeth . .	5	22	540	,,
19th year .	Nîsân . . .	6	6	0	Ṣêdeḳ.
	Tammuz . .	6	13	540	,,
	Tishrî . .	6	21	0	Ma'adhîm.
	Têbeth . .	0	4	540	,,

Column of the Solar Cycle.	The Months of the Four Teḳûfôth.	The intervals between the Teḳûfôth and the beginning of the Night of Sunday.			*The Masters of the hours* in which the Teḳûfôth occur.
20th year .	Nîsân . . .	0	12	0	Ḥammâ.
	Tammuz . .	0	19	540	„
	Tishrî . .	1	3	0	Nôgah.
	Têbeth . .	1	10	540	„
21st year .	Nîsân . . .	1	18	0	Kôkhabh Ḥammâ.
	Tammuz . .	2	1	540	„
	Tishrî . .	2	9	0	Lebhênâ.
	Têbeth . .	2	16	540	„
22nd year .	Nîsân . . .	3	0	0	Shabbethâî.
	Tammuz . .	3	7	540	„
	Tishrî . .	3	15	0	Ṣêdeḳ.
	Têbeth . .	3	22	540	„
23rd year .	Nîsân . . .	4	6	0	Ma'adhîm.
	Tammuz . .	4	13	540	„
	Tishrî . .	4	21	0	Ḥammâ.
	Têbeth . .	5	4	540	„
24th year .	Nîsân . . .	5	12	0	Nôgah.
	Tammuz . .	5	19	540	„
	Tishrî . .	6	3	0	Kôkhabh Ḥammâ.
	Têbeth . .	6	10	540	„
25th year .	Nîsân . . .	6	18	0	Lebhênâ.
	Tammuz . .	0	1	540	„
	Tishrî . .	0	9	0	Shabbethâî.
	Têbeth . .	0	16	540	„
26th year .	Nîsân . . .	1	0	0	Ṣêdeḳ.
	Tammuz . .	1	7	540	„
	Tishrî . .	1	15	0	Ma'adhîm.
	Têbeth . .	1	22	540	„
27th year .	Nîsân . . .	2	6	0	Ḥammâ.
	Tammuz . .	2	13	540	„
	Tishrî . .	2	21	0	Nôgah.
	Têbeth . .	3	4	540	„
28th year .	Nîsân . . .	3	12	0	Kôkhabh Ḥammâ.
	Tammuz . .	3	19	540	„
	Tishrî . .	4	3	0	Lebhênâ.
	Têbeth . .	4	10	540	„

Names of the Planets and the Signs of the Zodiac.—The names p. 192
of the planets which we have mentioned in the *Table of the Teḳûfôth* are Hebrew names, in which form they are used by them. Each nation, however, if they want to mention the planets, must call them by the names of their own language. Therefore here follows a table exhibiting the names of the planets in various languages. The reader will find here the Hebrew names which we have mentioned as well as the names in other languages.

TABLE

OF THE NAMES OF THE SEVEN PLANETS.

in Arabic	Zuḥal	Almushtarî	Almirrîkh	Alshams	Alzuhara	ʻUṭârid	Alḳamar
in Greek	Kronos	Zeus	Ares	Helios	Aphrodite	Hermes	Selene
in Persian	Kaiwân	Hurmuzd	Bahrâm	Mihr, Khurshîd	Nâhîd	Tîr	Mâh
in Syriac	ܟܐܘܢ	ܒܠ	ܢܪܝܓ	ܫܡܫܐ	ܐܣܬܪܐ ܒܠܬܝ	ܢܒܘ	ܣܗܪܐ
in Hebrew	Shabbethâî	Ṣêdeḳ	Ma'adhîm	Ḥammâ	Nôgah	Kôkhab Ḥammâ	Lebhênâ
in Sanscrit	Çanaiçcara	Vṛhaspati	Maṅgala	Âditya	Çukra	Budha	Sôma
in Chorasmian	——	ريمزد	اريهر	اخير	ناهيج	چىرى	ماه

And now natural relationship (between the planets and the signs of the zodiac) demands, although it is not necessary in this place of our book, nor is it requisite, that we should do the same with regard to the signs of the zodiac which we have done for the planets, *i.e.* construct a table containing all that we know of their names in various languages. For he who wants this for the planets, wants something of the same kind for the signs of the zodiac.

Here follows the table containing the names of the signs of the zodiac in various languages. p. 193

Arabic.	Greek.	Persian.	Syriac.	Hebrew.	Sanscrit.	Chorasmian.
Alḥamal Alkabsh	Κριός	Bara	ܐܡܪܐ	טלה	Mêsha	ورن
Althaur	Ταῦρος	Gâu	ܬܘܪܐ	שׁור	Vṛsha	غاو
Aljauzâ Altau'amân	Δίδυμοι	Dûpaikar	ܬܐܡܐ	תאמים	Mithuna	الدوبهرقريك
Alsaraṭân	Καρκῖνος	Karzang	ܣܪܛܢܐ	סרטן	Karkaṭa	خرچنك
Al'asad	Λέων	Shîr	ܐܪܝܐ	ארי	Siṅha	سرغ
Alsunbula Al'adhrâ	Παρθένος	Khôsha	ܫܒܠܬܐ ܒܬܘܠܬܐ	בתולה	Kanyâ	وولينى
Almîzân	Ζυγός	Tarâzû	ܡܣܐܬܐ	מוזנים	Tulâ	ترازك
Al'akrab	Σκορπίος	Kazhdum	ܥܩܪܒܐ	עקרב	Vṛśćika	درمچيك
Alkaus Alrâmî	Τοξευτής	Nîmasp	ܩܫܬܐ ܫܕܝ ܓܐܪܐ	קשׁת	Dhanu	ذتيك
Aljady	Ἀιγόκερως	Bahî	ܓܕܝܐ	גדי	Makara	ثارليك
Aldalw	Ὑδροχόος	Dôl	ܕܘܠܐ	דלי	Kumba	دور
Alḥût Alsamaka	Ἰχθύες	Mâhî	ܢܘܢܐ	דג	Mîna	كيب

p. 194 **The Author criticizes the Jewish computation of the Tekufoth.**— We return to our subject and say: The calculation and tables, given in the preceding, enable the student to find the week day on which the Teḳûfâ falls; the corresponding day of the Syrian month, however, to which they bring us, differs from real time to an intolerable extent.

Let us *e.g.* take the *Æra Adami* for the 1st of Tishrî, the môlêd of which falls on Sunday the 1st of Îlûl in the year 1311 of Alexander. The number of complete years of the *Æra Adami* is

4759

or 8 *great cycles* ($8 \times 532 = 4256$), 26 *small cycles* ($26 \times 19 = 494$), and 9 *complete years*, arranged according to the *Ordo Intercalationis* בהזיגוח so that six out of these nine years are common-years and three leap-years.

If we convert this sum of cycles and years into days, we get the sum of

1,738,200d. 7h. 253^H

This is the interval between the môlêd of the first year of the *Æra Adami* and the môlêd of the present above-mentioned year (*A. Adami* 4759).

We have already stated before that according to Jewish dogma the Teḳûfôth-Tishrî, *i.e.* the autumnal equinox, occurred at the beginning of the *Æra Adami*, 5 days and 1 hour after the môlêd of the year.

If we subtract these 5d. 1h. from the sum we have got, we get as remainder the interval between the Teḳûfath-Tishrî of the first year of the era and the môlêd of the present year.

If we divide this interval by $365\frac{1}{4}$d, we get

4,758 years

and a remainder of

$335\frac{3}{4}$ days.

Till this Solar year is complete, and night and day are again equal, 29d. 11h. 827^H more are required. If we add this number of days, hours, and Ḥalâḳîm to the môlêd of the present year, *i.e.* to Sunday 7h. 253^H of daytime, we advance as far as the night of Tuesday 9h. on the 1st day of the month *Tishrîn Primus*.

Now, this Teḳûfâ falls by 14 days later than the equinox as determined by astronomical observation. Such a difference, even if it be much less, is quite intolerable, although popular use may be based upon it. This popular use we have illustrated by our table according to the theory of the Jews.

If we, further, take this interval between the first Teḳûfâ and the môlêd of the present year, *i.e.*

1,738,195d. 6h. 253^H,

and multiply it by

98,496,

which is the number of fractions of one day of their Solar year (of R. Addâ), we get the sum of

171,280,305

(*Great lacuna.*)

Methods showing how to find the beginning of a year of any era. p. 195

TABLE OF THE BEGINNINGS OF THE SYRIAC AND GREEK MONTHS.

Column of the Solar Cycle.	October. Tishrîn I.	November. Ti[illegible]n II.	December. Kânûn I.	January. Kânûn II.	February. Shubâṭ.	March. Adhâr.	April. Nîsân.	May. 'Iyyâr.	June. Ḥazîrân.	July. Tammûz.	August. Âbh.	September. Îlûl.	Leap-years.
1	2	5	7	3	6	6	2	4	7	2	5	1	
2	3	6	1	4	7	7	3	5	1	3	6	2	
3	4	7	2	5	1	2	5	7	3	5	1	4	L
4	6	2	4	7	3	3	6	1	4	6	2	5	
5	7	3	5	1	4	4	7	2	5	7	3	6	
6	1	4	6	2	5	5	1	3	6	1	4	7	
7	2	5	7	3	6	7	3	5	1	3	6	2	L
8	4	7	2	5	1	1	4	6	2	4	7	3	
9	5	1	3	6	2	2	5	7	3	5	1	4	
10	6	2	4	7	3	3	6	1	4	6	2	5	
11	7	3	5	1	4	5	1	3	6	1	4	7	L
12	2	5	7	3	6	6	2	4	7	2	5	1	
13	3	6	1	4	7	7	3	5	1	3	6	2	
14	4	7	2	5	1	1	4	6	2	4	7	3	
15	5	1	3	6	2	3	6	1	4	6	2	5	L
16	7	3	5	1	4	4	7	2	5	7	3	6	
17	1	4	6	2	5	5	1	3	6	1	4	7	
18	2	5	7	3	6	6	2	4	7	2	5	1	
19	3	6	1	4	7	1	4	6	2	4	7	3	L
20	5	1	3	6	2	2	5	7	3	5	1	4	
21	6	2	4	7	3	3	6	1	4	6	2	5	
22	7	3	5	1	4	4	7	2	5	7	3	6	
23	1	4	6	2	5	6	2	4	7	2	5	1	L
24	3	6	1	4	7	7	3	5	1	3	6	2	
25	4	7	2	5	1	1	4	6	2	4	7	3	
26	5	1	3	6	2	2	5	7	3	5	1	4	
27	6	2	4	7	3	4	7	2	5	7	3	6	L
28	1	4	6	2	5	5	1	3	6	1	4	7	

p. 196 If we want to know the same for the *Æra Augusti* (*i.e.* to find the week-day on which a year of this era commences), we take its complete years and add thereto $\frac{1}{4}$ of them. To this sum we add 6 and divide the whole by 7. Thereby we get the *Signum* (of the week-day) of the 1st of Thôt.

To this *Signum* we add 2 for each complete month that has elapsed before the date you want to find, and the sum we divide by 7. Thereby we find the *Signum* of the month we seek.

The leap-years are in this era ascertained in this way, that we add 1 to the number of the complete years and divide the sum by 4. If there is a remainder, the current year is not a leap-year; if there is no remainder, it is a leap-year.

If we want to know the same for the *Æra Antonini*, we increase its complete years by $\frac{1}{4}$ of them, and to the sum we add $4\frac{3}{4}$. Then we make the same calculation (as for the *Æra Augusti*).

The leap-years in this era are ascertained in this way, that we add 3 to its complete years and divide the sum by 4. If there is no remainder, the year is a leap-year; if there is a remainder, it is a common-year.

As regards the *Æra Diocletiani*, we add to its years $\frac{1}{4}$ of them, and to the sum we add $4\frac{1}{4}$. With the remainder, and in order to find the beginnings of the single months, we reckon in the same way as we have done for the *Æra Alexandri* according to the Greek system.

The leap-year in the *Æra Diocletiani* is ascertained in this way, that we add 2 to its complete years and divide the sum by 4. If there is no remainder, the year is a leap-year; if there is a remainder, it is a common-year.

If we want to learn the beginnings of the years and months of the *Æra Fugæ* by chronological computation, we take its complete years and write them down in three places. The first we multiply by 354 days, the second by 22 minutes, and the third by 1 second. To the number of minutes we add 34 minutes. Then we convert the three sums in the three places into wholes. If the minutes are more than 15, we add them as one whole; if they are less, we drop them. The sum we get represents the time which has elapsed between the beginning of the *Æra Fugæ* and the beginning of the year in question, consisting of days. We add 5 to them and divide the sum by 7. Now, the remainder of less than 7 is the *Signum* of Muḥarram.

If we want to learn the *Signum* of another month, we take for the months, which have elapsed before the month in question, alternately for one month 2 days, for the other 1 day, and the sum we add to the *Signum* of Muḥarram. The whole we divide by 7, and the remainder is the *Signum* of the month in question, as determined by chronological computation which is based upon the mean motion of the moon.

The computation according to the appearance of new-moon is a subject the exposition of which would be both of great length and difficulty and would require difficult calculations and numerous tables. It is sufficient to know what on this subject is said in the *Canon* of Muḥammad b. Jâbir Albattânî, and in that one of Ḥabash the mathematician. In case of necessity the student may consult them.

The same principle we have explained has been adopted by the sect who claim to have esoteric doctrines and represent themselves as the party of the Family (of 'Alî). So they have produced a calculation which they maintain to be one of the mysteries of prophecy. It is this: p.197.

If you want to know the beginning of Ramaḍân, take the complete years of the Hijra, multiply them by 4 and add to the sum $\frac{1}{5}$ and $\frac{1}{6}$ (*i.e.* $\frac{11}{30}$) of the number of years. If in both these portions (in $\frac{1}{5}$ and $\frac{1}{6}$ of the year of the Hijra) you get a fraction, add it as one complete day to the other days, if one of them or both together are more than half the denominator of either of the two fractions ($\frac{1}{5}$ and $\frac{1}{6}$). Then add to the sum 4 and divide the whole by 7. The remainder beyond 7, which you get, is the *Signum Ramaḍâni.*

This calculation is based upon what we have mentioned. For if you divide the days of each Lunar year, *i.e.* 354 days, by 7, you get as remainder 4. If therefore the years of the Hijra are multiplied by 4, it is the same as if the days of each year and the remainders (*i.e.* the 4 days which remain, if you divide 354 by 7) were converted into weeks.

Further, to take $\frac{1}{5}$ and $\frac{1}{6}$ of the years of the Hijra is the same as if you would take $\frac{1}{5}$ day and $\frac{1}{6}$ day for each single year. So this method of taking $\frac{1}{5}$ and $\frac{1}{6}$ of the years comes to the same thing as if you multiplied each year by $\frac{1}{5}$ and $\frac{1}{6}$ day and divided the products by the denominators of the two fractions (*i.e.* $5 \times 6 = 30$).

If, therefore, the whole is divided by 7 and the remainder is counted from Friday, which is the beginning of the *Æra Fugæ*, we come to the *Signum Muḥarrami.* And if we add thereto 6 and count the sum from Sunday, the matter comes to the same result.

Further, the reason why those people add 4 is this, that you get—by alternately taking 2 days for one month and 1 day for the next one—till the beginning of Ramaḍân the sum of 5 days. If you add these to the *Signum Muḥarrami,* you get the *Signum Ramaḍâni.* Having already added 6 for Muḥarram and combining with it the 5 days, which are necessary for the time till Ramaḍân, you get a total of 11 days. Subtract 7 and you get as remainder 4; this is what remains of the sum of the two additions (*i.e.* the addition of 6 days for the purpose of postponing the epoch of the era from Friday to Sunday, and the addition of 5 days for the purpose of converting the *Signum Muḥarrami* into the *Signum Ramaḍâni*).

The two computations, the one which is counted from Friday, the other—mentioned shortly before—which is counted from Thursday, agree with each other, for this reason, that in the former case the 34 minutes are summed to one day, whilst in the latter case none of the fractions are raised to a whole.

This and similar modes of computation have been adopted by the followers of this new theory in this sect, who are known in Khwârizm as the *Baghdâdiyya* sect, so called from their founder, a Shaikh who lives in Baghdâd. I have found that one of their leaders has taken the *Jadwal-Mujarrad* (*i.e.* the pure table, divested of any accessory), which was constructed by Ḥabash in his *Canon* for the purpose of correcting the method of dating employed in astronomical calculations. Now this sectarian has added to each number of the table, *i.e.* the *Signum Muḥarrami*, 5, for the reason just mentioned; further he has altered the shape of the table, giving it—instead of the perpendicular form of a table—the form of a screw-like train, similar to a wound-up serpent, as some people in Ṭabaristân have given it the form of a circle, in which the beginning and the end of the numbers meet together.

He has also followed the example of the people (of the same sect) in composing a book in which he abuses those who want to find the new-
p.198. moon by observation; he attacks them and blames them, saying that for both Christians and Jews it is rendered superfluous by their tables to observe new-moon for the determination of their fast-days and the beginnings of their months, whilst Muslims trouble themselves with a subject of so dubious a character (as the observation of new-moon). But if he had read farther (in the book of Ḥabash) beyond that place where the *Jadwal-Mujarrad* occurs, as far as the chapter of the astronomical methods for the observation of new-moon, if he had acquainted himself with their nature and with the real character of the practices of both Jews and Christians, he would have learned that that which they have adopted is obscurity itself.

Perhaps he who is acquainted with our preceding explanations will find out the truth of this. For astronomers agree that the assumed measures in the most difficult parts of the practice of the observation of new-moon are certain distances which cannot be ascertained except by experiment. Besides, the observations themselves are subject to certain circumstances of a geometrical nature, in consequence of which that which is observed by the eye differs in greatness and smallness. A man who considers astronomical affairs with an unbiassed mind could not decide against the necessity of the observation of new-moon nor against its possibility, particularly when new-moon occurs near the end of that distance which has been assumed.

Here follows the screw-figure (here given in the form of a common table) which has been transformed out of the *Jadwal-Mujarrad*.

TABLE showing on what WEEK-DAYS the SINGLE YEARS of the CYCLE of 210 LUNAR YEARS commence.

A. The Single Years of the Cycle of 210 Years.	B. The Week-Days on which the Single Years commence.	A	B	A	q	A	B	A	B	A	B	A	B
1	IV	31	II	61	VII	91	V	121	III	151	I	181	VI
2	I	32	VI	62	IV	92	II	122	VII	152	V	182	III
3	VI	33	IV	63	II	93	VII	123	V	153	III	183	I
4	III	34	I	64	VI	94	IV	124	II	154	VII	184	V
5	VII	35	V	65	III	95	I	125	VI	155	IV	185	II
6	V	36	III	66	I	96	VI	126	IV	156	II	186	VII
7	II	37	VII	67	V	97	III	127	I	157	VI	187	IV
8	VI	38	IV	68	II	98	VII	128	V	158	III	188	I
9	IV	39	II	69	VII	99	V	129	III	159	I	189	VI
10	I	40	VI	70	IV	100	II	130	VII	160	V	190	III
11	V	41	III	71	I	101	VI	131	IV	161	II	191	VII
12	III	42	I	72	VI	102	IV	132	II	162	VII	192	V
13	VII	43	V	73	III	103	I	133	VI	163	IV	193	II
14	V	44	III	74	I	104	VI	134	IV	164	II	194	VII
15	II	45	VII	75	V	105	III	135	I	165	VI	195	IV
16	VI	46	IV	76	II	106	VII	136	V	166	III	196	I
17	IV	47	II	77	VII	107	V	137	III	167	I	197	VI
18	I	48	VI	78	IV	108	II	138	VII	168	V	198	III
19	V	49	III	79	I	109	VI	139	IV	169	II	199	VII
20	II	50	I	80	VI	110	IV	140	II	170	VII	200	V
21	VII	51	V	81	III	111	I	141	VI	171	IV	201	II
22	V	52	III	82	I	112	VI	142	IV	172	II	202	VII
23	II	53	VII	83	V	113	III	143	I	173	VI	203	IV
24	VI	54	IV	84	II	114	VII	144	V	174	III	204	I
25	IV	55	II	85	VII	115	V	145	III	175	I	205	VI
26	I	56	VI	86	IV	116	II	146	VII	176	V	206	III
27	V	57	III	87	I	117	VI	147	IV	177	II	207	VII
28	III	58	I	88	VI	118	IV	148	II	178	VII	208	V
29	VII	59	V	89	III	119	I	149	VI	179	IV	209	II
30	IV	60	II	90	VII	120	V	150	III	180	I	210	VI

In the original Arabic this table is arranged in the form of a screw. In the longitudinal fields of the screw there is a steady progression of both numbers, the numbers of the years rising by 21, the numbers of the week-days rising by 1. For instance, in the field of the first years the years rise in this way :—

1. 22. 43. 64. 85. 106. 127. 148. 169. 190. (1. 22. 43. etc.) ;

and the week-days rise in this way :—

IV. V. V. VI. VII. VII. I. II. II. III. (IV. V. etc.).

Considering that in the *Jadwal Mujarrad* produced by Ḥabash the sage in his canon known as the *Canon Probatus* (lacuna). This man whom we have mentioned, transferred thence the screw-figure (into his work), adding five in places where Ḥabash had added the fractions as a whole day to the other days, which he ought not to have done. His method is the same for the *Tabula Mediorum*, so that by this he was preserved from error.

Let him who wants to ascertain the truth of our words compare this screw-figure—for it is the *Jadwal-Mujarrad* itself, only increased by 5 so as to represent the *Signum Ramaḍâni*—with the *Corrected Table* which we have computed for the *Signum Muḥarrami*. The fractions following after the whole days we have also noticed, wishing that they should come under occular inspection, and so afford a help also for other things.

If you use this corrected table, subtract always 210 from the years of the Hijra, including the current year, if their number be more than 210. With the remainder compare the column of the numbers and take the days and minutes which you find opposite in the corresponding square. Add to the minutes 5 days and 34 minutes, and convert them into whole days. Eliminate the 7, if the number is more than 7, and you get the *Signum* for the 1st of Muḥarram. If you add thereto 5, you get the *Signum* of Ramaḍân.

The result of this computation compare with the screw-figure. For in some dates there is a difference on account of the conversion of the minutes under 60 into days.

It will be clear to the reader why the table has been constructed for 210 years, and not for a less or larger number of years, if he studies the subject thoroughly.

God is all-wise. He is our sufficiency and our help!

The Corrected Table.

pp.199, 200.

Column of numbers.	Days.	Minutes.	Column of numbers.	Days.	Minutes.	Column of numbers.	Days.	Minutes.	Column of numbers.	Days.	Minutes.	Column of numbers.	Days.	Minutes.	Column of numbers.	Days.	Minutes.	Column of numbers.	Days.	Minutes.
1	4	22	31	2	22	61	0	22	91	5	22	121	3	22	151	1	22	181	6	22
2	1	44	32	6	44	62	4	44	92	2	44	122	0	44	152	5	44	182	3	44
3	6	6	33	4	6	63	2	6	93	0	6	123	5	6	153	3	6	183	1	6
4	3	28	34	1	28	64	6	28	94	4	28	124	2	28	154	0	28	184	5	28
5	0	50	35	5	50	65	3	50	95	1	50	125	6	50	155	4	50	185	2	50
6	5	12	36	3	12	66	1	12	96	6	12	126	4	12	156	2	12	186	0	12
7	2	34	37	0	34	67	5	34	97	3	34	127	1	34	157	6	34	187	4	34
8	6	56	38	4	56	68	2	56	98	0	56	128	5	56	158	3	56	188	1	56
9	4	18	39	2	18	69	0	18	99	5	18	129	3	18	159	1	18	189	6	18
10	1	40	40	6	40	70	4	40	100	2	40	130	0	40	160	5	40	190	3	40
11	6	2	41	4	2	71	2	2	101	0	2	131	5	2	161	3	2	191	1	2
12	3	24	42	1	24	72	6	24	102	4	24	132	2	24	162	0	24	192	5	24
13	0	46	43	5	46	73	3	46	103	1	46	133	6	46	163	4	46	193	2	46
14	5	8	44	3	8	74	1	8	104	6	8	134	4	8	164	2	8	194	0	8
15	2	30	45	0	30	75	5	30	105	3	30	135	1	30	165	6	30	195	4	30
16	6	52	46	4	52	76	2	52	106	0	52	136	5	52	166	3	52	196	1	52
17	4	14	47	2	14	77	0	14	107	5	14	137	3	14	167	1	14	197	6	14
18	1	36	48	6	36	78	4	36	108	2	36	138	0	36	168	5	36	198	3	36
19	5	58	49	3	58	79	1	58	109	6	58	139	4	58	169	2	58	199	0	58
20	3	20	50	1	20	80	6	20	110	4	20	140	2	20	170	0	20	200	5	20
21	0	42	51	5	42	81	3	42	111	1	42	141	6	42	171	4	42	201	2	42
22	5	4	52	3	4	82	1	4	112	6	4	142	4	4	172	2	4	202	0	4
23	2	26	53	0	26	83	5	26	113	3	26	143	1	26	173	6	26	203	4	26
24	6	48	54	4	48	84	2	48	114	0	48	144	5	48	174	3	48	204	1	48
25	4	10	55	2	10	85	0	10	115	5	10	145	3	10	175	1	10	205	6	10
26	1	32	56	6	32	86	4	32	116	2	32	146	0	32	176	5	32	206	3	32
27	5	54	57	3	54	87	1	54	117	6	54	147	4	54	177	2	54	207	0	54
28	3	16	58	1	16	88	6	16	118	4	16	148	2	16	178	0	16	208	5	16
29	0	38	59	5	38	89	3	38	119	1	38	149	6	38	179	4	38	209	2	38
30	5	0	60	3	0	90	1	0	120	6	0	150	4	0	180	2	0	210	0	0

p.201. Further, I have found with 'Aḥmad b. Muḥammad b. Shihâb, who was counted among the leaders of the Ḥarûriyya-sect and one of the greatest of their missionaries, the following table, which, he says, is to be used in this way: Take the complete years of the *Æra Fugæ*, add thereto 4 and divide the sum by 8. The remainder under 7 you compare with the column of numbers, and opposite you find the week-day of the beginning of whatever month you like.

TABLE OF THE MONTHS.

Column of Numbers.	Almuharram.	Safar.	Rabî' I.	Rabî' II.	Jumâdâ I.	Jumâdâ II.	Rajab.	Sha'bân.	Ramadân.	Shawwâl.	Dhû-alka'da.	Dhû-alhijja.
1	3	5	6	1	2	4	5	7	1	3	4	6
2	7	2	3	5	6	1	2	4	5	7	1	3
3	5	7	1	3	4	6	7	2	3	5	6	1
4	2	4	5	7	1	3	4	6	7	2	3	5
5	6	1	2	4	5	7	1	3	4	6	7	2
6	4	6	7	2	3	5	6	1	2	4	5	7
7	1	3	4	6	7	2	3	5	6	1	2	4
8	6	1	2	4	5	7	1	3	4	6	7	2

This table, too, is certainly derived from the *Jadwal-Mujarrad*. If the student would consider the *Octaeteris* on which this table is based, he would find that the new-year-days of the years of this cycle return to the same day of the week, that they, however, fall short (of a complete revolution and return to the same day) by a fraction of 4 minutes. Therefore this table does not differ from the corrected *Jadwal-Mujarrad*, except when the Octaeteris in the course of time recurs many times. In this case the minus-difference of 4 minutes causes a very disagreeable confusion.

This same trickster of a missionary relates that this table was the work of Ja'far b. Muḥammad Alṣâdiḳ at the time when he—so that man says—explained the difference of opinion and the uncertainty that exists among Muslims regarding the month Ramaḍân. According to him Ja'far said: "I swear by him who in truth has sent Muḥammad
p.202. as a prophet, that *He* (the prophet) did not leave his people, before he had disclosed before our eyes both the past and the future till the end of the world. And the least of this is the knowledge of fasting for every year and every day." Further, he is reported to have said:

"Sha'bân has never been more—and Ramaḍân has never been less—than 30 days."

This malefactor has invented tales about that wise Lord, the noblest of the nobles, the wisest of the Imâms—God's blessing be upon their names!—by making him responsible for something that is inconsistent with the religion of his ancestor (*i.e.* 'Alî). It has been proved that the contrary of these assertions is the truth. That pious Imâm was far from sullying himself by traditions like those, and never dreamt that he would be defiled by their insolence in referring them to his authority —God's blessing be upon him!—

There are two methods for the finding of the *Signum Muḥarrami*, mentioned by 'Abû-Ja'far Alkhâzin in his *Great Introduction to Astronomy:*

I. Take for each complete 30 years of the *Æra Hijræ* which have elapsed, 5 days. As regards the remainder of less than 30 years, take for each 10 years $1\frac{2}{3}$ days, *i.e.* 1 day 16 hours. For each 5 years of the further remainder take 20 hours, and for each complete single year take 4 days $8\frac{4}{5}$ hours. To the sum you get in this way add 5 or subtract 2. The remainder divide by 7, and the remainder you get is the *Signum Muḥarrami.*

This method is correct, and proceeds in the same way as the before-mentioned methods. For the days and fractions of days that are taken for certain numbers of years are the remainders which you get, if you convert those years into days and divide them by 7, as the *Corrected Table* shows.

To the sum we add 5, in order to make the days begin with Sunday, as we have mentioned before. It is the same whether you add 5 or subtract 2, *i.e.* 7 minus 5, as long as you use the hebdomadal cycle, which must be adhered to.

If you want the *Signum* of any other month (*but* Muḥarram), add to the *Signum Muḥarrami* 2 days for each month whose number in the order of months is an *odd* one, and 1 day for each month whose number is an *even* one. Divide the sum by 7, and the remainder is the *Signum* of the month in question.

II. The second method is this: Take half of the number of years, if it is an *even* number; if it is an *odd* one, subtract 1 therefrom, and keep in mind for it 4d. 22min. (*i.e.* make a mental note of it). Then take the half of this remainder of years and put it into two different places. Multiply this number in one place by 3, and divide it by 4. So you get days. In the other place multiply it by 8, and add the sum to the number of days, with the addition of 5. From the sum subtract a number of *day-minutes* which is equal to half the number of the years. With the remainder combine that which you have kept in mind (4d. 22.), if the years are without a fraction. But if there is a fraction of more than 30 minutes, count it as a whole; if it is less, omit it. Divide the sum by 7, and the remainder is the *Signum Muḥarrami.*

This method, too, is correct, and based on the circumstances we have mentioned.

That which you keep in mind (4d. 22′) is the intercalatory portion of the year which you subtract from the total sum of years, the remainder
p.203. which you get after having divided 354d. 22′ by 7.

To multiply the half of the remaining years (*i.e.* after the subtraction of 1, in case the number of years be an *odd* one) by 8, is the same as to divide the whole by 4. These 4 days are the whole days which you get by dividing the Lunar year by 7 (354d. 22′ : 7, remainder 4).

Finally you take $\frac{1}{5}$ and $\frac{1}{6}$ day, *i.e.* $\frac{11}{30}$ or $\frac{22}{60}$ day for each year. However, the half of $\frac{3}{4}$ (*i.e.* $\frac{3}{8}$) of any number is more than $\frac{1}{5}+\frac{1}{6}$ (*i.e.* $\frac{11}{30}$) of the whole number by a measure (a quantity) which amounts for the *whole* number to the same as a corresponding number of sixtieth parts (or minutes) for half the number (*i.e.* for the whole number x this plus-difference is $\frac{1}{120}x$, which is the same as [illegible] of $\frac{1}{2}x$). If you, therefore, multiply half of the number of years by 3 and divide the product by 4, you get $\frac{3}{4}$ of the number, which is more than $\frac{1}{5}+\frac{1}{6}$ ($\frac{11}{30}$) of the whole number of years by a number of minutes which is equal to half the number of years. If they, now, are counted in 60th parts, *i.e.* in minutes, and you subtract them from the sum, you get $\frac{1}{5}$ and $\frac{1}{6}$ ($\frac{11}{30}$) of the years. The analogy of the other parts of this calculation with what we have before mentioned is evident.

If we want to find the *Signum* of the new-year's day of a year of the *Æra Yazdagirdi*, we take the number of complete years and add thereto always 3. The sum we divide by 7, and the remainder of this division is *Signum* of Farwardîn-Mâh.

If we want to know the *Signum* of another month, we take for each of the complete months, that have passed, 2 days, except Âbân-Mâh, for which we take nothing. The sum we add to the *Signum* of Farwardîn-Mâh, and subtract 7, if the number be more than seven. The remainder is the *Signum* of the month in question.

For the *Æra Magorum*, the epoch of which is the death of Yazdagird, we add always 5 to the number of complete years, and the remainder we compute in the same way as we have done for the preceding era, in case we use for this era the Persian months. But if we use the months of the Ṣughdians or Khwârizmians, we always add 3 to the number of complete years, and divide the sum by 7. As remainder we get the *Signum* of *Nausard* or *Nâusârjî*. For each following month we add 2 days to the *Signum Nausardi*. In this way we find the *Signum* of the month in question.

If we want to know the intercalation, as practised by the Persians before the decline of their empire, we take the Persian years from the end of the reign of Yazdagird, which event is the epoch of the *Æra Magorum*, and add thereto 70, for the reason which we have mentioned

in the first part of this book. The sum we divide by 120. The quotient is the number of intercalations that ought to have been carried out since the time when they commenced to neglect intercalation. Now we take for the total sum of the years of the era a number of months corresponding to the number of intercalations. If, then, these months make up complete years, without giving a remainder, the year is a leap-year approximately, for there is confusion in their chronology. But p.204.
if there is a remainder of months, the year is a common year. Thereupon we add the leap-months we have got to the beginning of the year in question, and we find Naurôz on that day to which this calculation brings us. So Naurôz comes again to be there, where it used to be in the time of the Kisrâs, when it used to coincide with the summer-solstice as calculated by their astronomical tables.

For the *Æra of Almu'taḍid* we find the *Signum* of Farwardîn-Mâh by adding to the complete years $\frac{1}{4}$ of them, and to the sum $4\frac{1}{4}$. The whole we divide by 7, and the remainder is the *Signum* of Farwardîn-Mâh. Knowing the *Signum* of New-year's day of a year, and wishing to find the *Signum* of some other month, we add for each month that has passed 2 days, except Âbân-Mâh, for which we take 1 in a leap-year and nothing in a common year. The sum we divide by 7, and the remainder is the *Signum* of the month in question.

The leap-years of this era you find by dividing its complete years by 4. If there is no remainder the year is a leap-year; if there is a remainder it is a common-year.

Now we think that this long exposition will be sufficient. Much praise be unto God, as is due to Him!

CHAPTER VIII.

ON THE ERAS OF THE PSEUDO-PROPHETS AND THEIR COMMUNITIES WHO WERE DELUDED BY THEM, THE CURSE OF THE LORD BE UPON THEM!

We shall explain the method of dating the eras by the pseudo-prophets. For in the intervals between the prophets and kings whom we have mentioned, pseudo-prophets came forward, the number and history of whom it would be impossible to detail in this book. Some of them perished without having gained adherents, not leaving anything behind them but a place in history. Whilst others were followed by a community who kept up their institutes and used their method of dating. It is necessary, therefore, to mention the eras of the most notorious among them, for this affords a help, also, for the knowledge of their history.

Budhasaf.—The first mentioned is *Bûdhâsaf*, who came forward in India after the 1st year of Ṭahmûrath. He introduced the Persian writing and called people to the religion of the *Ṣâbians*. Whereupon many people followed him. The Pêshdâdhian kings and some of the Kayânians who resided in Balkh held in great veneration the sun and moon, the planets and the primal elements, and worshipped them as holy beings, until the time when Zarâdusht appeared thirty years after the accession of Bishtâsf.

The remnants of those Ṣâbians are living in Ḥarrân, their name (*i.e. Alḥarrâniyya*) being derived from their place. Others derive it from Hârân b. Teraḥ, the brother of Abraham, saying that he among their chiefs was the most deeply imbued with their religion and its most
p.205. tenacious adherent. *Ibn Sankilâ* (Syncellus), the Christian, relates in his book which he, intending to refute their creed, stuffed with lies and

futile stories, that Abraham left their community simply because leprosy appeared on his foreskin, and that everybody who suffered from this disease was considered impure, and excluded from all society. Therefore he cut off his foreskin, *i.e.* he circumcised himself. In this state he entered one of their idol-temples, when he heard a voice speaking to him: "O Abraham, you went away from us with one sin, and you return to us with two sins. Go away, and do not again come to us." Thereupon Abraham, seized by wrath, broke the idols in pieces, and left their community. But, after having done it, he repented and wished to sacrifice his son to the planet Saturn, it being their custom to sacrifice their children, as that author maintains. Saturn, however, on seeing him truly repentant, let him go free with the sacrifice of a ram.

Also 'Abd-almasîḥ b. 'Isḥâḳ Alkindî, the Christian, in his reply to the book of 'Abdallâh b. 'Ismâ'îl Alhâshimî, relates of them, that they are notorious for their sacrificing human beings, but that at present they are not allowed to do it in public.

All, however, we know of them is that they profess monotheism and describe God as exempt from anything that is bad, using in their description the *Via Negationis*, not the *Via Positionis*. *E.g.* they say "he is indeterminable, he is invisible, he does *not* wrong, he is *not* unjust." They call him by the *Nomina Pulcherrima*, but only metaphorically, since a real description of him is excluded according to them. The rule of the universe they attribute to the celestial globe and its bodies, which they consider as living, speaking, hearing, and seeing beings. And the fires they hold in great consideration.

One of their monuments is the cupola over the *Miḥrâb* beside the *Maḳṣûra* in the great Mosque of Damascus. It was their place of worship as long as Greeks and Romans professed their religion; afterwards it passed into the hands of the Jews, who made it their synagogue. Then it was occupied by the Christians, who used it as their church till the time of the rising of Islam, when the Muslims made it their Mosque.

They had temples and images, called by the names of the sun, the forms of which are known, and the like of which are mentioned by 'Abu-Ma'shar Albalkhî in his book on the houses of worship. For instance, the temple of Ba'al-bek was sacred to the idol of the Sun. The city of Ḥarrân was attributed to the moon, it being built in the shape of the moon like a *Ṭailasân*. Close to Ḥarrân there are another place called *Selemsin*, its ancient name being *Ṣanam-sin*, *i.e. imago Lunæ*, and another village called *Tera'-'ûz*, *i.e.* Porta Veneris. People say, too, that the Ka'ba and its images originally belonged to them, and that the worshippers of those images belonged to their community, and that *Allât* was called Zuḥal and *Al'uzzâ*, Alzuhara.

They have many prophets, most of whom were Greek philosophers, *e.g.* Hermes the Egyptian, Agathodæmon, *Wâlis*, Pythagoras, Bâbâ, and Sawâr the grandfather of Plato on the mother's side, and others. Some of them did not allow themselves to eat fish, fearing it might be a *Silurus Electricus*, nor chickens because they are always feverish, nor garlic because it produces headache and burns the blood or the *sperma genitale* on which the existence of the world depends, nor peas because they stupify and impair the intellect and originally grew in the skull of man.

p.206. They have three prayers in writing, one for the time of sunrise with eight inclinations, the second immediately before the sun leaves the centre of heaven (the meridian) with five inclinations, the third at sunset with five inclinations. Each of the *inclinations* at their prayer consists of three *prostrations*. Besides, they have voluntary prayers, one in the second hour of the day, another in the ninth hour of the day, and a third one in the third hour of the night. Their prayer is preceded by purification and washing. They also wash themselves after a pollution. They do not circumcise themselves, not being ordered to do so, as they maintain.

Most of their regulations about women and their penal law are similar to those of the Muslims, whilst others, relating to pollution caused by touching dead bodies, etc., are similar to those of the Thora.

They offer offerings to the stars, their images and temples, and practise sacrifices carried out by their priests and seducers. By this means they elicit the knowledge of the future of the man who offers the offering, and the answer to his inquiries.

Idrîs, who is mentioned in the Thora as Henokh, they call Hermes, whilst according to others Hermes is identical with Bûdhâsaf.

Again, others maintain that the Ḥarrânians are not the real Ṣabians, but those who are called in the books *Heathens* and *Idolaters*. For the Ṣabians are the remnant of the Jewish tribes who remained in Babylonia, when the other tribes left it for Jerusalem in the days of Cyrus and Artaxerxes. Those remaining tribes felt themselves attracted to the rites of the Magians, and so they *inclined* (were *inclined*, *i.e.* Ṣâbî) towards the religion of Nebukadnezzar, and adopted a system mixed up of Magism and Judaism like that of the Samaritans in Syria.

The greatest number of them are settled at Wâsiṭ, in Sawâd-al'irâḳ, in the districts of Ja'far, Aljâmida, and the two Nahr-alṣila. They pretend to be the descendants of Enos the son of Seth. They differ from the Ḥarrânians, blaming their doctrines and not agreeing with them except in few matters. In praying, even, they turn towards the north pole, whilst the Ḥarrânians turn towards the south pole.

Some of those to whom God has given a divine code (Jews and Christians) say that Methuselah had another son besides Lamech, who called himself *Ṣâbi'*, and that the Ṣabians derive their name from him.

Before the first establishment of their rites and the appearance of Bûdhâsaf people were Σαμαναῖοι, inhabiting the eastern part of the world and worshipping idols. The remnants of them are at present in India, China, and among the *Toghazghar*; the people of Khurâsân call them *Shamanân*. Their monuments, the *Bahâras* of their idols, their *Farkhâras* are still to be seen on the frontier countries between Khurâsân and India. They believe in the eternity of time and the migration of souls; they think that the globe of the universe is flying in an infinite *vacuum*, that therefore it has a rotatory motion, since anything that is round, when thrown off its place, goes downward in a circular motion, as they say. But others of them believe that the world has been created (within time), and maintain that its duration is one million of years, which they divide into four periods, the first of four hundred thousand years, the *Aurea Ætas*.

(*Great lacuna*. The end of the chapter on Bûdhâsaf, the whole chapter on Zaradûsht, and the beginning of the chapter on Bardaiṣân are missing.)

So he gets the sum of 3,457. We think they will dispute with us on the astronomical interpretation we propose, for we, as well as themselves, p.207.
are familiar with the science of the subject. Therefore any arguing on the subject and any interpretation are altogether devoid of sense.

What we have just mentioned regarding the division is a proof in favour of the Egyptians in the matter of the *Termini*. For according to them the duration of the *Terminus* of Venus in Pisces is 400 years, whilst Ptolemy reckons it as 266 years. We have already said before that the time between Alexander and Ardashîr is longer than 400 years, and have endeavoured to settle this question of chronology.

We return now to our subject, and go on to state that the Persians adhered to the Magian religion of Zarâdusht, that they had no schism or dissension in it till the time came when Jesus rose, and his pupils spread through all the world preaching the Gospel. When they thus spread through the countries, one of them came to Persia, and both Bardaiṣân and Marcion were among those who followed his call and heard the word of Jesus. Part they took from him, part from what they had heard from Zarâdusht. So each of them derived from both systems a separate doctrine, containing the dogma of the eternal existence of the two *Principia*. Each of them produced a gospel, the origin of which he traced back to the Messiah, and declared everything else to be a lie. Ibn-Daiṣân maintained that the *Light* of God was residing in his own heart.

The difference, however, did not go so far as to separate them and their followers from the bulk of the Christians, nor were their gospels in all matters different from that of the Christians; in some regards they contained more, in others less. God knows best!

Mani.—After Bardaiṣân and Marcion, *Mâni* the pupil of Fâdarûn came forward. On having acquainted himself with the doctrines of the Magians, Christians, and Dualists, he proclaimed himself to be a prophet. In the beginning of his book called *Shâbûrkân*, which he composed for Shâpûr b. Ardashîr, he says: "Wisdom and deeds have always from time to time been brought to mankind by the messengers of God. So in one age they have been brought by the messenger, called Buddha, to India, in another by Zarâdusht to Persia, in another by Jesus to the West. Thereupon this revelation has come down, this prophecy in this last age through me, Mânî, the messenger of the God of truth to Babylonia." In his gospel, which he arranged according to the twenty-two letters of the alphabet, he says that he is the Paraclete announced by Messiah, and that he is the seal of the prophets (*i.e.* the last of them).

His doctrines regarding the existence and the form of the world are contradicted by the results of scientific arguments and proofs. He preached of the empire of the worlds of light, of the Πρῶτος Ἄνθρωπος, and of the spirit of life. He taught that light and darkness are without beginning and end. He absolutely forbade his followers to slaughter animals and to hurt them, to hurt the fire, water, and plants. He established laws which are obligatory only for the *Ṣiddiḳs*, *i.e.* for the saints and ascetics among the Manichæans, viz. to prefer poverty to riches, to suppress cupidity and lust, to abandon the world, to be
p.208. abstinent in it, continually to fast, and to give alms as much as possible. He forbade them to acquire any property except food for one day and dress for one year; he further forbade sexual intercourse, and ordered them continually to wander about in the world, preaching his doctrines and guiding people into the right path.

Other laws he imposed upon the *Sammâ'* (laymen), *i.e.* their followers and adherents who have to do with worldly affairs, viz. to give as alms the tithe of their property, to fast during the seventh part of life-time, to live in monogamy, to befriend the *Ṣiddiḳs* (saints), and to remove everything that troubles and pains them.

Some people maintain that he allowed pederasty, if a man felt inclined, and as proof of this they relate that every Manichæan used to be accompanied by a young, beardless and hairless servant. I, however, have not found in what I have read of his books a word indicating anything of this kind. Nay, even his life proves the contrary of this assertion.

Mânî was born in a village called Mardînû on the upper canal of Kûthâ, according to his own statement in his book Shâbûrḳan, in the chapter on the coming of the prophet, in the year 527 of the era of the Babylonian astronomers, *i.e.* the *Æra Alexandri*, in the 4th year of the king Adharbân. He received the first divine revelation in his 13th year, *Anno Astronomorum Babyloniæ* 539, in the 2nd year of Ardashîr, the King of Kings. This part of chronology we have already tried to correct in the chapter preceding that of the duration of the rule of the Ashkânians and the *Mulûk alṭawâ'if*.

According to Yaḥyâ b. Alnu'mân, the Christian, in his book on the Magians, Mânî was called by the Christians *Corbicius the son of Patecius.*

When he came forward, many people believed in him and followed him. He composed many books, his gospel, the Shâbûrkân, *Kanz-al'iḥyâ (Thesaurus Revivicationis)*, the Book of the Giants, the Book of Books, and many treatises. He maintained that he had explained *in extenso* what had only been hinted at by the Messiah.

Manichæism increased by degrees under Ardashîr, his son Shâpûr and Hurmuz b. Shâpûr, until the time when Bahrâm b. Hurmuz ascended the throne. He gave orders to search for Mânî, and when he had found him, he said: "This man has come forward calling people to destroy the world. It will be necessary to begin by destroying him, before anything of his plans should be realized."

It is well known that he killed Mânî, stripped off his skin, filled it with grass, and hung it up at the gate of Gundîsâpûr, which is even still known as the "Mânî-gate." Hurmuz also killed a number of the Manichæans.

Jibrâ'îl b. Nûḥ, the Christian, says in his reply to Yazdânbakht's refutation of the Christians, that one of Mânî's pupils composed a book, in which he relates the fate of Mânî, that he was put in prison on account of a relative of the king who believed that he was possessed by the devil; Mânî had promised to cure him, but when he could not effect it, he was chained hand and foot, and died in prison. His head was exposed before the entrance of the royal tent, and his body was thrown into the street, that he should be a warning example to others. p.209.

Of his adherents, some remnants that are considered as Manichæan are still extant: they are scattered throughout the world and do not live together in any particular place of Muhammadan countries, except the community in Samarkand, known by the name of *Ṣâbians.* As regards non-Muhammadan countries, we have to state that most of the eastern Turks, of the people of China and Thibet and some of the Hindus, adhere still to his law and doctrine.

Regarding their prophet Mânî they hold two different opinions, one party maintaining that he never worked a mircle, and relating that he only informed people of the signs and wonders indicative of the coming of the Messiah and his companions, whilst the other party maintains that he in fact worked signs and miracles, and that the king Shâpûr came to believe in him when he had ascended with him towards heaven, and they had been standing in the air between heaven and earth. Mânî, thereby, made him witness a miracle. Besides, they relate that he sometimes used to rise to heaven from among his companions, to stay there for some days, and then to redescend to them.

I have heard the Ispahbadh Marzubân ben Rustam say that Shâpûr banished him out of his empire, faithful to the law of Zarâdusht which demands the expulsion of pseudo-prophets from the country. He imposed upon him the obligation never to return. So Mânî went off to India, China, and Thibet, and preached there his gospel. Afterwards h returned, was seized by Bahrâm and killed for having broken the stipulation, as he had thereby forfeited his life.

Mazhdak.—Thereupon came forward a man called *Mazhdak ben Hamadâdân*, a native of Nasâ. He was *Maubadhân-Maubadh*, *i.e.* chief-justice during the reign of Kobâdh ben Fêrôz. He preached Dualism and opposed Zarâdusht in many points. He taught that both property and women belonged in common to all. So he found innumerable followers.

Kobâdh, too, believed in him. But some of the Persians maintain that his adhesion was a compulsory one, since his reign was not safe against the mass of the followers of Mazhdak. According to others, again, this Mazhdak was a cunning sort of man, who managed to concoct this system, and to come forward with it simply because he knew that Kobâdh was charmed by a woman who was the wife of his cousin; and that for this reason Kobâdh hastened to adopt it. Mazhdak ordered him to abstain from sacrificing cattle before the natural term of their life had come. Kobâdh said: "Your enterprise shall not succeed unless you make me master of the mother of Anûshirwân, that I may enjoy her." Mazhdak did as he wished, and ordered her to be handed over

(*Lacuna.* Missing, the end of Mazhdak and beginning of Musailima.)

Musailima.—"To Muḥammad the Prophet of God. Greeting unto thee! etc. God has made me partake with thee in the rule. One half of the earth belongs to us and one half to Ḳuraish But the Ḳuraish are evil-doing people." This letter he sent off with two messengers. To these the Prophet said: "What is it you are speaking?" They answered: "We are speaking just as He spoke." Thereupon the Prophet said: "If it was not the custom not to kill messengers, I should behead both of you." Then he gave them his answer: "From Muḥammad the Prophet of God to Musailima the liar. Greeting unto those who follow the right guidance! etc. The earth belongs to God, he gives it as an inheritance to whomsoever of his servants he pleases. And the end will be in favour of the pious."

The people of Yamâma let themselves be deluded by him by such tricks as introducing an egg that had been soaked in vinegar into a glass-bottle, by fitting together the wings of birds, which he had previously
p.210. cut off, by means of similar feathers; and by such-like humbug and swindle.

The Banû Ḥanîfa kept possession of Yamâma until Musailima was killed by Khâlid b. Alwalîd in the year when 'Abû-Bakr Alṣiddîḳ succeeded. Then they lamented his death in verses; one of the Banû Ḥanîfa says:

"Alas for thee, o' Abû-Thumâma!
[Thou wast] like the sun beaming forth from a cloud."

Before Musailima in the time of heathendom the Banû Ḥanîfa had got an idol of *Ḥais* (*i.e.* a mixture of dates, butter, and dried curd), which they worshipped for a long time. But once, being pressed by hunger, they devoured it. So a poet of the Banû Tamîm said:

"The Banû Ḥanîfa have eaten their Lord for hunger,
From which they were suffering already a long time, and from want."

Another said:

"The Ḥanîfas have eaten their Lord
At the time of want and hunger.
They did not guard against the punishment,
Which their Lord might inflict upon them."

Bahafirid b. Mahfurûdhin.—Thereupon in the days of 'Abû-Muslim, the founder of the 'Abbâside dynasty, came forward a man called *Bahâfirîd ben Mâhfurûdhîn* in Khwâf, one of the districts of Nîshâpûr, in a place called Sîrâwand, being a native of Zûzan. In the beginning of his career he disappeared and betook himself to China for seven years. Then he returned, and brought with him among other Chinese curiosities a green shirt which, when folded up, could be held in the grasp of a human hand; so thin and flexible was it. He went up to a temple during the night, and when he thence descended in the morning, he was observed by a peasant who was ploughing part of his field. This man he told that he had been in heaven during his absence from them, that heaven and hell had been shown unto him, that God had inspired him, had dressed him in that shirt, and had sent him down upon earth in that same hour. The peasant believed his words, and told people that he had witnessed him descending from heaven. So he found many adherents among the Magians, when he came forward as a prophet and preached his knew doctrine.

He differed from the Magians in most rites, but he believed in Zarâdusht, and claimed for his followers all the institutes of Zarâdusht. He maintained that he secretly received divine revelations, and he established seven prayers for his followers, one in praise of the *one* God, one relating to the creation of heaven and earth, one relating to the creation of the animals and to their nourishment, one relating to death, one relating to the resurrection and last judgment, one relating to those in heaven and hell and what is prepared for them, and one in praise of the people of paradise.

He composed for them a book in Persian. He ordered them to worship the substance of the sun, kneeling on one knee, and in praying

always to turn towards the sun wherever he might be, to let their hair and looks grow, to give up the *Zamzama* at dinner, not to sacrifice small cattle except they be already decrepit, not to drink wine, not to eat the flesh of animals that have died a sudden death as not having
p.211. been killed according to prescription, not to marry their mothers, daughters, sisters, nieces, not to exceed the sum of four hundred dirhams as dowry. Further, he ordered them to keep roads and bridges in good condition by means of the seventh part of their property and of the revenue of their labour.

When 'Abû-Muslim came to Nîshâpûr, the Maubadhs and Herbadhs assembled before him telling him that this man had infected Islâm as well as their own religion. So he sent 'Abdallâh b. Shu'ba to fetch him. He caught him in the mountains of Bâdaghîs and brought him before 'Abû-Muslim, who put him to death, and all his followers of whom he could get hold.

His followers, called the *Bahâfirîdiyya*, keep still the institutes of their founder and strongly oppose the *Zamzamîs* among the Magians. They maintain that the servant of their prophet had told them that the prophet had ascended into heaven on a common dark-brown horse, and that he will again come down to them in the same way as he ascended and will take vengeance on his enemies.

Almukanna'.—Thereupon came forward Hâshim b. Ḥakîm, known by the name of *Almuḳanna'*, in Marw, in a village called Kâwakímardân. He used to veil himself in green silk, because he had only one eye. He maintained that he was God, and that he had incarnated himself, since before incarnation nobody could see God.

He passed the river Oxus and went to the districts of Kash and Nasaf. He entered into correspondence with the Khâkân and solicited his help. The sect of the *Mubayyiḍa* and the Turks gathered round him, and the property and women (of his enemies) he delivered up to them, killing everybody who opposed him. He made obligatory for them all the laws and institutes which Mazhdak had established.

He scattered the armies of Almahdî, and ruled during fourteen years, but finally he was besieged and killed A.H. 169. Being surrounded on all sides he burned himself, that his body might be annihilated, and, in consequence, his followers might see therein a confirmation of his claim of being God. However, he did not succeed in annihilating his body; it was found in the oven, and his head was cut off and sent to the Khalîf Almahdî, who was then in Ḥalab.

There is still a sect in Transoxiana who practise his religion, but only secretly, whilst in public they profess Islâm. The history of Almuḳanna' I have translated from the Persian into Arabic; the subject has been exhaustively treated in my history of the Mubayyiḍa and the Ḳarmaṭians.

Alhallaj.—Thereupon came forward a Ṣûfî of Persian origin, called

Alḥusain ben Mânṣûr Alḥallâj. He was the first to preach the coming of Almahdî, maintaining that he would come from Ṭâlaḳân in Dailam. They seized upon him and led him into Baghdâd, parading him through the streets. Then he was put into prison, but he contrived to get out of it again. He was a juggler and artful sort of man, mixing himself up with every human being according to his belief and his views. Further, he preached that the Holy Ghost was dwelling in him, and he called himself God. His letters to his followers bore the following superscription: "From the *He*, the eternal, the first *He*, the beaming and shining light, the original origin, the proof of all proofs, the Lord of the Lords, who raises the clouds, the window from which the light shines, the Lord of the Mountain (Sinai), who is represented in every shape—to his slave N. N." And his followers began their letters to him in this way: "Praise unto thee, O being of all beings, the perfection of all delights, O great, O sublime being, I bear witness p.212. that thou art the eternal creator, the light-giver, who reveals himself in every time and age, and in this our time in the figure of Alḥusain b. Manṣûr! Thy slave, thy wretched and poor one, who seeks help with thee, who flies for refuge to thee, who hopes for thy mercy, O thou who knowest all mysteries!—speaks thus and thus."

He composed books on the subject of his preaching, *e.g.* the *Kitâb-Nûr-Al'aṣl*, the *Kitâb-Jamm-Al'akbar* and the *Kitâb-Jamm-Al'aṣghar.*

A.H. 301, the Khalîf Almuḳtadir-billâh laid hands upon him; he ordered the executioner to give him a thousand lashes, to cut off his hands and feet and to behead him; then they besprinkled him with nafta and burnt his body, and threw the ashes into the Tigris. During the whole execution he did not utter a syllable nor distort his face nor move his lips.

A remnant of his followers who are called after him is still extant; they preach the coming of Almahdî, and say that he will issue from Tâlaḳân. Of this same Mahdî it is said in the *Kitâb-Almalâḥim* that he will fill the earth with justice as it heretofore has been filled with injustice. Somewhere in the book it is said that he will be Muḥammad b. 'Abdallâh, elsewhere that he will be Muḥammad b. 'Alî. Nay, when Almukhtâr b. 'Abî-'Ubaid Althaḳafî called people to rally round Muḥammad b. Alḥanafiyya, he produced as a testimony an authentic tradition, and maintained that this was the predicted Mahdî.

Even in our time people expect the Mahdî to come, believing that he is alive and resides in the mountain Raḍwâ. Likewise the Banû-'Umayya expect the coming of Alsufyânî who is mentioned in the *Malâḥim*. In that book it is also mentioned that Aldajjâl, the seducer, will issue from the district of Isfahân, whilst astrologers maintain that he will issue from the island of *Bartâ'il* four hundred and sixty-six years after Yazdagird ben Shahryâr. Also in the Gospel you find mentioned the signs that will foreshow his coming. In Greek, in Christian books, he is called 'Αντίχριστος, as we learn from Mâr Theodorus, the Bishop of Mopsueste, in his commentary on the Gospel.

Historians relate that 'Umar ben Alkhaṭṭâb on entering Syria was met by the Jews of Damascus. They spoke thus: "Greeting to thee, O Fârûk! Thou art the Lord of Ælia. We adjure thee by God, do not return until you conquer it." He asked them as to Aldajjâl, whereupon they answered: "He will be one of the tribe of Benjamin. By God, you, O nation of the Arabs, you will kill him at a distance of ten to twenty yards from the gate of Lydda."

In the times after Alḥallâj the Ḳarmatians rose into power. 'Abû-Ṭâhir Sulaimân b. 'Abî-Sa'îd Alḥasan b. Bahrâm Aljannâbî marched out and reached Makka A.H. 318; he killed in an atrocious way the people who were passing round the circuit of the Ka'ba, and threw the corpses into the well Zamzam; he carried off the garments and the golden implements of the Holy House, and destroyed its aqueduct; he took away the black stone, smashed it, suspended it afterwards in the Mosque of Kûfa, and then he returned home.

p.213. **Ibn Abi-Zakariyya.**—On the 1st Ramaḍân A.H. 319 came forward *Ibn 'Abî-Zakariyyâ*, a native of Ṭamâm, a young man of bad character, a male prostitute. He called upon people to recognise him as the Lord, and they followed him. He ordered them to cut open the stomachs of the dead, to wash them and to fill them with wine; to cut off the hand of everybody who extinguished the fire with his hand, the tongue of everybody who extinguished it by blowing; to have intercourse with young men,—but with this restriction, *ne justo magis penem immitterent.* If anybody infringed this rule, he should be dragged on his face over a distance of forty yards. Those who would not practise pederasty were killed by the butcher. He ordered them to worship and honour the fires, he cursed all the prophets of former times and their companions, for they were "artful deceivers and on the wrong path," and more of that sort, which I have sufficiently related in my history of the Mubayyiḍa and the Ḳarmaṭians.

In such a conditon they remained during eighty days, till God gave him into the power of that man who had originally brought him forward. He slaughtered him, and so their schemes turned back upon their own necks.

If, now, this be the time which Jâmâsp and Zarâdusht meant, they are right as far as chronology is concerned. For this happened at the end of Æra Alexandri 1242, *i.e.* 1,500 years after Zarâdusht. They are wrong, however, as regards the restoration of the empire to the Magians. Likewise 'Abû-'Abdallâh Al'âdî has been mistaken, a man who is stupidly partial to Magism and who hopes for an age in which

Alkâ'im is to appear. For he has composed a book on the cycles and conjunctions, in which he says that the 18th conjunction since the birth of Muḥammad coincides with the 10th millennium, which is presided over by Saturn and Sagittarius. Now he maintains that then a man will come forward who will restore the rule of Magism; he will occupy the whole world, will do away with the rule of the Arabs and others, he will unite all mankind in one religion and under one rule; he will do away with all evil, and will rule during 7½ conjunctions. Besides he asserts that no Arabian prince will rule after that one who is ruling in the 17th conjunction.

That time which this man indicates must of necessity refer to Almuktafî and Almuḳtadir, but it has not brought about those events which, according to his prophecies, were to have taken place after their time.

People say that the Sasanian rule existed during *fiery* conjunctions. Now, the rule over Dailam was seized by 'Alî b. Buwaihi called 'Imâd-aldaula during fiery conjunctions. This is what people used to promise each other regarding the restoration of the rule to the Persians, although the doings of the Buwaihi family were not like those of the ancient kings.

I do not know why they preferred the Dailamite dynasty, whilst the fact of the transitus into a *fiery Trigonon* is the most evident proof indicative of the Abbaside dynasty, who are a Khurâsânî, an eastern dynasty. Besides, both dynasties (Dailamites as well as Abbasides) are alike far from renewing the rule of the Persians and farther still from restoring their ancient religion.

Before the appearance of that youth (Ibn 'Abî Zakariyyâ) the
Ḳarmaṭians believed in some dogmas of the Esoterics, and they were p.214.
considered as adherents of the family of the blessed House (of 'Alî). They promised each other the coming of him who is expected to come during the 7th conjunction under a *fiery Trigonon,* so that 'Abû-Ṭâhir Sulaimân b. Alḥasan says on this subject:

"The most glorious benefit I bestow on you will be my return to Hajar.
Then, after a while, verily the news will reach you.
When Mars rises from Babylonia,
When the *Two Stars* have left him, then beware, beware!
Is it not I who is mentioned in all the scriptures?
Is it not I who is described in the Sûra *Alzumar?*
I shall rule the people of the earth, east and west,
As far as the Ḳairawân of the Greeks, to the Turks and Chazars.
And I shall live until the coming of Jesus the son of Mary.
Then he will praise my exploits and approve of what he ordered.
Then, no doubt, my dwelling-place will be in paradise,
Whilst the others will burn in fire and hell."

Thereupon came forward a man called *Ibn 'Abî-Al'azâkir* b. 'Alî b. Shalmaghân. He maintained that the Holy Ghost was dwelling in him, and composed a book which he called *The 6th Sense*, relating to the abrogation of the rites.

(The end of this chapter and the beginning of the following are missing.)

CHAPTER IX.

ON THE FESTIVALS IN THE MONTHS OF THE PERSIANS. p.215.

(Farwardîn-Mâh.)

(1. Naurôz.)

. . . and he divided the cup among his companions, and said, "O that we had Naurôz every day!"

A philosopher of the Ḥashwiyya-school relates that when Solomon the son of David had lost his seal and his empire, but was reinstated after forty days, he at once regained his former majesty, the princes came before him, and the birds were busy in his service. Then the Persians said, "*Naurôz âmadh*," i.e. the new day has come. Therefore that day was called Naurôz. Solomon ordered the wind to carry him, and so it did. Then a swallow met him, and said, "O king, I have got a nest with little eggs in it. Please, turn aside and do not smash them." So Solomon did, and when he again descended to earth the swallow came bringing some water in his beak, which he sprinkled before the king, and made him a present of the foot of a locust. This is the cause of the water-sprinkling and of the presents on Naurôz.

Persian scholars say that in the day of Naurôz there is an hour in which the sphere of Fêrôz is driven on by the spirits for the purpose of renovating the creation.

The happiest hours of this day are the hours of the sun. On its morning, dawn is the shortest possible, and it is considered as a good omen to look at this dawn. It is a "*preferable*" day because it is called Hurmuz, which is the name of God who has created, formed, produced, and reared the world and its inhabitants, of whose kindness and charity nobody could describe even a part.

Sa'îd b. Alfaḍl relates: On the mountain Damâ in Fârs every night of Naurôz there is observed a far-spreading and strong-shining light.

ning, whether the sky be clear or covered with clouds, in every state of the weather.

Still more curious than this are the fires of Kalwâdhâ, although one does not feel inclined to believe the thing without having seen it. 'Abû-alfaraj Alzanjâni, the mathematician, told me that he had witnessed it together with a number of other people who went to Kalwâdhâ in that year when 'Adud-aldaula entered Baghdâdh, and that there are innumerable fires and lights which appear on the west side of the Tigris, opposite Kalwâdhâ, in that night with the morning of which Naurôz begins. The Sulṭân had there posted his guards to find out the truth in order not to be deceived by the Magians. All, however, they found out was this, that as soon as they came nearer to the fires they went farther off, and as soon as they went away the fires came nearer. Now I said to 'Abû-Alfaraj, "The day of Naurôz recedes from its proper place in consequence of the Persians neglecting intercalation. Why, then, does not this phenomenon remain back behind Naurôz? Or if it is not necessary that it should remain behind, did it then fall earlier at the time when they practised intercalation?" Upon which he could not give a satisfactory answer.

p.216. The charm-mongers say: He who thrice sips honey on Naurôz in the morning before speaking, and perfumes his room with three pieces of wax, will be safe against all diseases.

One Persian scholar adduces as the reason why this day was called Naurôz, the following: *viz.* that the Ṣâbians arose during the reign of Tahmûrath. When, then, Jamshîd succeeded, he renovated the religion, and his work, the date of which was a Naurôz, was called *New-Day.* Then it was made a feast day, having already before been held in great veneration.

Another account of the reason why it was made a feast day is this, that Jamshîd, on having obtained the carriage, ascended it on this day, and the Jinns and Dêws carried him in one day through the air from Dabâwand to Babel. Now people made this day a feast day on account of the wonder which they had seen during it, and they amused themselves with swinging in order to imitate Jamshîd.

Another report says that Jam was going about in the country,—that he, when wishing to enter Âdharbaijân, sat on a golden throne and was thus carried away by the men on their necks. When, then, the rays of the sun fell on him and people saw him, they did homage to him and were full of joy and made that day a feast day.

On Naurôz it was the custom for people to present each other sugar. According to Âdharbâdh, the Maubadh of Baghdâdh, the reason is this, that the sugar-cane was first discovered during the reign of Jam on the day of Naurôz, having before been unknown. For Jam on seeing a juicy cane which dropped some of its juice, tasted it, and found that it had an agreeable sweetness. Then he ordered the juice of the sugar-cane to be pressed out and sugar to be made thereof. It was ready on the fifth day, and then they made each other presents of sugar. The same was also the custom on Mihrjân.

They have adopted the time of the summer-solstice as the beginning of the year for this reason in particular, that the two solstitial-points are easier to be ascertained by the help of instruments and by observation than the equinoctial points, for the former are the beginning of the advance of the sun towards one of the two poles of the universe and of his turning away from the same pole. And if the perpendicular shadow at the summer-solstice is observed, and the level shadow at the winter-solstice, in whatsoever place of the earth the observation be made, the observer cannot possibly mistake the day of the solstice, though he may be entirely ignorant in geometry and astronomy, because a variation of the level shadow takes place notwithstanding the small amount of declination, if the *Height* is considerable. On the other hand the two equinoctial days cannot be ascertained, unless you have found beforehand the latitude of the place and the *General Declination*. And this nobody will find out unless he studies astronomy and has profited something thereby, and knows how to place and how to use the instruments of observation.

Therefore the solstitial points are better adapted for marking the beginning of the year than the equinoctial points. And as the summer-solstice is nearer to the zenith of the northern countries, people preferred it to the winter-solstice; for this reason, moreover, that it is the time of the ripening of the corn. Therefore it is more proper to p.217.
gather the taxes at this time than at any other.

Many of the scholars and sages of the Greeks observed the horoscope at the time of the rising of Sirius and commenced the year at that time, not with the vernal equinox, because the rising of Sirius coincided in bygone times with this solstice, or occurred very near it.

This day, I mean Naurôz, has receded from its original proper place, so that in our time it coincides with the sun's entering the sign of Aries, which is the beginning of spring. Whence it has become the custom of the princes of Khurâsân on this day to dress their warriors in spring—and summer—dresses.

On the 6th of Farwardîn, the day Khurdâdh, is the Great Naurôz, for the Persians a feast of great importance. On this day—they say—God finished the creation, for it is the last of the six days, mentioned before. On this God created Saturn, therefore its most lucky hours are those of Saturn. On the same day—they say—the *Sors Zarathustræ* came to hold communion with God, and Kaikhusrau ascended into the air. On the same day the happy lots are distributed among the people of the earth. Therefore the Persians call it "*the day of hope.*"

The charm-mongers say: He who tastes sugar on the morning of this day before speaking, and anoints himself with oil, will keep off all sorts of mishap during the greater part of this same year.

On the morning of this day, a silent person with a bundle of fragrant flowers in his hand is seen on the mountain Bûshanj; he is visible for one hour and then disappears, and does not reappear until the same time of the next year.

Zâdawaihi says that the cause of this was the rising of the sun from the southern region, *i.e. Afâhtar.* For the cursed 'Iblîs had deprived eating and drinking of their beneficial effect, so that people could not satisfy their hunger nor quench their thirst; and he had prevented the wind from blowing. So the trees withered up and the world was near to utter decay. Then came—by the command and under the guidance of God—Jam to the southern region. He marched towards the residence of 'Iblîs and of his followers, and remained there for some time until he had extinguished that plague. Then people returned into a state of justice and prosperity and were freed from that trial. Under such circumstances Jam returned to the world (*i.e.* Eran) and rose on that day like the sun, the light beaming forth from him, as though he shone like the sun. Now people were astonished at the rising of two suns, and all dried-up wood became green. So people said *rôz-i-nau*, *i.e.* a new day. And everybody planted barley in a vessel or somewhere else, considering it as a good omen. Ever since, it has been the custom on this day to sow around a plate seven kinds of grain on seven columns, and from their growth they drew conclusions regarding the corn of that year, whether it would be good or bad.

On the same day Jamshîd issued a proclamation to those who were present, and wrote to those who were absent, ordering them to destroy the old temples and not to build a new one on that day.

His behaviour towards the people was such as pleased God, who rewarded him by delivering his people from diseases and decrepitude,
p.218. from envy and frailty, and sorrows and disasters. No being was sick or died, as long as he ruled—until the time when Bêwarasp, his sister's son, appeared, who killed Jam and subdued his realm. In the time of Jam the population increased at such a rate that the earth could no longer contain them; therefore God made the earth thrice as large as it had been before. He (Jam) ordered people to wash themselves with water in order to clean themselves of their sins, and to do so every year that God might keep them aloof from the calamities of the year.

Some people maintain that Jam ordered channels to be dug, and that the water was led into them on this day. Therefore people rejoiced at their prosperity, and washed themselves in the water that was sent them (by the channels), and in this respect the later generations have considered it a good omen to imitate the former ones.

Others, again, maintain that he who let the water into the channels was Zû, after Afrâsiâb had ruined all the dwellings of Erânshahr.

According to another view, the cause of the washing is this—that this day is sacred to Harûdhâ, the angel of the water, who stands in relation to the water. Therefore people rose on this day early, at the rising of dawn, and went to the water of the aqueducts and wells. Frequently, too, they drew running water in a vase, and poured it over themselves, considering this a good omen and a means to keep off hurt.

On the same day people sprinkle water over each other, of which the cause is said to be the same as that of the washing. According to another report, the reason was this—that during a long time the rain was withheld from Erânshahr, but that they got copious rain, when Jamshîd, having ascended the throne, brought them the good news of which we have spoken. Therefore they considered the rain a good omen, and poured it over each other, which has remained among them as a custom.

According to another explanation, this water-sprinkling simply holds the place of a purification, by which people cleansed their bodies from the smoke of the fire and from the dirt connected with attending to the fires. Besides it serves the purpose of removing from the air that corruption which produces epidemic and other diseases.

On the same day Jam brought forward all kinds of measures; therefore the kings considered his way of counting as of good omen. On the same day they used to prepare all the necessary paper and the hides on which their despatches to the provinces of the empire were written, and all the documents to which the royal seal was to be applied were sealed. Such a document was called *Espîdânuwisht.*

After the time of Jam the kings made this whole month, *i.e.* Farwardîn-Mâh, one festival, distributed over its six parts. The first five days were feast days for the princes, the second for the nobility, the third for the servants of the princes, the fourth for their clients, the fifth for the people, and the sixth for the herdsmen.

The man who connected the two Naurôz with each other is said to have been Hormuz ben Shâpûr the Hero, for he raised to festivals all the days between the two Naurôz. Besides he ordered fires to be kindled on high places, because he considered it a good omen, and for the purpose of purifying the air, since they consume all unwholesome elements in the air and dissolve and scatter those miasmata that produce corruption.

In these five days it was the custom of the Kisrâs that the king opened the Naurôz and then proclaimed to all that he would hold a session for them, and bestow benefits upon them. On the second day the session was for men of high rank, and for the members of the great families. On the third day the session was for his warriors, and for the p.219.
highest Maubadhs. On the fourth day it was for his family, his relations and domestics, and on the fifth day it was for his children and clients. So everybody received the rank and distinction he was en-

titled to, and obtained those remunerations and benefits which he had deserved. When the sixth day came and he had done justice to all of them, he celebrated Naurôz for himself and conversed only with his special friends and those who were admitted into his privacy. Then he ordered to be brought before him the whole amount of presents, arranged according to those who had presented them. He considered them, distributed of them what he liked, and deposited what he liked in his treasury.

The 17th is the day of Serôsh, who first ordered the *Zamzama*, *i.e.* expressing yourself by whispering, not by clear speech. For they said prayers, praised and celebrated God, whilst handing to each other the food; now, speaking not being allowed during prayer, they express themselves by whispers and signs. Thus I was told by the geometrician Âdharkhûrâ. According to another authority, the *Zamzama* is intended to prevent the breath of the mouth from touching the food.

This day is a blessed day in every month, because Serôsh is the name of that angel who watches over the night. He is also said to be Gabriel. He is the most powerful of all angels against the Jinns and sorcerers. Thrice in the night he rises above the world; then he smites the Jinns and drives off the sorcerers; he makes the night shine brilliantly by his appearance. The air is getting cold, the water sweet; the cocks begin crowing, and the lust of sexual intercourse begins to burn in all animals. One of his three risings is the rising of dawn, when the plants begin to thrive, the flowers to grow, and the birds to sing; when the sick man begins to rest, and the sorrowful to feel somewhat relieved; when the traveller travels in safety; when the time is agreeable; when such dreams occur as will be fulfilled one day; and when all angels and demons enjoy themselves.

On the 19th, or Farwardîn-Rôz, there is a feast called *Farwardagân* on account of the identity of the name of the day and of the month in which it lies. A similar feast-day they have got in every month.

Ardîbahisht-Mâh.

On the 3rd, or Ardîbahisht-Rôz, there is a feast, *Ardîbahishtagân*, so called on account of the identity of the name of the month and the day. The word Ardîbahisht means "*truth is the best,*" or according to another explanation, "*the utmost of good.*"

Ardîbahisht is the genius of fire and light; both elements stand in relation to him. God has ordered him to watch over these elements; to remove the weaknesses and diseases by drugs and nourishments; to distinguish truth from falsehood, the true man from the liar, by means of those oaths that are manifest in the Avastâ.

The 26th, or Ashtâdh-Rôz, is the first day of the third Gahanbâr; it lasts five days, the last of which is the last day of the month. In these days God created the earth. This Gahanbâr is called *Paitishahîm-*

Gâh. The six Gahanbârs, each of which lasts five days, have been established by Zoroaster.

Khurdâdh-Mâh. p.220.

The 6th day, or Khurdâdh-Rôz, is a feast Khurdâdhagân, so called on account of the identity of the name of the month and the day. The meaning of the name is "the stability of the creation." Harûdhâ is the genius instructed to watch over the growth of the creation, of the trees and plants, and to keep off all impure substances from the water.

The 26th, or Ashtâdh-Rôz, is the first day of the fourth Gahanbâr, the last day of which is the last of the month. During this Gahanbâr God created the trees and plants. It is called *Ayathrema-Gâh.*

Tîr-Mâh.

On the 6th, or Khurdâdh-Rôz, there is a feast called *Cashn-i-nîlûfar,* considered to be of recent origin.

On the 13th, or Tîr-Rôz, there is a feast Tîragân, so called on account of the identity of the name of the month and the day. Of the two causes to which it is traced back, one is this, that Afrâsiâb after having subdued Erânshahr, and while besieging Minôcihr in Ṭabaristân, asked him some favour. Minôcihr complied with his wish, on the condition that he (Afrâsiâb) should restore to him a part of Erânshahr as long and as broad as an arrow-shot. On that occasion there was a genius present, called Isfandârmadh; he ordered to be brought a bow and an arrow of such a size as he himself had indicated to the arrow-maker, in conformity with that which is *manifest* in the Avastâ. Then he sent for Arish, a noble, pious, and wise man, and ordered him to take the bow and to shoot the arrow. Arish stepped forward, took off his clothes, and said: "O king, and ye others, look at my body. I am free from any wound or disease. I know that when I shoot with this bow and arrow I shall fall to pieces and my life will be gone, but I have determined to sacrifice it for you." Then he applied himself to the work, and bent the bow with all the power God had given him; then he shot, and fell asunder into pieces. By order of God the wind bore the arrow away from the mountain of Rûyân and brought it to the utmost frontier of Khurâsân between Farghâna and Ṭabaristân; there it hit the trunk of a nut-tree that was so large that there had never been a tree like it in the world. The distance between the place where the arrow was shot and that where it fell was 1,000 Farsakh. Afrâsiâb and Minôcihr made a treaty on the basis of this shot that was shot on this day. In consequence people made it a feast-day.

During this siege Minôcihr and the people of Erânshahr had been suffering from want, not being able to grind the wheat and to bake the bread because the wheat was late in ripening; finally they took the wheat and the fruits, unripe as they were, ground them and ate them. Thence it has become a rule for this day to cook wheat and fruits.

According to another report, the arrow was shot on this day, *i.e.* Tîr-Rôz, and the festival of this day is the small Tîragân; on the other hand the 14th, or Gôsh-Rôz, is the great Tîragân, that day on which the news arrived that the arrow had fallen.

On Tîr-Rôz people break their cooking-vessels and fire-grates, since on this day they were liberated from Afrâsiâb and everybody was free to go to his work.

The second cause of the feast Tîragân is the following: The *Dahûfadhiyya,* which means "the office of guarding and watching over the world and of reigning in it," and the *Dahkana,* which means "the office of cultivating the world, of sowing in it, and of distributing it"—these two are twins on whom rest the civilization of the world, and its
p.221. duration, and the setting right of anything that is wrong in it. The *Kitâba* (the office of writer) follows next to them and is connected with both of them.

The *Dahûfadhiyya* was founded by Hôshang, the Dahkana by his brother Waikard. The name of this day is Tîr or Mercury, who is the star of the scribes. Now Hôshang spoke in praise of his brother on this same day, and gave to him as his share the *Dahkana,* which is identical with the *Kitâba.* Therefore people made this day a feast in praise and honour of him (Waikard). On this day he (Hôshang) ordered people to dress in the dress of the *Scribes* and *Dihkâns.* Therefore the princes, Dihkâns, Maubadhs, etc., continued to wear the dress of the *Scribes* until the time of Gushtâsp, in praise and honour of both the *Kitâba* and *Dahkana.*

On the same day the Persians used to wash themselves, of which the reason is this—that Kaikhusrau, on returning from the war against Afrasiâb, passed on this day through the territory of Sâwa. He went up the mountain which overhangs the town, and sat down at a fountain quite alone at some distance from his encampment. There an angel appeared unto him, whereby he was so terrified that he swooned. About that time Wîjan ben Jûdarz arrived, when the king had already recovered himself; so he sprinkled some of that water on his face, leaned him against a rock, and said ماندیش *i.e. do not be afraid.* Thereupon the king ordered a town to be built around that fountain, and called it *Mandîsh,* which afterwards was altered and mutilated into *Andîsh.* Ever since, it has been the custom of people to wash themselves in this water and in all fountain-waters, this being considered a good omen. The inhabitants of Âmul go out to the *Bahr-alkhazar,* play in the water, and make fun, and try to dip each other on this day the whole day long.

Murdâdh-Mâh.

On the 7th, or Murdâdh-Rôz, there is the feast Murdâdhagân, so called on account of the identity of the name of the month and the day. The meaning of the word Murdâdh is "*the everlasting duration of the world without death and destruction.*" Murdâdh is the angel appointed to guard the world and to produce vegetable food and drugs that are remedies against hunger, misery, and disease. God knows best!

Shahrêwar-Mâh.

On the 7th, or Shahrêwar-Rôz, is the feast Shahrêwaragân, so called on account of the identity of the name of the month and the day. Shahrêwar means sperma and love. It is the angel who is appointed to watch over the seven substances, gold, silver, and the other metals, on which rests all handicraft, and in consequence all the world and its inhabitants.

Zâdawaihi relates that this feast was called *Âdhar-ćashn*, *i.e.* the feast of the fires that are found in the human dwelling-places. It was the beginning of winter, therefore people used to make great fires in their houses, and were deeply engaged in the worship and praise of God; also they used to assemble for eating and merriment. They maintained that this was done for the purpose of banishing the cold and dryness that arises in winter-time, and that the spreading of the warmth would keep off the attacks of all that which is obnoxious to the plants in the world. In all this, their proceeding was that of a man who marches out p.222.
to fight his enemy with a large army.

According to the Maubadh, Khurshêd Âdhar-ćashn was the first day of this month, and only a feast for the nobility. It does, however, not belong to the feast-days of the Persians, although it was used in their months. For it is one of the feast-days of the people of Tukhâristân, and is a custom of theirs based on the fact that about this time the season altered and winter set in. In this our time the people of Khurâsân have made it the beginning of autumn.

This day, *i.e.* Mihr-Rôz, is the first day of the fifth Gahanbâr, the last of which is Bahrâm-Rôz. During this Gahanbâr God created the cattle. It is called *Maidhyâirîm-Gâh.*

Mihr-Mâh.

On the 1st of it, or Hurmuzd-Rôz, falls the *Second Autumn*, a feast for the common people, agreeably with what has been before mentioned.

On the 16th, or Mihr-Rôz, there is a feast of great importance, called Mihrajân. The name of the day is identical with that of the month; it means "*the love of the spirit.*" According to others, Mihr is the name of the sun, who is said to have for the first time appeared to the world on this day; that therefore this day was called Mihr. This is indicated by the custom of the Kisrâs of crowning themselves on this day with a crown on which was worked an image of the sun and of the wheel on which he rotates. On this day the Persians hold a fair.

People maintain that the special veneration in which this day is held is to be traced to the joy of mankind when they heard of Frêdûn's coming forward, after Kâbî had attacked Alḍaḥḥâk Bêvarasp, expelled him and called upon people to do homage to Frêdûn. Kâbî is the same whose standard the Persian kings adopted, considering it a good omen; it was made of the skin of a bear, or, as others say, of that of a lion; it was called *Dirafsh-i-Kâbiyân*, and was in later times adorned with jewels and gold.

On the same day the angels are said to have come down to help Frêdûn. In consequence it has become a custom in the houses of the kings, that at the time of dawn a valiant warrior was posted in the court of the palace, who called at the highest pitch of his voice: "O ye angels, come down to the world, strike the Dêws and evil-doers and expel them from the world."

On the same day, they say, God spread out the earth and created the bodies as mansions for the souls. In a certain hour of this day the sphere of *Ifranjawî* breathes for the purpose of rearing the bodies.

On the same day God is said to have clad the moon in her splendour and to have illuminated her with her light, after He had created her as a black ball without any light. Therefore, they say, on Mihrajân the moon stands higher than the sun, and the luckiest hours of the day are those of the moon.

Salmân Alfârisî has said: In Persian times we used to say that God has created an ornament for his slaves, of rubies on Naurôz, of emeralds on Mihrajân. Therefore these two days excel all other days in the same way as these two jewels excel all other jewels.

Alêrânshahrî says: God has made the treaty between Light and Darkness on Naurôz and Mihrajân.

Sa'îd b. Alfaḍl used to say: Persian scholars relate, that the top of
p.223. the mountain Shâhîn appears always black during the whole length of summer, whilst on the morning of Mihrajân it appears white as if covered with snow, whether the sky be clear or clouded, in any weather whatsoever.

Alkisrawî relates:—I heard the Maubadh of Almutawakkil say: On the day of Mihrajân the sun rises in Hâmîn, in the midst between light and darkness. Then the souls die within the bodies; therefore the Persians called this day *Mîragân*.

The charm-mongers say: He who eats on the day of Mihrajân a piece of pomegranate and smells rose-water, will be free from much mishap.

The Persian theologians have derived various symbolic interpretations from these days. So they consider Mihrajân as a sign of resurrection and the end of the world, because at Mihrajân that which grows reaches its perfection and has no more material for further growth, and because animals cease from sexual intercourse. In the same way they make Naurôz a sign for the beginning of the world, because the contrary of all these things happens on Naurôz.

Some people have given the preference to Mihrajân by as much as they prefer autumn to spring. In their arguments they chiefly rely upon what Aristotle said in reply to Alexander, when he was asked by him regarding them: "O king, in spring the reptiles begin growing, in autumn they begin to die away. From this point of view autumn is preferable."

This day used in former times to coincide with the beginning of winter. Afterwards it advanced, when people began to neglect intercalation. Therefore it is still in our time the custom of the kings of Khurâsân, that on this day they dress their warriors in autumn—and winter—dresses.

On the 21st, or Râm-Rôz, is the *Great Mihrajân* in commemoration of Frêdûn's subduing and binding Al-Ḍaḥḥâk. People say, that when he was brought before Frêdûn he spoke: "Do not kill me in retaliation for thy ancestor." Upon which Frêdûn answered, refusing his entreaty, "Do you want to be considered as equal to Jam b. Wîjahân in the way of retaliation? By no means. I shall punish you for an ox, that was in the house of my ancestor." Thereupon he put him in fetters and imprisoned him in the mountain Dubâwand. Thereby people were freed from his wickedness, and they celebrated this event as a feast. Frêdûn ordered them to gird themselves with *Kustîks*, to use the *Zamzama* (speaking in a whispering tone) and to abstain from speaking loud during dinner, as a tribute of thanks to God for having again made them their own masters with regard to their whole behaviour and to the times of their eating and drinking, after they had been living in fear so long as 1,000 years. This has come down to posterity as a rule and custom on the day of Mihrajân.

All the Persians agree that Bêvarasp lived 1,000 years, although some of them say that he lived longer and that the 1,000 years are only the time of his rule and tyranny. People think that the Persian mode of salutation, according to which the one wishes the other to live as long as 1,000 years—I mean the words "*Hazâr sâl bazî*"—comes down from that time, because they thought it was allowed and possible (that a man should live 1,000 years) from what they had seen of Al-Ḍaḥḥâk. God knows best!

Zarâdusht has ordered that both Mihrajân and Râm-Rôz should be p.224.
held in equal veneration. In consequence, they celebrated both days as feast-days, until Hurmuz b. Shâpûr, the Hero, connected the two days with each other, and raised to feast-days all the days between them, as he had done with the two Naurôz. Afterwards the kings and the people of Êrânshahr celebrated as feast-days all the days from Mihrajân till thirty days afterwards, distributing them over the several classes of the population in the same way as we have heretofore explained regarding Naurôz. Each class celebrated its feast for five days.

Âbân-Mâh.

On the 10th, or Âbân-Rôz, there is a feast *Abânajân*, so called on account of the identity of the name of the month and the day. On this day Zau b. Tahmâsp ascended the throne; he ordered the channels to be dug and to be kept in good preservation.

On the same day the news reached all the seven κλίματα of the world that Frêdûn had put in fetters Bêvarasp; that he had assumed the royal dignity; that he had ordered people to take possession of their houses, their families and children, and to call themselves *Kadhkhudâ*, *i.e.* master of this house; that he ruled over his family, his children, and his empire with supreme authority; whilst before that, in the time of Bêvarasp, they had been in a deserted state, and Dêws and *rebels* had alternately been haunting their houses, without their being able to keep them off. This institute (that of a *Kadhkhudâ*) has been abolished by *Alnâsir Al'uṭrûsh*, who made again the *rebels* partake of the Kadhkhudâdom together with the people.

The last five days of this month, the first of which is Ashtâdh, are called Farwardajân. During this time people put food in the halls of the dead and drink on the roofs of the houses, believing that the spirits of their dead during these days come out from the places of their reward or their punishment, that they go to the dishes laid out for them, imbibe their strength and suck their taste. They fumigate their houses with juniper, that the dead may enjoy its smell. The spirits of the pious men dwell among their families, children, and relations, and occupy themselves with their affairs, although invisible to them.

Regarding these days there has been among the Persians a controversy. According to some they are the last five days of the month Âbân, according to others they are the Andergâh, *i.e.* the five *Epagomenæ* which are added between Âbân and Âdhâr-Mâh. When the controversy and dispute increased, they adopted all (ten) days in order to establish the matter on a firm basis, as this is one of the chief institutes of their religion, and because they wished to be careful, since they were unable to ascertain the real facts of the case. So they called the first five days the first Farwardajân, and the following five days the second Farwardajân; the latter, however, is more important than the former.

The first day of these *Epagomenæ* is the first day of the sixth Gahanbâr, in which God created man. It is called *Hamaçpaṭmaêdhaêm-gâh.*

The reason of the Farwardajân is said to be this—that when Cain had killed Abel, and the parents were lost in grief, they implored God to restore his soul to him. God did so on the day Ashtâdh of Âbân-Mâh, and the soul remained in him for ten days. Abel was sitting erect and

looking at his parents, but it was not allowed to him to speak. Then his p.225.
parents collected—(*Missing, the end of Âbân-Mâh*).

[*Âdhâr-Mâh.*]

[1. *Bahâr-ćashn*, the feast of the *Riding of Alkausaj*. This day was the beginning of spring at the time of the Kisrâs. Then a thin-bearded (Kausaj) man used to ride about, fanning himself with a fan to express his rejoicing at the end of the cold season and the coming of the warm season. This custom is in Persis still kept up for fun.]

Its most lucky hours are those during which Aries is the horoscope. People consider the hour of morning as of good omen—I mean the charm-mongers—and they maintain that everything that is mentioned during this hour exists absolutely. Besides they say that he who tastes a quince and smells an orange in the morning of this day before speaking will be happy during that same year.

According to Ṭâhir b. Ṭâhir, the Persians, in old times, used to drink honey on this day if the moon happened to stand in a fiery station, and to drink water if it stood in a watery station, always adapting themselves to the character of the stations of the moon.

Alêrânshahrî says: I heard a number of Armenian learned men relate that on the morning of the *Fox-day* there appears on the highest mountain, between the *Interior* and the *Exterior* country, a white ram that is not seen at any other time of the year except about this time of this day. Now the inhabitants of that country infer that the year will be prosperous if the ram bleats; that it will be sterile if he does not bleat.

On the morning of the *Fox-day* the Persians thought it to be a good omen to look at the clouds; and from the fact whether they were clear or dark, thin or dense, they drew conclusions as to whether the year would be prosperous or not, fertile or barren.

On the 9th, the day of Âdhar, is a feast called *Âdharćashn*, so called on account of the identity of the name of the day and the month. On that day people want to warm themselves by the fire, for this is the end of the winter months, when the cold, at the end of the season, is most biting and the frost is most intense. It is the feast of the fire, and is called by the name of the genius who has to watch over all the fires.

Zarâdusht has given the law that on this day people should visit the fire-temples, and that they should there offer offerings and deliberate on the affairs of the world.

Dai-Mâh, also called Khûr-Mâh.

The first day of it is called *Khurram-Rôz*. This day and the month are both called by the name of God, *i.e.* (Hormuzd), *i.e.* a wise king, gifted with a creative mind.

On this day the king used to descend from the throne of the empire, to dress in white dresses, to sit on white carpets in the plain, to suspend for a time the duties of the chamberlains and all the pomp of royalty, and exclusively to give himself up to the consideration of the affairs of the realm and its inhabitants. Whosoever, high or low, wanted to speak to him in any matter, went into his presence and addressed him, nobody preventing him from doing so. Besides, he held a meeting with the Dihkâns and agriculturists, eating and drinking with them, and then he used to say: "To-day I am like one of you. I am your brother; for the existence of the world depends upon that culture which is wrought by your hands, and the existence of this culture depends upon government; the one cannot exist without the other. This being the case, we are like twin brothers, more particularly as this (royalty and agriculture) proceeds from twin brothers, from Hôshang and Waikard."

This day is also called *Nuwâd-Rôz* (90 *days*), and is celebrated as a
p.226. feast, because there are 90 days between this day and Naurôz.

The 8th, 15th, and 23rd days of this month are feast-days on account of the identity of the names of these days with that of the month, as we have heretofore explained.

The 11th, or Khûr-Rôz, is the first day of the (first?) Gahanbâr; its last day is the 15th, or Dai-ba-Mihr. This Gahanbâr is called Maidhyô-zaremaya-gâh. During it God created heaven.

On the 14th, or Gôsh-Rôz, there is a feast called Sîr-sawâ, when people eat garlic and drink wine, and cook the vegetables with pieces of meat, by which they intend to protect themselves against the devil. The original purpose of the thing was to rid themselves of their affliction when they were oppressed in consequence of Jamshîd's being killed, and were in sorrow and swore that they would never touch any fat. This has remained as an usage among them. By that dish they cure themselves of the diseases which they attribute to the influence of the evil spirits.

The 15th, or Dai-ba-Mihr-Rôz, is called بسكان, when they used to make a human-like figure of paste or clay and posted it at the gateways. This, however, was not practised in the houses of the kings. At present this custom has been abolished on account of its resemblance to idolatry and heathendom.

The night of the 16th, or Mihr-Rôz, is called, درامزينان, and also كاكثل. Its origin is this, that Êrânshahr was separated and liberated from the country of the Turk, and that they drove their cows, which the enemy had driven away, back to their houses. Further: when Frêdûn had put Bêvarasp out of the way, he let out the cows of *Athfiyân* (Âthwyâna) that had been hidden in some place during the siege, whilst Athfiyân defended them. Now they returned to his house. Athfiyân was a man of high standing and noble character, a benefactor of the poor, busying himself with the affairs of the poor and taking care of them, and liberal towards all who applied to him. When Frêdûn had freed his property, people celebrated a feast in hope of his gifts and presents.

On the same day the weaning of Frêdûn took place. It was the first day when he rode on the ox in a night when the ox appears which drags the carriage of the moon. It is an ox of light, with two golden horns and silver feet, which is visible for an hour and then disappears. The wish of him who looks at the ox when it is visible will be fulfilled in the same hour.

In the same night there appears on the highest mountain, as they maintain, the spectre of a white ox, that bellows twice if the year is to be fertile, and once (if the year is to be barren). (*Here follows a lacuna*).

[23. Feast of the third day, Dai.]

Bahman-Mâh.

[2. **Bahmanja.**

5. **Barsadhak, or Nausadhak.**

10. **The Night of Alsadhak.**]

They fumigate their houses to keep off mishap, so that finally it has become one of the customs of the kings to light fires on this night and to make them blaze, to drive wild beasts into them, and to send the birds flying through the flames, and to drink and amuse themselves round the fires.

May God take vengeance on all who enjoy causing pain to another being, gifted with sensation and doing no harm!

After the Persians had neglected intercalation in their months, they hoped that the cold would cease at this time, as they reckoned as the p.227.
beginning of winter the 5th of Âbân-Mâh, and as the end the 10th of Bahman-Mâh. The people of *Karaj* called this day شبِ گرزه, *i.e.* the biting night, on account of its being so cold.

Another report accounts for the lighting of the fires during this night in the following way: When Bêvarasp had ordered people to provide him every day with two men, that he might feed his two serpents with their brains, he commissioned immediately after his arrival a man called Azmâ'îl to attend to this. Now, this man always used to set free one of the two, giving him food, and ordering him to settle in the western part of mount Dunbâwand and there to build himself some sort of house, whilst he fed the two serpents with the brains of a ram instead of that prisoner whom he had set free, mixing them with the brains of the other victim who was killed. When Frêdûn had conquered Bêvarasp, he ordered Azmâ'îl to be fetched and punished in revenge for those whom he had killed. Thereupon Azmâ'îl told him the tale of those whom he had set free, speaking the truth, and asked the king to send out a messenger with him that he might show them to him. So the king did, and Azmâ'îl ordered those whom he had set free to light fires on the roofs of their houses, in order that their number might be seen.

This happened in the 10th night of Bahman. Therefore the messenger said to Azmâ'îl: "What a number of them thou hast set free! May God give thee a good reward!" He returned to Frêdûn and brought him his report. Frêdûn exceedingly rejoiced at the matter, and set out himself for Dunbâwand to see the thing himself. Thereupon he conferred great honour upon Azmâ'îl, he gave him Dunbâwand as a fief, made him sit on a golden throne, and called him *Masmaghân*.

Regarding the two serpents of Bêvarasp, people say that they came out of his shoulders, feeding upon brains; whilst according to another view, they were two painful wounds which he besmeared with brains, hoping to get relief from them.

The two serpents are something wonderful—possible, indeed, but hardly likely. For worms are produced out of flesh, and in flesh lice and other animals are living. Further, there are other animals that do not entirely leave their birthplace, like that one of which people relate that it, living in India, peeps out of the womb of its mother to eat grass and then to return, that it does not leave the womb of its mother entirely until it has grown strong and thinks itself able to run faster than its mother, even if the mother should run after it; then it jumps out and runs away. People say that the young animal fears the tongue of its mother, which is the roughest thing imaginable. For the mother, if she finds the young one, licks it continually, until the flesh is severed from the bones. And out of the hair of the head that has been torn out together with its white root which originally is fixed in the flesh, snakes grow, in case the hair falls into water or some wet place in the midst of summer, growing within the time of three weeks or less.

This fact cannot be denied, since it has been witnessed, and the formation of other animals out of other materials has also been witnessed.

'Abû-'Uthmân Aljâḥiẓ relates, that he saw at 'Ukbarâ a piece of clay,
p.228. one half of which was a part of the body of a field-mouse, whilst the other half was still a common and unchanged piece of clay. I have heard this also from a number of people in Jurjân who had observed something similar in that country.

Aljaihânî relates that in the Indian Ocean there are the roots of a tree which spread along the sea-coast in the sand, that the leaf is rolled up and gets separated from the tree, and that it then changes into a king-bee and flies away.

The formation of scorpions out of figs and mountain-balm, that of bees from the flesh of oxen, that of wasps from the flesh of horses, is well known to all naturalists. We ourselves have observed many animals, capable of propagating their species, that had originally grown out of plants and other materials by a clear process of formation, and who afterwards continued their species by sexual intercourse.

The 22nd, or Bâdh-Rôz, is called by this name (*lacuna*).

On that day certain usages are practised in Ḳumm and neighbourhood that have a likeness to those festive customs of drinking and making fun which are practised at Ispahân in the days of Naurôz, when people hold a fair and celebrate a feast. At Ispahân people call it کژین. However, Bâdh-Rôz is only *one* day, whilst کژین lasts a whole week.

The 30th, or Anêran, is called *Âfrijagân* at Ispahân, which means "pouring out the water." Its origin is this: that once in the time of Fêrôz, the grandfather of Anôshirwân, the rain was kept back, and people in Êrânshahr suffered from barrenness. Therefore Fêrôz remitted them the taxes of these years, opened the doors of his storehouses, borrowed money from the properties of the fire-temples, and gave all to the inhabitants of Êrânshahr, taking care of his subjects as a parent does for his children; and the consequence was that during those years nobody died of hunger. Now, Fêrôz went to the famous fire-temple in *Âdharkhûrâ* in Fârs; there he said prayers, prostrated himself, and asked God to remove that trial from the inhabitants of the world. Then he went up to the altar and found there the ministers and priests standing before it. They, however, did not greet him as is due to kings. So he felt that there was something the matter with the priests. Then he went near the fire, turned his hand and arms round the flame, and pressed it thrice to his bosom, as one friend does with another when asking after each other's health; the flame reached his beard, but did not hurt him. Thereupon Fêrôz spoke: "O my Lord, thy names be blessed! if the rain is held back for my sake, for any fault of mine, reveal it to me that I may divest myself of my dignity; if something else is the cause, remove it, and make it known to me and tc the people of the world, and give them copious rain." Then he descended from the altar, left the cupola, and sat down on the دسکا made of gold, similar to a throne, but smaller. It was a custom for a famous fire-temple to have a golden دسکا for the purpose that the king should sit upon it when he came to the temple. Now the ministers and priests came near him and greeted him as is due to kings. The king spoke to them: "What has hardened your hearts, what has offended you and made you suspicious, that you did not greet me before?" They replied: "Because we were standing before another king more sublime than you. We were not allowed to greet you whilst p.229
standing before him." The king believed them and made them presents. Then he started from the town Âdharkhûrâ in the direction of the town Dârâ. But having come as far as the place where is now the village called Kâm-Fêrôz in Fârs—it was at that time an uncultivated plain—a cloud rose and brought such copious rain as had never been witnessed before, till the water ran into all the tents, the royal tent as well as the other ones. Fêrôz recognized that God had granted his prayer; he praised God, and ordered that on that spot his tents should be pitched. He gave alms, made liberal presents, held assemblies, and was full of joy. He did not leave this place before he had built the famous village which

he called Kâm-Fêrôz. Fêrôz is his name, and *kâm* means "*wish*;" so it signifies "*that he had obtained his wish.*" In the joy which everybody felt over this event, they poured the water over each other. In consequence this has become a custom in Êrânshahr ever since. In every town they celebrate this feast on that day when they got the rain, and the people of Ispahân got the rain on this day.

Isfandârmadh-Mâh.

On the 5th, or Isfandârmadh-Rôz, there is a feast on account of the identity of the names of the month and the day. The word means "intelligence" and "ripeness of mind."

Isfandârmadh is charged with the care of the earth and with that of the good, chaste, and beneficent wife who loves her husband. In past times this was a special feast of the women, when the men used to make them liberal presents. This custom is still flourishing at Ispahân, Rai, and in the other districts of Fahla. In Persian it is called Muzhdgîrân.

This day is famous for the inscribing of pieces of paper. For on this day common people eat sun-raisins and the kernels of pomegranates unmoistened and not kneaded with water, but pulverized, believing that to be an antidote against the bite of the scorpions, and, besides, they write in the time between dawnrise and sunrise upon square pieces of paper the following charm: "In the name of God the gracious, the merciful—Isfandârmadhmâh and Isfandârmadhrôz—I have bound (by the charm) the going and coming—below and above—except the cows—in the name of the Yazatas and in the name of Jam and Frêdûn—in the name of God—(I swear) by Adam and Eve, God alone is sufficient unto me!" Three such paper pieces they fix on this day on three walls of the house, whilst they leave unmarked the wall opposite to the front of the house, believing that if they fix something also on this fourth wall the reptiles get bewildered and do not find an outlet, and raise their heads towards the window, preparing to leave the house. Sometimes you find places influenced by some charm where scorpions do not bite, as, *e.g.* Dînâr-Râzî in Jurjân, ten miles beyond the frontier towards Khorâsân. For there you find under every stone a number of large black scorpions,
p.230. which people touch and play with, and which do not bite. But when they are taken away and brought over the frontier of that district, which is a bridge not farther off than a bowshot, then they bite, causing instantaneous death.

In the district of Ṭûs there is said to be a village where the scorpions do not bite. And 'Abû-alfaraj Alzanjânî has told me that in the city of Zanjân there are scorpions only in one place, called the "*Cemetery of the Ṭabaristânîs,*" and that a man when he goes there at night and gathers some of them in a pot and leaves the pot somewhere else, finds that they hurriedly return to their former places.

Now, as regards these pieces of paper we have mentioned, they are evidently useless, because the power of the incantation cannot affect the object of incantation, though its influence be strong, because the planetary cycles do not agree with the Persian year, and because the conditions of talismans are not fulfilled in them. Perhaps we shall speak of the incantations, charms, and talismans in the *Book of physical and technical wonders and curiosities*, giving such explanations as will plant certain persuasions in the minds of intelligent men and remove doubt from the minds of those who seek for information, if God will mercifully postpone the end of my life and by His grace remove mental calamities. He has the power to do so.

The 11th, or Khûr-Rôz, is the first day of the second Gahanbâr, the last of which is Dai-ba-mihr-Rôz. It is called *Maidhyôshema-gâh*. During this Gahanbâr God created the water.

The next following day, the 16th, or Mihr-Rôz, is called *Misk-i-tâza* (fresh musk).

The 19th, or Farwardîn-Rôz, is called *Naurôz of the rivers and of all running waters*, when people throw perfumes, rose-water, &c. into them.

The Zoroastrians have no fasting at all. He who fasts commits a sin, and must, by way of expiation, give food to a number of poor people. They have fairs in the days of the months we have mentioned, but as they differ in different places, we cannot fix them, as little as we can the watercourses of a torrent, it being impossible to count them.

'Aḍud-aldaula has founded two feast-days, each of which is called Ćashn-i-Kard-i-Fanâkhusrau. The one is the day Serôsh in Farwardîn-Mâh, when the water of the aqueduct coming from a distance of four farsakh reached the town, which he had built one farsakh below the citadel of Shîrâz, and which he had called Kard-i-Fanâkhusra. The other is the day Hormuz in Âbân-Mâh, the day when he commenced building that same town, A. Yazd. 333. On both days people hold fairs of seven days duration, and they assemble for merriment and drinking.

The Persians divide all the days of the year into preferable and lucky days and into unlucky and detested ones. Besides they have other days, bearing names which are common to them in every month, which are festival days for one class of the people to the exclusion of the other.

Further, they have certain rules regarding the appearance of snakes on the different days of the month, which we unite in the following *Jadwal-alikhtiyârât* (Table of Selections) :—

pp. 231, 232.

Names of the days of the months.	Far-wardin-Mâh.	Ardîba-hisht-Mâh.	Khur-dâdh-Mâh.	Tîr-Mâh.	Murdâdh-Mâh.	Shahrê-war-Mâh.	Mihr-Mâh.	Âbân-Mâh.	Âdhar-Mâh.	Dai-Mâh.	Bahman-Mâh.	Isfandâr-madh-Mâh.	What the appearance of a snake in one of the days of the month signifies.
Hormuz 1 - -	preferable, because it bears the name of God					-	-	-	-	-	-	-	Before noon : Sultân.
Bahman - -	middling	middling	middling	middling	middling	middling	middling	middling	middling	middling	unlucky	middling	Illness and disease.
Ardîbahisht -	"	lucky	lucky	lucky	"	unlucky	"	"	lucky	"	middling	lucky	Death or any loss in the family.
Shahrêwar -	"	"	"	middling	lucky	lucky	"	"	middling	"	"	middling	Something useful and a help coming from the people of his place.
Isfandârmadh 5	"	"	"	lucky	middling	"	"	lucky	lucky	"	"	unlucky	Reputation and praise.
Khurdâdh -	"	middling	"	"	lucky	"	"	"	middling	lucky	"	lucky	A very useful journey.
Murdâdh - -	"	"	middling	middling	"	"	"	"	"	"	"	"	Illness and disease.
Dai-ba-Âdhar -	"	"	"	"	middling	middling	"	middling	"	middling	"	middling	Coming to the Sultân.
Âdhar - -	unlucky	"	lucky	lucky	lucky	lucky	"	"	"	lucky	lucky	unlucky	Like the day before.
Âbân 10 - -	middling	"	middling	middling	middling	middling	"	unlucky	"	middling	middling	middling	Matchmaking and marrying.
Khûr - - -	"	"	unlucky	"	"	"	"	middling	"	"	"	"	Money without exertion.
Mâh - - -	preferable, because it bears the name of the moon					-	-	-	-	-	-	-	Good—till noon, bad—afterwards.
Tîr - - -	lucky	lucky	middling	unlucky	middling	middling	middling	middling	middling	middling	middling	middling	An increase of wealth.
Gôsh - - -	middling	"	"	lucky	lucky	"	"	"	"	"	"	"	Nourishment alone from quadrupeds.
Dai-ba-Mihr 15 -	"	unlucky	"	middling	middling	"	"	lucky	"	"	"	"	Illness followed by convalescence.
Mihr - - -	"	lucky	lucky	"	"	"	lucky	middling	"	"	"	lucky	The acquisition of something you had not got before.
Srôsh - -	"	middling	middling	"	"	"	middling	lucky	lucky	"	lucky	middling	Journey and return.
Rashn - -	"	"	"	"	"	lucky	lucky	middling	middling	"	middling	"	Journey and illness during it.
Farwardîn - -	lucky	"	"	lucky	"	middling	middling	"	lucky	"	"	"	An increase of wealth.
Bahrâm 20 - -	middling	"	"	middling	unlucky	"	"	"	unlucky	unlucky	"	lucky	Some one of the family dies.
Râm - - -	lucky	lucky	"	"	lucky	lucky	lucky	lucky	middling	lucky	lucky	"	Journey and victory over the adversaries.
Bâdh - - -	unlucky	middling	"	"	"	"	middling	middling	"	"	unlucky	middling	Suspicion of theft.
Dai-ba-Dîn -	middling	"	"	"	middling	middling	"	lucky	"	middling	middling	"	Illness and disease.
Dîn - - -	lucky	"	"	"	"	"	"	middling	"	lucky	lucky	lucky	The acquisition of money.
Ard 25 - -	"	lucky	"	"	"	"	lucky	"	"	middling	"	"	Bad and blamable.
Ashtâdh - -	middling	middling	"	lucky	"	lucky	middling	"	"	lucky	middling	middling	Building a new house.
Âsmân - -	"	lucky	lucky	"	"	"	"	"	"	"	"	"	Being accused of lying.
Zâmyâd - -	"	middling	middling	middling	"	middling	"	"	"	"	"	"	A calamity in the property and family.
Mârasfand -	"	"	"	"	"	"	"	"	"	middling	"	"	A short journey.
Anêrân 30 - -	"	"	unlucky	"	"	"	"	"	"	"	"	"	Punishment for fornication.

The day Mâh they consider to be a *preferable* day from its being called p.233. by the name of the moon, which God created for the purpose of distributing what is good and agreeable over the world. Therefore the waters increase, and animals, trees, and plants grow from new-moon till the time when the moon begins to wane.

The two days of conjunction and opposition they hold to be *unlucky* days.

On the day of conjunction the Demons and Satans feel the lust of intermingling viciously with the things in the world. Then madness and epilepsy are brought about. The seas begin to ebb, the waters to decrease, the male turtle-doves are suffering from epilepsy. The sperma which on this day settles in the uterus is born as a child of imperfect structure; hair which is torn out of the body will be replaced only sparsely; everything that is planted will only produce scanty fruit, more particularly so if there be an eclipse on the same day. If a hen sits hatching her eggs at new-moon, the eggs will be bad; at new-moon a narcissus is sure to wither.

Al-Kindî says: Conjunction is *detested* because then the moon is being burned, who is the guide of all bodies; and therefore people dread destruction and ruin for them.

At the time of opposition, people say, the Ghûls and sorcerers feel the lust to mix with impure spirits. In consequence there is much epilepsy. The seas begin to flow, the waters to increase; the she-turtle-doves are becoming epileptical. The sperma which settles in the uterus on the day of opposition is born as a child of more than common structure. The hair which is torn out will be replaced abundantly. All that is planted on this day will produce worm-eaten fruits and will be very impure, more particularly so if there be an eclipse on the same day.

Al-Kindî says: Full-moon is *detested* because then the light of the moon requires help from the light of the sun, who is the guide of the spirits. Therefore people fear lest the spirits should leave the bodies.

CHAPTER X.

ON THE FESTIVALS IN THE MONTHS OF THE SUGHDIANS.

The months of the inhabitants of Sogdiana were likewise distributed over the four quarters of the year. The first day of the Sughdian month Nausard was the first day of summer. There was no difference between them and the Persians regarding the beginning of the year and the beginnings of *some* of the months, but there was a difference regarding the place of the five *Epagomenæ*, as we have heretofore explained. And they did so for no other reason but this, that they honoured their kings to such a degree that they would not do the same things which the kings did. They preferred to use as new-year that moment when Jam returned successful, whilst the kings preferred as new-year that moment when Jam started (set out).

Some people maintain that these two different new-years were to be traced to a difference that was discovered in the astronomical observations. For the ancient Persians used a solar year of 365 days 6 hours 1 minute, and it was their universal practice to reckon these 6 hours *plus* the 1
p.234. minute as a unit (*i.e.* to disregard the 1 minute in reckoning).

But afterwards when Zoroaster appeared and introduced the religion of the *Magi*, when the kings transferred their residence from Balkh to Persis and Babel and occupied themselves with the affairs of their religion, they ordered new observations to be made, and then they found that the summer-solstice preceded by five days the beginning of the year, which was the third year after intercalation. In consequence, they gave up their former system and adopted what astronomical observation had taught them, whilst the people of Transoxiana kept the old system and disregarded the state of that same year (*i.e.* its deviation from real time), on which their calendar was based. Hence the difference of the beginnings of the Persian and Sughdian years.

Other people maintain that originally both the Persian and Sughdian years had the same beginning, until the time when Zoroaster appeared. But when after Zoroaster the Persians began to transfer the five *Epagomenæ* to each of the leap-months, as we have before mentioned, the Sughdians left them in their original place and did not transfer them. So they kept them at the end of the months of their year, whilst the Persians, after they began to neglect intercalation, retained them at the end of Âbân-Mâh. God knows best!

The Sughdians have many festivals and famous memorial days in the same way as the Persians. What we have learned of them, regarding this subject, is the following:—

Nausard. The 1st day is their Naurôz, which is the *Great Naurôz*.

The 28th is a feast for the Magians of Bukhârâ, called *Râmush-Âghâm*, during which they assemble in a fire-temple in the village Râmush. These Âghâms are the most important of their festivals, which they celebrate alternately in each village, assembling in the house of each chieftain, eating and drinking.

Jirjîn. Nothing mentioned.

Nîsanaj. The 12th is the first Mâkhîraj.

Basâkanaj. The 7th is the بكح Âghâm, a feast of theirs at Baikand, where they assemble.

The 12th is the second Mâkhîraj.

The 15th is the feast عمس *Khwâra*, when they eat leavened bread after abstaining from eating and drinking and from everything that is touched by the fire except fruits and vegetables.

Ashnâkhandâ. The 18th is the feast Bâba-Khwâra, also called Bâmî-Khwâra, *i.e.* drinking the good, pure must.

The 26th is Karm-Khwâra.

Mazhîkhandâ. The 3rd is the feast Kishmîn, when they hold a fair in the village كمجكث. On the 15th they hold a fair in Al-ṭawâwîs. There the merchants of all countries gather and hold a fair of seven days duration.

Faghakân. The 1st is called *Nîm-sarda*, *i.e.* the half of the year.

The 2nd is a feast called مس عيد Khwâra, when they assemble in their p.235.
fire-temples and eat a certain dish which they prepare of the flour of millet, of butter and sugar. Some people put *Nîm-sarda* five days earlier, *i.e.* on the 1st of Mihr-Mâh, to make it agree with the Persian calendar, whilst, in fact, the middle of the year ought to be celebrated when after its beginning 6 months and $2\frac{1}{2}$ days have passed.

The 9th is the feast تسيبس Âghâm.

The 25th is the first day of Karm-Khwâra.

Abhânaj. The 9th is the last day of Karm-Khwâra.

Fûgh. Nothing mentioned.

Marsâfûgh. From the 5th till the 15th of this month they have a feast. After that the Muhammadans hold a fair of seven days in Alshargh.

Zhîmadânaj. The 24th is the Bâdh-Amghâm.

Khshûm. On the last day of this month the Sughdians cry over those who died in past times, they lament over them and cut their faces. They lay out for them dishes and drinks, as the Persians do in Farwardajân. For the five days, which are the ἡμέραι κλοπιμαῖαι to the Sughdians, they fix at the end of this month, as we have mentioned before.

Besides, they hold fairs in the villages in the districts of Bukhârâ and Ṣughd on those days that have only one name in every month (*i.e.* the 8th, 15th, 23rd, which are called *Dast*).

CHAPTER XI.

ON THE FESTIVALS IN THE MONTHS OF THE KHWÂRIZMIANS.

The Khwârizmians agree with the Sughdians regarding the beginnings of the year and the months, and they disagree with the Persians in the same subjects. The cause of this is the same which we have described when speaking of the Sughdians. Their usages in their months are similar to those of the Sughdians. The beginning of their summer was the 1st of Nâusârjî. They had festivals in their months which they celebrated before the time of Islâm. They maintain that God Almighty ordered them to celebrate those festivals. Besides they celebrate other days in commemoration of the deeds of their ancestors. But at the present time there are only very few of the Magians among the Khwârizmians left, who do not particularly care for their religion; they know nothing of it except its outward forms, and they do not inquire into its spirit and real meaning. In consequence, they regulate their festivals by the knowledge of their distances from each other, not according to their real places which they occupy in the single months.

Those, now, of their days and festivals that are not connected with their religion are the following:

Nâusârjî. The 1st day is the feast of new-year, the new-day, as we have already mentioned.

Ardiwisht. Nothing mentioned. p.236.

Harûdâdh. The 1st day is called اريجها سوان. In ante-Muhammadan times this day was the time of extreme heat; therefore, they say, it was originally called اريجهاس چوزان which means: *the dress will be put off*, signifying that it was the time for baring and undressing themselves.

In our time this day coincides with the time of the sowing of sesame and what is sown together with it. So people have come to use it as an epoch.

Ćîrî. The 15th is called *Ajghâr*, which means: *the firewood and the flame.* In bygone times it was the beginning of that season when people felt the need of warming themselves at the fire, because the air was changing in autumn. In our time it coincides with the middle of summer. From this day they count 70 days, and then commence sowing the autumn wheat.

Hamdâdh. Nothing mentioned.

Ikhsharêwarî. The 1st day is called فغبرية; but originally, they say, it was called *Faghrubah*, i.e. *the exitus of the Shâh.* For about this time the kings of Khwârizm used to march out, because the heat was then decreasing and the cold drawing near; then they went into winter-quarters outside their residence, driving away the Ghuzz-Turks from their frontiers and defending the limits of their empire against their inroads.

Ûmrî. The 1st day is the feast *Azâd Kand Khwâr*, i.e. *the day of eating the bread prepared with fat.* On that day they sought protection from the cold, and assembled for the purpose of eating the *bread prepared with fat*, around the burning fire-grates.

The 13th is the feast *Ćîrî-Rôj*, which the Khwârizmians hold in the same veneration as the Persians their Mihrajân.

The 21st is likewise a feast, called *Râm-Rôj.*

Yânâkhun. Nothing mentioned in this month.

Adû. Nothing mentioned.

Rîmazhd. The 11th is called *Nîmkhab.* People say that it was originally called *Mînać Akhîb*, which was then wrongly altered for the sake of easier pronunciation, as it was frequently used. It means: *the night of Mîna.* Now, some of them maintain that Mîna was one of their queens or chieftains, that she left her castle intoxicated, dressed in a silk dress, at spring time. She fell down outside the castle and lost all self-control; she fell asleep, was smitten by the cold of the night, and died. Now people were astonished that the cold had killed a human being about this time in spring. So they used it as an epoch for some-
p.237. thing miraculous, extraordinary, that does not happen at its proper time.

This day has been advancing beyond its proper time to such a degree, that now-a-days people consider it as the middle of winter.

On this day and about this time the people of Khwârizm use perfumes and incense, and they make the smells rise up from the dishes which they lay out for the purpose of keeping off all the injuries of the demons and evil spirits.

This proceeding is necessary, by way of careful precaution, if some spiritual matters are connected with it. I mean charms, incantations, and prayers, which the most distinguished philosophers have acknowledged and allowed, after having witnessed their effects, *e.g.* Galenus, and others like him, though they are few. These precautions are likewise to be recommended if people in doing so derive some help from astronomical occurrences, as, *e.g.* the *Tempora Parata* and the *Tempora Selecta*, with the constellations that are mentioned for such purposes. We cannot help taking notice of those who try to prove that all such precaution is futile and false by no other arguments but by mockery, derision, and sneers.

The existence of jinns and demons has been acknowledged by the most famous philosophers and scholars, *e.g.* by Aristotle, when he describes them as beings of air and fire and calls them "*human beings.*" Likewise Yaḥyâ Grammaticus and others have acknowledged them, describing them as the impure parts of the erring souls, after they have been separated from their bodies, who (the souls) are prevented from reaching their primal origin, because they did not find the knowledge of the truth, but were living in confusion and stupefaction. Something similar to this is what Mânî indicates in his books, although his indications are expressed in subtle words and phrases.

Akhamman. Nothing mentioned in this month.

Ispandârmajî. The 4th is called *Khêzh, i.e.* the rising.

The 10th is a feast called *Wakhsh-Angâm.* Wakhsh is the name of the angel who has to watch over the water and especially over the river Oxus.

The 20th is called ابجهه, which means: houses that are built close together.

Besides they have other festivals which they want for the affairs of their religion; they are the following six:—

I. The first is called بنخجاچی ريد on the 11th of Nâusarjî. Common people call it *Nâusârjakânîk* by the month in which it occurs.

II. The second is called ميث سخن ريد on the 1st of Ćîrî. It is also called *Jâwardamînik,* i.e. القرعى and *Ajghârminik,* so called from the month Ajghâr, because it falls 15 days before that feast (on the 15th of Ćîrî).

III. The third is called مذبان ريد on the 15th of Hamdâdh. It is also called انجمردكانيك.

IV. The fourth is called ميث زرمى ريد on the 15th of Ûmrî, also called p.238.
خيرروچكانيك.

V. The fifth is called (*lacuna*) on the 1st of Rîmazhd, also called بكجدريكانيك.

VI. The sixth is called ارثمين ريد on the 1st of Akhamman, also called ارثمين دكانيك.

In the five last days of Ispandârmajî and the following five *Epagomenæ* they do the same which the Persians do in Farwardajân, *i.e.* they lay out food in the temples for the spirits of the dead.

(The Lunar Stations with the Chorasmians).—They were in the habit of using the stations of the moon and deriving from them the rules of astrology. The names of the stations in their language they have preserved, but those who made use of them, who knew how to observe them and how to draw conclusions from them, have died out. Their using the lunar stations is clearly proved by the fact that in the Khwârizmî dialect an astronomer is called *Akhtar-wênîk*, i.e. *looking to the lunar stations*, for *Akhtar* means a station of the moon.

They used to distribute these stations over the twelve signs of the Zodiac, for which they also had special names in their language. They knew them (the signs of Zodiac) even better than the Arabs, as you learn by the fact that their nomenclature of them agrees with the names given to them by the original designer of their figures, whilst the names of the Arabs do not agree, and they represent these signs as quite different figures.

For instance, they count Aljauzâ among the number of the Zodiacal signs instead of Gemini, whilst Aljauzâ is the figure Orion. The people of Khwârizm call this sign (Gemini) *Adhûpaćkarîk*, i.e. *having two figures*, which means the same as Gemini.

Further, the Arabs represent the figure of Leo as composed of a number of figures. In consequence, Leo extends in longitude over something more than three signs, not to mention its extension in latitude. For they consider the two heads of Gemini as his outstretched forefoot, and the nebula, in the foremost part of which is Cancer, I mean *Alnathra*, as his nose. The breast of Virgo, I mean *Al'awwâ*, they consider as his two loins; the hand of Virgo, I mean *Alsimâk Al'a'zal*, as one of his shanks; and Alrâmiḥ as his other shank. According to their opinion, the figure of Leo extends over the signs Cancer, Leo, Virgo, and part of Libra, and a number of constellations both of the northern and southern hemispheres, whilst in reality the matter is not what they assume.

If you, likewise, inquire into the names of the Arabs for the fixed stars, you will see that they were very far from an accurate knowledge of the Zodiacal signs and the star-figures, although 'Abû-Muḥammad 'Abdallâh b. Muslim b. Ḳutaiba Aljabalî used to make a great to-do and to be very verbose in all his books, and specially in his book on the superiority of the Arabs over the Persians, maintaining that the Arabs were the best-informed nation regarding the stars and the times of their rising and setting. I do not know whether he was really ignorant, or only pretended to be ignorant, of what the agriculturists and peasants in every place and district have got in the way of knowledge regarding the beginning of the agricultural works and other things, and of knowledge of the proper times for similar subjects. For he whose roof is heaven, who has no other cover, over whom the stars continually rise and set in one and the same course, p.239. makes the beginnings of his affairs and his knowledge of time depend upon them. But the Arabs had, moreover, one advantage in which others did not share; this is the perpetuation of what they knew or believed, right or wrong, praise or blame, by means of their poetry (Ḳaṣîdas), by Rajaz poems, and by compositions in rhymed prose. These things one generation inherited from the other, so as to remain among them and after them. If you study those traditions in the *'Anwâ* books, and specially his book which he called "*The Science of the Appearance of the Stars,*" part of which we have communicated to the reader at the end of this book, you will find that the Arabs had no particular knowledge on this subject beyond that which is familiar to the peasants of every country. The man (*i.e.* 'Abdallâh b. Muslim Aljabalî), however, is extravagant in the subject into which he plunges, and not free from Jabalî (*i.e.* mountaineer) character, as far as obstinacy of opinion is concerned. The style of his book which we have mentioned shows that there must have been enmities and grudges between him and the Persians. For he is not satisfied at exalting the Arabs at the expense of the Persians, but he must needs make the Persians the meanest, vilest, and most degraded of all nations, attribute to them even more want of belief and obstinacy against Islâm than God attributes to the Arab Bedouins in the Sûra *Altauba* (Sûra ix. 98), and heap upon them all that is abominable. If he had only taken a moment's consideration and had called to mind the first period of those whom he preferred to the Persians, he would have given the lie to himself in most of what he says about both parties from sheer want of moderation and equity.

In the following we give the names of the lunar stations in the dialect both of the Sughdians and the Khwârizmians. Afterwards we shall describe the constellations in which they appear, when we speak of the times of their rising and setting.

TABLE OF THE LUNAR STATIONS. p.240.

Their Names in Arabic.	In Sogdian.	In Chorasmian.
1. Althurayyâ	پروی	پروی
Aldabarân	باپرو	باپرو
Alhaḳ'a	مرازنه	اخماه
Alhan'a	رهنوند	خویا
5. Aldhirâ'	غنف	غوثف
Alnathra	هنب	جیری
Alṭarf	خمشریش	خمشیش
Aljabha	مغ	اچیر
Alzubra	وذه	امغ

Their Names in Arabic.	In Sogdian.	In Chorasmian.
10. Alṣarfa	ويذو	ويذيو
Al'awwâ	فستشى	أفسست
Alsimâk	هغار	اخشفرن
Alghafr	سرو	هوهك
Alzubaniyân	فسرو	سرافسريو
15. Al'iklîl	هنوند	اغنونه
Alḳalb	بهنوند	بهنوند
Alshaula	مغن سدويس	داريند
Alna'â'im	بسشم	سرذيو
Albalda	وزريك	مرخشيك
20. Sa'd aldhâbiḥ	ونند	خجمن
Sa'd bula'	يوغ	يوغ
Sa'd alsu'ûd	هدمشير	سدمسيج
Sa'd al'akhbiya	هوهت	مشتوند
Alfargh almuḳaddam	فرهت باث	فرخشبيث
25. Alfargh almu'akhkhar	بر فرهت	وبير
Baṭn alḥût	ريوند	زداك
Alsharaṭân	بشيش	ريوند
Albuṭain	ترو	فرخند

CHAPTER XII. p.241.

ON KHWARIZM-SHAH'S REFORM OF THE KHWÂRIZMIAN FESTAL CALENDAR.

'Abu-Sa'îd 'Ahmad b. Muḥammad b. 'Irâḳ followed the example of Almu'taḍid-billâh regarding the intercalation of the Chorasmian months. For on having been freed from his fetters at Bukhârâ, and having returned to his residence, he asked the mathematicians at his court regarding the feast Ajghâr, whereupon they pointed out to him its place in the calendar. Further, he asked with what day of Tammûz it corresponded, and this also they told him. This date he kept in memory, and when seven years later at the same time of the year he again came to think of it, he rejected this sort of calculation. He was not as yet acquainted with the intercalations and all matters connected with them. Then he ordered Alkharâjî and Alḥamdakî and other astronomers of his time to be brought before him, and asked them as to what was the reality of the case. These scholars then gave him a minute explanation and told him how the Persians and Chorasmians had managed their year. Thereupon he said: "This is a system which has become confused and forgotten. The people rely upon these days (*i.e.* certain feast-days, *Ajghâr*, *Nîmkhab*, etc.), and thereby they find the cardinal points of the four seasons, since they believe that they never change their places in the year; that *Ajghâr* is always the middle of summer, *Nîmkhab* the middle of winter; certain distances from these days they use as the proper times for sowing and ploughing. Something like this (*i.e.* the deviation of the Chorasmian year from proper time) is not perceived except in the course of many years. And this is one of the reasons why they disagree among each other regarding the fixing of those distances, so that some maintain that 60 days after Ajghâr is the proper time for sowing the wheat, whilst others put this time earlier or later. The proper thing would be that we should find some means to fix those things uniformly and to invariable times of the year, so that the proper times for these things should never differ."

Now, the scholars told him that the best way in this matter would be to fix the beginnings of the Chorasmian months on certain days of the Greek and Syrian months—in the same way as Almu'taḍid had done—and after that to intercalate them as the Greeks and Syrians do. This plan they carried out A. Alex. 1270, and they arranged that the 1st of Nâusârjî should fall on the third of the Syrian Nîsân, so that *Ajghâr* would always fall in the middle of Tammûz. And accordingly they regulated the times of agricultural works, *e.g.* the time of gathering grapes for the purpose of making raisins is 40–50 days after *Ajghâr*; the time of gathering grapes for the purpose of hanging them up, and the time of gathering pears, is 55–65 days after *Ajghâr*. In the same way they fixed all the times for sowing, for the impregnation of the palm-trees, for planting and binding together, etc. If the Greek year is a leap-year, the *Epagomenæ* at the end of Ispandârmajî are six days. If people had made this reform of Khwârizm-Shâh the epoch of an era, we should have added it to the other eras which we have before mentioned.

---

p.242. Regarding the festivals in the non-intercalated months of the Egyptians, although it is likely that they had similar ones with the other nations, we have not received any information. Likewise we have no information regarding their festivals in the intercalated months which they use now-a-days, except this, that people say that new-year of the Egyptians is the 1st of Thôth, and that the water of the Nile begins to swell and to increase on the 16th of Payni, according to another report on the 20th of Payni. It is likely that they would celebrate the same festivals as the Greeks and Syrians, because Egypt lies in the midst between them and because they all use the same kind of year. Some matters, however, are quite peculiar to the Egyptians, *e.g.* their country, Egypt, has certain peculiarities, in which no other country shares—appearances of the water, the air, the rain, etc.

The famous days of the Greeks and Syrians are of two kinds, one for the affairs of any sort of secular life, for certain aerial appearances, etc., as we have already mentioned, and another kind for the matters of their religion, which is Christianity. We shall describe in its proper place as much as we have learned about both kinds, and as has been reported to us, if God permits!

CHAPTER XIII.

ON THE DAYS OF THE GREEK CALENDAR AS KNOWN BOTH AMONG THE GREEKS AND OTHER NATIONS.

The Greek year agrees with the solar year; its seasons retain their proper places like the natural seasons of the solar year; it revolves parallel with the latter, and its single parts never cease to correspond with those of the latter, except by that quantity of time (the *Portio Intercalationis*) which, before it becomes perceptible, is appended to the year and added to it as one whole day (in every fourth year) by means of intercalation. Therefore the Greeks and Syrians and all who follow their example fix and arrange by this kind of year all annual, consecutive occurrences, and also the meteorological and other qualities of the single days that experience has taught them in the long run of time, which are called *'Anwâ* and *Bawârih*.

Regarding the cause of these *'Anwâ*, scholars do not agree among each other. Some derive them from the rising and setting of the fixed stars, among them the Arabs. (Some poet says):

"Those are my people (a bad set) like the Banât-Na'sh,
Who do not bring rain like the other stars;"

i.e. they are good-for-nothing people like the *Banât-Na'sh*, whose rising and setting do not bring rain.

Others, again, derive them from the days themselves, maintaining that they are peculiarities of them, that such is their nature, at least, on an average, and that besides they are increased or diminished by other causes. They say, for instance: The nature of the season of summer is heat, the nature of the season of winter is cold, sometimes in a higher degree, sometimes less. The excellent Galenus says: "To decide between p.243.

these parties is only possible on the basis of experiment and examination. But to examine this difference of opinion is not possible except in a long space of time, because the motion of the fixed stars is very little known and because in a short space of time we find very little difference in their rising and setting."

Now, this opinion has filled Sinân b. Thâbit b. Ḳurra with surprise. He says in his book on the *'Anwâ*, which he composed for the Khalif Almu'taḍid: "I do not know how Galenus came to make such a mistake, skilled as he was in astronomy. For the rising and setting of the stars differ greatly and evidently in different countries. *E.g. Suhail* rises at Baghdâd on the 5th of Îlûl, at Wâsiṭ two days later, at Baṣra somewhat earlier than at Wâsiṭ. People say: '*the 'Anwâ* differ in different countries.' But that is not the case. On the contrary, they occur always on one and the same day (everywhere); which proves that the stars and their rising and setting have nothing to do with this matter."

Afterwards he has given the lie to himself, though it is correct what he said, viz., that the rising and setting of the stars are not to be considered as forming one of the causes of the 'Anwâ, if you limit his assertion by certain conditions and do not understand it in that generality in which he has proclaimed it.

Further he (Sinân b. Thâbit) says: "The *'Anwâ* of the Arabs are mostly correct for Alḥijâz and the neighbourhood, those of the Egyptians for Egypt and the coasts of the sea, those of Ptolemy for Greece and the neighbouring mountains. If anybody would go to one of those countries and examine them there, he would find correct what Galenus says regarding the difficulty of an examination of the *'Anwâ* in a short space of time." In this respect he (Sinân) is right. Galenus mentions and believes only what he considers as a truth, resting on certain arguments, and keeps aloof from everything that is beset with doubt and obscurity.

Sinân relates of his father, that he examined the *'Anwâ* in 'Irâḳ about thirty years with the view of finding certain principles with which to compare the *'Anwâ* of other countries. But fate overtook him before he could accomplish his plan.

Whichever of the two theories may be correct, whether the *'Anwâ* are to be traced back to the days of the year or to the rising and setting of the *Lunar Stations*, in any case there is no room for a third theory. To each of these theories, whichever you may hold to be correct, certain conditions attach, on which the correctness of the *'Anwâ* depends, *i.e.* to prognosticate the character of the year, the season, the month, whether it will be dry or moist, whether it will answer to the expectations of people or not, to prognosticate it by means of the signs and proofs, of which the astronomical books on meteorology are full. For if the *'Anwâ* agree with those signs and proofs, they are true and will be fulfilled in their entire extent; if they do not agree, something different will occur. Thus the matter stands between these two theories.

Sinân b. Thâbit prescribes that we should take into regard whether the Arabs and Persians agree on a *Nau'*. If they do agree, its probability is strengthened and it is sure to take place; if they do not agree, the contrary is the case.

I shall mention in this book the comprehensive account of Sinân in his book on the *'Anwâ* and the proper times for secular affairs occurring in the Greek months. Of the rising and setting of the Lunar Stations I shall speak in a special chapter at the end of this book. For since the astronomers have found that their rising and setting proceed according to one and the same uniform order in these months, they have assigned them to their proper days, in order to unite them and prevent p.244.
them from getting into confusion. God lends support and help!

Tishrîn I. (October.)

1. People expect rain (Euctemon and Philippus); turbid air (Egyptians and Callippus).

2. Turbid winterly air (Callippus, Egyptians, and Euctemon); rain, (Eudoxus and Metrodorus).

3. Nothing mentioned.

4. Wearing wind (Eudoxus); winterly air (Egyptians).

5. Winterly air (Democritus); beginning of the time of sowing.

6. North wind (Egyptians).

7. South wind (Hipparchus).

8. Nothing mentioned. Winterly air, according to Sinân.

9. 'Επισημαίνει (Eudoxus); east wind (Hipparchus); west wind (Egyptians).

10. Nothing mentioned.

11. Episemasia (Eudoxus and Dositheus).

12. Rain (the Egyptians).

13. Unsteady wind, Episemasia, thunder, and rain (Callippus); north wind or south wind (Eudoxus and Dositheus). Sinân attests that this is frequently true. On this day the waves of the sea are sure to be in great commotion.

14. Episemasia and north wind (Eudoxus).

15. Change of the winds (Eudoxus).

16. Nothing mentioned.

17. Rain and Episemasia (Dositheus); west wind or south wind (Egyptians).

18. Nothing mentioned.

19. Rain and Episemasia (Dositheus); west wind or south wind (Egyptians).

20, 21. Nothing mentioned.

22. Unsteady, changing winds (Egyptians). On this day the air begins to get cold. It is no longer time for drinking medicine and for

phlebotomy except in case of need. For the *Favourable Times* for such things are always then, when you intend thereby to preserve the health of the body. For if you are compelled to use such means, you cannot wait for a night or day, for heat or cold, for a lucky or unlucky day. On the contrary, you use it as soon as possible, before the evil takes root, when it would be difficult to eradicate it.

23. Episemasia (Eudoxus); north wind or south wind (Cæsar).

24. Episemasia (Callippus and Egyptians).

25. Episemasia (Metrodorus); change in the air (Callippus and Euctemon).

26. Nothing mentioned.

p.245. 27. Winterly air (Egyptians).

28. Nothing mentioned. It is a favourable day for taking a warm bath and for eating things that are of a sharp, biting taste, nothing that is salt or bitter.

29. Hail or frost (Democritus); continual south wind (Hipparchus); tempest and winterly air (Egyptians).

30. Heavy wind (Euctemon and Philippus). The kites, the white carrion-vultures (*vultur percnopterus*), and the swallows migrate to the lowlands, and the ants go into their nest.

31. Violent winds (Callippus and Euctemon); wind and winterly air (Metrodorus and Cæsar); south wind (Egyptians). God knows best!

Tishrîn II. (November.)

1. Clear (*lit.* unmixed) winds (Eudoxus and Conon).

2. Clear air with cold north wind and south wind.

3. South wind blows (Ptolemæus); west wind (Egyptians); north or south wind (Eudoxus); rain (Euctemon, Philippus, and Hipparchus).

4. Episemasia (Euctemon); rain (Philippus).

5. Winterly air and rain (Egyptians).

6. South or west wind (Egyptians); winterly air (Dositheus). Sinân says that this is borne out by practical experience.

7. Rain with whirlwind (Meton); cold wind (Hipparchus). This is the first day of the rainy season, when the sun enters the 21st degree of Cancer. Astrologers take the horoscope of this time and derive therefrom an indication as to whether the year will have much rain or little. Herein they rely upon the condition of Venus at the times of her rising and setting. I believe, however, that this is only peculiar to the climate of 'Irâḳ and Syria, not to other countries, for very frequently it rains with us in Khwârizm even before this time. 'Abû-alḳâsim 'Ubaid-Allâh b. 'Abdallâh b. Khurdâdhbih relates in his *Kitâb-almasâlik walmamâlik* that in Ḥijâz and Yaman it rains during Ḥazîrân, Tammûz, and part of Îlûl. I myself have been dwelling in Jurjân during the summer months, but there never passed ten consecutive days during which the sky was clear and

free from clouds, and when it did not rain. It is a rainy country. People relate that one of the khalifs, I think it was Almaʼmûn, stayed there during forty days whilst it rained without any interruption. So he said: "Lead us out of this pissing, splashing country!"

The nearer a district is to Ṭabaristân, the more its air is moist, the more rainy it is. The air of the mountains of Ṭabaristân is so moist that if people break and pound garlic on the tops of the mountains, rain is sure to set in. As the cause of this subject, the vice-judge, Alâmulî, the author of the *Kitâb-Alghurra*, mentions this, that the air of the country is moist and dense with stagnant vapours. If, now, the smell of p.246. garlic spreads among these vapours, it dissolves the vapours by its sharpness and compresses the density of the air, in consequence of which rain follows.

Granted, now, that this be the cause of this appearance produced by the pounding of garlic, how do you, then, account for the famous well in the mountains of Farghâna, where it begins to rain as soon as you throw something dirty into this well?

And how do you account for the place called "*the shop of Solomon the son of David*," in the cave called Ispahbadhân in the mountain of Ṭâḳ in Ṭabaristân, where heaven becomes cloudy as soon as you defile it by filth or by milk, and where it rains until you clean it again?

And how do you account for the mountain in the country of the Turks? For if the sheep pass over it, people wrap their feet in wool to prevent their touching the rock of the mountain. For if they touch it, heavy rain immediately follows. Pieces of this rock the Turks carry about, and contrive to defend themselves thereby against all evil coming from the enemy, if they are surrounded by them. Now, those who are not aware of these facts consider this as a bit of sorcery on the part of the Turks.

Of a similar character is a fountain called "*the pure one*" in Egypt in the lowest part of a mountain which adjoins a church. Into this fountain sweet, nicely-smelling water is flowing out of a source in the bottom of the mountain. If, now, an individual that is impure through pollution or menstruation touches the water, it begins at once to stink, and does not cease until you pour out the water of the fountain and clean it; then it regains its nice smell.

Further, there is a mountain between Herât and Sijistân, in a sandy country, somewhat distant from the road, where you hear a clear murmur and a deep sound as soon as it is defiled by human excrements or urine. These things are natural peculiarities of the created beings, the causes of which are to be traced back to the simple elements and to the beginning of all composition and creation. And there is no possibility that our knowledge should ever penetrate to subjects of this description.

There are other districts of quite another character from that of the mountains of Ṭabaristân, *e.g.* Fusṭâṭ in Egypt, and the adjacent parts, for there it rains very seldom. And if it rains, the air is infected, becomes pestilential and hurts both animals and plants. Such things (*i.e.* such climatical differences) depend upon the nature of the place and its situation, whether it lies in the mountains or on the sea, whether it is a place of great elevation or a low country; further, upon the degree of northern or southern latitude of the place.

8. Rain and winterly air (Euctemon); winterly air and whirlwinds (Metrodorus); south wind or εὖρος, *i.e.* south-east wind (Euctemon); east wind (Egyptians).

9. Nothing mentioned.

10. Winterly air and whirlwinds (Euctemon and Philippus); north wind, or cold south wind and rain (Hipparchus).

11. Episemasia (Callippus, Conon, and Metrodorus). Sinân says that this is borne out by experience.

12. Winterly air (Eudoxus and Dositheus).

13. Episemasia (Eudoxus); winterly air on land and sea (Democritus). Ships that are at sea on this day put in to shore, and navigation to Persia and Alexandria is suspended. For the sea has certain days when
p.247. it is in uproar, when the air is turbid, the waves roll, and thick darkness lies over it. Therefore navigation is impracticable. People say that at this time there arises the wind at the bottom of the sea that puts the sea in motion. This they conclude from the appearance of a certain sort of fishes which then swim in the upper regions of the sea and on its surface, showing thereby that this storm is blowing at the bottom.

Frequently, people say, this submarine storm rises a day earlier. Every sailor recognizes this by certain marks in his special sea. For instance, in the Chinese sea this submarine storm is recognized by the fishing-nets rising of themselves from the bottom of the sea to its surface. On the contrary, they conclude that the sea bottom is quiet if a certain bird sits hatching her eggs—for they hatch in a bundle of chips and wood on the sea, if they do not go on land nor sit down there. They lay their eggs only at that time when the sea is quiet.

Further, people maintain that any wood which is cut on this day does not get worm-eaten, and that the white ant does not attack it. This peculiarity perhaps stands in connection with the nature of the mixture of the air on this special day.

14. Winterly air (Cæsar); south wind or Eurus, *i.e.* south-east wind (Egyptians).

15. Nothing mentioned.

16. Winterly air (Cæsar).

17. Rain (Eudoxus); winterly air (Cæsar); north wind during night and day (Cæsar).

18. Nothing registered.

19. Sharp winterly air (Eudoxus).

20. North wind (Eudoxus); severe winterly air (Egyptians). People say

that on this day all animals that have no bones perish. This, however, is different in different countries. For I used to be molested by the gnats, *i.e.* animals without bones, in Jurjân, whilst the sun was moving in the sign of Capricorn.

21. Winterly air and rain (Euctemon and Dositheus).

22. Very winterly air (Eudoxus). On this day people forbid to drink cold water during the night, for fear of the *Yellow Water*.

23. Rain (Philippus); winterly air (Eudoxus and Conon); continual south wind (Hipparchus and Egyptians). On this day falls the feast of gathering the olives, and the fresh olive-oil is pressed.

24. Light rain (Egyptians).

25, 26. Nothing mentioned.

27. In most cases a disturbance of the air on land and sea (Democritus); Episemasia (Dositheus); south wind and rain (Egyptians).

28. Nothing mentioned. People say that on this day the waves of the sea roll heavily and that there is very little fishing.

29. Winterly air (Eudoxus and Conon); west or south wind and rain (Egyptians).

30. Nothing mentioned by the authorities hitherto quoted, nor by others.

Kânûn I. (December.) p.248.

1. Winterly air (Callippus, Eudoxus, and Cæsar). On this day people hold a fair in Damascus, which is called "the fair of the cutting of the ben-nut," *i.e. Nux unguentaria*.

2. Pure winds (*lit.* not mixed) (Euctemon and Philippus); sharp, winterly air (Metrodorus).

3. Winterly air (Conon and Cæsar); light rain (Egyptians).

4. (Missing.)

5. Winterly air (Democritus and Dositheus). The same is confirmed by Sinân.

6. Winterly air (Eudoxus); vehement north wind (Hipparchus).

8. Nothing mentioned.

9. Winterly air and rain (Callippus, Euctemon, and Eudoxus).

10. Sharp winterly air (Callippus, Euctemon and Metrodorus); thunder and lightning, wind and rain (Democritus).

11. South wind and Episemasia (Callippus); winterly air and rain (Eudoxus and Egyptians). According to Sinân this is borne out by practical experience. Continued sexual intercourse on this day is objected to, which I do not quite understand. For sexual intercourse is not approved of in autumn, in the beginning of winter, and at the times of epidemic disease; on the contrary, at such times it is most noxious and pernicious to the body. Although we must say that the conditions of sexual intercourse depend upon a great many other things, as, *e.g.* age,

time, place, custom, character, nourishment, the fulness or emptiness of the stomach, the desire, the female genitals, etc.

12. Winterly air (Egyptians).

13. Vehement south wind or north wind (Hipparchus).

14. Winterly air (Eudoxus); rain and wind (Egyptians).

15. Cold north wind or south wind and rain (Egyptians).

16. Winterly air (Cæsar).

17. Nothing mentioned. People forbid on this day to take of the flesh of cows, of oranges, and mountain balm, to drink water after you lie down to sleep, to smear the camels with *Nûra* (a depilatory unguent made of arsenic and quick-lime), and to bleed anybody except him whose blood is feverish. The reason of all this is the cold and the moistness of the season. This day people call the "*Great Birth*," meaning the winter-solstice. People say that on this day the light leaves those limits within which it decreases, and enters those limits within which it increases, that human beings begin growing and increasing, whilst the demons begin withering and perishing.

Ka'b the Rabbi relates that on this day the sun was kept back for Yosua the son of Nûn during three hours on a clouded day. The same story is told by the simpletons among the Shî'a regarding the prince of the believers, 'Alî b. 'Abî Ţâlib. Whether, now, this story have any foundation or not, we must remark that those who are beset by calamity find its duration to be very long and think that the moment of liberation is very slow in coming. So, *e.g.* 'Alî b. Aljahm said in a sleepless night, when he had gone out to war against the Greeks, oppressed by wounds and fatigue:

p.249. "Has a stream swept away the morning,
Or has another night been added to the night?"

Afterwards on being released he indulged in hallucinations and lying reports.

Something similar frequently happens on fast-days, if heaven during the latter part of them be clouded and dark; then people break their fast, whilst shortly afterwards, when the sky or part of it clears up, the sun appears still standing above the horizon, having not yet set. The charm-mongers say that it is a good omen on this day to rise from sleeping on the right side, and to fumigate with frankincense in the morning before speaking. It is also considered desirable to walk twelve consecutive steps towards the east at the moment of sunrise.

Yaḥyâ b. 'Alî, the Christian writer of 'Anbâr, says that the rising-place of the sun at the time of the winter-solstice is the true east, that he rises from the very midst of paradise; that on this day the sages lay the foundations of the altars. It was the belief of this man that paradise is situated in the southern regions. But he had no knowledge of the difference of the zeniths. Besides, the dogma of his own religion proves his theory to be erroneous, for their law orders them to turn in praying towards the *east* (*i.e.* the rising-place of the sun), whilst he told them that the sun rises in paradise (*i.e.* in the *south* according to *his* theory). Therefore the Christians turn to no other rising-place but to that one of the equator, and they fix the direction of their churches accordingly.

This theory is not more curious than his view of the sun. For he maintains that the degrees through which the sun ascends and descends are 360 in number, corresponding to the days of the year; that during the 5 days which are the complement of the year the sun is neither ascending nor descending. Those are $2\frac{1}{2}$ days of Ḥazîrân and $2\frac{1}{2}$ days of Kânûn I.

A similar idea hovered in the mind of 'Abû-al'abbâs Alâmulî when he said in his book *On the Proofs for the Ḳibla* that the sun has 177 rising and setting places, thinking evidently that the solar year has got 354 days. He, however, who undertakes what he does not understand, incurs ignominy. Those crotchets of his are brought into connection with the argument regarding the 5 supernumerary days of the solar year and the 6 deficient days of the lunar year, of which we have already spoken.

18. Nothing mentioned.
19. South wind (Eudoxus, Dositheus, and Egyptians).
20. Winterly air (Eudoxus).
21. Episemasia (Egyptians).
22. Nothing mentioned.
23. Nothing mentioned.
24. Winterly air (Cæsar and Egyptians); Episemasia and rain (Hipparchus and Meton).
25. Middling winterly air (Democritus).
26. (Missing.)
27. Nothing mentioned.
28. Winterly air (Dositheus).
29. Episemasia (Callippus, Euctemon, and Democritus). People forbid p.250. on this day the drinking cold water after rising from sleep. They say that the demons vomit into the water, and that therefore he who drinks of it is affected by stupidity and phlegm. This serves as a warning to people against that which they dread most. The cause of all this is the coldness and moisture of the air.
30. Winterly air on the sea (Egyptians).
31. Winterly air (Euctemon).

Kânûn II. (January.)

1. Nothing mentioned by the *Parapegmatists*.
2. Episemasia (Dositheus). Some people say that wood which is cut on this day will not soon get dry.

3. Changeable air (Egyptians).

4. Episemasia (Egyptians); south wind (Democritus), which observation is confirmed by Sinân.

5, 6. Nothing mentioned. People say that on the 6th there is an hour during which all salt water of the earth is getting sweet. All the qualities occurring in the water depend exclusively upon the nature of that soil by which the water is enclosed, if it be standing, or over which the water flows, if it be running. Those qualities are of a stable nature, not to be altered except by a process of transformation from degree to degree by means of certain *media*. Therefore this statement of the waters getting sweet in this one hour is entirely unfounded. Continual and leisurely experimentation will show to any one the futility of this assertion. For if the water were sweet it would remain sweet for some space of time. Nay, if you would place—in this hour or any other—in a well of salt water some pounds of pure dry wax, possibly the saltishness of the water would diminish. This has been mentioned by the experimenters, who go so far as to maintain that if you make a thin vase of wax and place it in sea water, so that the mouth of the vase emerges above the water, those drops of water which splash over into the vase become sweet. If all salt water were mixed with so much sweet water as would overpower its nature, in that case their theory would be realized (*i.e.* all salt waters would become sweet). An example of this process is afforded by the lake of Tinnîs, the water of which is sweet in autumn and winter in consequence of the great admixture of the water of the Nile, whilst at the other seasons it is salt, because there is very little admixture of Nile water.

7. Winterly air (Eudoxus and Hipparchus).

8. South wind (Callippus, Euctemon, Philippus, and Metrodorus); south wind and west wind and winterly air on the sea (Egyptians).

9. Violent south wind and rain (Eudoxus and Egyptians).

The authors of talismans say that if you draw the figures of grapes on a table, between the 9th and the 16th of the month, and place it among the vines as a sort of offering at the time of the setting of the Tortoise, *i.e. Alnasr Alwâḳi'*, the fruit will not be injured by anything.

p.251. 10. Violent south wind and Episemasia (Cæsar and Egyptians).

11. South wind (Eudoxus and Dositheus); mixed winds (Hipparchus).

12. Nothing mentioned.

13. Winterly air (Hipparchus); a north wind or a south wind blows (Ptolemy).

14. Nothing mentioned.

15. East wind (Hipparchus).

16. Nothing mentioned.

17. Violent wind (Cæsar).

18. Winterly air (Euctemon and Philippus); change of the air (Metrodorus).

19. Winterly air (Eudoxus and Cæsar); suffocating air (Egyptians).

20. Clear sky (Euctemon and Democritus); north wind (Hipparchus) winterly air and rain (Egyptians).

21. Middling winterly air (Eudoxus).

22. Episemasia (Hipparchus); rain (Egyptians).

23. Nothing mentioned. On this day people do not smear the camels with *Nûra* (a depilatory unguent of arsenic and quicklime), nor bleed anybody except in cases of special need.

24. Clear sky (Callippus and Euctemon); middling winterly air (Democritus). Besides, the rule of the preceding day as regards the use of *Nûra* and phlebotomy refers also to this day.

25. East wind (Hipparchus).

26. Rain (Eudoxus and Metrodorus); winterly air (Dositheus).

27. Severe winter (Egyptians).

28. South wind blows and Episemasia (Ptolemy).

29. Nothing mentioned.

30. South wind (Hipparchus).

31. Nothing mentioned.

Shubâṭ. (February.)

It is the leap-month. It appears to me that the following is the reason —but God knows best!—why people have shortened this month in particular so that it has only 28 days, and why it has not had assigned to it 29 or 30 or 31 days: If it were assigned 29 days and were then to be increased by the leap-day, it would have 30 days and would no longer be distinguishable from the other months in a leap-year. The same would be the case if it had 30 days, whether the year be a leap-year or not. Likewise if it had 31 days, the same similarity with the other months in all sorts of years would exist. For this reason the leap-month has been assigned 28 days, that it might be distinguished from the other months both in leap and common years.

For the same reason it was necessary that in the Greek year two months of more than 30 days should follow each other. For at the beginning they intentionally gave to each month 30 days and took away 2 from Shubâṭ. So they got 7 supernumerary days (*i.e.* the 5 Epagomenæ and the 2 days of Shubâṭ), which they had to distribute over 11 months, because Shubâṭ had to be left out. Now, it was not possible to p.252
distribute the complete months of 30 days so as to fall each of them between two months of 31 days, for the latter (*i.e.* the months of 31 days) are more in number than the former. Therefore it was necessary to let several months of more than 30 days follow each other. But the most important subject of their deliberation was to add them in the places which would be the most suitable to them, so that the sum of the days of both spring and summer is more than the sum of the days of

autumn and winter, a fact which is the result of both ancient and modern observations.

Further, their months are proportional to each other in most cases; I mean to say: the sum of each month and of the seventh following one is 61 days, which is nearly equal to the time of the sun's mean motion through two signs of the zodiac. However, the sum of the days of Âb and Shubâṭ is 59 days. This could not have been otherwise, for the reason we have mentioned for Shubâṭ. For if Âb had been assigned more than 31 days, it would have been different from all the other months, and people would have thought that this in particular was the leap-month. As for Tammûz and Kânûn the Last, the sum of their days is 62. This, again, was necessary, because the number of the months of more than 30 days is greater than that of the months of 30 days. Wherever the supernumerary day is placed the circumstances are always the same. And, further, intercalation has been applied to Shubâṭ to the exclusion of the other months only for this reason, that Adhâr I., which is the leap-month in the Jewish leap-year, falls on Shubâṭ and near it.

1. Rain (Eudoxus). The cold decreases a little.

2. West wind or south wind intermixed with hail (Egyptians). Sinân says that this is frequently the case.

3. Clear sky and frequently the west wind blows (Eudoxus).

4. Clear sky and frequently the west wind blows (Dositheus); severe winterly air, rain and unmixed winds (Egyptians).

5. Nothing mentioned. People say that the four winds are in uproar.

6. Rain (Cæsar); winds (Egyptians); the west wind begins blowing (Democritus).

7. Beginning of the blowing of the west wind, frequently the air is winterly (Eudoxus and Egyptians). On this day the first *Coal* falls, called the minor one.

8. The time of the blowing of the west wind (Callippus, Metrodorus, and Hipparchus); rain (Eudoxus and Egyptians). This is confirmed by Sinân as borne out by his observations.

9, 10. Nothing mentioned.

11. Winterly air (Philippus and Metrodorus); west wind (Eudoxus and Egyptians).

12. North and east wind (Hipparchus); east wind alone (Egyptians).

13, 14. Nothing mentioned. On the 14th falls the second *Coal*, called
p.253. the middle one. As the poet says:

"When Christmas has passed and Epiphany after it,
And ten days and ten days and five complete days,
And five days and six and four of Shubâṭ,
Then, no doubt, the greatest cold vanishes.
That is the time of the falling of the two *Coals*; afterwards
The cold remains only a few nights."

15. Winterly air (Euctemon, Philippus, and Dositheus); changing wind (Egyptians); south wind (Hipparchus). This day is cold (Arabs), during which the coal is kindled. The Persians say: "The Summer has put his hands into the water." On this day the moisture of the wood is flowing from the lowest parts of the trees to the highest, and the frogs begin croaking.

16. A change in the winds and rain (Egyptians). People say that on this day the interior of the earth is getting warm. In Syria the mushrooms are coming forth; those which stand near the root of the olive-tree are deadly poison, as people maintain. This may be true, for it is not approved of to take much of the mushroom and fungus, nor of that which is prepared from them. Its pharmacological treatment is mentioned in most of the medical compilations in the chapter of preparing poisons from these materials.

17. Nothing mentioned.

18. West wind, and hail falls, or rain (Egyptians).

19. Cold north wind (Hipparchus).

20. Winds (Egyptians).

21. Nothing mentioned. On this day the third *Coal* falls, called the great one. Between the falling of each of the two *Coals* there is an interval of one complete week. They were called *Coals* because they are days characterised by the spreading of the heat from the interior of the earth to the surface, according to those who hold this theory. According to those who hold the opposite view, this change is brought about by the air's receiving heat instead of cold from the body of the sun, for the body of the sun and the near approach of a column of rays are the first cause of the heat. With this subject also the question is connected why the earthen jars or pipes of which subterranean channels are formed, and the water of wells, are warm in the winter and cold in the summer.

Between 'Abû-Bakr b. Zakariyyâ Alrâzî and 'Abû-Bakr Ḥusain Altammâr several questions and answers, expostulations and refutations have been exchanged that will satisfy the curiosity of the reader and inform him of the truth.

The Arabs used these three days (the so-called *Coals*) in their months until they got into confusion, as we have mentioned, and these days no longer fell at their proper times. Thereupon they were transferred into (*i.e.* fixed on certain days of) the Greek months which keep always their proper places. On the first day, people say, the 1st and 2nd *κλίματα* are getting warm, on the second the 3rd and 4th, on the third the remaining *κλίματα*. Further, they say that on the *Coal*-days vapours are rising p.254.
from the earth which warm the earth on the 1st *Coal*-day, the water on the 2nd, and the trees on the 3rd.

According to another view, they are days noticeable for the rising of Lunar Stations, or some special parts of them; whilst other subtle people maintain that they are the *termini* of the cold in winter, and serve to denote the differences in the beginning of heat and cold as known in the different countries. Some inconsiderate and over-zealous people of our ancestors have introduced these *Coal*-days into Khwârizm, so that the first fell on the 21st of Shubâṭ, the second a week later, and the third two weeks after the second one.

22. A cold north-east wind begins blowing and the swallows appear (Euctemon and Hipparchus).

23. Winds are blowing and the swallows appear (Callippus, Philippus, and the Egyptians); rain at the time of the appearance of the swallows, north-east wind during four days (Eudoxus, Conon, Callippus, and Philippus).

24. Cold north wind and west wind (Hipparchus); north-east wind with other winds (Egyptians); days with changeable air (Democritus).

25. Winterly air (Cæsar and Dositheus).

26, 27. Nothing mentioned.

28. Cold north wind (Hipparchus).

In this month fall the *Days of the Old Woman, i.e.* seven consecutive days beginning with the 26th; if the year is a leap-year, four days fall into Shubâṭ and three into Adhâr; if it is a common year, three fall into Shubâṭ and four into Adhâr. They are called by the Arabs by special names; the 1st is called *Al-ṣinn, i.e.* the severity of the cold, the 2nd is called *Al-ṣinnabr, i.e.* a man who leaves things as *Ṣanbara, i.e.* as something that is coarse and thick. The *Nûn* in this word is not radical, the same as in بلنصى *balanṣâ*, the plural of بَلَصُوص *balaṣûs*. The third is their brother *Al-wabr*, so called from the verb وَبَرَ, i.e. *he followed the trace of these days.* The 4th is called *Alâmir* (*commanding*), because he *commands* people to beware of him. The 5th is *Al-mu'tamir, i.e.* he has an impulse of doing harm to mankind. The 6th is *Al-mu'allil, i.e.* he diverts people by some relief which he affords.

The 7th is *Muṭfi'-aljamr* (the extinguisher of coals), the most severe of them, when the coals used to be extinguished. It is also called *Mukfi'-alḳidr* (who turns the kettle upside down) in consequence of the cold wind of this day. Some poet has connected these names in a *versus memorialis* in this way:

"The winter is closed by seven dusty (days),
Our *Old Woman's Days* of the month;
When her days come to an end,
Ṣinn, Ṣinnabr, and *Wabr,*
Âmir, and his brother *Mu'tamir,*
Mu'allil, and *Muṭfi'-aljamr,*
p.255. Then the cold retires, passing away with the end of the month,
And a burning (wind) comes to thee from the beginning of the next month."

The 6th day is also called *Shaibân*, and the 7th *Milḥân*. These days are scarcely ever free from cold and winds, the sky being dark and variously coloured. Mostly during these days the cold is most vehement, because it is about *to turn away* (*i.e.* to cease). And hence the Lunar Station *Alṣarfa* has got its name, because its setting occurs about this time.

Nobody need be astonished at the fact that the cold towards its end, when it is about to cease, is the most severe and vehement. Quite the same is the case with the heat, as we shall mention hereafter. Similar observations you may make in quite common physical appearances. *E.g.* if the lamp is near the moment of extinction, because there is no more oil, it burns with an intense light, and flickers repeatedly, like the quivering (of human limbs). Sick people furnish another example, specially those who perish by hectic fever or consumption, or the disease of the belly, or similar diseases. For they regain power when they are near death; then those who are not familiar with these things gain new hope, whilst those who know them from experience despair.

I have seen a treatise of Ya'ḳûb b. 'Isḥâḳ Alkindî on the cause of this appearance in these days (*i.e.* of the vehemence of the cold during them). His whole argument comes to this, that the sun then reaches the quadrature of his apogee, which is the place of all changes, and that the sun's influence upon the atmosphere is greater than that of anything else. In that case it would be necessary that that change which the sun effects in its own sphere should be proportional to that one which it effects in the atmosphere, and that this effect should on an average continue as long as the moon stands in that quarter (of her own course) in which the effect commenced, and in that quarter of the sun in which the effect took place.

I have been told that 'Abdallâh b. 'Alî, the mathematician, in Bukhârâ, on having become acquainted with this treatise of Alkindî, transferred these days into the calendar of his people in conformity with the amount of the progression of the apogee. Therefore they were called the *Days of the Old Woman of 'Abdallâh.*

[*Lacuna.*]

Regarding the reason why these days were called the *Days of the Old Woman*, the ancients relate the following: They are the days which God mentions in his Book (Sûra lxix. 7), "*seven nights and eight days, unlucky ones*," and the people of 'Âd perished by their cold wind, their whirlwinds, and the other terrors which happened during them. Of all of them only one old woman remained, lamenting the fate of her nation. Her story is well known. Therefore these days are said to have been called the *Days of the Old Woman.*

People say that the wind which destroyed them was a west wind, for the prophet says: "I have been assisted by the east wind—viz. on the

Yaum-alkhandaḳ—and 'Âd has been destroyed by the west wind." A poet says:

"The west wind has destroyed the sandy tracts of 'Âd;
So they perished, thrown down like the trunks of palm-trees."

Further, people say that the *unlucky days* mentioned in the Coran (Sûra xli. 15) coincide, each set of four of them, with a day of the month in the date of which there is a 4, *i.e.* the 4th, or the 14th, or the 24th from beginning or end of the month.

p.256. Some people maintain that the *Days of the Old Woman* received their name from this, that an old woman, thinking that it was warm, threw off her *Miḥsha'* (a sort of garment) and perished in the cold of these days.

Some Arabs maintain that the *Days of the Old Woman* (*Al'ajûz*) were given this name because they are the *'Ajuz*, i.e. *pars postica*, of the winter.

We find that the Arabs have names for the five *Epagomenæ* between Âbân-Mâh and Âdhâr-Mâh like those of the *Days of the Old Woman.* The 1st is called *Hinnabr*, the 2nd *Hinzabr*, both words meaning the injury from cold; the 3rd is called *Ḳâlib-alfihr* (*i.e.* turning the braying-stone upside down), viz. through the vehemence of the wind; the 4th, *Ḥâliḳ-alẓufr* (*i.e.* cutting the nail), for they mean that the wind is so sharp as *e.g.* to cut the nail; the 5th is called *Mudaḥrij-alba'r* (whirling about the dung), viz. in the plains, so that the vehemence of the wind carries it to human habitations. Somebody has brought them into a verse in this way:

"The first of them is *Hinnabr*, an excessive day,
After him comes *Hinzabr*, one who strikes with the fore-foot,
Striking till he comes who exercises justice.
And *Ḳâlib-alfihr* is justly called thus;
And *Ḥâliḳ-alẓufr* who evidently cuts
And splits the rocks by the cold.
After them the last of them, the fifth,
Mudaḥrij-alba'r, the biting and licking one.
There is no sixth name after it."

Adhâr.

1. Nothing mentioned by the Parapegmatists. People say that on this day the locusts and all creeping animals come forth, and that *the heat of heaven and the heat of the earth meet each other.* This is a somewhat hyperbolical expression for the beginning of the heat, its increase and spreading, and for the air's preparing itself for the reception of the heat. For the heat is nothing but the rays of the sun detached from

the body of the sun towards the earth or from the warm body which touches the inside of the Lunar sphere, which is called *Fire*.

Regarding the rays of the sun many theories have been brought forward. Some say that they are fiery particles similar to the essence of the sun, going out from his body. Others say that the air is getting warm by its being situated opposite to the sun, in the same way as the air is getting warm by being opposite to the fire. This is the theory of those who maintain that the sun is a hot, fiery substance.

Others, again, say that the air is getting warm by the rapid motion of the rays in the air, which is so rapid as to seem *timeless, i.e.* without time ("*aeitlos*"). This is the theory of those who maintain that the nature of the sun has nothing in common with the natures of the four elements.

Further, there is a difference of opinion regarding the motion of the rays. Some say this motion is *timeless,* since the rays are not bodies. Others say this motion proceeds in very short time; that, however, there is nothing more rapid in existence by which you might measure the degree of its rapidity. *E.g.* the motion of the sound in the air is not so fast as the motion of the rays; therefore the former has been compared with the latter, and thereby its time (*i.e.* the degree of its rapidity) has been determined.

As the reason of the heat which exists in the rays of the sun, people assign the acuteness of the angles of their reflexion. This, however, is not the case. On the contrary, the heat exists in the rays (is inherent in them).

Regarding the body that touches the inside of the sphere, *i.e.* the fire, people maintain that is a simple element like earth, water, and air, and that it is of a globular form. According to my opinion, the warmth of the air is the result of the friction and violent contact between the sphere, moving rapidly, and his body, and that its shape is like a body p.257.
which you get by making a crescent-like figure revolve around its chord. This explanation is in conformity with the theory, viz. that none of the existing bodies is in its natural place, that all of them are where they are only in consequence of some force being employed, and that force must of necessity have had a beginning.

On this subject I have spoken in a more suitable place than this book is, specially in my correspondence with the youth 'Abû-'Alî Alḥusain b. 'Abdallâh b. Sînâ, consisting of discussions on this subject.

Both sorts of heat are brought to bear upon the earth in an equal manner during the four seasons. The heat of the earth consists either of the solar rays that are reflected from its surface, or of the vapours that are produced—according to one theory—by the heat of the interior of the earth, or—according to another—by that heat which accidentally comes to the earth from outside, for the motion of the vapours in the air causes them to get warm.

The heat of the fire (*i.e.* the body touching the inside of the sphere) remains always at the same distance (from us, *i.e.* is always of the same degree), because the rotation of the celestial sphere proceeds always at the same rate. And the reflected rays are not to be referred to the earth (*i.e.* the earth is not to be considered as their source), and the vapours reach only to a certain limit which they do not go beyond.

The author of this theory, I think, must believe that within the earth heat is contained which proceeds from the interior to the outside, whilst the air has become warm through the rays of the sun. *Thus the two sorts of heat meet each other.* This, at all events, *is* a theory, if there is any; one must accept it.

2. Cold north wind (Hipparchus); south wind and fall of hail (Egyptians).

3. Nothing mentioned.

4. Cold north wind (Euctemon). Sinân says that this is mostly true.

5. Winterly air (Egyptians). Beginning of the Χελιδονίαι (Cæsar): they blow during ten days.

6. Troubled air (Egyptians). Beginning of the cold ὀρνιθίαι, which blow during nine days (Democritus).

7. Nothing mentioned. Some people say that a change of the violent winds takes place.

8. Episemasia and cold north wind (Euctemon, Philippus, and Metrodorus); swallows and kites appear (Eudoxus). On the same day is the feast of the Small Lake of Alexandria.

9. North wind (Euctemon and Metrodorus); violent south wind (Hipparchus); light rain (Egyptians); the kites appear (Dositheus).

10. Nothing mentioned.

11. The ancients do not mention an apparent change on this day. Sinân says that there is frequently winterly air.

12. Moderate north wind (Callippus). People say that on this day the traces of the winter disappear, and that phlebotomy is advisable.

p.258. 13. Ὀρνιθίαι begin blowing; the kite appears (Euctemon and Philippus).

14. Cold north wind (Euctemon and Hipparchus); west or south wind (Egyptians); ὀρνιθίαι begin blowing (Eudoxus).

15. Cold north wind (Euctemon and Egyptians).

16. North wind (Callippus). This Sinân confirms from his experience.

17. Nothing mentioned. People say that on this day it is agreeable to go out on the sea. The snakes open their eyes, for during the cold season, as I have found them myself in Khwârizm, they gather in the interior of the earth and roll themselves up one round the other so that the greatest part of them is visible, and they look like a ball. In this condition they remain during the winter until this time.

On this day (the 17th) in a leap-year, and on the 18th in a common year, takes place the equinox, called the first equinox. It is the first day of the Persian spring and of the Chinese autumn, as we have mentioned. This, however, is impossible, for spring and autumn or winter and summer cannot at one and the same time alternately exchange their places except in countries north or south of the equator. And China, having only few degrees of latitude, does not lie south, but north of the equator, in the farthest end of the inhabited world towards the east.

The country south of the Line is not known, for the equatorial part of the earth is too much burned to be inhabitable. Parts of the inhabited world do not reach nearer the equator than to a distance of several days' journey. There the water of the sea is dense, because the sun so intensely vaporises the small particles of the water, that fishes and other animals keep away from it. Neither we nor any of those who care for those things have ever heard that any one has reached the Line or even passed the Line to the south.

Some people have been beguiled by the expressions "*Æquator Diei*" and "*Linea Æquitatis*," so as to think that there the air is *equal* (moderate), just as day and night there are equal. So they have made the equator the basis of their fictions, describing it as a sort of paradise and as being inhabited by creatures like angels.

As to the country beyond the Line, someone maintains that it is not inhabitable, because the sun, when reaching the perigee of his eccentric sphere, stands nearly in its utmost southern declination, and then burns all the countries over which he culminates, whilst all the countries of 65 degrees of southern latitude have the climate of the middle zone of the north. From that degree of latitude to the pole the world is again inhabitable. But the author of this theory must not represent this as necessary, because excessive heat and cold are not alone the causes which render a country uninhabitable, for they do not exist in the second quarter of the two northern quarters, and still that part of the world is not inhabited. So the matter is (and will be), because the apogee and perigee of the eccentric sphere, the sun's greater and less distance from the earth, are necessitated exclusively by the difference in the sun's rotation.

'Abû-Ja'far has designed a figure different from the eccentric sphere p.259
and the epicycle, in which the sun's distance from the earth, notwithstanding the difference of its rotation, is always identical. Thereby he gets two regions, a northern and a southern one, equal to each other in heat and cold.

The day of the equinox, as calculated by the Hindûs according to their *Canon*,—of which they are impudent enough to pretend that it is eternal, without beginning and end, whilst all the other *Canons* are derived therefrom,—is their Naurôz, a great feast among them. In the first hour of the day they worship the sun and pray for happiness and bliss to the spirits (of the deceased). In the middle of the day they worship the sun again, and pray for the resurrection and the other world. At the end of the day they worship the sun again, and pray for health and happiness for their bodies. On the same day they make presents to each other, consisting of precious objects and domestic animals. They maintain that the winds blowing on this day are spiritual beings of great use for mankind. And the people in heaven and hell look at each other affectionately, and light and darkness are equal to each other. In the hour of the equinox they light fires in sacred places.

The omina of this day are the following, viz.: to rise from sleep lying on the back, the tree *Salix Ægyptia* and to fumigate with its wood before speaking. For he who performs this will be free from all sorts of pain.

People say that a man who has no children, on looking to the star *Al-Suhâ* in the night of this day and then having intercourse with his wife, will get children.

Muḥammad b. Miṭyâr maintains that in the hour when this day begins to decline, (*i.e.* after noon,) the shadow of everything is half its size. This, however, is only partially the case, not in general. It is true only for such places of which the latitude is about 27 degrees.

On this day the crocodile in Egypt is thought to be dangerous. The crocodile is said to be the water-lizard when it has grown up. It is an obnoxious animal peculiar to the Nile, as the thesking is peculiar to other rivers. People say that in the mountains of Fusṭâṭ there was a talisman made for that district. Around this talisman the crocodile could not do any harm. On the contrary, when it came within its limits, it turned round and lay on its back, so that the children could play with it. But on reaching the frontier of the district it got up again and carried all it could get hold of away to the water. But this talisman, they say, has been broken and lost its power.

18. Winterly air and cold winds (Democritus and the Egyptians).

19. North wind (Hipparchus); winds, and cold in the morning (Egyptians).

20. North wind (Cæsar).

21. North wind (Eudoxus).

22. Nothing mentioned.

23. North wind (Cæsar); rain (Hipparchus).

24. Rain and mizzle (Callippus, Euctemon, and Philippus); Episemasia (Hipparchus); thunder and Episemasia (Egyptians). On this day people like to purify the children by circumcision. The fecundating winds are said to blow.

p.260. 25. North wind (Eudoxus); Episemasia (Meton, Conon, and the Egyptians).

26. Rain and snow-storm (Callippus); wind (Egyptians).

27. Rain (Callippus, Eudoxus, and Meton).

Of the rest of the month nothing is mentioned. Sinân says that the 30th frequently brings an Episemasia. God knows best!

Nîsân.

1. Rain (Callippus, Euctemon, Meton, and Metrodorus).

2. Nothing mentioned.

3. Wind (Eudoxus); rain (Egyptians and Conon).

4. West wind or south wind; hail falls. Sinân says that this is frequently the case.

5. South wind and changing winds (Hipparchus).

6. Episemasia (Hipparchus and Dositheus). This is confirmed by Sinân.

7. Nothing mentioned.

8. Rain (Eudoxus); south wind (Egyptians).

9. Rain (Hipparchus); unmixed winds (Egyptians).

10. Unmixed winds (Euctemon and Philippus); rain (Hipparchus and Egyptians). The raining is confirmed by the experience of Sinân.

11. West wind and mizzle (Eudoxus).

12. Nothing mentioned.

13. Rain (Cæsar and Dositheus).

14. South wind, rain, thunder, and mizzle (Egyptians). Sinân says that this is frequently the case.

15. Rain and hail (Euctemon and Eudoxus); unmixed winds (Egyptians).

16. West wind (Euctemon and Philippus); hail falling (Metrodorus).

17. West wind and rain (Eudoxus and Cæsar); hail falling (Conon and Egyptians).

18. Winds and mizzle (Egyptians).

19. Nothing mentioned.

20. Wind, south wind or another one, the air unmixed (Ptolemy).

21. Cold south wind (Hipparchus). Sinân maintains that this is frequently the case. The water begins to increase.

22. Rain (Eudoxus); winterly air (Cæsar and Egyptians). People fear for the ships at sea.

23. South wind and rain (Egyptians). People hold a fair at Dair-'Ayyûb. 'Abû-Yaḥyâ b. Kunâsa says that the Pleiades disappear under the rays of the sun during 40 days, and this fair is held when the Pleiades appear. So the Syrians make them rise 15 days earlier than in reality they rise, because they are in a hurry to settle their affairs. This fair lasts 7 days. Then they count 70 days until the fair of Buṣrâ. Through these fairs, that are held alternately in certain places, the com- p.261.
merce of the people of these countries has been promoted and their wealth been increased. They have proved profitable to the people, to both buyers and sellers.

24. Frequently hail falls (Callippus and Metrodorus); Episemasia (Democritus); south wind, or a wind akin to it, and rain (Egyptians). The Euphrates begins to rise.

25. Mizzle and rain (Eudoxus and Egyptians).

26. Rain and frequently hail (Callippus and Euctemon); Episemasia and west wind (Egyptians).

27. Dew and moisture (Cæsar); winds (Egyptians).

28. Wind (Egyptians); rain (Eudoxus). Sinân confirms the rain from his own observations. On this, they say, the south wind blows, and then the streams and rivers begin to rise. This increase of the water, however, does not apply to all streams and rivers uniformly; on the contrary, they greatly differ from each other in this respect. *E.g.* the Oxus has high water when there is little water in the Tigris, Euphrates, and other rivers. The fact is this, that those rivers the sources of which are situated in cold places, have more water in summer and less in winter. For the greatest part of the ordinary volume of their water is gathered from springs, and an increase and decrease of them exclusively depend upon the fall of dew in those mountains where the rivers originate or through which they flow; thereupon the springs pour their volumes into the rivers. Now it is well known that dew-fall is more frequent in winter and beginning of spring than at any other season. In the countries far up to the north, where the cold is intense, this dew-water freezes at those seasons. But when the air is getting warm and the snow melts, at that time the Oxus rises.

As for the water of the Tigris and Euphrates, their sources are not so high up in the north. Therefore they have high water in winter and spring, because the dew that falls flows instantaneously into the rivers, and that portion of water that may have been frozen melts away in the beginning of spring.

The Nile, again, has high water when there is low water in both Tigris and Euphrates, because its source lies in the *Mons Lunæ*, as has been said, beyond the Abyssinian city Assuan in the southern region, coming either exactly from the equator or from countries south of the equator. This is, however, a matter of doubt, because the equatorial zone is not inhabited, as we have before mentioned. It is evident that in those regions there is no freezing of moist substances at all. If, therefore, the high water of the Nile is caused by falling dew, it is evident that the dew does not stay where it has fallen, but that it directly flows off to the Nile. But if the high water is caused by the springs, these have the most abundant water in spring. Therefore the Nile has high water in summer, for when the sun is near us and our zenith, it is far distant from the zenith of those places whence the Nile originates, and which in consequence have winter.

p.262 As to the question why the springs have the most copious water in winter, we must observe: the all-wise and almighty Creator, in creating the mountains, destined them for various purposes and uses. Some of them have been mentioned by Thâbit b. Ḳurra in his book on the reason why the mountains were created. It is the same cause which renders complete the intention (of the Creator) which he had in making the sea-water salt.

Evidently more wet falls in winter than in summer, in the mountains more than in the plains. When, now, the wet falls and part of it flows away in the torrents, the remaining part sinks down into the channels in the mountain caves, and there it is stored up. Afterwards it begins to flow out through the holes, called springs. Therefore the springs have the most copious water in winter, because the substance by which they are nourished is then most copious. If, further, these mountain caves are clean and pure, the water flows out just as it is, *i.e.* sweet. If that is not the case, the water acquires different qualities and peculiarities, the causes of which are not known to us.

The bubbling of the fountains and the rising of the water to a certain height are to be explained in this way, that their reservoirs lie higher than they themselves, as is the case with artificial well-springs, for this is the only reason why water rises upward.

Many people who attribute to God's wisdom all they do not know of physical sciences (*i.e.* who excuse their ignorance by saying "*Allah is all-wise!*"), have argued with me on this subject. In support of their view they relate that they have observed the water rise in rivers and other watercourses, that the water the more it flows away (from its source) the more it rises. This they assert in complete ignorance of the physical causes and because they do not sufficiently distinguish between the higher and lower situations (of the springs of rivers and of the rivers themselves). The matter is this, that they observed water flowing in mountain streamlets, the bed of which was going downward at the rate of 50–100 yards and more for the distance of one mile. If the peasants dig a channel somewhere in this terrain, and this channel is made to incline a little towards the country (*i.e.* if the channel is rising), at first the water flows only very little, until it rises to an enormous height above the water of the river; (then it commences to flow strongly).

If, now, a man who has no training in these things believes that the natural direction of the river is to flow in a horizontal line or with a small inclination (upwards), he must of necessity imagine that the river is rising in height. It is impossible to free their mind from this illusion unless they acquaint themselves with the instruments by which pieces of soil are weighed and determined, and by which rivers are dug and excavated—for if they weigh the earth through which the water flows, the reverse of what they believe becomes evident to them;—or unless they study physical sciences, and learn that the water moves towards the centre of the earth and to any place which is nearest to the centre. There is no doubt that the water may rise to any place where you want to have it, even if it were to the tops of the mountains, if previously it descends to a place which is lower than its maximum of ascent (which it ultimately reaches), and if you keep away from it any substance which might occupy the place instead of the water when it finds the place empty. Now, the water in its natural function is only assisted by the co-operation of something forcible which acts like an instrument, and that is the air. This has frequently been carried out in canals, in the midst of which there were mountains which it was impossible to perforate.

An illustration of this principle is the instrument called *Water-thief*, *κλεψύδρα*. For if you fill it with water and put both its ends into two
p.263. vessels, in both of which the water reaches to the same level, then the water in the *κλεψύδρα* stands still even for a long time, not flowing off into either of the two vessels. For the one vessel is not nearer (to the water) than the other, and it is impossible that the water should flow off equally into both vessels, for in that case the instrument would get empty. Now, emptiness is either a *non-ens*, as most philosophers suppose, or it is an *ens* which attracts bodies, as others believe. If, now, the *vacuum* cannot exist, the matter is impossible, or *if* it is something which attracts bodies, it keeps back the water and does not let it flow off, except its place be occupied by some other body. But if you then place the one end of the *κλεψύδρα* a little lower (than the other), the water flows immediately off into that direction. For if its place has once become lower, it has come nearer to the centre of the earth, and so it flows towards it, flowing continually in consequence of the adhesion and connection of the water-atoms amongst each other. It flows so long until the water of that vessel, whence the water is drawn, is finished, or until the level of the water in the vessel where it flows is equal to the level of the water in the vessel whence it is drawn. So the question returns to its original condition. On this principle people have proceeded in the mountains.

Sometimes even the water rises in artificial fountains out of wells, in case they have got springing water. For one sort of well-water, which is gathered from droppings from the sides, does not rise at all; it is taken from neighbouring masses of water, and the level of the water which is gathered in this way is parallel to the level of those waters by which it is nourished. On the other hand there is one kind of water which bubbles (springs) already at the bottom. Of this water people hope that it may rise to the earth and flow on over its surface. This latter kind of water is mostly found in countries near to mountains, in the midst of which there are no lakes or rivers with deep water. If the source of such water is a reservoir much above the level of the earth, the water rises springing, if it is confined (to a narrow bed or channel); but if its reservoir be lower, the water does not succeed in rising to the earth. Frequently the reservoir is higher by thousands of yards in the mountains; in that case the water may rise up to the castles, and, *e.g.*, to the tops of the minarets.

I have been told that people in Yaman often dig until they come to a certain rock under which they know that there is water. Then they knock upon this rock, and by the sound of the knocking they ascertain the quantity of the water. Then they bore a small hole and examine it; if it is all right, they let the water bubble out and flow where it likes. But if they have some fear about the hole, they hasten to stop it up with gypsum and quicklime and to close it over repeatedly. For frequently they fear that from such a hole a spring similar to the *Torrent of Al'arim* might originate.

As to the water on the top of the mountain between Abrashahr and Ṭûs, a small lake of one farsang in circumference, called Sabzarôd, one of the following three things must be the case:

1. Either its material is derived from a reservoir much higher than the lake itself, although it may be far distant, and the water flows into it in such a quantity as corresponds to that which the sun absorbs and vaporises. Therefore the water of the lake remains in the same condition, quietly standing.

2. Or its material is derived from a reservoir which lies on the same level with the lake, and therefore the water of the lake does not rise above that of the reservoir.

3. Or, lastly, the condition of its sources in some way resembles that of the water of the instrument called *Al-daḥj*, and the *self-feeding lamp*. p.264.
The case is this: You take a water-jug, or an oil-vase; in several places of the edge or lip of the vase you make fine splits, and you bore a narrow hole in it deeper than the mouth by so much as you wish the water to remain in the jug and the oil in the vase (*i.e.* the hole is to represent the line to which people wish the water or oil to rise). Thereupon you turn the jug upside down in the cup and the vase in the lamp. Then both water and oil flow out through the splits, until they reach the level of the hole. When, then, so much has been consumed as the hole allows to pass, then comes forth that which lies next to the hole. In this way both oil and water keep the same level.

Similar to this little lake is a sweet-water well in the district of the Kîmâk in a mountain called Mankûr, as large as a great shield. The surface of its water is always on a level with its margin. Frequently a whole army drinks out of this well, and still it does not decrease as much as the breadth of a finger. Close to this well there are the traces of the foot, two hands with the fingers, and two knees of a man who had been worshipping there; also the traces of the foot of a child, and of the hoofs of an ass. The Ghuzzî Turks worship those traces when they see them.

Moreover, similar to this is a small lake in the mountains of Bâmiyân, one mile square, on the top of the mountain. The water of the village which lies on the slope of the mountain comes down from that lake through a small hole in such a quantity as they require; but they are not able to make it flow more copiously.

Frequently the springing (rising of water) occurs also in a plain country which gets its water from a reservoir in a high situation. If the rising power of the water were kept down by an obstacle, and then this obstacle is removed, the water begins at once to spring (rise). *E.g.* Aljaihânî has mentioned a village between Bukhârâ and Alķarya Alḥadîtha, where there is a hill that was perforated by diggers for hidden treasures. Suddenly they hit upon water which they were unable to keep back, and it has been flowing ever since till this day.

If you are inclined to wonder, you may well wonder at a place called Fîlawân (Failawân) in the neighbourhood of Almihrjân. This place is like a portico dug out in the mountain, from the roof of which water is always dropping. If the air gets cold, the water freezes and hangs down in long icicles. I have heard the people of Almihrjân maintain that they frequently knock the place with pickaxes, and that in consequence the spot which they knock becomes dry; but the water never increased, whilst reason would demand that it should always remain in the same condition if it does not increase.

More wonderful even than this is what Aljaihânî relates in his *Kitab-Almamâlik wal-masâlik* of the two columns in the grand mosque of Ķairawân, the material of which people do not know. People maintain that on every Friday before sunrise they drop water. It is curious that this should take place just on a Friday. If it occurred on any week-day in general, it would be combined with the moon's reaching such and such a place of the sun's orbit, or with the like of it. This, however, is not admissible, since Friday is a *conditio sine quâ non* of this occurrence. The Greek king is said to have sent to buy them. He said: "It is better for the Muslims to utilize their prize than to have two stones in the mosque." But the people of Ķairawân refused, saying: "We shall not let them pass out of the house of God into that of the devil."

Still more marvellous than this is the self-moving column in Alķaira-
p.265. wân. For it inclines towards one side. People put something underneath when it inclines, and this you can no longer take away if the column again stands erect; if glass is put underneath, you hear the sound of breaking and crushing. This is no doubt a got-up piece of artifice, as also the place where the column stands seems to indicate.

We return to our subject, and say:

29. Winterly air (Cæsar); winds, or moisture of the ground, and rain (Egyptians).

30. Episemasia (Egyptians); winds and dew, moisture and mizzle (Callippus and Euctemon).

Ayyâr.

1. Mizzle (Egyptians).
2. Nothing mentioned.
3. Wind, mizzle, dew, moisture, and thunder (Egyptians).

4. Rain (Eudoxus), mizzle (Egyptians).

5. Rain (Dositheus). Sinân says that this is frequently the case and that it brings a strong episemasia.

6. Wind (Egyptians), rain (Eudoxus), mizzle and episemasia. (*Lacuna.*) Some people extend the rainy season as far as this day. It is the time when the sun passes the (first) 20 degrees of Leo. In this respect the matter stands as we have explained it at the beginning of the rainy season, when the sun moves in Cancer.

7. Winds (Egyptians). Sinân says that this is frequently the case, more particularly so if on the preceding day heaven has a rainy appearance.

8. Gushes of rain (Eudoxus and Dositheus), rain (Egyptians).

9. Rain (Egyptians).

10. Episemasia and wind (Callippus and Euctemon), rain (Egyptians).

11. Episemasia (Dositheus). Sinân says that it is true.

12. Episemasia (Eudoxus, Metrodorus, and Hipparchus); rain (Cæsar); west-wind (Egyptians). People say that on this and the following day there is no fear of frost doing harm to the fruits. This remark can, however, only apply to one particular place; it cannot be meant in general.

13. Rain (Eudoxus); north wind and hail (Egyptians).

14. Episemasia (Callippus, Euctemon, and Egyptians).

15. Rain (Cæsar).

16. Episemasia (Cæsar). People say that on this day the first *Samûm* is blowing.

17. South wind or east wind and rain (Hipparchus and Egyptians).

18. Episemasia (Eudoxus); rain and thunder (Egyptians).

19. Episemasia and mizzle (Hipparchus and Egyptians).

20. Nothing mentioned.

21. Episemasia (Cæsar); south wind (Dositheus), west wind (Egyptians).

22, 23. Nothing mentioned.

24. Episemasia (Callippus, Euctemon, and Philippus); winds (Egyp- p.266.
tians).

25. Episemasia (Euctemon, Philippus, and Hipparchus).

26. Episemasia (Callippus and Euctemon); cold north wind (Egyptians).

27. Dew and moisture (Callippus and Euctemon); episemasia (Egyptians).

28. Rain (Metrodorus and Egyptians).

29. South wind or west wind (Hipparchus).

30. South wind (Cæsar).

31. Nothing mentioned.

Ḥazirân.

1. Dew and moisture (Eudoxus and Dositheus); west wind (Egyptians).

2. West wind (Egyptians).

3. Wind and mizzle (Egyptians), and thunder.

4. Rain (Cæsar).

5. Mizzle (Egyptians). Confirmed by Sinân.

6, 7, 8. Nothing mentioned.

9. West wind and thunder (Egyptians).

10, 11, 12. Nothing mentioned. The 11th is the Naurôz of the Khalif, when people in Baghdâd splash in the water, strew about dust, and play other games, as is well known.

12. Sinân says that frequently a change of the weather takes place.

13. West wind and mizzle (Egyptians).

14. Nothing mentioned.

15. Mizzle (Egyptians).

16. Nothing mentioned. People say that on this day the water sinks into the earth, whilst the Nile begins to rise. The reason of this is, as we have mentioned before, the difference of their sources and of other circumstances, those of the Nile standing in direct opposition to those of all other rivers.

On this day in a leap-year, and on the 17th in a common year, the *Plenitudo Maxima* takes place, which is celebrated by Arabs and Persians. They call it *Mîrîn*, which means *the Sun's getting full*, *i.e.* the summer-solstice. On this day light subdues darkness. The light of the sun is falling into the wells, as Muḥammad b. Miṭyâr mentions; but this is only possible in countries the latitude of which is like the greatest declination, over which, therefore, the sun culminates.

The *Ḥayawâniyya*-sect maintains that on this day the sun takes breath in the midst of heaven; that, therefore, the spirits recognise each other in the greatest heat. It is considered as a good omen to look into the intense heat. People eat pomegranates before having eaten anything else, and Hippocrates is said to have taught that he who eats a pomegranate on this day before having eaten anything else, enlightens his constitution and his χυμός is pure during forty mornings.

People relate, on the authority of Ḥanna the Hindû, that Kisra Parwîz has said: "Sleeping in the shadow of a pomegranate cures a man of bad disease and makes him safe from the demons."

It belongs to the *omina* of this day to rise in the morning from sleep on the left side, and to fumigate with saffron before speaking.

p.267. 17. Episemasia (Dositheus); heat (Egyptians).

18. West wind and heat (Egyptians).

19. Rain (Egyptians).

20. West wind, rain, and thunder (Egyptians).

21. Nothing mentioned

22. Episemasia (Democritus).

23. South wind or west wind (Hipparchus).

24. Nothing mentioned. People say that on this day the *Samûms* begin blowing during fifty-one days. The Oxus rises and frequently injures the shores and their inhabitants.

25. West wind and heat (Egyptians).

26. West wind (Democritus and Egyptians).

27. Nothing mentioned.

28. Episemasia (Eudoxus); west wind and south wind and rain (Democritus); then the north wind begins to blow during seven days.

29. Nothing mentioned. People say that practical obververs examine on this day the dew; if it is copious, the Nile rises; if it is not copious, the Nile does not rise, and they get a barren year.

30. Winds (Egyptians) and unmixed air.

31. Nothing mentioned.

Tammûz.

1, 2. Nothing mentioned by our authorities.

3. South wind and heat (Cæsar and Egyptians).

4. Wind (Egyptians); frequently it rains in their country.

5. South wind (Callippus, Metrodorus, and Hipparchus); west wind and thunder (Egyptians).

6. South wind (Callippus and Metrodorus); west wind and thunder (Egyptians).

7. Episemasia (Ptolemy). According to Sinân the weather frequently changes.

8. Dew and moisture, according to Meton, in his country.

9. Dew (Euctemon and Philippus); west-by-west wind (Egyptians).

10. Bad air (Egyptians). On this day they begin to hold the fair of Buṣrâ during 25 days; in the time of the Banû-'Umayya this fair used to last 30–40 days.

11. Nothing mentioned.

12. West wind (Metrodorus); winds (Egyptians).

13. Unmixed winds (Hipparchus). According to Sinân the weather frequently changes.

14. Heavy wind (Cæsar); the north wind begins to blow (Hipparchus); heat (Egyptians).

15. Nothing mentioned.

16. Frequently it rains in rainy countries (Ptolemy); rain and whirl- p.268.
winds (Democritus); heavy wind (Egyptians).

17. Dew and heat (Dositheus and Egyptians).

18. The Etesian winds (ἐτησίαι) begin to blow (Hipparchus). According to the general consent of seamen and peasants, and all those who

have experience in this subject, this is the first day of the dog-days, *i.e.* seven consecutive days, the last of which is the 24th of this month. On each of these days they draw conclusions from certain changes of the weather regarding the months of the autumn and winter and part of spring; these changes mostly occur in the evening and morning. People maintain that these days are to the year what the *critical* days are to acute diseases, when their *criteria* appear, in consequence of which people conceive either hope or fear as to the end in which they will issue. Both words *bâḥûr* and *buḥrân* in the Greek and Syriac languages are derived from a word which means the decision of the rulers (*v.* κρίσις and κρίσιμος ἡμέρα). According to another view, *buḥrân* is derived from *baḥr* (the Arabic for *sea*), because the *critical* state of a sick person resembles the motion of the sea, called ebb and flow. This derivation is very likely correct, because of both appearances the motions of the moon, her cycles and phases, are the cause, whether the moon revolves in a *Great Circle*, as it is in the case of the flow, for the flow sets in when the moon reaches the western and eastern point of the horizon. The same is the case with the ebb, for it sets in when the moon reaches the sphere of the meridian of noon and midnight. Or whether it be that the moon revolves from one certain point of her cycle back to the same, or from the sun to that point. So the flow is the strongest in the first half of the lunar month, the weakest in the second half. Besides, also, the sun has an influence upon this. It is curious what people relate of the *Western Sea*, viz. that there is flow from the side of Andalusia always at sunset, that then the sea decreases at the rate of about 5–6 farsang in one hour and then it ebbs. And this appearance takes place always precisely at this time.

If on the evening of the 18th there is a cloud on the horizon, people expect cold and rain at the beginning of Tishrîn I. If the same is the case at midnight, the cold and rain will come in the middle of Tishrîn I.; and if it is the case towards morning, the same will come in the end of that month. The matter is the same, if you observe a cloud on the horizon during *daytime;* however, the changes of the sky in the night are more evident. And if you observe those changes on all four sides of the compass, the same, too, will occur in Tishrîn I. Herein the nights are counted after the days, as we ha. ntioned in the beginning of this book, in consequence of which those who count the nights before the days think that the night of the 18th is the 19th; therefore they consider the 19th as the first of the dog-days and the 25th as the last of them.

The 1st of these seven days serves to prognosticate the character of Tishrîn I., the 2nd that of Tishrîn II., the 3rd that of Kânûn I., etc. etc., and lastly, the 7th, that of Nîsân.

Practical observers prescribe the following: Take a plate some time before the dog-days, sow upon it all sorts of seeds and plants, and let it stand until the 25th night of Tammûz, *i.e.* the last night of the dog-days; then put the plate somewhere outside at the time when the stars rise and set, and expose it uncovered to the open air. All seeds, then, that will grow in the year will be yellow in the morning, and all whose growth will not prosper will remain green. This experiment the Egyptians used to make. p.269.

Practical observers have produced many contrivances for the purpose of prognosticating the character of the year by help of these (the dog) days; they have even gone as far as to use incantations and charms. So some people maintain that if you take the leaves of twelve different olive-trees, and write upon each leaf the name of some Syrian month, if you then put them, in the night we have mentioned, somewhere in a wet place, you will find that, if a leaf has dried up in this night, the month which was written upon it will be rainless.

According to others, you learn whether the year will have much rain or little, by this method: You look out for a level place, around which there is nothing that might keep off the dew, wind, and light rain; then you take two yards of a cotton dress, you weigh it and keep in mind its weight. Then you spread it over that place and leave it there during the first four hours of the night. Thereupon you weigh it a second time; then each *Mithkâl* which it weighs more the second time than the first time signifies one rainy day in that month which stands in relation with this particular dog-day of which we have heretofore spoken.

These dog-days are the time of the rising of Sirius (*Kalb-aljabbâr* or *Alshi'râ Alyamâniya Al'abûr*). Hippocrates, in his book of the seasons, forbids taking hot drugs and bleeding twenty days before and after the rising of this star, because it is the hottest time of summer and the heat reaches its maximum, and because summer time by itself warms, dissolves, and takes away all moist substances. However, Hippocrates does not forbid those things if you take but very little of them. Afterwards, when autumn comes with its cold and dryness, you cannot be sure whether the natural warmth may not be entirely extinguished.

Some people who have no practice in physical sciences and no knowledge of the *μετέωρα*, think that the influence we have mentioned must be attributed to the body of this star, to its rising and revolution. They go even as far as to make people imagine that the air is warmed by its great mass; that, therefore, it is necessary to indicate and to explain its proper place and to determine the time of its rising. The same opinion is indicated by the verse of 'Abû-Nu'âs:

> "Îlûl has gone and the hot night-wind passed away,
> And Sirius has extinguished his fire."

For this reason 'Alî b. 'Alî, the Christian secretary, maintains that the first of the dog-days is the 22nd of Tammûz, suggesting that the dog-

days have changed their place along with the star itself, whilst I maintain that Sirius always revolves during the whole year in one and the same orbit parallel to the equator. Hippocrates, however, meant by this time the central portion of the summer, the period when the heat is greatest
p.270. in consequence of the sun's being near to our zenith, whilst he at the same time begins in his eccentric sphere to descend from the apogee of his orbit. And this event was in the time of Hippocrates contemporaneous with the rising of Sirius. Therefore he has only said in general *at the time when Sirius rises,* knowing that no scientific man could misunderstand the truth. For if Sirius changed its place so as to advance even as far as the beginning of Capricorn or Aries, the time during which he forbids taking drugs would not therefore advance in the same way.

Sinân says in his *Kitâb-al'anwâ* that the shepherds have seven special days of their own, beginning with the 1st of Tammûz, which they use like the dog-days, drawing from them conclusions regarding the single winter months. They are known as "*the dog-days of the shepherds.*" The weather of these days is always different from that of the time immediately preceding and following. During all or at least some of them heaven is never free from a speck of clouds.

19. West wind or heat (Egyptians). The *water dogs* are getting strong and do much damage.

20. West wind or a similar one (Egyptians). Practical observers say that on this day frequent cases of inflammation of the eyes occur.

21. The Etesian winds are blowing (Euctemon); the heat begins (Callippus, Euctemon, and Metrodorus).

22. Bad air (Euctemon); beginning of the heat (Hipparchus); west wind and heat (Egyptians).

23. Winterly air on sea, winds (Philippus and Metrodorus); beginning of the blowing of the Etesian winds (Egyptians). On this day 'Abû-Ja'far Almanṣûr began to build Baghdâd, that part which is called *Manṣûr's-town,* on the western side of the Tigris in the present Baghdâd. This was A. Alexandri, 1074. Astrologers are obliged to know dates like this, and must date from such an epoch by means of their knowledge of the *Permutationes, Terminationes, Cycles,* and *Directiones,* until they find the horoscopes of those people who were born at those times. It was Naubakht who determined the time (for the commencement of building). The constellation which heaven showed at the time, and the stations of the planets which appeared on heaven, were such as are indicated in the following figure.

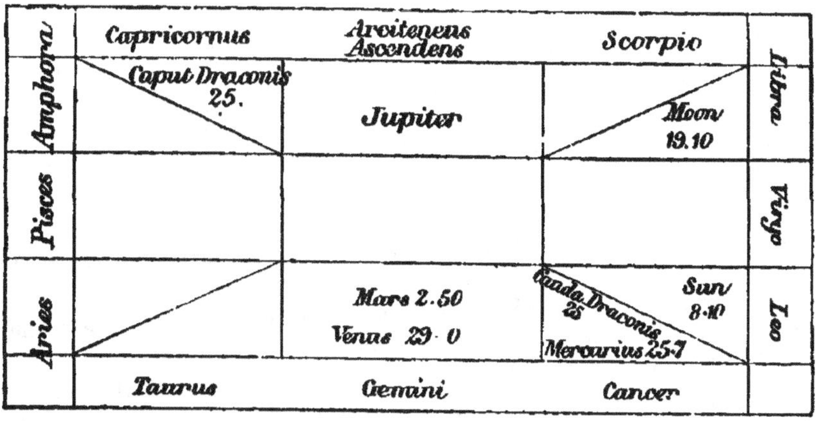

p.271

24. Winds (Philippus and Metrodorus); the Etesian winds blow (Hipparchus).

25. South wind (Eudoxus and Cæsar); west or south wind (Egyptians). Sexual intercourse and all exertion are forbidden, because it is the time of the greatest heat. The river Oxus begins to rise.

26. South wind and heat (Philippus, Meton, Metrodorus, Democritus, and Hipparchus).

27. Dew and wet, and oppressive air (Euctemon and Dositheus). This oppressive air mostly occurs when heaven is covered and the air is in perfect repose. But often, too, this is peculiar to a place where this cause does not exist, *e.g.* to the region beyond that bridge which, according to Aljaihânî, was in old times built by the Chinese, reaching from the top of one mountain to that of another on the road that leads from Khotan to the region of the residence of the Khâḳân. For those who pass this bridge come into an air which makes breathing difficult and the tongue heavy, in consequence of which many travellers perish there, whilst others are saved. The Tibetans call it the "*poison-mountain.*"

28. Nothing mentioned.

29. Beginning of the Etesian winds (Dositheus); heat (Egyptians). They hold the fair of Buṣrâ for a whole month, and that of Salamiyya p.272.
for two weeks.

30. The Etesian winds blow (Eudoxus); west wind and heat (Egyptians).

31. South wind (Cæsar).

Âb.

1. Heat (Hipparchus).
2. Nothing mentioned.
3. Dew falls (Eudoxus and Dositheus); episemasia (Cæsar).
4. Great heat (Eudoxus).
5. Heat, still and oppressive air, then blowing of winds (Dositheus

and Egyptians). They hold a fair at 'Adhri'ât during fifteen days, also in Al'urdunn, and in several districts of Palestine.

6, 7. Nothing mentioned.

8. The air is still and oppressive (Callippus); wind, and intense heat (Egyptians). According to Sinân, frequently there occurs a change of the air.

9. Heat and still air (Euctemon and Cæsar); south wind and turbid air (Egyptians).

10. Heat and still air (Eudoxus, Metrodorus, and Dositheus); episemasia (Democritus). At this time the heat is very intense.

11. The northerly winds cease to blow (Callippus, Euctemon, and Philippus); heavy wind (Eudoxus); different winds blow together (Hipparchus); thunder (Egyptians). According to Sinân there is always a change of the weather on this day. He says: I do not know whether we, I and all those who make meteorological observations, are correct in describing a day like this. On this day there is almost always a change of the weather for the better. It is the first day when the air of Al'irâḳ begins to be agreeable. Sometimes this change is most evident, whilst at other times it is only slightly perceptible. But that the day should be free from such a change, almost never occurs.

Some of the ancients consider this day as the beginning of the autumnal air, whilst others take as such the following day.

Sinân says: Thâbit used to say: If in a rare year that which we have described does not take place on this day it is not likely to take place on the 13th or 14th, but rather in the middle of Âb. If it takes place on the 11th, a season of agreeable air is sure to return about the middle of the month, though it may only be short.

12. Heat (Euctemon and Egyptians).

13. Episemasia and still air (Cæsar). Sinân says that on this day an irregular change of the air frequently occurs.

14, 15. Nothing mentioned.

16. Episemasia (Cæsar).

17. Episemasia (Eudoxus).

18. Nothing mentioned. The Samûms are said to cease.

19. Episemasia, rain, and wind (Democritus); west wind (Egyptians).

p.273. 20. Episemasia (Dositheus); heat and density in the air (Egyptians).

21. Nothing mentioned.

22. West wind and thunder (Eudoxus); episemasia and bad air (Cæsar and Egyptians).

23. West wind (Egyptians).

24. Episemasia (Eudoxus and Metrodorus). The heat relaxes a little at the time when the sun passes the first 6 degrees of Virgo.

25. Episemasia (Eudoxus); south wind (Hipparchus); heat (Egyptians).

26. Rotating winds (Hipparchus). Between this day and the first of

the *Days of the Old Woman* (*i.e.* 26 Shubâṭ) lies one half of a complete year. On this day the heat, at the time when it is about to disappear, returns once more with renewed force, as does also the cold at the time when it is about to disappear. It is a time of seven days, the last of which is the 1st of Îlûl, called by the Arabs *Waḳdat-Suhail* (*i.e.* the burning of Suhail). It is the time of the winds that accompany the rising of *Aljabha* (*Frons Leonis*, the 10th Lunar Station), but as Suhail rises in its neighbourhood, it has become the prevailing use to call the time by *Suhail* and not by *Aljabha*. The heat of these days is more intense than at any time before or afterwards. But after this time the nights begin to be agreeable. This is an occurrence generally known among people, which scarcely ever fails. Muḥammad b. 'Abd-almalik Alzayyât says:

"The water had become cold and the night long,
And the wine was found to be sweet;
Ḥazîrân had left you, and Tammûz and Âb."

27. Episemasia (Philippus).

28. West wind (Egyptians).

29. Rain and thunder; the Etesian winds are about to cease (Eudoxus and Hipparchus).

30. Episemasia (Hipparchus).

31. The Etesian winds are about to cease (Ptolemæus); changing winds (Eudoxus); winds, rain, and thunder (Cæsar); east wind (Hipparchus).

Îlûl.

1. Episemasia and the Etesian winds are getting quiet (Callippus). A fair is held at Manbij (Mabbug)

2. Density in the air (Metrodorus). Conon says that on this day the Etesian winds cease.

3. Wind, thunder, and density in the air (Eudoxus); wet and dew (Hipparchus); fog, heat, rain, and thunder (Egyptians). On this day people begin to light their fires in cold countries.

4. Dense and changing air (Callippus, Euctemon, Philippus, and Metrodorus); rain, thunder, and changing wind (Eudoxus).

5. Changing winds and rain, and the Etesian winds are getting quiet (Cæsar); rains and winterly air at sea, and south wind (Egyptians). On this day midsummer ends, and a time comes which is good for bleeding p.274.
and for taking drugs during forty days.

6. West wind (Egyptians).

7. Density in the air (Philippus); episemasia (Dositheus).

8. West wind and episemasia (Egyptians).

9. Nothing mentioned.

10. The air is not troubled (mixed) (Dositheus).

11. The north winds are ceasing (Cæsar).

12. South wind (Eudoxus).

13. Episemasia (Callippus and Conon).

14. The north winds are ceasing (Eudoxus); episemasia (Democritus and Metrodorus). After this time no swallow is seen.

15. Wet and dew (Dositheus); rains and episemasia (Egyptians).

16. Density in the air, and rain at sea (Hipparchus).

On the 16th in a common year and on the 17th in a leap-year occurs the second equinox, which is the first day of the Persian autumn and the Chinese spring, as people maintain. But we have already explained that this is impossible.

The winds, now, blowing on this day are said to be of a psychical nature. To look towards the clouds that rise on this day emaciates the body and affects the soul with disease. I think the reason of this is that people conceive fear on account of the cold and the disappearance of the agreeable time of the year.

It is one of the *omina* of this day to rise from sleep in a worshipping attitude, and to fumigate with tamarisks before speaking.

People say that if a woman who is sterile looks on this day at the star *Alsuhâ* and then has intercourse with her husband, she is sure to conceive.

Further, they say, that in the night of this day the waters are getting sweet. We have already heretofore shown the impossibility of such a thing.

This second equinox is, according to the *Canon Sindhind,* a great festival with the Hindûs, like the Mihrjân with the Persians. People make each other presents of all sorts of valuable objects and of precious stones. They assemble in their temples and places of worship until noon. Then they go out to their pleasure-grounds, and there they assemble in parties, showing their devotion to the (Deity of) Time and humbling themselves before God Almighty.

17. Rain at sea and density in the air (Metrodorus).

18. West, then east wind (Egyptians).

19. Wet and dew (Eudoxus); west wind, mizzle, and rain (Egyptians). On this day the water returns from the upper parts of the trees to the roots.

20, 21. (Missing.)

22. Nothing mentioned.

23. Rain (Eudoxus); west wind or south wind (Hipparchus).

24. Nothing mentioned. On this day the fair of Thu'âliba is held. Practical observers say that people mark on this day what wind is constantly blowing until night or until the time when the sun begins to decline; for this will be the most constant of all the winds of the year. This day they called the *Turning of the winds.* The white-and-black crows appear on this day in most countries.

25. Episemasia (Hipparchus and Eudoxus); west wind or south wind p.275. (Egyptians).

26, 27, 28. (Missing.)

29. Episemasia (Euctemon and Eudoxus); west wind or south wind (Hipparchus).

30. Nothing mentioned by the ancients, either about the air or anything else.

This, now, is the calendar used by the Greeks, to which we have added all that Sinân has mentioned in his *Kitâb-al'anwâ*. This is the concise summary of his book. We have not kept back anything which we have learned regarding the days of the calendar. We quote them by the names of the Syrians (*i.e.* as the 1st of Tishrîn, Kânûn, etc.) only, because they are generally known among people, and because this serves the same purpose (as if we were to call them by the Greek names).

Next we shall speak of the memorable days in the months of the Jews, if God Almighty permits!

CHAPTER XIV.

OF THE FESTIVALS AND FAST-DAYS IN THE MONTHS OF THE JEWS.

AFTER having explained the method how to learn the beginning of the year of the Jews, and its character,—after having solved this problem by the help both of computation and tables,—after having shown the arrangement of the months according to their beginnings and to the number of their days,—we hold it now to be necessary to explain their festivals and memorable days. For getting acquainted with them we shall at the same time learn the reason why they, even New-Year's Day itself, are not allowed to fall on certain days of the week. We begin with the first month, *i.e.*

Tishrî.

It has 30 days and only one *Rôsh-Ḥôdesh.* As we have explained before, the 1st Tishrî cannot fall on a Sunday, Wednesday, or Friday אדו. When, according to calculation, it ought to fall on one of these days, it is disregarded, and New-Year's Day is either the following day, if it is a *Dies licita,* or the preceding day, in case the following one is not a *Dies licita* according to the conditions that have been laid down in the *Tabula Terminorum* in the first part of this book. This proceeding of theirs they call דְּחִי. The 1st is the feast of New-Year, when they blow the trumpots and trombones, which are rams-horns. All work ceases on this day as on Sabbath. On this day, they maintain, Abraham offered his son Isaac, but then Isaac was ransomed by means of a ram. According to Jews and Christians, the person offered was Isaac, whilst there is a passage in the Coran in the Sûra *Wal-ṣâffât* (Sûra xxxvii. 99–113), showing that it was Ishmael. And, according to tradition, the Prophet is reported to have said: "I am the son of the two sacrificed ones," meaning 'Abdallâh b. Almuṭṭalib and Ishmael However, the discussion of this question is a subject of great extent. God knows best!

3. Fasting of Gedalyâ b. 'Aḥîḳâm, the governor of Nebucadnezar over Jerusalem. On this day he was killed, together with eighty-two p.276.
people, in a cistern in which the water collected until it rose above their heads. In consequence the Israelites were stricken with sorrow, and have ever since fasted on the day of his death.

5. Fasting of 'Akîbhâ. People wanted to compel him to worship the idol; he, however, did not submit. So they put him into a cage where he died of hunger, surrounded by twenty fellow prisoners.

7. *Fasting of punishment.* Its origin is this, that David, on having counted the Israelites, rejoiced in their number, and people themselves were puffed up on account of their great number, so as to go astray. Therefore God became angry with them, and sent the prophet Nathan to David and the assembly of the tribes to threaten them with the sword, with famine, and sudden death. His threatening was fulfilled. So they were stricken with fright, and have ever since fasted on this day.

On the same day the Israelites killed each other on account of the worship of the calf. They say that it was Aaron who made the calf, and so it is related in the Thora.

The Jew Ya'ḳûb b. Mûsâ Alniḳrisî (*i.e.* the physician) told me in Jurjân the following: Moses wanted to leave Egypt together with the Israelites, but Joseph the prophet had ordered that they should take his coffin along with them. As he, however, was buried in the bottom of the Nile and the water flowed over him, Moses could not get him away. Now, Moses took a piece of a paper and cut it into the figure of a fish; over this he recited some sentence, breathed upon it, wrote something upon it, and threw it into the Nile. Waiting for the result he stayed there, following the course of the river, but nothing appeared. So Moses took another piece of paper and cut it into the figure of a calf, wrote upon it, recited over it, breathed upon it, but then, when he was just about to throw it into the water, as he had done the first time, the coffin appeared. So he threw away the figure of the calf which he just held in his hand, but it was taken up by one of the bystanders.

Afterwards, when Moses disappeared on the mountain to speak with the Lord, and when the Israelites became anxious at his staying there so long, they pressed Aaron and demanded of him that he should give them a viceregent instead of Moses. Aaron, no doubt, did not know what to do; so he said: "Bring me all the precious ornaments of your women." So he spoke in order to gain time, knowing that the women would not be in a hurry to part with their ornaments. Possibly Moses might return before that. But it happened that the women gave up their ornaments most speedily. They fetched Aaron and *he* melted the ornaments and poured them into a mould; but the result was nothing but broken pieces of ingots. The same work he repeated in a hurry, hoping for the return of Moses and for news of him. Now he happened to have with himself the figure of that calf (which Moses had cut out of paper). So he said to himself: "By the figure of the fish once a wonderful miracle has been wrought. Now, let me see what the figure of the calf will produce!" He took the figure and threw it into the molten gold; when then the liquid mass was poured into a mould, it was formed into a calf which roared. Thereby the people were at that time seduced from the true belief without Aaron's having intended it.

p.277. 10. Fasting of *Kippûr*, also called *Al-'âshûrâ*. This fast-day is obligatory, whilst all other ones are voluntary. *Kippûr*-fasting begins half an hour before sunset of the 9th and lasts until half an hour after sunset of the 10th during 25 hours. In this way, too, all the voluntary fast-days are held. Therefore it is impossible that two of their fast-days should immediately follow each other, because one hour would belong to both of them in common, and because there would be no possibility of breaking the fast between them. Ya'ḳûb, however, maintains that this is a peculiarity only of this fast-day, whilst in the case of all the other fast-days it is allowed to fast in the same way (*i.e.* the same length of time) that the Muslims do.

On this day God addressed Moses the son of Amram. The fasting of this day is an atonement for all sins that are committed by mistake. The Jewish law orders everybody to be killed who does not fast on this day. They recite five prayers on this day, prostrating themselves upon the earth, which is not the custom on the other festivals.

15. *The feast of Tabernacles*, lasting seven consecutive days, during which they rest under the shadow of willows and reeds and other branches on the roofs of their houses. This is obligatory only for him who dwells at home, not for the traveller. On these days all work ceases, as God says in the third book of the Thora (Levit. xxiii. 34–43): "And on the fifteenth of the seventh month is the feast of Tabernacles. Then you shall not work during seven days. You shall celebrate a feast before God and you shall sit in the tents, the whole house of Israel, during seven days, that your (future) generations should know that it was I who made the Israelites dwell in tabernacles, when I led them out of Egypt." This feast is celebrated by the whole Jewish nation, whilst 'Abû-'Îsâ Alwarrâk says in his *Kitâb-almaḳâlât* of the Samaritans that they do not celebrate it.

The last or seventh day of the feast of Tabernacles, the 21st of the month, is called *'Arâbhâ*. On this day the clouds stood over the heads of the Israelites in the desert Altîh.

On the same day is the *feast of the Congregation*, when the Jews assemble in *Hârharâ* of Jerusalem, carrying around in procession the Ark of the Covenant, which in their synagogues is like the pulpit (*Minbar*) in a mosque.

22. The *feast of Benediction*, by which this feast-time is completed. All work ceases. They maintain that on this day the communication of the Thora was finished, and that the Thora was handed over to their

chiefs to be deposited in their synagogues. On this day they take the Thora out of its shrine, they bless themselves by it, and try to derive auguries from unfolding and reading it.

Marḥeshwân.

It has always two *Rosh-Ḥodesh*, and it has 30 days in a Perfect year and 29 days in an Intermediate year or in an Imperfect one. On these two Rosh-Ḥodesh there is no feast.

6. *Fasting of Zedekia.* Its origin is this, that Nebukadnezar killed the children of Zedekia, whilst he stood before them, patient and enduring, not weeping nor manifesting any sign of despair. Then both his eyes p.278. were put out. Therefore the Israelites were stricken with sorrow, and have ever since fasted on this day.

Differing herefrom, other people fix this fast-day on the Monday falling between the 8th and the 13th of this month. This, however, is not like a method suitable to Jewish ways; it is rather like Christian theories. The generality of Jews fix their fast-days on certain dates in the months, not on week-days.

Kislew.

It has only one Rosh-Ḥodesh in a Perfect year. It has 30 days in a Perfect and Intermediate year; 29 in an Imperfect year.

8. A fast-day. Its origin is this, that Yehoyakim burned the papers, called קינות *i.e.* the *Lamentations.* They contained a promise of God, and were brought by the prophet Jeremia. They treated of the condition of the Israelites in future times and of the calamities that would befall them. Jeremia sent the book through Barukh b. Neriyya, but Yehoyakim threw it into the fire, and therefore there arose manifold lamentations.

Other people fix this fasting on the Thursday falling between the 19th and the 25th of this month.

25. Beginning of the feast *Ḥanukkâ*, i.e. *purification.* It lasts eight days, during which they light lamps at the door of the hall; on the first night one lamp for each inhabitant of the house, on the second night two lamps, in the third three, etc. etc., and finally eight lamps on the eighth night, by which they mean to express that they increase their thanks towards God from day to day by the *purification* and sanctification of Jerusalem. The origin is this: Antiochus, the king of the Greeks, had subdued and maltreated them during a long period. It was his custom to violate the women, before they were led to their spouses, in a subterranean vault. From this vault two cords led outside, where two bells were fixed at their ends. When, now, he wanted a woman, he rung the right bell, and the woman entered; when he had done with her, he rung the left bell and dismissed her. Further, there was an Israelite who had eight sons, and one daughter whom another Israelite had demanded in marriage. Now, wanting to marry her, the father of his

bride said: "Give me time; for I stand between two things. If we lead my daughter to you, she will be dishonoured by the cursed tyrant, and she then is no longer a lawful wife for you. And if she does not submit to him, he will make me perish." For this state of things he blamed and reviled his sons, who became greatly excited and angry. But the youngest of them jumped up, dressed like a woman, hid a dagger in his garments, and went to the gate of the king, behaving like the whores. Now, the tyrant rang the right bell, and he was ushered into his presence; there, being alone with him, he killed him and cut off his head; then he rang the left bell and was let out, and stuck up the head (somewhere). Therefore the Israelites celebrate a feast on that and the following days (*i.e.* seven days), corresponding to the number of the brothers of this youth. God knows best!

p.279.

Ṭêbeth.

It has one Rosh-Ḥodesh in an Imperfect year, two in a Perfect and Intermediate year. It has 29 days.

5. *First appearance of darkness.* Ptolemy, the king of the Greeks, had asked them for the Thora, compelled them to translate it into Greek, and deposited it in his treasury. They maintain that this is the version of the *Seventy*. In consequence darkness spread over the world during three days and nights.

8. A fast-day, the last of the three *Dark* days, so called for the reason just mentioned.

9. A fast-day which they are ordered to keep, the origin of which they are ignorant of.

10. A fast-day, the day on which Nebukadnezar arrived before Jerusalem and laid siege to it.

Shebâṭ.

It has only one Rosh-Ḥodesh and 30 days.

5. A fast-day on account of the death of the saints in the time of Josua b. Nûn. Other people fix this fast-day on the Monday between the 10th and 15th of this month.

23. *Fasting of the Rebellion.* Its origin is this: The tribe of Benjamin were a godless and lawless set of people, who behaved like the people of Lot. Now, there came a man who wanted to pass through their country with his wife and maid-servant, making his pilgrimage to Jerusalem. Some countryman of his received him in his house; but scarcely had darkness fallen when the people of the place surrounded the door of his house, demanding his guest for their lust. Now, the master of the house offered to them his own daughter; but they said: "We do not want her." Then he gave up to them the servant-girl of his guest, and then they raped her the whole night. The girl expired towards dawn. Then her master cut her into pieces (12) according to the number of the

tribes of Israel; and to each tribe he sent one of her limbs, in order to rouse their wrath. Now, they assembled and made war upon that tribe, but they could not conquer them. Thereupon they fasted on this day and humiliated themselves before God. Finally He gave them victory over Benjamin; forty thousand men of this tribe were killed and seventy thousand of the others.

Adhâr I.

It is the leap-month in the leap-year. It does not exist in common years, and is not counted among their months. It has two Rosh-Ḥodesh and 30 days. There is not fast or feast day in this month.

Adhâr II.

This is the original Adhâr, which is called so in general (without the addition of I. or II.) in common years. There cannot be any ambiguity about what we just mentioned, speaking of another Adhâr preceding this one (because this only relates to leap-years). It has two Rosh-Ḥôdesh and 29 days.

7. A fast-day, because on this day Moses b. Amram died, and because with his death the manna and the quails ceased to appear.

9. A *fast-day which the Israelites established* for themselves at the time when the war between the people of Shammâi and of Hillel took p.280.
place, in which twenty-eight thousand men were killed.

Others fix this fast-day on the Monday between the 10th and 15th of this month.

13. *The fasting of Albûrî* (Pûrim), *i.e.* casting lots. Its origin is this: Once a man called Haman, a man of no importance, travelled to Tustar in order to undertake some office. But on the way thither he met with an obstacle which prevented him from reaching the end of his journey, and this happened on the identical day on which the offices (in Tustar) were bestowed. So he missed this opportunity and fell into utter distress. Now, he took his seat near the temples and demanded for every dead body (that was to be buried) $3\frac{1}{3}$ dirhams. This went on until the daughter of King Ahashwerosh died. When people came with her body, he demanded something from the bearers, and on being refused he did not allow them to pass, until they yielded and were willing to pay him what he asked for. But then he was not content with his first demand; he asked for more and more, and they paid him more and more, till at last it reached an enormous sum. The king was informed of the matter, and he ordered them to grant him his desire. But after a week he ordered him into his presence, and asked him: "Who invested you with such an office?" But Haman simply answered this: "And who forbade me to do so?" When the king repeated his question, Haman said: "If I am now forbidden to do so, I shall cease and give it up, and I shall give you with the greatest pleasure so and so many ten

thousands of denars." The king was astonished at the great sum of money which he mentioned, because he with all his supreme power had nothing like it. So he said: "A man who gathered so much money from the rule over the dead, is worthy to be made wazîr and councillor." So he entrusted him with all his affairs, and ordered his subjects to obey him.

This Haman was an enemy of the Jews. He asked the *Haruspices* and *Augures* which was the most unlucky time for the Jews. They said: "In Adhâr their master Mûsâ died, and the most unlucky time of this month is the 14th and 15th." Now Haman wrote to all parts of the empire, ordering people on that day to seize upon the Jews and to kill them. The Jews of the empire prostrated themselves before him, and appeared before him, crossing their hands upon their breasts, except one man, Mordekhai, the brother of Ester, the king's wife. Haman hated her, and planned her destruction on that day; but the king's wife understood him. Now she received (in her palace) the king and his wazîr, entertaining them during three days. On the fourth day she asked the king permission to lay before him her wishes. And then she asked him to spare *her* life and that of her brother. The king said: "And who dares to attempt anything against you both?" She pointed to Haman. Now the king rose from his seat in great wrath; Haman dashed towards the queen, prostrating himself before her, and then kissing her head, but she pushed him back. Now the king got the impression that he wanted to seduce her; so he turned towards him and said: "Hast thou in thy impudence come so far as to raise thy desire to her?" So the king ordered him to be killed, and Ester asked him to have him crucified on the same tree which he had prepared for her brother. So the king did, and wrote to all parts of the empire to kill the partisans of Haman. So they were killed on the same day on which he had intended to kill the Jews, *i.e.* on the 14th. Therefore there is great joy over the death of Haman on this day.

This feast is also called the *Feast of Megillâ*, and further *Hâmân-Sûr*.
p.281. For on this day they make figures which they beat and then burn, imitating the burning of Hâmân. The same they practise on the 15th.

Nîsan.

It has only one Rosh-Ḥodesh and 30 days.

1. Fasting over the death of Nadab and Abihu, the sons of Aaron, who died because they introduced foreign fire into the temple of God.

10. Fasting over the death of Maryam, the daughter of Amram, and over the sinking and disappearing of the water, a miracle which occurred on account of her death, as the manna and the quails ceased to appear in consequence of the death of Moses b. Amram. Some people fix this day on Monday between the 5th and the 10th of the month.

15. Passover-feast, of which we have already treated at such length

that there is no necessity for a repetition. This day is the first of the *Days of Unleavened Bread*, during which they are not allowed to eat leavened bread. For such is the command of God in the third book of the Thora (Levit. xxiii. 6), where He says: "On the fifteenth of this month is the feast of the unleavened bread unto God. Then you shall eat unleavened bread during seven days, and you shall not work during them." These days end with sunset of the 21st. On this day God drowned Pharao; it is also called المكس.

Iyâr.

It has two Rosh-Ḥodesh and 29 days.

10. *Fasting over the Ark.* It is the day when the Israelites were deprived of the ark, and when thirty men of them were killed. The priest Eli then managed their affairs. His gall-bladder split, and he fell dead from his seat, when he heard the news. Others fix this fasting on the Thursday between the 6th and 11th of the month.

28. Fasting, because on this day the prophet Samuel died.

Siwan.

It has only one Rosh-Ḥodesh and 30 days.

6. *The Feast of the Congregation*, a great festival, and one of the חגים of the Israelites. On this day their elders were present at Mount Sinai, where they heard the voice of God from the mountain speaking to Moses, ordering and forbidding, promising and threatening. They were ordered to celebrate a feast on this day as a thanksgiving to God for having preserved them from all mishap in their country, and their crops from thunder, cold, and rain. God says in the second book of the Thora: "And you shall make a pilgrimage to me thrice in every year: first, at the time of the unleavened bread; secondly, when the Thora was sent down, this is the pilgrimage of the *Feast of the Congregation*; and the third time, at the end of the year, when you bring in your fruit from the fields. Your feasting and your devotion to God shall be in sacred houses."

On this day they offer the first-fruits. Then they read prayers over them and invoke the blessing of God upon them.

Between the first of the *Days of the Unleavened Bread* and the *Feast of the Congregation* there are fifty days. These are the celebrated weeks during which they received their commandments, when their law was completed, and they were taught all knowledge relating to God.

Fasting on the Monday between the 9th and 14th. p.282.

23. Fasting. They say that on this day Jerobeam b. Nebaṭ ordered the ten tribes to worship two golden calves, and that they obeyed him. His children ruled over them about two hundred and fifty years, until Salmân

Al'a'shar, the king of Mosul, conquered them and led them into captivity. Then they were united with the other tribes in the time of Hizkia.

Yerobeam b. Nebâṭ was one of the slaves of Solomon, the son of David; he fled from his master, and the Israelites made him their king. Then he kept them from making pilgrimages to Jerusalem by the worship of these two calves, knowing that if they went to Jerusalem they would come to consider why they had made him their king; they would learn the reality of his case, and would depose and kill him.

25. Fasting over the death of Simeon, Samuel, and Ḥananyâ.

27. Fasting, for this reason: one of the Greek kings wanted to force Rabbâ Ḥananyâ b. Teradhyôn to worship the idol; he, however, did not yield. Therefore the king ordered a Thora to be wrapped round him, and him to be burned in it. Besides he put in prison Rabbâ 'Akîbâ, and forbade people to follow him, and he strove to abolish the Sabbath.

Tammuz.

It has two Rosh-Ḥodesh and 29 days. It has no feast.

17. Fasting, for on this day Moses broke the tables, and the fortifications of Jerusalem began to be destroyed at the time when Nebukadnezar besieged them. Further, on this day they put up an idol for worship in Jerusalem, and placed it in the altar-place of the temple, from sheer insolence and rebellion against God. On this day the Thora was burned, and the sacrifices ceased to be practised.

Âbh.

It has only one Rosh-Ḥodesh and 30 days.

1. Fasting, because on this day Aaron b. Amram died, and the cloud was raised as a miracle in his honour.

9. Fasting, because on this day they were told in the desert that they should not enter Jerusalem, and were sorry in consequence. On this day Jerusalem was conquered and entered by Nebukadnezar, who destroyed it by fire. On this day it was destroyed the second time, and its soil ploughed over.

25. Fasting, because the fire was extinguished in the temple. On this day Nebukadnezar left Jerusalem, and the conflagration of its storehouses and temples was put an end to.

28. Fasting, because the lamp of the temple was extinguished in the days of the prophet Ahas, which was a sign of God's wrath against them.

Elûl.

It has two Rosh-Ḥodesh and 29 days, but no feast.

7. *Fasting of the Spies.* On this day the spies returned to Moses, and brought him the report of the giants. Therefore the Israelites were sorry, but Josua b. Nûn refuted them. For this reason the fast-day was p.283.
established. Other Jews, however, place this fast-day on the Monday or Thursday which falls within the last seven days before the beginning of the next year.

(On the דחיות of the Jewish Calendar.)—The reason why they did not allow that—

The first of Tishr should ever be אדו (I. IV. VI. days of the week),
„ Kippûr be אגו (I. III. VI.),
„ Purim or Haman Sûr בדז (II. IV. VII.),
„ Passover בדו (II. IV. VI.),
„ 'Aṣereth גהז (III. V. VII.),

was this, that they wanted to prevent a day for any work falling on a Sabbath; for in that case they would not have been able to celebrate it, since they are not allowed to work on a Sabbath. For God says in the second book (Exod. xxxv. 2): "He who works on a Sabbath shall be killed." And in the fourth book (Num. xv. 32–36) it is related that they found a man of the Israelites in the desert working on a Sabbath and gathering wood. He was brought before Moses and Aaron, and they put him in prison. But God said to Moses, "Kill him," and so he was stoned to death.

A second reason (why they did not allow the feasts to fall on the days mentioned) was this, that they wished to prevent a Sabbath and another day on which all work ceases following each other.

As for Sunday, א, they did not allow it to be New-Year's Day, because God says in the third book (Levit. xxiii. 24–25): "On the *first* day of the *seventh* month you shall have rest, and a memorial of blowing of trumpets. Then you shall not work on that day, but you shall offer sacrifices." If, now, this day follows a Sabbath, the Jew gets two consecutive days of rest; the means of his maintenance are getting scanty, and he is brought to a condition in which it is difficult for him to make good the deficiency. In this case, *'Arâbhâ* falls on a Sabbath, and almsgiving, and the other works prescribed for this day, could not be carried out. For the same reason *Kippûr* could not fall on a Tuesday, nor the preceding Passover on a Friday, nor the preceding *'Aṣereth* on a Sabbath, because if this were the case the 1st of Tishrî would fall on a Sunday.

The reason why they do not allow New-Year's Day to be a Wednesday (ד), is that God says in the third book (Levit. xxiii. 27–32): "On the tenth day of the seventh month shall be remission. On this day you shall not do the least work from the evening of the ninth of the month till the (next) evening." Therefore all work is suspended on Kippûr (the 10th, in this case a Friday), and the following Sabbath is likewise a day of rest. So Kippûr cannot fall on a Friday, nor the preceding Passover on a Monday, nor the preceding *'Aṣereth* on a Tuesday.

The reason why they do not allow New-Year's Day to fall on a Friday (ו) is this, that Friday is followed by Sabbath, and because in that case Kippûr would fall on a Sunday, following upon a Sabbath, and the *Feast of Benediction* would fall on a Friday preceding a Sabbath, an order of days which is forbidden by the law. For the same reason Kippûr cannot fall on a Sunday, nor the preceding Passover on a Wednesday, nor the preceding *'Aṣereth* on a Thursday, because all this would necessitate
p.284. New-Year's Day being a Friday, and thence would result those consequences which we have mentioned.

Therefore people endeavoured to construct the calendar in such a way as to prevent two days of rest following each other, and *'Arâbhâ* falling on a Sabbath, because on this day they must give alms and must make a pilgrimage around the pulpit, which they call 'Ârôn, ארון, or *Kilwâdh*. Further, they had to prevent Pûrîm falling on a Sabbath, which would keep them from burning Haman and uttering their joy thereat. And lastly, they had to prevent *'Aṣereth* falling on a Sabbath, because in that case they could not bring their seeds and their first-fruits, and other things that are prescribed for this day.

'Abû-'Îsâ Alwarrâḳ speaks in his *Kitâb-Almaḳâlât* of a Jewish sect called the Maghribîs, who maintain that the feasts are not legal unless the moon rises in Palestine as a full moon in the *night* of Wednesday, which follows after the *day* of Tuesday, at the time of sunset. Such is their New-Year's Day. From this point the days and months are counted, and here begins the rotation of the annual festivals. For God created the two great lights on a Wednesday. Likewise they do not allow Passover to fall on any other day except on Wednesday. And the obligations and rites prescribed for Passover they do not hold to be necessary, except for those who dwell in the country of the Israelites. All this stands in opposition to the custom of the majority of the Jews, and to the prescriptions of the Thora.

The Ananites fix the beginning of the months by the observation of the appearance of new moon, and settle intercalation by that sort of prognostication which we have mentioned. They do not mind on what days of the week the feasts fall, except as regards Sabbath. For in this case they postpone the feasts to the following Sunday. This postponement they call דְּחִיָּא. On a Sabbath they do not touch any work whatsoever; even the circumcision of the children they postpone till the following day, in opposition to the practice of the Rabbanites.

With the suspension of work on a Sabbath certain curious affairs are connected. In the first instance God says in the Coran (Sûra vii. 163): "Then their fishes appearing on the surface of the water come to them on the day when they celebrate Sabbath; but on a day on which they do not celebrate Sabbath the fishes do not come to them." Further, Aljaihânî relates in his *Liber Regnorum et Viarum*, that eastward of Tiberias lies the city of Balinas (Apollonias?), where the Jordan has its source. There the river drives mills, that stand still on a Sabbath and do not work, because the water disappears beneath the earth until the end of Sabbath. For this occurrence I am unable to find a physical explanation, because its repetition and revolution is based upon the days of the week. Annual occurrences are accounted for by the sun and his rays, monthly occurrences by the moon and her light, as *e.g.* the altar in Greece which of itself burned the sacrifices on one certain day of the year, under the influence of the reflected solar rays which were concentrated on a certain spot of the altar, etc.

'Abû-'Îsâ Alwarrâḳ relates in his *Kitâb-Almaḳâlât* that a Jewish sect, the Alfâniyya (*Millenarii*), reject the whole of the Jewish feasts, and p.285. maintain that they cannot be learned except through a prophet, and that they keep no other feast-day but Sabbath.

The following table, the *Tabula Argumentationis*, illustrates what we have stated before regarding the feasts, and shows that New-Year's Day cannot fall on the days mentioned, *i.e.* the days of the sun, of Mercury and Venus. The red ink indicates a *Dies illicita*, the black ink a *Dies licita*. If, now, the transversal line of numbers which correspond to the feasts mentioned at the tops of the single columns is black from beginning to end, all these numbers signify *Dies licitæ*; if, however, some of those numbers, or all of them, are written in red ink, these *some* or all of them are *Dies illicitæ*. Opposite the numbers we have placed a special column for the terms "*Necessary*," "*Possible*," and "*Impossible*." The terms *necessary* and *impossible* do not need an explanation. The term *possible* means that if New-Year's Day falls on a *Dies licita*, but some of the numbers indicating the single feast-days in the transversal line are written in red ink, those days are *Dies illicitæ* in common years, whilst they are *Dies licitæ* in a leap-year of the same quality, and *vice versâ*. This table shows clearly why some of the (three) kinds of Jewish years can follow each other, whilst others cannot, as we have mentioned before. For if Rosh-hashshânâ of a year following after a year of a certain quality (כ or ש) is such as could not be the beginning of a year of another quality, these two kinds may follow each other; in any other case they cannot follow each other. From this rule, however, we must except the *Imperfect* years (ח), because the fact that two years ח cannot follow each other rests on another ground; hereof we have already spoken in the preceding part.

p.286.

Tabula Argumentationis I.*

1. Primary qualities.	2. Secondary qualities.	3. Divisions.	4. 1st of Tishrî, upon which the other feasts depend.	5. Kippûr, 10th of Tishrî.	6. 'Arâbhâ, 21st of Tishrî.	7. Pûrîm, 14th of Adhâr.	8. Passover, 15th of Nîsan.	9. 'Aṣereth, 6th of Sîwan.	10. Beginning of the following year, 1st of Tishrî.
Common years.	Imperfect -	Impossible	I.	III.	VII.	VII.	II.	III.	IV.
	”	Necessary	2	4	1	1	3	4	5
	”	Impossible	3	5	2	II.	IV.	V.	VI.
	”	Impossible	IV.	VI.	3	3	5	6	7
	”	Possible	5	7	4	IV.	VI.	VII.	I.
	”	Impossible	VI.	I.	5	5	7	1	2
	”	Necessary	7	2	6	6	1	2	3
	Intermediate	Impossible	I.	III.	VII.	1	3	4	5
	”	Impossible	2	5	1	II.	IV.	V.	VI.
	”	Necessary	3	4	2	3	5	6	7
	”	Impossible	IV.	VI.	3	IV.	VI.	VII.	I.
	”	Possible	5	7	4	5	7	1	2
	”	Impossible	VI.	I.	5	6	1	2	3
	”	Impossible	7	2	6	VII.	II.	III.	IV.
	Perfect -	Impossible	I.	III.	VII.	II.	IV.	V.	VI.
	”	Necessary	2	4	1	3	5	6	7
	”	Possible	3	5	2	IV.	VI.	VII.	I.
	”	Impossible	IV.	VI.	3	5	7	1	2
	”	Necessary	5	7	4	6	1	2	3
	”	Impossible	VI.	I.	5	VII.	VI.	III.	IV.
	”	Necessary	7	2	6	1	3	4	5

* The *Dies Licitæ*, in the Arabic original written in black ink, are here written in Arabic numerals, whilst the *Dies Illicitæ*, written in the original in red ink, are here written in Latin numerals.

Tabula Argumentationis II.

p.287.

	1. Primary qualities.	2. Secondary qualities.	3. Divisions.	4. 1st of Tishrî, upon which the other feasts depend.	5. Kippûr, 10th of Tishrî.	6. 'Arûbhâ, 21st of Tishrî.	7. Pûrîm, 14th of Adhâr.	8. Passover, 15th of Nisan.	9. 'Aṣereth, 6th of Siwan.	10. Beginning of the following year, 1st of Tishrî.
	Leap-years.	Imperfect -	Impossible	I.	III.	VII.	II.	IV.	V.	VI.
		,,	Necessary	2	4	1	3	5	6	7
		,,	Impossible	3	5	2	IV.	VI.	VII.	I.
		,,	Impossible	IV.	VI.	3	5	7	1	2
		,,	Possible	5	7	4	6	1	2	3
		,,	Impossible	VI.	I.	5	VII.	II.	III.	IV.
10		,,	Necessary	7	2	6	1	3	4	5
		Intermediate	Impossible	I.	III.	V.	3	5	6	7
		,,	Impossible	2	4	1	IV.	VI.	VII.	I.
		,,	Necessary	3	5	2	5	7	1	2
		,,	Impossible	IV.	VI.	3	6	1	2	3
		,,	Possible	5	7	4	VII.	II.	III.	IV.
		,,	Impossible	VI.	I.	5	1	3	4	5
		,,	Impossible	7	2	6	II.	IV.	V.	VI.
		Perfect -	Impossible	I.	III.	VII.	IV.	VI.	VII.	I.
		,,	Necessary	2	4	1	5	7	1	2
20		,,	Possible	3	5	2	6	1	2	3
		,,	Impossible	IV.	VI.	3	VII.	II.	III.	IV.
		,,	Necessary	5	7	4	1	3	4	5
		,,	Impossible	VI.	I.	5	II.	IV.	V.	VI.
		,,	Necessary	7	2	6	3	5	6	7

CHAPTER XV.

p.288. ON THE FESTIVALS AND MEMORABLE DAYS OF THE SYRIAN CALENDAR, CELEBRATED BY THE MELKITE CHRISTIANS.

THE Christians are divided into various sects. The first of them are the Melkites (*Royalists*), *i.e.* the Greeks, so called because the Greek king is of their persuasion. In Greece there is no other Christian sect beside them.

The second sect are the Nestorians, so called after Nestorius, who brought forward their doctrine between A. Alex. 720 and 730.

The third sect are the Jacobites.

These are their principal sects. They differ among each other on the dogmas of their faith, as *e.g.* on the persons (τὰ πρόσωπα in Christ), on the divine nature, the human nature, and their union (ἕνωσις). There is another sect of them, the *Ariani*, whose theory regarding Christ comes more near that of the Muslims, whilst it is most different from that of the generality of Christians. Besides there are many other sects, but this is not the place to enumerate them. This subject has been exhaustively treated and followed up into its most recondite details in the books treating of philosophical and religious categories and doctrines, and which at the same time refute those sects.

The most numerous of them are the Melkites and Nestorians, because Greece and the adjacent countries are all inhabited by Melkites, whilst the majority of the inhabitants of Syria, 'Irâk and Khurâsân are Nestorians. The Jacobites mostly live in Egypt and around it.

Certain days of the Syrian months are celebrated among them; on some of them they agree, on others they differ. The reason of their agreeing is this, that those days were spread through the Christian world before the schism in their doctrines was brought about. The reason of the difference is this, that some days belong to one sect and to one province in particular.

Besides they have other days depending upon their Great Fast, and weeks depending upon the most famous days. On this category of days, as on the former, they partly agree, partly disagree.

I shall now enumerate the calendar days of the Melkites in Khwârizm according to the Syrian calendar. You find very rarely that the Christians, Jews, and Zoroastrians in different countries agree among each other in the use of festivals and memorial days. Only regarding the greatest and most famous feasts they agree, whilst generally on all others they differ.

Secondly, I shall speak of their fasting, and all the days connected with it, on which the various sects agree.

Lastly, I shall treat of the feasts and memorable days of the Nestorians, if God permits!

Tishrîn I.

1. Commemoration (*μνήμη*) of the bishop and martyr Ananias, the pupil of St. Paul. It is a Christian custom on these commemoration days to celebrate the memory of the saint to whom the day is dedicated; they pray to God for him, and praise him, and humble themselves before God in his name. To every child which is born on this day or later, until the next commemoration day, they give the name of the saint of the day. Frequently, too, they give each other the names of two commemorations, so as to say χ, also called N from the commemoration of the saint N. When this commemoration comes, people assemble in his house, and he receives them as his guests, and gives them a repast.

2. Arethas (Ḥârith) of Najrân, martyr with the other martyrs.

3. Mary the nun, who wore man's dress; she lived like a monk, and concealed her sex before the monks. Being accused of fornication with a woman, she bore this wrong patiently, and her sex did not become known before her death. When they, then, wanted to wash her body, and saw the genitals, they found out the reality of the case, and—her innocence. p.289.

4. Dionysius, the bishop and astronomer, the pupil of St. Paul.

These titles (like bishop, etc.) indicate clerical degrees, of which they in their religion have nine:—

1. Cantor, *ψάλτης*.
2. Reader, ܩܪܘܝܐ
3. Hypodiaconus.
4. Diaconus, in Arabic *Shammâs*.
5. Presbyter, in Arabic *Ḳass*.
6. Bishop, in Arabic *'Usḳuf*. He stands under the *metropolita*.
7. *Metropolita*, who stands under the *catholicus*. The residence of the metropolita of the Melkites in Khurâsân is Marw.
8. Catholicus, in Arabic *Jâtheliḳ*. The residence of the catholicus of

the Melkites in Muhammadan countries is Baghdâd. He stands under the Patriarch of Antiochia. The Nestorian catholicus is appointed by the Khalif on the presentation of the Nestorian community.

9. Patriarch, in Arabic *Baṭrîḳ*. This dignity exists only among the Melkites, not among the Nestorians. There are always four patriarchs in Christendom; as soon as one dies, at once a successor is created, being chosen by the remaining patriarchs, the *catholici*, and by the other dignitaries of the Church. One patriarch resides in Constantinople, another in Rome, the third in Alexandria, and the fourth in Antiochia. These towns are called *θρόνοι*.

There is no degree beyond that of the patriarch, and none below that of the *cantor*. Frequently they count only from the diaconus upwards, and do not reckon the singers and altar-servants among the officials of the Church. To each degree attach certain rules, usages, and conditions, on which this is not the proper place to enlarge.

'Abû-Alḥusain 'Aḥmad b. Alḥusain Al'ahwâzî, the secretary, reports in his *Book of the Sciences of the Greeks*, what he himself has learned in Constantinople of the degrees of the service both of Church and State. His report is this:—

I. Patriarch, highest Church dignitary, supreme authority throughout the empire.
II. Χρνσχς (?) the prefect of the greatest monastery.
III. 'Επίσκοπος, *i.e.* bishop.
IV. Μητροπολίτης, *i.e.* the governor [or ruler].
V. 'Ηγούμενος, prefect of a monastery, highly revered by them.
VI. Καλόγηρος. His degree comes near to that of the *Hegoumenos*.
VII. Πάπας, in Arabic *Ḳass*.
VIII. Διάκονος, in Arabic *Shammâs*.

However, the more trustworthy account of the matter is the one given above. Because 'Abû-Alḥusain has mixed up with the men of the official degrees other people, who, although important personages, are not exactly dignitaries of such and such degree; or perhaps they belong to one of those degrees, but then his description does not fit.

The laic degrees of the State service are the following:—

I. Βασιλεύς, *i.e.* Cæsar, king of the Greeks.
p.290. II. Λογοθέτης, his vazîr and dragoman.
III. Παρακοιμώμενος, the first of the chamberlains.
IV. Δομέστικος, commander of the army.
V. Ακσιωrs (?), a man in the king's special confidence in the army, similar to the *domesticus*, both being of the same rank.
VI. Αρχντιρχν (?), the head of the πατρίκιοι.
VII. Πατρίκιος, in Arabic *baṭrîḳ*. These baṭrîḳs are in the army something like chief-commander, not to be confounded with the

batrîks whom we have mentioned as clerical dignitaries. Those who fear the ambiguity of the words call the clerical dignitary *baṭrak*.

VIII. Ῥογάτωρ, who has to review the army and to pay the stipends of the soldiers.

IX. Στράτηγος. His rank is half that of a Πατρίκιος.

X. Πρωτοσπαθάριος, a man in the king's confidence in the army of the Πατρίκιος, whom the Πατρίκιος consults in every affair.

XI. Μαγλαβίτης, the officer of the royal whip (*Præfectus lictorum*).

XII. Ἔξαρχος, an officer over 1,000 men.

XIII. Ἑκατοντάριος, a commander of 100 men.

XIV. Πεντηκοντάριος, a commander of 50.

XV. Τεσσαρακοντάριος, a commander of 40.

XVI. Τριαντάριος, a commander of 30.

XVII. Ἐικοσιτάριος, a commander of 20.

XVIII. Δέκαρχος, a commander of 10.

Now we return to our subject.

5. Commemoration of the Seven Sleepers of Ephesus, who are mentioned in the Coran. The Khalif Almu'taṣim had sent along with his ambassador another person who saw the place of the Seven Sleepers with his own eyes, and touched them with his own hands. This report is known to everybody. We must, however, observe that he who touched them, *i.e.* Muḥammad b. Mûsâ b. Shâkir himself, makes the reader rather doubt whether they are really the corpses of those seven youths or other people,—in fact, some sort of deception.

'Alî b. Yaḥyâ, the astronomer, relates that on returning from his expedition, he entered that identical place, a small mountain, the diameter of which at the bottom is a little less than one thousand yards. At the outside you see a subterranean channel which goes into the interior of the mountain, and passes through a deep cave in the earth for a distance of three hundred paces. Then the channel runs out into a sort of half-open hall in the mountain, the roof being supported by perforated columns. And in this hall there is a number of separate compartments. There, he says, he saw thirteen people, among them a beardless youth, dressed in woollen coats and other woollen garments, in boots and shoes. He touched some hairs on the forehead of one of them, and tried to flatten them, but they did not yield. That their number is more than seven, which is the Muhammadan, and more than eight, which is the Christian tradition, is perhaps to be explained in this way, that some monks have been added who died there in the same spot. For the corpses of monks last particularly long, because they torture themselves to such a degree that finally all their moist substances perish, and between bones and skin only very little flesh remains. And therefore their life is extinguished like a lamp when it has no more oil Frequently they remain for generations in the same posture, leaning on their sticks. Such a thing you may witness in regions where monks live.

According to the Christians these youths slept in their cave three hundred and seventy-two years; according to us (Muslims) three hundred solar years, as God says in the Coran in the chapter which specially treats of their history (*i.e.* Sûra xviii.). As for the addition of nine years (Sûra xviii. 24: "And they remained in their cave three hundred years and nine years more"), we explain them as those nine years which you must add if you change the three hundred solar years into lunar years. To speak accurately, this addition would be

9 years, 75 days, $16\frac{4}{5}$ hours.

However, according to the way in which people reckoned at that time,
p.291. they counted the 300 years as 15 *Minor Cycles* (of 19 years) plus 15 years of the 16th cycle. The number of months that were to be intercalated for such a space of time was 110 according to anyone of the *Ordines Intercalationis* which they may have applied to the rest of the (15) years. And 110 months amount to 9 years and 2 months. Such fractions, however, (as 2 months or $\frac{1}{6}$ year) are neglected in a historical account.

7. Commemoration of Sergius and Bacchus.

10. Commemoration of the prophet Zacharias. On this day the angels announced to him the birth of his son John, as it is mentioned in the Coran, and in greater detail in the Gospel.

11. Cyprianus, the bishop, the martyr.

14. Gregory of Nyssa, the bishop.

17. Cosmas and Damianus, the physicians, the martyrs.

18. Lucas, author of the third Gospel.

23. Anastasia, the martyr.

26. Commemoration of the sepulture of the head of John the son of Zacharias.

Tishrîn II.

1. Cornutus, martyr.

11. Menas (Μηνᾶς), martyr.

15. Samonas, Gurias, and Abibos, the martyrs.

16. Beginning of the fasting for the nativity of Jesus, the son of Mary, Messiah. People fast forty consecutive days before Christmas (16 Nov.–25 Dec.).

17. Gregorius Thaumaturgos.

18. Romanus, the martyr.

20. Isaac, and his pupil Abraham, the martyrs.

25. Petrus, bishop in Alexandria.

27. Jacob, who was cut to pieces.

30. Andreas, martyr, and Andreas the apostle.

Kânûn I.

1. Jacob, the first bishop of Ælia.

3. Johannes, the *Father*, who collected in a book the rites and laws of Christianity. To address a man by the title of "*Father*," is with them the highest mark of veneration, because thereupon (upon the veneration towards their spiritual fathers?) their dogmas are based. There is no original legislation in Christianity; their laws are derived and developed by their most venerated men from the canonical sayings of Messiah and the apostles. So they represent the matter themselves.

4. Barbara and Juliana, the martyrs.

5. Sâbâ, abbot of the monastery in Jerusalem.

6. Nicolaus, patriarch of Antiochia.

13. The five martyrs.

17. Modestus, patriarch of Ælia.

18. Sîsîn, the catholicus of Khurâsân.

20. Ignatius, third patriarch of Antiochia. p.292.

22. Joseph of Arimathia ὁ βουλευτής, who buried the body of the Messiah in a grave which he had prepared for himself, as is related towards the end of all four Gospels. Almaʻmûn b. Aḥmad Alsalamî Alharawî maintains that he has seen it in the Church of the Resurrection in Jerusalem, in a vault as a grave cut into the rock in a gibbous form, and inlaid with gold. To this grave attaches a curious story, which we shall mention when speaking of the Christian Lent. People say that the king does not allow the Greeks to visit the grave.

23. Gelasius, martyr.

25. In the night after the 25th of this month, *i.e.* in the night of the 25th according to the Greek system, is the feast of the Nativity (ܝܠܕܐ), the birth of the Messiah, which took place on a Thursday night. Most people believe that this Thursday was the 25th, but that is a mistake; it was the 26th. If anybody wants to make the calculation for this year by means of the methods mentioned in the preceding part, he may do so. For the 1st of Kânûn I. in that year was a Sunday.

26. David, the prophet, and Jacob, the bishop of Ælia.

27. Stephanus, head of the deacons.

28. Herodes killed the chilren and infants of Bethlehem, searching for the Messiah, and hoping to kill Him among the others, as is related in the beginning of the Gospel.

29. Antonius, martyr. Christians believe him to be identical with *'Abû-Rûḥ*, the cousin of Hârûn Alrashîd. He left Islâm, and became a convert to the Christian Church, wherefore Hârûn crucified him. They tell a long and miraculous tale about him, the like of which we never heard nor read in any history or chronicle. Christians, however, on the whole are very much inclined to accept and to give credit to such things, more particularly if they relate to their creeds, not at all endeavouring by

the means at their disposal to criticise historical traditions, and to find out the truth of bygone times.

Kânûn II.

1. Basilius, also feast of the *Calendæ* (*Calendas*). Calendas means "*may it be good*" (καλόν plus?). On this day the Christian children assemble and go round through the houses, crying with the highest voice and some sort of melody "*Calendas.*" Therefore they receive in every house something to eat, and a cup of wine to drink. As the reason of this custom some people assert that this is the Greek New-Year's Day, *i.e.* one week after Mary had given birth to Christ. Others relate as its reason the following story: Arius on having come forward with his theory, and having found adherents, took possession of one of the Christian churches, but the people of that church protested against it. Finally they arranged with each other, and came to this agreement: That they would shut the door of the church for three days; then they would proceed together to the church, and read before it alternately. That party, then, to whom the door would open of itself, should be its legal owner. So they did. The church door did not open of itself to Arius, but it opened to the other party. So they say. Therefore their children do
p.293. such things in imitation of the lucky message which they received at that time.

2. Silvester, the metropolitan, through whom the people of Constantinople became christianized.

5. Fasting for the feast of Epiphany.

6. Epiphany (ܕܢܚܐ) itself, the day of baptism, when John, the son of Zacharias, baptized Messiah, and made him dive under the baptismal water of the river Jordan, when the Messiah was thirty years of age. The Holy Ghost came over him in the form of a dove that descended from heaven, according to the relation of the Gospel.

The same, now, Christians practise with their children when they are three or four years of age. For their bishops and presbyters fill a vessel with water and read over it, and then they make the child dive into it. This being done, the child is christened. This is what our Prophet says: "Every child is born in the state of original purity, but then its parents make it a Jew, or Christian, or Magian."

'Abû-Alḥusain Al'ahwâzî describes in his book of the Sciences of the Greeks the process of christening. First they read prayers for the child in the church during seven days, early and late; on the seventh day it is undressed, and its whole body anointed with oil. Then they pour warm water into a marble vessel which stands in the middle of the church. On the surface of the water the priest makes five dots with oil in the figure of a cross, four dots and one in the middle. Then the child is raised, its feet are placed so as completely to cover the dot in the middle, and it is put into the water. Then the priest takes a handful of water from one side, and pours it over the head of the child. This he does four times, taking the water successively from all four sides corresponding to the four sides of the cross. Then the priest steps backward, and that person comes who wants to take the child out of the water, the same who has placed it there. Then the priest washes it, while the whole congregation of the church is praying. Then it is definitively taken out of the water, is adorned with a shirt, and carried away to prevent its feet from touching the ground, whilst the whole church cries seven times: κύριε ἐλεησόν, *i.e.* "*O Lord, have mercy upon us!*" Then the child is completely dressed, always being borne in the arms; then it is put down. Thereafter either it remains in the church, or it goes there again and again during seven days. On the seventh day the priest washes it again, but this time without oil, and not in the baptismal vessel.

11. Theodosius, the monk, who tortured himself, and loaded himself with chains.

13. End of the feast of Epiphany. On this day the noble saints on Mount Sinai were killed.

15. Petrus, Patriarch of Damascus.

17. Antonius, the first of the monks, and their head.

20. Euthymius, the monk, the teacher.

21. Maximus, the anchorite.

22. Cosmas, author of Christian canons and laws.

25. Polycarpus, the bishop, the martyr, who was burned with fire. p.294.

25. Johannes, called Chrysostomus. 'Ιωάννης is the Greek form for John.

31. Johannes and Cyrus, the martyrs.

Shubâṭ.

1. Ephraem, the teacher.

2. *Wax Feast*, in recollection of Mary's bringing Jesus to the temple of Jerusalem, when he was forty days of age. This is a Jacobite feast, held in great veneration among them. People say that on this day the Jews introduce their children into the temples, and make them read the Thora. If this is the case, it is in Shebaṭ (the Jewish form of the name) not in Shubâṭ (the Syrian form), since the Jews do not use the Syrian names.

Between the 2nd Shubâṭ and the 8th Adhâr the beginning of their Lent varies, of which we shall speak hereafter. When fasting they never celebrate the commemoration-days we mention, except those that fall on a Sabbath; those and only those are celebrated.

3. Belesys, martyr, killed by the Magians.

5. Sîs Catholicus, who first brought Christianity down to Khurâsân.

14. Commemoration in recollection of the finding of the head of the Baptist, *i.e.* John, the son of Zacharias.

Adhâr.

9. The forty martyrs who were tortured to death by fire, cold, and frost.

11. Sophronius, Patriarch of Jerusalem.

25. *Annuntiatio Sanctissimæ Deiparæ.* Gabriel came to Mary announcing to her the Messiah. From this day until the day of His birth is a little more than 9 months and 5 days, which is the natural space of time for a child's sojourn in the mother's womb. Jesus, though he had no human father, and though supported by the Holy Ghost, was in His earthly life subject to the laws of nature. And so it is only proper that also His sojourn in the womb of His mother should have been in agreement with nature.

The mean place of the moon at noon of this day, Monday the 25th Adhâr A. Alex. 303, for Jerusalem, was about 50 minutes in the first degree of Taurus. Those, now, who follow in the matter of the *Numûdhâr* (*i.e.* a certain method of investigation for the purpose of finding the *ascendens* or horoscope under which a child is born) the theory of Hermes the Egyptian, must assume the last part of Aries and the beginning of Taurus as the *ascendens* of the Messiah. However, Aries and Taurus were *ascending* at the time Christ was born, during the *daytime,* because the mean place of the sun for Jerusalem for noon of Thursday following after the night in which Christ was born, is about 2 degrees and 20 minutes of Capricorn. The above-mentioned time of
p.295. Christ's sojourning in His mother's womb (9 months 5 days) is, according to their theory, a *conditio sine quâ non* for every child that is born in the night of Christmas, when the moon is standing under the earth at a distance of $\frac{1}{10}$ circumference from the degree of the horoscope. Now, knowing so much about the moon's place on the day of the Annunciation, we find that the horoscope (of the hour of Christ's birth) was near 24 degrees of Pisces. And if we compute the mean place of the moon for the 25th of Kânûn I. for the time when she stands under the earth at the distance of $\frac{1}{10}$ circumference, we find the horoscope to have been nearly 20 degrees in Aries.

Both calculations, however, (that of the astrologers and Albîrûnî's own) are worthless, because those who relate the birth of Christ relate that it occurred at night, whilst our calculations would lead to the assumption that it occurred in the day. This is one of the considerations which clearly show the worthlessness of the *Numûdhârs.* We shall dedicate a special book to the *genera* and *species* of the *Numûdhârs,* where we shall exhaust the subject and not conceal the truth, if God permits me to live so long as that, and if He by His mercy delivers me from the remainder of pain and illness.

Nísân.

1. Mary the Egyptian, who fasted 40 consecutive days without any interruption. As a rule, this commemoration-day is celebrated on the first Friday after breaking fast; therefore, Friday being a *conditio sine quâ non*, it falls on the 1st of Nîsân only four times in a *Solar Cycle*, viz. in the 4th, 10th, 15th, and 21st years, if you count the cycles from the beginning of the *Æra Alexandri*, the current year included.

15. The 150 martyrs.

21. The six synods. *Synod* means a meeting of their wise men, of their priests, bishops, and other church dignitaries, for the purpose of anathematizing some innovation, and for something like cursing each other, or for the consideration of some important religious subject. Such synods are not convoked except at long intervals, and if one takes place, people keep its date in memory and frequently celebrate the day, hoping to obtain a blessing thereby, and wanting to show their devotion.

1. The first of the six synods was that of the 318 bishops at Nicæa, A.D. 325, under the king Constantine, convoked on account of Arius, who opposed them in the question of the *Persons*, and for the purpose of perpetuating the dogma which they all agreed upon regarding the two Persons of the Father and the Son, and their agreement regarding this subject that *Fast-breaking* should always fall on Sunday after the resurrection of the Messiah; for there had come forward some people proposing to break the fast on the 14th of the Jewish Passover month (Τεσσαρεσκαιδεκατῖται, or *Quartodecimani*).

2. Synod of the 150 bishops in Constantinople, A.D. 381, under the king Theodosius, son of Arcadius the Elder, convoked on account of a man called "*enemy of the Spirit*" (*πνευματόμαχος*), because he opposed the Catholic Church in the description of the Holy Ghost, and for the purpose of perpetuating their dogma regarding this *Third Person*.

3. Synod of the 200 at Ephesus, A.D. 431, under the king Theodosius Junior, convoked on account of Nestorius, the Patriarch of Constantinople, p.296.
the founder of Nestorian Christianity, because he opposed the Catholic Church regarding the Person of the Son.

4. Synod of the 630 at Chalcedon, A.D. 451, under the king Marcianus, on account of Eutyches, because he taught that the body of the Lord Jesus consisted, *before* the *ἕνωσις*, of two natures, afterwards only of *one* nature.

5. Synod under Justinian I., A.D. 553, convoked for the purpose of condemning the bishops of Mopsuestia, of Edessa, and others, who opposed the Church in its fundamental dogmas.

6. Synod of 187 bishops in *Constantinople*, A.D. 680, under Constantine (Pogonatus) the Believer, convoked on account of Cyrus and Simon Magus.

23. Mâr Georgios, the martyr, tortured repeatedly and by various tortures, till he died.

24. Marcus, author of the second Gospel.

25. Elias, Catholicus of Khurâsân.

27. Christophorus.

30. Simeon b. Ṣabbâ'ê Catholicus, killed in Khûzistân, together with other Christians.

Ayyâr.

1. Jeremia, the prophet.

2. Athanasius, the patriarch.

3. *The Feast of Roses* according to the ancient rite, as it is celebrated in Khwârizm. On this day they bring Jûrî-roses to the churches, the reason of which is this, that Mary presented on this day the first roses to Elizabeth, the mother of John.

6. Hiob, the prophet.

7. Feast of the Apparition of the Cross in Heaven. Christian scholars relate:—In the time of Constantine the Victorious there appeared in heaven the likeness of a cross of fire or light. Now people said to the king Constantine, "Make this sign your emblem, and thereby you will conquer the kings who surround you." He followed their advice, he conquered, and therefore became a Christian. His mother Helena he sent to Jerusalem to search for the wood of the Cross. She found it, but together with the two crosses on which, as they maintain, the two robbers had been crucified. Now they were uncertain, and did not know how to find out which was the wood of the Cross of Christ. Finally they placed each cross upon a dead body: when, then, it was touched by the wood of the Cross of Christ, the dead man became alive again. Thereby, of course, Helena knew that this cross was the right one.

Other Christians, who are not learned people, speak of the cross in the constellation of the Dolphin, which the Arabs call *Ḳa'ûd* (riding-camel), *i.e.* four stars close to *Alnasr Alwâḳi'*, the situation of which is like the angles of a quadrangle. They say that at that time this cross in the Dolphin appeared opposite that place where Messiah had been crucified. Now, it is very strange that those people should not reflect a little and find out that there are nations in the world who consider it as their business to observe the stars and to examine everything connected with the stars for ages and ages, one generation inheriting from the
p.297. other at least this knowledge, that the stars of the Dolphin are *fixed stars*, which in this quality of theirs had long ago been recognized by their ancestors who cared for such things.

And more than this. This Christian sect indulges *in majorem Crucis gloriam* in all sorts of tricks and hallucinations, *e.g.* God ordered the Israelites to make a serpent of brass and to hang it on a beam, which was to be erected, for the purpose of keeping off the injury done by the serpents when they had become very numerous among them in the desert. Now from this fact they infer and maintain that it was a prophecy and a hint indicative of the Cross (of Christ).

Further they say that the *sign* of Moses (*i.e.* the divine gift by which he wrought miracles) was his staff, and a staff is a longitudinal line. Now when Christ came, He threw His staff over such a line, and a cross was formed, which is to be indicative of the fact that the law of Moses was *completed* (finished) by Christ. But I should think that that which is perfect in itself does not admit of any increase or decrease, which you might prove in this way, that if you threw a third staff over the cross, from whatever side you like, you get the lines of the word لا (*no*), which means *no* increase and *no* decrease.

This is certainly the same sort of hallucination frequently occurring among those Muslims who try to derive mystical wisdom from the comparison of the name of Muhammad (محمد) with the human figure. According to them the Mîm is like his head, the Ḥâ like his body, the second Mîm like his belly, and the Dâl like his two feet. These people seem to be completely ignorant of lineaments, if they compare the measure of the head and the belly (both expressed by the same letter Mîm) and the quantity of the limbs which project out of the mass of the body, forgetting at the same time the means for the perpetuation of our race. Perhaps, however, they meant individuals of the feminine, not of the masculine sex. I should like to know what they would say of such names as in their outward form, but for the addition or omission of one letter, resemble the form of the name of Muḥammad, for instance, حميد or مجيد (*Ḥamîd* or *Majîd*), and others. If you would compare some of them according to their method, the matter would simply become ridiculous and ludicrous.

More curious still than this is the fact that this Christian sect, in the matter of the Cross and its verification, refers to the wood of *Pæonia*. For, if you cut this wood, you observe in the plane of the cut lineaments which resemble a cross. They go even so far as to maintain that this fact originated at the time when Christ was crucified. This wood is frequently used in this way, that a piece of it is attached to a man who suffers from epilepsy, being considered as a symbol of the resurrection of the dead. Now, I should like to know whether they never study medical books and never hear of those authors who lived long before Christ, and on whose authority the excellent Galenus gives the description of this wood. Those who use the works of soul and nature as arguments regarding physical appearances, from whatever theory they start, and how widely soever discordant their theories may be, will always manage to find that the starting-point of their argument agrees with that which they maintain, and that their first sentence resembles that at which they aim. However, such arguments can never be accepted, unless there be a reason which properly connects that which is

measured with that by which you measure, the proof with that which is to be proved. There exist, *e.g.* double formations or correlations in things opposite to each other (*e.g.* black and white, &c.), triple formations in many leaves of plants and in their kernels, quadruplications in the motions of the stars and in the fever days, quintuplications in the bells of the flowers and in the leaves of most of their blossoms, and in their veins; sextuplications are a natural form of cycles, and occur also in bee-
p.298. hives and snow-flakes. So all numbers are found in physical appearances of the works of soul and life, and specially in flowers and blossoms. For the leaves of each blossom, their bells and veins, show in their formation certain numbers (numerical relations) peculiar to each species of them. Now, if anybody wants to support his theory by referring to one of these species, he *can* do so (*i.e.* there is material enough for doing so), but who will believe him?

Also in minerals you find sometimes wonderful physical peculiarities. People relate, *e.g.* that in the *Maksûra* (altar-place) of the Mosque in Jerusalem there is a white stone, with a nearly-obliterated inscription to this effect: "*Muhammad is the prophet of God, may God be merciful to him!*" And behind the Kibla there is another white stone with this obliterated inscription: "In the name of God the clement, the merciful! Muḥammad is the prophet of God, Hamza is his help." Further, stones for rings, with the name '*Alî*, the Prince of the Believers, are of frequent occurrence, because the figure of the name '*Alî* is frequently found in the veins of mountains.

To this category, too, belong certain forgeries, *e.g.* some Shî'a preacher once asked me to teach him something which he might utilize. So I produced to him from the *Kitâb-altalwîḥ* of Alkindî the recipe of (an information how to make) an ink composed of various pungent materials. This ink you drop upon an agate and write with it; if you then hold the stone near the fire the writing upon the stone becomes apparent in white colour. Now, in this manner he wrote (upon stones) the names Muḥammad, 'Alî, etc., even without doing the thing very carefully or understanding it particularly well, and then he proclaimed that these stones were formations of nature and had come from such and such a place. And for such forgeries he got much money from the Shî'a people.

Among the peculiarities of the flowers there is one really astonishing fact, viz. the number of their leaves, the tops of which form a circle when they begin to open, is in most cases conformable to the laws of geometry. In most cases they agree with the chords that have been found by the laws of geometry, not with conic sections. You scarcely ever find a flower of 7 or 9 leaves, for you cannot construct them according to the laws of geometry in a circle as isoscele (triangles). The number of their leaves is always 3 or 4 or 5 or 6 or 18. This is a matter of frequent occurrence. Possibly one may find one day some species of flowers with 7 or 9 leaves, or one may find among the species hitherto known such a number of leaves; but, on the whole, one must say nature preserves its *genera* and *species* such as they are. For if you would, *e.g.* count the number of seeds of one of the (many) pomegranates of a tree, you would find that all the other pomegranates contain the same number of seeds as that one the seeds of which you have counted first. So, too, nature proceeds in all other matters. Frequently, however, you find in the functions (actions) of nature which it is her office to fulfil, some fault (some irregularity), but this only serves to show that the Creator who had designed something deviating from the general tenor of things, is infinitely sublime beyond everything which we poor sinners may conceive and predicate of Him.

Now we return to our subject.

8. Commemoration of John, author of the fourth Gospel, and of Arsenius, the monk.

9. Icsaia, the prophet. Dâdhîshû', in his commentary on the Gospel, p.299.
calls him شعيا. God knows best (which is the right form).

10. Dionysius, the bishop.

12. Epiphanius, the archbishop.

13. Julianus, martyr.

15. *Feast of Roses* according to the new rite, (postponed to this date) because on the 4th the roses are still very scarce. On the same date it is celebrated in Khurâsân, not on the original date.

16. Zacharias, the prophet.

20. Cyriacus, the anchorite.

22. Constantine the Victorious. He was the first king who resided in Byzantium and surrounded it with walls. The town was after him called Constantinople; it is the residence of his successors.

24. Simeon, the monk, who wrought a great miracle.

Ḥazîrân.

1. *Feast of Ears*, when people bring ears of the wheat of their fields, read prayers over them, and invoke the blessing of God for them.

On the same day commemoration of John the son of Zakaria, through which they purpose gaining the favour of God for their wheat. This feast they celebrate instead of the Jewish *'Azereth*.

3. Commemoration of Nebukadnezar's burning the children, 'Azaryâ, Ḥananyâ, and Michael. Also commemoration of the renovation of the temple.

5. Athanasius, the patriarch.

8. Cyrillus, the patriarch, who drove Nestorius, the author of Nestorianism out of the Church, and excommunicated him.

12. Matthew, Mark, Luke, and John, the four evangelists.

18. Leontius, martyr.

21. Berekhyâ, the presbyter, who brought Christianity to Marw about two hundred years after Christ.

22. Gabriel and Michael, the archangels. Their commemoration they consider as a means to gain the favour of God, and they ask God to protect the creation from any injury done by the heat.

25. Birth of John b. Zachariah. Between the annunciation of his birth and his birth itself there elapsed 258 days, *i.e.* 8 months and 18 days.

26. Febronia, the martyr, who was tortured to death.

29. Death of Paul, the teacher, the apostle of Christianity.

30. Peter, *i.e.* Simeon Kephas, the head of the messengers, *i.e.* apostles.

Tammûz.

1. The twelve apostles, the pupils of Christ.

2. Thomas, the apostle, who did not believe in Christ when he had returned after His crucifixion, until he touched the ribs of His side. There he felt the trace of the wound, where the Jews had pierced Him. He is the same apostle through whom India was Christianized.

p.300. 5. Dometius, martyr.

7. Procopius, martyr.

8. Martha, the mother of Simeon Thaumaturgus.

9. Commemoration of Nebukadnezar's burning the three children. They assert that, if they did not keep this commemoration, they would suffer from the heat of Tammûz.

10. The forty-five martyrs.

11. Phocas, martyr.

13. Thuthael, martyr.

14. John of Marw, the younger, who was killed in our time.

15. Cyricus, and his mother, Julitta. Cyricus is said to have argued, when a child of three years, with decisive arguments against some king. Through him fourteen thousand men were converted to Christianity.

20. *Feast of the Grapes*, when they bring the first grapes, and pray to God that He may give blessing and increase, rich thriving and growing.

21. Paphnutius, martyr.

26. Panteleémon, martyr, the physician.

27. Simeon Stylites, the monk.

30. The seventy-two disciples of Christ.

Âb.

1. Fasting on account of the illness of Mary, the mother of Christ; it lasts fifteen days, and the last day is the day of her death.

On the same day, commemoration of Solomonis the Makkabean. The magians killed her seven children, and roasted them in roasting-pans.

5. Moses, the son of Amram.

6. Feast of Mount Tabor, regarding which the Gospel relates that once the prophets, Moses, the son of Amram, and Elias, appeared to Christ on Mount Tabor, when three of His disciples, Simeon, Jacob, and John, were with Him, but slept. When they awoke and saw this, they were frightened, and spoke: "May our Lord, *i.e.* Messiah, permit us to build three tents, one for Thee, and the other two for Moses and Elias." They had not yet finished speaking when three clouds standing high above them covered them with their shadow; then Moses and Elias entered the cloud and disappeared. Moses was dead already a long time before that, whilst Elias was alive, and is still living, as they say; but he does not show himself to mankind, hiding himself from their eyes.

7. Elias, the ever-living, whom we mentioned just now.

8. Elisha, the prophet, disciple of Elias.

9. Rabûla, the bishop.

10. Mamas, martyr.

15. Feast in commemoration of the death of Mary. The Christians make a difference between "*Commemoration*" and "*Feast*"; the latter is an affair of more importance than the former.

16. Iesaia, Jeremia, Zakaria, and Hezekiel, the prophets.

17. The martyrs Seleucus and his bride Stratonice. p.301.

20. Samuel, the prophet.

21. Lucius, martyr.

26. Saba, the monk, weak from age.

29. Decapitation of John. Alma'mûn b. 'Aḥmad Alsalamî Alharawî relates that he saw in Jerusalem some heaps of stones at a gate, called *Gate of the Column;* they had been gathered so as to form something like hills and mountains. Now people said that those were thrown over the blood of John the son of Zacharias, but that the blood rose over them, boiling and bubbling. This went on till Nebukadnezar killed the people, and made their blood flow over it; then it was quiet.

Of this story there is nothing in the Gospel, and I do not know what I am to say of it. For Nebukadnezar came to Jerusalem nearly four hundred and forty-five years before the death of John; and the second destruction was the work of the Greek kings, Vespasian and Titus. But it seems that the people of Jerusalem call everybody who destroyed their town Nebukadnezar,; for I have heard some historian say that in this case is meant Jûdarz b. Shâpûr b. Afkûrshâh, one of the Ashkanian kings.

30. Commemoration of all the prophets.

Ilûl.

1. *Festum coronæ anni.* They pray and invoke God's blessing for the end of the year, and the beginning of the new one because with this month the year reaches its end.

3. Commemoration of the seven martyrs killed in Nîshâpûr.

8. Ḥanna, mother of Mary, and Joyakim, the father of Mary.

13. Feast of the renovation of the temple, with prayers. On this day they renovate their churches.

14. Feast in recollection of Constantine and Helena his mother finding the Cross, which they seized out of the hands of the Jews It was buried in Jerusalem, but on this subject we have spoken already.

15. Commemoration of the Six Synods.

16. Euphemia, martyr.

20. The martyrs Eustathius, his wife, and mother.

23. Vitellius, martyr.

24. Thecla, martyr, who was burned to death. On the same day, the feast of the *Church of the Sweepings* (*i.e.* Church of Resurrection) in Jerusalem.

25. The martyrs Sabinianus, Paulus, and Ṭaṭṭâ.

28. Chariton, the monk.

29. Gregorius, the bishop, the apostle of the Armenians.

This, now, is all we know of the commemorations and feasts of the Melkites, in some of which they agree with the Nestorians. Of these we p.302. shall treat in a special chapter, but first we shall give an explanation of Lent as something which lies in the midst between both sects, being common to both of them.

CHAPTER XVI.

ON THE CHRISTIAN LENT, AND ON THOSE FEASTS AND FESTIVE DAYS WHICH DEPEND UPON LENT AND REVOLVE PARALLEL WITH IT THROUGH THE YEAR, REGARDING WHICH ALL CHRISTIAN SECTS AGREE AMONG EACH OTHER.

Heretofore we have explained in such a manner as will suffice for every want, and more than that, all the particulars relating to the Passover of the Jews, its conditions, the mode in which it is calculated, and the reason on which this calculation rests. Christian Lent is one of the institutions dependent on Passover, and is in more than one way connected with it. We now present such information regarding Lent as corresponds to the purpose for which the practices of Lent are intended —by the help of God and His mercy.

Christian Lent always lasts forty-eight days, beginning on a Monday and ending on a Sunday, the forty-ninth day after its beginning. The last Sunday before the end of Lent (or Fast-breaking), is that one which they call Sa'ânîn (*i.e.* Hosanna or Palm Sunday).

Now, one of the conditions which they have established is this, that Passover (Easter) must always fall in the time between Palm Sunday and Fast-breaking, *i.e.* in the last week of Lent. It cannot fall earlier than Palm Sunday, nor later than the last day of Lent.

The limits within which the Jewish Passover revolves, we have already heretofore mentioned. Regarding these the Christians do not agree with them, nor regarding the beginning of the cycles (*Gigal*). The word *Jijal*, or cycle, is an Arabized Syriac word, in Syriac *Gigal*, meaning the same as the Jewish *Maḥzór*. But it is only proper that we should mention the *Termini* peculiar to each nation. So they call the *Great Cycle*, *Indictio* (*sic*); but as it is troublesome to pronounce this word so frequently in our discourse, we shall use the term *Great Jijal* (*i.e.* Great Cycle).

The difference regarding the cycles has this origin: According to the Jews the first year of the *Æra Alexandri* is the tenth year of the Cycle (*Enneadecateris*), whilst according to the Christians it is the 13th year. For some of them count the interval between Adam and Alexander as 5069 years, others as 5180 years. The majority uses the latter number; it is also well known among scholars (of other nations). It occurs *e.g.* in the following verses of Khâlid b. Yazîd b. Mu'âwiya b. 'Abî-Sufyân, who was the first philosopher in Islâm; people say even that the source of his wisdom was that learning which Daniel had derived from the *Treasure-Cave*, the same one where Adam the father of mankind had deposited his knowledge.

"When 10 years had elapsed besides other 3 complete years,
And further 100 single years, which were joined in right order to 6 times 1000,
He manifested the religion of his lord, Islâm, and it was consolidated and established by the *Flight* (Hijra);" i.e. *Anno Adami* 6113.

p.303. The Hijra occurred *A. Alexandri* 933. If you subtract this from the just mentioned 6113 years of the *Æra Mundi*, you get as remainder

5180 years

(as the interval between Adam and Alexander). Now they converted this number of years into *Small Cycles*, and got as remainder

12 years,

i.e. at the beginning of the *Æra Alexandri* 12 years of the current Enneadecateris had already elapsed.

Further they arranged the years of the *Enneadecateris* according to the *Ordo Intercalationis* בדהזיגוח (*i.e.* 2. 5. 7. 10. 13. 16. 18.), because this arrangement stands by itself, as not requiring you to subtract anything from the years of the era.

In the first year of the cycle they fixed Passover on the 25th of Adhâr, because in the year when Christ was crucified it must have fallen on this date. Starting from this point they arranged the Passovers of all the other years. Its earliest date is the 21st Adhâr, its latest date the 18th Nîsân. So the *Terminus Paschalis* extends over 28 days.

Therefore the earliest date of Passover falls always by two days later than the vernal equinox as observed by eye-sight (*i.e.* the 19th Adhâr). And this is to serve as a help and precaution against that which is mentioned in the 7th Canon of the *Canones Apostolorum*: "Whatever bishop, or presbyter, or diaconus celebrates the feast of Passover before the equinox together with the Jews, shall be deposed from his rank."

If the *Fast-breaking* (Fiṭr) of the Christians were identical with their Passover, or if it fell always at one and the same invariable distance from Passover, both would revolve through the years either on the same days, or parallel with each other on corresponding days. Since, however, Fast-breaking can never precede Passover, its earliest possible date falls by *one* day later than the earliest possible date of Passover, *i.e.* on the 22nd Adhâr (the 21st Adhâr being the earliest date of Passover). And the latest date of Fast-breaking falls by *one* week later than the latest date of Passover; because if one and the same day should happen (to be Fast-breaking and Passover, *i.e.* a Sunday), Fast-breaking would fall on the next following Sunday. In this case it would fall by one week later than Passover. If, therefore, Passover falls on its latest possible date (18th Nîsân), Fast-breaking also falls on its latest possible date, *i.e.* on the 25th Nîsân.

Therefore the days within which Fast-breaking varies are 35. And for the same reason the beginning of fasting varies parallel with Fast-breaking on the corresponding days, the earliest being the 2nd Shubâṭ, the latest the 8th Adhâr. Accordingly the greatest interval between the beginning of Lent and Passover is 49 days, the smallest interval 42 days.

Between the full moon of Passover and the new moon of Adhâr in a common year, of Adhâr Secundus in a leap-year, is an interval of

44 days, 7 hours, 6 minutes.

This new moon falls always between the beginning of the smallest interval and the greatest interval (between the beginning of Lent and Passover), and falls near the beginning of Lent. And this new moon p.304.
has been made the basis of the whole calculation in this way: You observe the new moon of Shûbâṭ and consider which *Monday* is the nearest to it, the preceding one or the following. If this Monday lies within the *Terminus Jejunii*, *i.e.* between the 2nd Shubât and the 8th Adhâr, it is the beginning of Lent. If, however, this Monday does not reach the *Terminus Jejunii*, and lies in the time before it, that new moon is disregarded, and you repeat the same consideration with the following new moon. In this way you find the beginning of Lent.

As we have mentioned already, Passover may proceed towards the beginning of the year as far as the 21st Adhâr, which is its earliest possible date. If full moon falls on this day and it is a Sabbath, the year is a common year, the new moon by which you calculate falls on the 4th Shubâṭ and the preceding Monday, which is the nearest Monday to this date, and therefore the beginning of the *Terminus Jejunii* is the 1st Shubâṭ, if the year be a leap-year, but the 2nd Shubâṭ if the year is a common year. This date lies within the *Terminus Jejunii*, and so it is the beginning of Lent.

The latest possible date of Passover is the 18th of Nîsân. If full moon falls on this day, and it is a Sunday, the year is a leap-year, the new moon by which you calculate, *i.e.* the new moon of *Adhâr Secundus,* falls on the 5th of the Syrian Adhâr, and the 8th of the same month is that Monday which follows after this new moon and falls the nearest to it, because in this case the 1st of the Syrian Adhâr is a Monday. Therefore the beginning of Lent is the 8th Adhâr, which is at the same time the latest possible date for the beginning of the *Terminus Jejunii.*

If we were to go back upon the new moon of Adhâr Primus, we should find that it falls on the 5th Shubâṭ in a common year, whilst the 1st Shubâṭ is a Sunday. In that case the preceding Monday would be nearest to it (the 2nd Shubâṭ), which is the beginning of the *Terminus Jejunii.* Now, this day would be suitable to be the beginning of Lent, if it also corresponded to all the other conditions (but that is not the case); viz. if we make this day the beginning of Lent, Fast-breaking would fall about one month earlier than Passover; and this is not permitted, according to a dogma of theirs. And if the year were a leap-year, new moon would fall on the 4th Shubâṭ, and then the preceding Monday, being the nearest to it, would be the 1st Shubâṭ, and this date does not lie within the *Terminus Jejunii* (2nd Shubâṭ—8th Adhâr). Therefore we must disregard this new moon and fall back upon the following one.

The followers of Christ wanted to know before-hand the Passover of the Jews, in order to derive thence the beginning of their Lent. So they consulted the Jews, and asked them regarding this subject, but the Jews, guided by the enmity which exists between the two parties, told them
p.305. lies in order to lead them astray. And besides, the eras of both parties differed. Finally, many of the Christian mathematicians took the work in hand and made calculations with the various cycles and different methods. Now, that method which they at last agreed to adopt, is the table called Χρονικόν, of which they maintain that it was calculated by Eusebius, Bishop of Cæsarea, and the 318 bishops of the Synod of Nicæa.

The *Chronicon* of the Christians.

pp. 306, 307.

Lunar Cycle.		1	2	3	4	5	6	7	8	9	10	11	12	13	14	15	16	17	18	19
Leap-years.	Solar Cycle.	Shubâț.	Shubâț, Adhâr.	Shubâț.	Shubâț.	Shubâț.	Shubâț.	Adhâr.	Shubâț.	Shubâț.	Shubâț, Adhâr.	Shubâț.	Shubâț.	Shubâț, Adhâr.	Shubâț.	Shubâț.	Shubâț.	Shubâț.	Adhâr.	Shubâț.
	1	11	25	18	4	25	11	4	25	11	4	18	11	25	18	4	25	11	4	18
	2	10	3	17	3	24	17	3	24	10	3	17	10	24	17	3	24	10	3	24
L.	3	9	1	16	9	23	16	8	23	9	1	23	9	1	16	9	23	16	1	23
	4	7	28	14	7	28	14	7	21	14	28	21	7	28	14	7	21	14	7	21
	5	6	27	28	6	27	13	6	20	13	27	20	6	27	13	6	27	13	6	20
	6	12	26	19	5	26	12	5	19	12	5	19	5	26	19	5	26	12	5	19
L.	7	11	3	18	4	25	18	3	25	11	3	18	11	25	18	4	25	11	3	25
	8	9	2	16	9	23	16	2	23	9	2	16	9	2	16	2	23	16	2	23
	9	8	1	15	8	22	15	8	22	8	1	22	8	1	15	8	22	15	1	22
	10	7	28	14	7	28	14	7	21	14	28	21	7	28	14	7	21	14	7	21
L.	11	13	27	20	6	27	13	5	20	13	5	20	6	27	20	6	27	13	5	20
	12	11	25	18	4	25	11	4	25	11	4	18	11	25	18	4	25	11	4	18
	13	10	3	17	3	24	17	3	24	10	3	17	10	24	17	3	24	10	3	24
	14	9	2	16	9	23	16	2	23	9	2	16	9	2	16	2	23	16	2	23
L.	15	8	29	15	8	29	15	7	22	15	29	22	8	29	15	8	22	15	7	22
	16	6	27	20	6	27	13	6	20	13	27	20	6	27	13	6	27	13	6	20
	17	12	26	19	5	26	12	5	19	12	5	19	5	26	19	5	26	12	5	19
	18	11	25	18	4	25	11	4	25	11	4	18	11	25	18	4	25	11	4	18
L.	19	10	2	17	10	24	17	2	24	10	2	17	10	2	17	3	24	17	2	24
	20	8	1	15	8	22	15	8	22	8	1	22	8	1	15	8	22	15	1	22
	21	7	28	14	7	28	14	7	21	14	28	21	7	28	14	7	21	14	7	21
	22	6	27	20	6	27	13	6	20	13	27	20	6	27	13	6	27	13	6	20
L.	23	12	26	19	5	26	12	4	26	12	4	19	12	26	19	5	26	12	4	19
	24	10	3	17	3	24	17	3	24	10	3	17	10	24	17	3	24	10	3	24
	25	9	2	16	9	23	16	2	23	9	2	16	9	2	16	2	23	16	2	23
	26	8	1	15	8	22	15	8	22	8	1	22	8	1	15	8	22	15	1	22
L.	27	7	28	21	7	28	14	6	21	14	28	21	7	28	14	7	28	14	6	21
	28	12	26	19	5	26	12	5	19	12	5	19	5	26	19	5	26	12	5	19

p.308. **Festivals depending upon Lent.**

[*Lacuna.*]

to give up their religion. Then they fled one night and perished to the last of them. This Friday they call also *The Small Hosanna.*

The first Sunday after Fast-breaking is called the *New Sunday*, on which day Messiah dressed in white. They use it as the commencement of all kinds of work, and as a date for commercial agreements and written contracts. For it is, as it were, the first Sunday, because the preceding one is specially known by a more famous name, *i.e.* Fast-breaking.

All Sundays are highly celebrated by the Christians, because Hosanna and Resurrection fall on Sundays. Likewise the Sabbaths are celebrated by the Jews because, as is said in the Thora, God rested on this day on having finished the creation. And, according to some scholars, Muslims celebrate their Friday because on that day the Creator finished the creation of the world and breathed His spirit into Adam. According to the astrologers, in all religions certain week-days are celebrated, because the horoscopes of their prophets and the constellations indicative of their coming stood under the influence of the planets that reign over these respective days.

Forty days after Fast-breaking is the feast of *Ascension*, always falling on a Thursday. On this day Messiah ascended to heaven from the Mount of Olives, and He ordered His disciples to stay in that room where He had celebrated Passover in Jerusalem, until He should send them the Paraclete, *i.e.* the Holy Ghost.

Ten days after Ascension is *Whitsun Day*, always on a Sunday. It is the day when the Paraclete came down and Messiah revealed Himself to His disciples, *i.e.* the Apostles. Then they began to speak different tongues; they separated from each other, and each party of them went to that country with the language of which they were inspired and which they were able to speak.

On the evening of this day the Christians prostrate themselves upon the earth, which they do not do between Fast-breaking and this day, for during this time they say their prayers standing erect, all in consequence of some biblical commandment to this effect. The same (prostration) is proclaimed for all the (other) Sundays by the last Canon of the first Synod.

The beginning of the *Fasting of the Apostles*, according to the Melkites, is a Wednesday, ten days after Whitsunday. It is broken always on a Sunday, 46 days after its beginning,

The third day of this fasting, a Friday, is called the *Golden Friday*. For on this day the Apostles passed a lame man in Jerusalem, who asked people for a gift. He invoked the name of God, asking them for alms. They answered: "We have neither gold nor silver. However, rise, carry away your bed, and go to your business. That is the best we can do for you." The man rose, free from pain, carried away his bed, and went to his business.

Most of these festivals are mentioned in the *Table of Fasting*, which is arranged in seven columns. If you find Fasting by this table, you find at the same time these festivals—if God permits!

p.309.

CHAPTER XVII.

ON THE FESTIVALS OF THE NESTORIAN CHRISTIANS, THEIR MEMORIAL AND FAST DAYS.

NESTORIUS, from whom this sect derives its origin and name, opposed the Melkites and brought forward a theory on the dogmas of Christianity which necessitated a schism between them. For he instigated people to examine and to investigate for themselves, to use the means of logic, syllogism, and analogy for the purpose of being prepared to oppose their adversaries, and to argue with them; in fact, to give up the *Jurare in verba magistri*. This was the method of Nestorius himself. He established as laws for his adherents those things in which he differed from the Melkites, differences to which he had been led by his investigation and unwearying study.

Now I shall proceed to propound all I have been able to learn regarding their festivals and memorial-days.

Nestorians and Melkites agree among each other regarding some memorial-days, whilst they disagree regarding others.

Those days, regarding which they differ, are of two kinds:

1. Days altogether abolished by the Nestorians.
2. Days not abolished by them, but celebrated at a time and in a manner different from that of the Melkites.

Further, such Nestorian festivals, not celebrated by the Melkites, which are derived from the feast-times common to both sects (Lent, Christmas, Epiphany).

Besides, there is a fourth class of Nestorian feast-days, not used by the Melkites, which are not derived from the (common) feast-times also used by the Melkites.

A. Feasts regarding which Nestorians and Melkites agree among each other: Christmas, Epiphany, *the Feast of Wax*, the beginning of the

Fasting, the Great Hosanna, the Washing of the Feet of the Apostles, the Passover of the Messiah, the Friday of Crucifixion, Resurrection, Fast-breaking, the New Sunday, Ascension, and Whitsunday, the fasting of Our Lady Mary, and some of the memorial-days which we have mentioned heretofore.

B. Feasts common to both sects, but celebrated by the Nestorians at a time and in a manner different from that of the Melkites :—

1. *Ma'al'thâ* (*Ingressus*). On this feast they wander from the naves of the churches up to their roofs, in commemoration of the returning of the Israelites to Jerusalem. It is also called ܩܘܕܫ ܥܕܬܐ (*Sanctification of the Church*). It is celebrated on the *first* Sunday of Tishrîn II., if the 1st of this month falls on a Wednesday, Thursday, Friday, Saturday, or Sunday; but if it falls on a Monday or Tuesday, the feast is celebrated on the *last* Sunday of Tishrîn I. The characteristic mark of the day, as I have heard John the Teacher say, is this, that it is the Sunday falling between the 30th of Tishrîn I. and the 5th of Tishrîn II.

2. *Subbâr* (*Annuntiatio*), Feast of the annunciation to Mary that she was pregnant with the Messiah, celebrated on the *first* Sunday in Kânûn I., if the first of the month falls on a Friday, Saturday, or Sunday; but if it falls on a Monday, Tuesday, Wednesday, or Thursday, the feast is celebrated on the *last* Sunday of Tishrîn II. In every case it is the 5th Sunday after the Sunday of *Ma'al'thâ*.

In the year when the Messiah was born, the 1st of Kânûn I. was a Sunday. Between this day and that of His birth there are 25 days. Now, Christians say: Messiah differs from mankind in so far as He has not originated through an act of begetting; likewise the period of p.310. His sojourning in the womb of His mother is contrary to the ways of human nature. The annunciation (of the pregnancy) may already have occurred at a time when the embryo (or growing child) was already settled in the womb; it may also have occurred earlier or later. I have been told that the Jacobites celebrate *Subbâr* on the 10th of the Jewish Nîsân; this day fell, in the year preceding the year of Christ's birth, on the 16th of the Syrian Adhâr.

3. *The Fasting of Our Lady Mary.* It begins on Monday after the Sunday of *Subbâr*, and it ends on Christmas-day.

4. *The Decollation of John the Baptist.* The Nestorians celebrate it on the 24th of Âb.

5. *Commemoration of Simeon b. Ṣabbâ'ê, i.e.* son of the dyers, on the 17th Âb.

6. *The Feast of the Cross*, celebrated by the Nestorians on the 13th Îlûl. For on this day Helena found the Cross, and she showed it to the people on the following day, the 14th. Therefore the Christians came to an agreement among each other, the Nestorians adopting the day of the finding, the others the day when it was shown to the people.

C. Feasts celebrated by the Melkites only, and fixed by them on certain dates of their own, are, *e.g.* :—

1. Commemoration of John of Kashkar, on the 1st Tishrîn I.
2. Commemoration of Mâr Phetion, on the 25th of Tishrîn I.
3. The feast of the Monastery of John, on the 6th of Kânûn I.
4. The feast of the Church of Mary in Jerusalem, on the 7th of Kânûn II.
5. Commemoration of Mâr لونيا, on the 25th of Ḥazîrân.
6. Beginning of the Feast of Revelation, on the 6th of Âb; it is the last day on which Christ appeared to men. On the same day the feast of *Dair-Alnâs.* The end of the Feast of Revelation is on the 16th Âb.
7. Feast of Mar Mârî, on the 12th Âb.
8. Commemoration of Crispinus and Crispinianus, on the 3rd Îlûl.

D. Feasts fixed by the Nestorians on certain week-days, regarding which the two sects have nothing in common. For instance:—

1. Commemoration of the monk Ḳûṭâ or Mâr Sergius, on the 7th Tishrîn I., if the 1st of the month is a Sunday; in any other case it is postponed to the Sunday following next after the 7th.
2. Commemoration of Solomonis, on the following Sunday, according to the practice of the Christians of Baghdâd.
3. The Feast of Dair-Abî Khâlid, on the first Friday in Tishrîn II.
4. Feast of the Monastery of Alḳâdisiyya, on the third Friday of Tishrîn II.
5. Feast of Dair-Alkaḥḥâi, on the fourth Friday of Tishrîn II.
6. Commemoration of برسفا (Mar Sâbâ ?), on the last Sunday of Îlûl.
7. Feast of Dair-Altha'âlib, on the last Sabbath of Îlûl; but if the 1st of Tishrîn I. of the next year be a Sunday, the feast is postponed to this day, and falls no longer in Îlûl. In that case the feast does not at all occur in the year in question, whilst it occurs twice in the following year, once at the beginning and once at the end.

p.311. E. Of those feasts, depending on certain days, which are common to both sects, there are three classes:—

I. Those depending on the Lent or Fast-breaking.
II. Those depending on Christmas.
III. Those depending on Epiphany.

I. Feasts depending on the beginning or end of Lent are, *e.g.* :—

1. The Friday of اهادر, the 12th day after beginning of Lent.
2. *Alfûrûḳa, i.e.* liberation, on Thursday, the 24th day after beginning of fasting.
3. Commemoration of Mar ترسيا and commemoration of Mâr Cyriacus, the Child who preferred death to apostasy, on Friday the 20th day after Fast-breaking.

4. Commemoration of Sûrîn and Dûrân the Armenians, who were killed by the king Shâpûr, on Sunday the 29th day after Fast-breaking.

5. Fasting of the Apostles, according to the Nestorians, always beginning on Monday, seven weeks after the Great Fast-breaking following after Whitsunday. It lasts during 46 days, and it is broken always on a Friday.

6. Commemoration of Mar Abdâ, the pupil of Mar Mârî, on Thursday, the 14th day after the end of the Fasting of the Apostles, which again depends on the Great Fast-breaking.

7. Commemoration of Mar Mârî on Friday, the 15th day after the end of the Fasting of the Apostles.

8. Fasting of Elias, beginning on Monday, 21 weeks after the Great Fast-breaking; it lasts during 48 days, and it ends on a Sunday.

Fasting of Ninive, on Monday, 22 days before the beginning of Lent, lasting three days. Tradition says that the people of the prophet Jona, after punishment had come upon them, and after God had again released them and they were in safety, fasted these three days.

10. The Night of *Almâshûsh* (the spy) is the night of a Friday, in which—as people say—they seek Messiah. There is, however, a difference; according to some it is the night of Friday, the 19th day after the Fasting of Elias; according to others it is the Friday on which Christ was crucified, called *Alṣalabût;* according to others it is the Friday of the Martyrs, one week after *Alṣalabût.* The preference we give to the first of these three opinions.

If, now, you know the beginning of Lent of a year in question, compare the column of the common year, if the year be a common year, or the column of the leap-year, if the year be a leap-year, and opposite, in the table of the feasts depending on Lent, you will find the date of every feast in question, and also the date of the Fasting of Ninive, which precedes Lent.

Here follows the table.

pp. 312, 313.

Table of the Feasts depending on Lent.

II.	II.	VI.		V.		VI.	I.	IV.	II.	VI.	V.	VI.	II.	VI.	II.	II.	
Beginning of Lent in a common year.	Beginning of Lent in a leap-year.	Friday of أجلار.		Alfâriṭa.		Com. of Mar برسوما and Mâr Cyriacus.	Com. of Suren and Dûrân.	Fast of the Apostles according to the Melkites.	Fast of the Apostles according to the Nestorians.	Golden Friday.	Commemoration of Mar 'Abdâ.	Commemoration of Mar Mârî.	Fast of Elias.	Night of the Spy.	Fast of Moses.	The Ninive-Fast preceding Lent.	
Shubâṭ.	Shubâṭ.	Shubâṭ.		Shubâṭ.		Nîsân.	Nîsân.	Ayyâr.	Ayyâr.	Ayyâr.	Tammûz.	Tammûz.	Âb.	Îlûl.	Tishrîn I.	Kânûn II.	
2	3	13	14	26	27	10	19	20	11	22	9	10	17	4	12	12	13
3	4	14	15	27	28	11	20	21	12	23	10	11	18	5	13	13	14
4	5	15	16	28	29	12	21	22	13	24	11	12	19	6	14	14	15
5	6	16	17	1 Adhâr.		13	22	23	14	25	12	13	20	7	15	15	16
6	7	17	18	2		14	23	24	15	26	13	14	21	8	16	16	17
7	8	18	19	3		15	24	25	16	27	14	15	22	9	17	17	18
8	9	19	20	4		16	25	26	17	28	15	16	23	10	18	18	19
9	10	20	21	5		17	26	27	18	29	16	17	24	11	19	19	20
10	11	21	22	6		18	27	28	19	30	17	18	25	12	20	20	21
11	12	22	23	7		19	28	29	20	31	18	19	26	13	21	21	22
12	13	23	24	8		20	29	30	21	1 Ḥazîrân.	19	20	27	14	22	22	23
13	14	24	25	9		21	30	31	22	2	20	21	28	15	23	23	24
14	15	25	26	10		22	1 Ayyâr.	1 Ḥazîrân.	23	3	21	22	29	16	24	24	25
15	16	26	27	11		23	2	2	24	4	22	23	30	17	25	25	26
16	17	27	28	12		24	3	3	25	5	23	24	31	18	26	26	27
17	18	28	29	13		25	4	4	26	6	24	25	1 Îlûl	19	27	27	28
18	19	1 Adhâr.		14		26	5	5	27	7	25	26	2	20	28	28	29
19	20	2		15		27	6	6	28	8	26	27	3	21	29	29	30
20	21	3		16		28	7	7	29	9	27	28	4	22	30	30	31
21	22	4		17		29	8	8	30	10	28	29	5	23	31	31	1 Shubâṭ.
22	23	5		18		30	9	9	31	11	29	30	6	24	1 Tishrîn II.	1	2
23	24	6		19		1 Ayyâr.	10	10	1 Ḥazîrân.	12	30	31	7	25	2	2	3
24	25	7		20		2	11	11	2	13	31	1 Âb.	8	26	3	3	4
25	26	8		21		3	12	12	3	14	1 Âb.	2	9	27	4	4	5
26	27	9		22		4	13	13	4	15	2	3	10	28	5	5	6
27	28	10		23		5	14	14	5	16	3	4	11	29	6	6	7
28	29	11		24		6	15	15	6	17	4	5	12	30	7	7	8
1	1	12		25		7	16	16	7	18	5	6	13	1 Tishrîn I.	8	8	9
2	2	13		26		8	17	17	8	19	6	7	14	2	9	9	10
3	3	14		27		9	18	18	9	20	7	8	15	3	10	10	11
4	4	15		28		10	19	19	10	21	8	9	16	4	11	11	12
5	5	16		29		11	20	20	11	22	9	10	17	5	12	12	13
6	6	17		30		12	21	21	12	23	10	11	18	6	13	13	14
7	7	18		31		13	22	22	13	24	11	12	19	7	14	14	15
8	8	19		1 Nîsân.		14	23	23	14	25	12	13	20	8	15	15	16

II. The feasts depending on Christmas are these:—The Feast of the Temple on Sunday after Christmas; the Commemoration of Our Lady Mary, *lit. Mârt Maryam—Mârt* means *mulier nobilis, domina*—on Friday after Christmas. If, however, Christmas falls on Thursday, it is postponed until the second Friday, for this purpose, that Christmas and this Commemoration should not follow each other immediately. For only the *night* of Thursday lies in the middle between the *day* of Thursday and the *day* of Friday (not one complete day). p.314.

III. Feasts depending on Epiphany:—The Fast of the Virgins on Monday after Epiphany; it lasts three days, and is broken on Thursday. It is also in use among the 'Ibâdites and the Arab Christians, who relate this story: Once the King of Al-ḥîra, before the time of Islâm, chose a number of women from among the virgins of the 'Ibâdites, whom he wanted to take for himself. Now, they fasted three days without any interruption, and at the end of them the king died without having touched them.

According to another report, this fast was kept by the Christian virgins among the Arabs as a thanksgiving to God for the victory which the Arabs gained over the Persians on the day of Dhû Ḳâr. So they were delivered from the Persians, who did not get into their power the virgin Al'anḳafîr, the daughter of Alnu'mân.

Frequently this fast is connected with the Ninive-Fast. For if Lent falls on its earliest date, the Monday after Epiphany is the Fast of the Virgins. Then there are twenty-two days between this fast and Lent. In that case this day is also the beginning of the Fast of Ninive. Both fasts (*Jejunium Virginum et Jejunium Niniviticum*) last three days.

Thereupon they celebrate the Commemoration of Mâr Johannes on Friday after Epiphany.

The Commemoration of Peter and Paul on the second Friday after Epiphany, that one which follows after the Commemoration of Mâr Johannes. Paulus was a Jew. Now, they maintain that Messiah worked a miracle in blinding his eyes and making them see again, whereupon he believed in Him. Then Messiah sent him as an apostle to the nations. Petrus is the same as Simeon Kephas.

The Commemoration of the Four Evangelists, on the third Friday.

The Commemoration of Stephanus, martyr, on the fourth Friday. Some people place it on Thursday, one day earlier.

The Commemoration of the Syrian Fathers, on the fifth Friday.

The Commemoration of the Greek Fathers, *i.e.* Diodorus, Theodorus, and Nestorius, the bishops, on the sixth Friday.

The Commemoration of Mâr Abbâ Catholicus, on the seventh Friday.

The Commemoration of the *Children of Adam*, *i.e.* of all mankind that have died up to that date, on the eighth Friday. But if there are not enough Fridays, and Lent is near, they drop the Commemoration of the Syrian Fathers, and celebrate instead the Commemoration of Mar

Abbâ Catholicus, and then they proceed according to the original order. During Lent they drop the Fridays, and on the evening of every Friday they have a *Ḳuddâs, i.e.* worship.

They have constructed for the days depending on Christmas and Epiphany and the week-days in question a table, indicating their dates in the Syrian months. If you want to use it, take the years of the *Æra*
p.315. *Alexandri*, including the current year, and change them into solar cycles. With the remainder compare the *Column of Numbers* in the table of the Nestorian festivals. There you find opposite the number each festival; if in red ink, its date in the month written in red ink at the top of the column; if in black ink, the date in the month written in black ink at the top of the column. Over the whole you find the week-day on which the feast always falls.

If we knew the system of the Jacobite Christians, we should explain it, as we have explained those of the other Christians. However, we never met with a man who belonged to their sect or knew their dogmas.

Here follows the table.

	I.	I.	I.	VI.	VI.	VI.	I.	II.	I.	VI.	II.	VI.	VI.	VI.	III.	VI.	VI.	VI.	VI.	VI.	I.	VII.
	Tishrîn I.	Tishrîn I.	Tishrîn I. Tishrîn II.	Tishrîn II.	Tishrîn II.	Tishrîn II.	Tishrîn II. Kânûn I.	Tishrîn II. Kânûn I.	Kânûn I. Kânûn II.	Kânûn I. Kânûn II.	Kânûn II.	Kânûn II.	Kânûn II.	Kânûn II.	Kânûn II. Shubât.	Kânûn II. Shubât.	Shubât.	Shubât.	Shubât.	Shubât. Âdhâr.	Ilûl.	Ilûl. Tishrîn I.
Column of the number.	Commemoration of Mâr Sergius.	Commemoration of Solomonis.	Ma'altha.	Feast of the Monastery of 'Abd-Khalil.	Feast of the Monastery of Alkadisiyya.	Feast of Dair-Alkahhil.	Feast of Annunciation.	Fast of Our Lady Mary.	Feast of the Temple.	Commemoration of Our Lady Mary.	Fast of the 'Ibâdites, or Fast of the Virgins.	Commemoration of Johannes.	Commemoration of Paul and Peter.	Commemoration of the Evangelists.	Commemoration of Mâr Johannes of Dailam.	Commemoration of Stephanus.	Commemoration of the Syrian Fathers.	Commemoration of the Greek Fathers.	Commemoration of Mar Abhâ Catholicus.	Commemoration of the Children of Adam.	Commemoration of Bar Safâ.	Feast of Dair-Althaʽâlib.
1	7	14 21	4	2	16	23	2	3	30	28	7	11	18	25	29	1	8	15	22	1	30	I. 29
2	13	20 27	3	1	15	22	1	2	29	3	13	10	17	24	4	31	7	14	21	28	29	28
*3	12	19 26	2	7	21	28	30	1	28	2	12	9	16	23	3	30	6	13	20	27	28	27
4	10	17 24	31	5	19	26	28	29	26	31	10	7	14	21	1	28	4	11	18	25	26	25
5	9	16 23	30	4	18	25	27	5	1	30	9	13	20	27	31	3	10	17	24	3	25	24
6	8	15 22	5	3	17	24	3	4	31	29	8	12	19	26	30	2	9	16	23	2	24	
*7	7	14 21	4	2	16	23	2	3	30	28	7	11	18	25	29	1	8	15	22	29	30	I. 29
8	12	19 26	2	7	21	28	30	1	28	2	12	9	16	23	3	30	6	13	20	27	28	27
9	11	18 25	1	6	20	27	29	30	27	1	11	8	15	22	2	29	5	12	19	26	27	26
10	10	17 24	31	5	19	26	28	29	26	31	10	7	14	21	1	28	4	11	18	25	26	25
*11	9	16 23	30	4	18	25	27	5	1	30	9	13	20	27	31	3	10	17	24	2	25	24
12	7	14 21	4	2	16	23	2	3	30	28	7	11	18	25	29	1	8	15	22	1	30	29
13	13	20 27	3	1	15	22	1	2	29	3	13	10	17	24	4	31	7	14	21	28	29	28
14	12	19 26	2	7	21	28	30	1	28	2	12	9	16	23	3	30	6	13	20	27	28	27
*15	11	18 24	1	6	20	27	29	30	27	1	11	8	15	22	2	29	5	12	19	26	27	26
16	9	16 23	30	4	18	25	27	5	1	30	9	13	20	27	31	3	10	17	24	3	25	24
17	8	15 22	5	3	17	24	3	4	31	29	8	12	19	26	30	2	9	16	23	2	24	
18	7	14 21	4	2	16	23	2	3	30	28	7	11	18	25	29	1	8	15	22	1	30	I. 29
*19	13	20 27	3	1	15	22	1	2	29	3	13	10	17	24	4	31	7	14	21	28	29	28
20	11	18 25	1	6	20	27	29	30	27	1	11	8	15	22	2	29	5	12	19	20	27	26
21	10	17 24	31	5	19	26	28	29	26	31	10	7	14	21	1	28	4	11	18	25	26	25
22	9	16 23	30	4	18	25	27	5	1	30	9	13	20	27	31	3	10	17	24	3	25	24
*23	8	15 22	5	3	17	24	3	4	31	29	8	12	19	26	30	2	9	16	23	1	24	
24	13	20 27	3	1	15	22	1	2	29	3	13	10	17	24	4	31	7	14	21	28	29	I. 28
25	12	19 26	2	7	21	28	30	1	28	2	12	9	16	23	3	30	6	13	20	27	28	27
26	11	18 25	1	6	20	27	20	30	27	1	11	8	15	22	2	29	5	12	19	26	27	26
*27	10	17 24	31	5	19	26	28	29	26	31	10	7	14	21	1	28	4	11	18	25	26	25
28	8	15 22	5	3	17	24	3	4	31	29	8	12	19	26	30	2	9	16	23	2	24	

pp. 316, 317.

p.318.

CHAPTER XVIII.

ON THE FEASTS OF THE ANCIENT MAGIANS AND ON THE FAST AND FEAST DAYS OF THE SABIANS.

The ancient Magians existed already before the time of Zoroaster, but now there is no pure, unmixed portion of them who do not practise the religion of Zoroaster. In fact, they belong now either to the Zoroastrians or to the *Shamsiyya* sect (sun-worshippers). Still, they have some ancient traditions and institutes, which they trace back to their original creed; but in reality those things have been derived from the laws of the sun-worshippers and the ancient people of Ḥarrân.

As regards the Sabians, we have already explained that this name applies to the real Sabians, *i.e.* to the remnants of the captive Jews in Babylonia, whom Nebukadnezar had transferred from Jerusalem to that country. After having freely moved about in Babylonia, and having acclimatized themselves to the country, they found it inconvenient to return to Syria; therefore they preferred to stay in Babylonia. Their religion wanted a certain solid foundation, in consequence of which they listened to the doctrines of the Magians, and *inclined* towards some of them. So their religion became a mixture of Magian and Jewish elements like that of the so-called Samaritans who were transferred from Babylonia to Syria.

The greatest part of this sect is living in Sawâd-al-'Irâḳ. These are the real Sabians. They live, however, very much scattered and nowhere in places that belong exclusively to them alone. Besides, they do not agree among themselves on any subject, wanting a solid ground upon which to base their religion, such as a direct or indirect divine revelation or the like. Genealogically they trace themselves back to Enos, the son of Seth, the son of Adam.

The same name is also applied to the Ḥarrânians, who are the remains of the followers of the ancient religion of the West, separated (cut off) from it, since the Ionian Greeks (*i.e.* the ancient Greeks, not the Ρωμαῖοι or Byzantine Greeks) adopted Christianity. They derive their system from Aghâdhîmûn (Agathodæmon), Hermes, Wâlîs, Mâbâ, Sawâr. They believe that these men and other sages like them were prophets. This sect is much more known by the name of Sabians than the others, although they themselves did not adopt this name before A. H. 228 under Abbaside rule, solely for the purpose of being reckoned among those from whom the duties of *Dhimma* (*μετοικία*) are accepted, and towards whom the laws of Dhimma are observed. Before that time they were called heathens, idolaters, and Ḥarrânians.

They call the months by the Syrian names and use them in a similar way to the Jews, whom they imitate, the Jews being the more ancient and having a greater claim to originality. To the names of the months they add the word *Hilâl* (new moon), so they say *Hilâl Tishrîn I.*, *Hilâl Tishrîn the Last*, etc.

Their New Year is Hilâl Kânûn the Last, but in counting the months they begin with Hilâl Tishrîn I. p.319.

Their day begins with sunrise, whilst all others, who use lunar months, make it begin with sunset.

Their lunar month begins with the second day after conjunction (new moon). If, now, conjunction precedes sunrise only by one minute, the third following day is the beginning of the month. But if conjunction coincides with sunrise or falls only a little later, the second day after conjunction is the beginning of the month.

When in the course of three years, one month and some days have summed up, they add this time as one month to their months after Hilâl Shubâṭ and call it *Hilâl Adhâr I.*

Muhammad b. 'Abd-Al'azîz Alhâshimî has given in his Canon called *Alkâmil* a short notice of the feasts of the Ṣabians, simply relating the facts without investigating and criticising their origin and causes. His report I have transferred into this chapter, adding thereto whatever I have learned from other sources. Regarding the more external part of this feast-calendar (*i.e.* the purely chronological part) I have made computations on my own account, only by way of induction, since I have not the same means to investigate this subject which I had for the others. God helps to what is right!

Hilâl Tishrîn I.

6. Feast of Al-Dhahbâna.
7. Beginning of the celebration of the feast.
13. Feast of Fûdî Ilâhî.
14. Feast of Ilâtî Fûdî.
15. Feast of the Lots (*Festum Sortium*).

Hilâl Tishrîn II.

1. The Great *Bakht* (i.e. *Fatum*).
2. Mâr Shelâmâ.

5. Feast of داءمو ملح for the shaving of the head.

9. ترسا the idol of Venus.

17. Feast of ترسا (Tarsâ). On the same day they go out of town to Baṭnæ.

18. Feast of Sarûg; it is the day of the renewal of the dresses.

According to 'Abû-alfarag Alzanjânî, they celebrate the *Feast of Tents* in this month, beginning with the 4th and ending on the 18th.

Hilâl Kânûn I.

7. Feast of the addressing (خطاب) to سان the idol of Venus.

10. Feast of the idols for Mars.

20. Feast of the Demons.

21. Beginning of the first fast, which is broken on the next following day of conjunction (new-moon's day). During this time they are not allowed to eat meat. At the time when they break their fast they are wont to practise almsgiving and charitable work.

28. Feast of the invocation of the Demons.

29. Feast of the *Fata* for the Demons.

30. Feast of consultation.

p.320. According to 'Abû-alfaraj Alzanjânî they celebrate on the 24th of this month the feast of the Nativity.

Hilâl Kânûn II.

All the invocations, fast and feast days of this month are sacred to the Demons.

1. Feast of New-Year's Day, like the calendar of the Greeks.

4. Feast of *Dair-aljabal*, and the feast of Baltî, *i.e.* Venus.

8. Fast of seven days: it is broken on the 15th.

12. Invocation of وحسرا.

20. They pray to the Bel of Ḥarrân.

25. Feast of the idol of Tirrathâ (Tir'athâ, Atergatis).

26. Feast of the nuptials (wedding) of the year.

Hilâl Shubâṭ.

9. Beginning of the minor fast; it lasts seven days and is broken on the 16th. During that time they do not taste any fat, nor anything of the feast-meals or what is taken from them.

10. Feast of the House of the Bridegroom for the Sun.

22. Feast of منطس for the Sun.

24. Feast of the *Venerable Old Man*, *i.e.* Saturn.

25. Feast of the nuptials of علاما.

Hilâl Adhâr.

1. Fast of اى; it lasts three days, and is broken on the 4th.

7. Feast of Hermes-Mercury.

8. Beginning of the Great Fast, during which only meat is forbidden. Its *Signum* is this, that they begin to lament on a day of this month, when the sun stands in the sign of Pisces (and the moon ?—*lacuna*). They continue their lamentations until the 31st day, when the sun stands in the sign of Aries, and the moon in the sign of Cancer, both standing in the same degree. The former day is the beginning of the fast, the latter is its breaking. Frequently this fast lasts only 29 days, when Hilâl Adhâr has less than 30 days.

10. Weaning of the children.

Hilâl Nîsân.

2. Feast of Damîs.
3. Feast of the Stibium.
4. Celebration of Πλοῦτος.
5. Feast of بليان, the idol of Venus.
6. Feast of سمار and of the *Living Being* of the Moon. On the same day is the feast of Dair-kâdhî.
8. The breaking of the Great Fast falls in most cases on the 8th of this month. On the same day is the feast of the birth of the spirits.
9. Feast of the Lords of the Hours.
15. Feast of the mysteries of Alsimâk (Spica).
20. Feast of the assembly at Dair-Kâdhî. p.321.
28. Feast of Dair-Sînî.

Hilâl Ayyâr.

2. Feast of Salûghâ, prince of the Satans.
3. Feast of a Baghdâdian house.
4. Feast of the vows.
8. Feast of اصيطلح, or feast of baptism,
7. Feast of Ḍaḥḍâk, the idol of the Moon.
11. Feast of Ḍaḥḍâḳ and جرورشيا.
12. Feast of جرشيا.
13. Feast of Barkhûshyâ.
15. Feast of Barkhurûshyâ.
17. Feast of *Bâb-altibn* (the straw-gate).
20. Feast of perfection for Ḍaḥḍâḳ, a blind idol. On the same day the feast of Tera'ûz.

Hilâl Ḥazîrân.

7. Commemoration of Tammûzâ with lamentation and weeping.
24. Feast of Alkurmûs or feast of genuflection.
27. Feast of the butcher's house.

Hilâl Tammûz.

15. Feast of the youths.
17. Feast of the nuptials of the elements.

18. Feast of the elements.
19. Also feast of the elements.

Hilâl Âb.

3. Feast of Dailafatân, the idol of Venus.
7. Also feast of Dailafatân.
24. Feast of bathing in the *Thermæ of Serûg.*
26. Another feast.
28. Feast of Kepharmîsâ.
30. End of the feast of bathing in the *Thermæ of Serûg.*

Hilâl Îlûl.

13. Feast of the Column of our Houses for the women, the end of a fasting.
14. Fasting of دنطا.
24. Feast of the Lords of the coming forth of the New Moons.
25. Feast of the candle on the hill of Ḥarrân.

In each of these months there is a fast of certain days which is obligatory for their priests. I think, either it lasts 14 days of each month, or it falls on the 14th. I cannot make out the truth.

One of those who record their doctrines says, that on the 17th of each month they celebrate a feast, the reason of which is the beginning of the deluge on the 17th of the month [*lacuna*]; further, that the days of the equinoxes and solstices are festivals with them, and that the winter-solstice is the beginning of their year.

p.322. This is all that Alhâshimî and others have related. We have collected these materials as we found them, simply transcribing the names as they were written. When we shall be in a position to hear these things from the people themselves (the Ḥarrânians), and to distinguish between what is peculiar to the Sabians, the Ḥarrânians, and the ancient Magians, we shall follow in this chapter the same method which we have followed elsewhere, if God permits!

(The author tries to form his information regarding the Harranian calendar into a system.)—Because their great fasting falls into the first phase (quadrature) of Hilâl Adhâr, whilst sun and moon stand in two *double-bodied* signs (Pisces and Gemini?), and because the end of the fasting falls into the first phase of Hilâl Nîsân, whilst sun and moon stand in certain two *inclining* signs (Aries and Cancer), their months must of necessity revolve in the solar year in a similar way to the Jewish months, that is to say: on an average. And between the causes of each of these two things there is a connection. For the Jewish Passover demands that sun and moon should stand in the *first* opposition in two signs of the equinoxes—for they may stand in opposition, and not only once, but twice—and the Ḥarrânian fast-breaking demands that which we have mentioned (in Hilâl Adhâr). Hence follows that the phase (quadrature) next preceding the Jewish Passover is the fast-breaking of the Ḥarrânians, and that the conjunction which falls next to the autumnal equinox is the beginning of their year, never falling beyond Îlûl.

If we compute these elements for a cycle of 19 years, we get a rough sort of computation, but only a rough one, for they themselves try to correct it by means of the time of the conjunction, as we have mentioned.

The methods of both Jews and Christians for the computation of Passover are based upon such motions of the luminaries, of which we have found out that they remain back behind real time, especially as regards the sun (the precession of the equinoxes having been neglected). If we examine the oppositions according to the motions that have been found by recent observations, we find that some of them precede the *Easter-limit* according to both Jewish and Christian systems; they, however, disregard this precession, whilst it is really the case, and we find that others of them (the oppositions) fall near the end of the *Easter-limit*; these latter oppositions they adopt and rely upon them, whilst they are utterly wrong; for the real time (or opposition) precedes that time already by one month.

Now, since it has been our object hitherto to point out scientific truth, to mediate between the two parties (Jews and Christians), and to adjust their differences, we have put forward the methods of each of the two sects according to their own theory as well as that of others, so as to show to each of them the *pro* and the *contra* of the case. And from our side we have proved that we candidly adopt their tradition and lean upon their theory, in order to make the truth clear to them. In all of which we are guided by the wish that both parties should dismiss from their minds the suspicion that we are partial to any side or try to mystify them; that their minds should not shrink back from our opposition, when we pass in review the (chronological) canons which they produce. For if they are left such as they are, they are not free from confusion and mistakes, most of which we have already pointed out.

Now we shall assume as the earliest date of the *Terminus Paschalis* the 16th of Adhâr; we shall let the day of opposition in reality fall into the two signs of the equinoxes; upon this basis we shall arrange the Passovers of the cycle that none of them precedes this terminus, and that each of them falls so that sun and moon stand in opposition to each other in the manner prescribed; the end of the terminus is to p.323.
be the 13th Nîsân, and within this space the sun must once have stood in opposition to the moon, although the sun may also after this terminus still stand in Aries without standing in opposition to the moon.

From these *corrected* Passovers we shall then derive the fast-breaking of the Sabians, and thence the beginning of the year, *i.e.* the conjunction in Hilâl Tishrîn I.

All this we have done and arranged in a table. Now if you take the years of the *Æra Alexandri*—the current year included—for the beginning of Tishrîn I., which follows after the conjunction of their New Year, and add thereto 16 or subtract therefrom 3; if you divide the sum by 19, if you neglect the quotient and compare the remainder with the column of the numbers in the *Table of the Corrected Cycles*, you find opposite their New Year, the end of their Great Fasting, the corrected Passover, and, hence derived, the mean fasting of the Christians, all fixed on the corresponding days of the Syrian months.

Here follows the *Table of the Corrected Cycle.*

p.324.

TABLE OF THE CORRECTED CYCLE.

Column of the Numbers.	Leap-years of the Cycle.	Date of the Sabian New Year in Îlûl and Tishrîn.	The Great Fast-breaking of the Sabians.	The month in which the Great Fast-breaking falls.	The corrected Passover.	The month in which the corrected Passover falls.	The mean fasting of the Christians as derived from the corrected Passover.	The month in which the mean fasting falls.	The first of Tishrîn following after this Passover.	The month in which the 1st of Tishrîn falls.
1		28	1	Nîsân	8	Nîsân	20	Shubâṭ	18	Îlûl
2	L.	17	21	Adhâr	28	Adhâr	9	Shubâṭ	7	Îlûl
3		6	9	Nîsân	16	Adhâr	28	Kânûn II.	26	Âb
4		25	29	Adhâr	4	Nîsân	16	Shubâṭ	14	Îlûl
5	L.	14	17	Adhâr	24	Adhâr	5	Shubâṭ	3	Îlûl
6		2	5	Nîsân	12	Nîsân	24	Shubâṭ	22	Îlûl
7	L.	21	25	Adhâr	1	Nîsân	13	Shubâṭ	11	Îlûl
8		10	13	Nîsân	21	Adhâr	2	Shubâṭ	31	Ab
9		29	2	Nîsân	9	Nîsân	21	Shubâṭ	19	Îlûl
10	L.	18	22	Adhâr	29	Adhâr	10	Shubâṭ	8	Îlûl
11		7	10	Nîsân	18	Adhâr	30	Kânûn II.	28	Âb
12		26	30	Adhâr	6	Nîsân	18	Shubâṭ	16	Îlûl
13	L.	15	19	Adhâr	26	Adhâr	7	Shubâṭ	5	Îlûl
14		4	7	Nîsân	15	Adhâr	27	Kânûn II.	25	Âb
15		23	27	Adhâr	3	Nîsân	15	Shubâṭ	13	Îlûl
16	L.	12	16	Adhâr	23	Adhâr	4	Shubâṭ	2	Îlûl
17		1	4	Nîsân	11	Nîsân	28	Shubâṭ	21	Îlûl
18	L.	20	24	Adhâr	31	Adhâr	12	Shubâṭ	10	Îlûl
19		9	12	Nîsân	20	Adhâr	1	Shubâṭ	30	Ab
1	2	3	4	5	6	7	8	9	10	11

CHAPTER XIX. p.325.

ON THE FESTIVALS OF THE ARABS IN THE TIME OF HEATHENDOM.

We have already mentioned that the Arabs had 12 months, that they used to intercalate them so as to make them revolve with the solar year in one and the same order, that the significations of the names of the months seem to indicate the reasons why they agreed among each other regarding this order, some of them indicating the corresponding times of the year, others indicating what the people did during them. We have already given the theory of some etymologists and historians of the Arabs regarding them; we shall now add another theory.

Al-Muḥarram, so called because four of their months were *Ḥurum*, *i.e. sacred* ones, one a *separate* one, *i.e.* Rajab, and three *consecutive* ones, *i.e.* Dhû-alka'da, Dhû-alḥijja and Almuḥarram, during which fighting was forbidden.

Ṣafar, so called on account of a contagious disease that used to befall them, when they became ill and their colour became *yellow*.

Rabî' Primus et *Postremus*; they fell into the season of autumn, which the ancient Arabs called *Rabî'*.

Jumâdâ Prima et *Postrema*, the time when the cold mornings, rime and hoar frost appeared, and when the water began to *freeze*,—the season of winter.

Rajab, so called because then people said *irjabû*, *i.e.* abstain from fighting and warlike expeditions, because it was a sacred month. According to others, so called because people immediately before it made haste, *being afraid* of it; for you say *rajibtuhu*, *i.e.* I was afraid of him.

Sha'bân, so called because then people *dispersed* to their camps and went out in search of booty.

Ramaḍân, the time when the heat commenced and the soil was *burning hot*. This month was held in high veneration in heathendom

Shawwâl, so called because then people said *shawwilû*, *i.e.* break up; according to another view: because about that time the she-camels *throw about* their tails, wanting to be covered. Therefore the Arabs did not like to marry their children in this month.

Dhû-alka'da, because then people said, *sit down* and abstain from fighting.

Dhû-alḥijja, so called because in this month they used to hold their pilgrimages.

(The seasons with different nations.)—Their months were distributed over the four seasons, beginning with autumn, which they called *Rabî'*; then winter; then spring, called *Ṣaif*, or by others *Rabî' Secundus*; then summer, called *Ḳaiẓ*. This nomenclature, however, has altogether been dropped and forgotten. Of the way in which they divided the seasons, we know only so much that the beginning of *Rabî'* or autumn fell on the 3rd Îlûl, the beginning of winter on the 3rd Kânûn I., the beginning of Ṣaif or spring on the 5th Adhâr, and the beginning of Ḳaiẓ or summer on the 4th Ḥazîrân. This you learn by the way in which they distribute the risings and settings of the lunar stations over the seasons.

Regarding the beginnings of these four seasons there has been a controversy. Ptolemy says, in his Introduction to the Spherical Art, that the ancient Greeks fixed their beginnings on the moments when the sun
p.326. enters the equinoctial and solstitial points, whilst the Chaldeans are said to have commenced the seasons 8 degrees after the equinoxes and solstices. The reason of this is, as it seems to me, that the computations in the Chaldean canons are back behind the computations to which the observations and canons of the ancient Greeks have led, and that just 8 degrees were assumed as the measure of this difference because they found such a difference in the progressive and retrograde motion of the sphere, the greatest extent of which is 8 degrees. But God knows best what they meant! The explanation of this motion you find in the *Zîj-alṣafâ'iḥ* of Abû-Ja'far Alkhâzin, and in the Book of the Motions of the Sun by Ibrâhîm b. Sinân, the best and most appropriate explanation possible.

The Byzantine Greeks and Syrians fixed the beginnings of the seasons earlier, one half sign (*i.e.* 15 degrees) before the equinoctial and solstitial points. In consequence, their seasons commence when the sun enters the middle of the signs that lie before the year-points. Therefore these signs were called the *corporeal* ones (Gemini, Virgo, Arcitenens, Pisces).

Sinân ben Thâbit relates two theories on this subject on the authority of the Egyptians and of Hipparchus, both nearly to this effect, that they fix the beginnings of the seasons one whole sign before the four year-points. The radicals among physical scholars make them precede the year-points by one sign and a half, and those of them who more than all deviate from the truth fix them on the times when the sun stands towards the equator at the half of his total inclination (15° Amphora, 15° Taurus, 15° Leo, 15° Scorpio). Such a division stands in direct opposition to common usage of mankind, and is in no way to be harmonized with the significations of the names of the seasons.

These theories in all their varieties are represented in the following table.

p. 327.

TABLE OF THE SEASONS ACCORDING TO THE DIFFERENT THEORIES.

The different theories.	The Byzantine Greeks, the Syrians, and the majority of Astronomers.		The ancient Greeks according to Ptolemy.		The Chaldeans according to Ptolemy.		The Arabs according to the *'Anwâ*-books.		The Egyptians according to Sinân b. Thâbit.		Hipparchus according to Sinân b. Thâbit.		The extreme natural philosophers who deviate from reality.		The natural philosophers who most of all deviate from reality.	
Seasons.	Dates.	Months.	Dates.	Months.	Dates.	Months.	Dates.	Months.	Dates.	Months.	Dates.	Months.	Dates.	Months.	Dates.	Months.
Spring	1	Adhâr	15	Adhâr	23	Adhâr	5	Adhâr	7	Shubâṭ	11	Shubâṭ	1	Shubâṭ	15	Kânûn II.
Summer	1	Ḥazîrân	16	Ḥazîrân	24	Ḥazîrân	4	Ḥazîrân	10	Ayyâr	12	Ayyâr	1	Ayyâr	15	Nîsân.
Autumn	1	Îlûl	17	Îlûl	25	Îlûl	3	Îlûl	15	Âb	16	Âb	1	Âb	17	Tammûz
Winter	1	Kânûn I.	15	Kânûn I.	23	Kânûn I.	3	Kânûn I.	11	Tishrîn II.	11	Tishrîn II.	1	Tishrîn II.	17	Tishrîn I.

p.328. **(On the fairs of the ancient Arabs.)**—The Arabs used to hold fairs in certain places and on certain dates of their months which were intercalated so as to agree with the solar year. Some of them have been mentioned by Abû-Ja'far Muḥammad ben Ḥabîb Albaghdâdî in the *Kitâb-Almujîr*. He says:

The fair of Dûmat-aljandal was held from the 1st of Rabî' I. till the middle of the month. There a bargain was concluded by *the throwing of a stone*, viz. if people gathered round an article of merchandise, he who liked to have it threw a stone. Now, frequently several people gathered around the same article, then the owner had to sell it to that man who threw the stone.

The fair of Almushaḳḳar commenced on the 1st of Jumâdâ II. There *the touching* was the mode of bargaining, viz. only to hint and to whisper, which they did for fear of swearing and lying.

The fair of Ṣuḥâr, from the 10th till 15th of Rajab.

The fair of Dabâ, on the last of Rajab. There the mode of bargaining was *Almusâwama* (*i.e.* chaffering).

The fair of Al-shiḥr, in the middle of Sha'bân. There the mode of bargaining was *the throwing of a stone.*

The fair of 'Adan, from the 1st till 10th of Ramaḍân.

The fair of Ṣan'â, from the middle of Ramaḍân till the end.

The fairs of Alrâbiya in Ḥaḍramaut, and of 'Ukâẓ in the highest part of Alnajd, not far from 'Arafât, fell on the same day, viz. the middle of Dhû-alka'da. The fair of 'Ukâẓ was one of the most important, being frequented by the tribes Ḳuraish, Hawâzin, Ghaṭafân, 'Aslam, 'Ukail, Almusṭaliḳ, the 'Aḥâbîsh, and by a motley crowd of other people. The fair was held from the middle of Dhû-alka'da till the end. As soon as the new moon of Dhû-alḥijja was observed, people went to Dhû-almajâz, a place in the neighbourhood of 'Ukâẓ. Then they held there a fair until the day of Altarwiya (the 8th of Dhû-alḥijja). Then they went up to Minâ.

The fair of Naṭâ in Khaibar and that of Ḥajr in Alyamâma were held from the 1st till the 10th of Almuḥarram. Since God has sent Islâm, most of these customs have been abandoned.

CHAPTER XX.

ON THE FESTIVALS OF THE MUSLIMS.

MUSLIMS use the months of the Arabs without any intercalation, for a reason which we have heretofore mentioned. They declared the four sacred months as sacro-sanct in consequence of the divine word (Sûra ix. 36): "Four of them are sacred ones (such is the right law). Therefore you shall not wrong yourselves in them."

The months Shawwâl, Dhû-alḳa'da, and the first ten days of Dhû-alḥijja they call the *Months of Pilgrimage*, of which God says (Sûra ii. 193): "Pilgrimage lasts for certain months. Therefore those on whom He has imposed the duty of pilgrimage shall not speak indecently, nor commit any wrong, nor quarrel during pilgrimage." They were called the Months of Pilgrimage because before this time the pilgrim is not allowed to enter the holy precincts. There are controversies regarding them between the lawyers of the four orthodox law-schools; they belong, however, to the science of law, and would swell this book too much if p.329. we were to propound them. These (two and one-third) months are named with the *Pluralis Paucitatis* (not dual), because the fraction, *i.e.* the third of a month, is added to the other months as one complete month.

The *Months of the Treaty*, which God describes in the following words (Sûra ix. 2): "Therefore ye shall go about on earth during four months," are the time from the *Day of Sacrifice* (the 10th of Dhû-Alḥijja) till the 10th of Rabî' II., for the Prince of the Believers ('Alî) recited this Sûra to the people (as a messenger of the Prophet) on the *Dies mactationis* (*i.e.* the 10th of Dhû-Alḥijja) on the fair.

The Arabs celebrate the following days of their calendar.

Almuḥarram.

The 1st is celebrated because it is the beginning and opening of the year.

The 9th is called *Tasû'â*, a word like *'Ashûrâ*. It is a day on which the devotees of the Shî'a say prayers.

The 10th is called *'Âshûrâ*, a most distinguished day. The Prophet is reported to have said: "O ye men, hasten to do good works on this day, for it is a grand and blessed day, on which God had mercy on Adam."

People celebrated this day until the murder of Alḥusain b. 'Alî b. 'Abî-Ṭâlib occurred on it, when he and his adherents were treated in such a way as never in the whole world the worst criminals have been treated. They were killed by hunger and thirst, through the sword; they were burned and their heads roasted, and horses were made to trample over their bodies. Therefore people came to consider this day as an unlucky one.

On the contrary, the Banû 'Umayya dressed themselves on this day in new garments, with various kinds of ornaments, and painted their eyes with *stibium;* they celebrated a feast, and gave banquets and parties, eating sweetmeats and various kinds of *confiseries.*

Such was the custom in the nation during the rule of the Banû 'Umayya, and so it has remained also after the downfall of this dynasty.

The Shî'a people, however, lament and weep on this day, mourning over the protomartyr (Alḥusain) in public, as, *e.g.* in Baghdâd and in other cities and villages; and they make a pilgrimage to the blessed soil (the tomb of Alḥusain) in Karbalâ. As this is a mourning-day, their common people have an aversion to renewing the vessels and utensils of the household on this day.

When the news of the murder of Alḥusain reached Medîna, the daughter of 'Aḳîl b. Abî-Ṭâlib came forward and said:

"What will you say, if once the Prophet speaks to you:
'What have you done, you, the last of all nations,
With my next relations and my family, if I inquire for them?'
One half of them are prisoners and one half tinged with blood.
It was not the proper reward for the advice I gave you,
That you, in playing the part of my successors, should bring woe over those who had sprung from my loins."

On the same day Ibrâhîm b. Al'ashtar, the helper of the Prophet's family, was killed.

People say that on this day God took compassion on Adam, that the ark of Noah stood still on the mountain Aljûdî, that Jesus was born, that Moses was saved (from Pharao), and Abraham (from the fire of Nebukadnezar), that the fire around him (which was to burn him) became cold. Further, on this day Jacob regained his eye-sight, Joseph was drawn out of the ditch, Solomon was invested with the royal power, the punishment was taken away from the people of Jona, Hiob was freed from his plague, the prayer of Zacharias was granted and John was given to him.

People maintain that the *Dies ornationis*, which is the time for the rendezvous of the sorcerers of Pharao, is this day 'Âshûrâ, especially the time after noon. p.330.

Although it be possible that all these events should have occurred on this day, we must state that all this rests only on the authority of popular story-tellers, who do not draw upon learned sources nor upon the agreement between the *owners of a divine writ* (*i.e.* Jews and Christians).

Some people say that 'Âshûrâ is an Arabized Hebrew word, viz. 'Âshûr, *i.e.* the 10th of the Jewish month Tishrî, in which falls the fasting Kippûr; that the date of this fasting was compared with the months of the Arabs, and that it was fixed on the 10th day of their *first* month, as it with the Jews falls on the 10th of *their first* month.

The Prophet gave orders to fast on this day in the first year of the Hijra, but afterwards this law was abrogated by the other law, to fast during the month of Ramaḍân, which falls later in the year. People relate that the Prophet of God on arriving in Medina saw the Jews fasting 'Âshûrâ. On inquiring of them, he was told that this was the day on which God had drowned Pharao and his people and had saved Moses and the Israelites. Then the Prophet said: "We have a nearer claim to Moses than they." In consequence he fasted on that day and ordered his followers to do the same. But when he afterwards issued the law regarding the fasting of Ramaḍân, he no longer ordered them to fast on 'Âshûrâ, but neither did he forbid them.

This tradition, however, is not correct, since scientific examination proves against it. For the 1st of Muḥarram in the year of the Hijra was a Friday, the 16th Tammûz, A. Alexandri 933. But if we compute the Jewish New-Year's Day for the same year, it was a Sunday, the 12th of Elûl, corresponding to the 29th of Ṣafar. Therefore the fasting 'Âshûrâ fell on Tuesday, the 9th of Rabî' I., and the flight of the Prophet occurred in the first half of Rabî' I.

When the Prophet was asked regarding the fasting of Monday, he said: "On this day I was born, I received my prophetical mission and divine revelation, and on this day I fled."

Further, it is a question on *which* Monday the flight occurred. According to some, it was the 2nd of Rabî' I., according to others the 8th, according to others the 12th of Rabî' I. However, according to the generally-adopted view, it was the 8th of Rabî' I. Both the 2nd and the 12th are excluded, since they were not Mondays, because the 1st of Rabî' I. of this year was a Monday (in consequence the 2nd was a Tuesday and the 12th a Friday). Now, for this reason the arrival of the Prophet in Medina (on Monday, the 8th of Rabî' I.) falls one day before the Jewish 'Âshûrâ (on Tuesday, the 9th of Rabî' I.), and 'Âshûrâ did not fall in Muḥarram, except at the time 3–10 years *before* the year of the flight, or 20–30 years *after* the year of the flight.

Therefore you could not maintain that the Prophet fasted 'Âshûrâ on account of its coinciding with the 10th in this year, unless you transfer 'Âshûrâ from the first of the Jewish months to the first of the Arabian months, so as to make them fall together. (In the first year of the flight the 1st of Muḥarram was a Friday, and therefore the 10th or 'Ashûrâ, Monday). Also in the second year of the flight the Jewish 'Âshûrâ and the date of Muḥammad's arrival in Medina cannot have coincided.

The assertion of the Jews that on this day God drowned Pharao is refuted by the Thora itself. For this took place on the 21st of Nîsân,
p.331. the seventh of the days of unleavened bread. Now, the beginning of the Jewish Passover after the arrival of the Prophet in Medina was a Tuesday, the 22nd Adhâr, A. Alex. 933, coinciding with the 17th Ramaḍân, and the day on which God drowned Pharao was the 23rd Ramaḍân. Therefore this tradition is altogether unfounded.

The 16th, Jerusalem was made the Ḳibla of the Muslims.

The 17th, the *Companions of the Elephants* (Ethiopians from the south of Arabia) arrived before Mekka.

Ṣafar.

1. The head of Alḥusain was brought to Damascus. Then he (Yazîd b. Mu'âwiya) placed it before himself, and with a stick in his hand he struck out the fore-teeth (the central four incisors), reciting these verses:

"I am not a descendant of Khindif, if I do not revenge
On the sons of Ahmad what he has done.
O that my chieftains in the battle of Badr had witnessed
The pain of Khazraj, caused by the hitting of the spears.
They would have praised God, and their faces would have beamed with joy,
And then they would say: 'O Yazîd, do not ask for anything more!
We have killed the generation of their chieftains;
We have tried to take vengeance on him for Badr, and we have got it.'"

On this day the Imâm Zaid b. 'Alî was killed and crucified on the border of the Euphrates; then his body was burned, and the ashes thrown into the water.

16. First appearance of the illness in the P .phet. This was the illness in which he died.

20. The head of Alḥusain was again laid to the body, and both were buried together.

On this day the pilgrimage of the forty men occurred, when they entered the holy district after their return from Syria.

23. Alma'mûn b. Alrashîd (the Abbaside Khalif) gave up again the green dress, after he had dressed in it during five and a half months.

He again adopted the black colours, the colours of the Abbaside party, after they had become excited against him.

24. Muḥammad left Mekka and concealed himself in a cave together with Abû-Bakr.

Rabî' I.

1. Death of the Prophet.
8. The Prophet arrives in Medina on the flight.
12. The Prophet is born on a Monday in the Year of the Elephants.

Rabî' II.

3. The Ka'ba was burned at the time when Alḥajjâj besieged 'Abdallâh b. Zubair.

15. Birth of 'Alî b. Abî-Ṭâlib.

Jumâdâ I.

3. The Battle of the Camel in Baṣra with 'Â'isha, Ṭalḥa, and Alzubair.
8. The death of the virgin Fâṭima, the Prophet's daughter.

Jumâdâ II.

2. Death of Abû-Bakr.
4. Fâṭima was born of Khadîja bint Khuwailid.

Rajab.

4. 'Alî and Mu'âwiya meet at Ṣiffîn.
26. God made Muḥammad His Prophet to all mankind.
27. Night of Ascension and the night-journey to Jerusalem.

Sha'bân.

3. Birth of Alḥusain b. 'Alî.

15. The great Liberation-night, also called *Lailat-alṣakk.*

15. The Ka'ba was made the Kibla instead of Jerusalem. The Ḥarrânians turn in praying towards the south pole, the Sabians towards the north pole. I believe that the Manichæans, too, turn towards the north pole, because this is, according to them, the middle of the dome of heaven and its highest place. I find, however, that the author of the *Book on Marriage,* who is a Manichæan and one of their missionaries, reproaches the people of the three religions with turning to one direction to the exclusion of another. With this he reproaches them, besides other things, and he seems to indicate that a man who prays to God does not need any Kibla at all.

Ramaḍân,

the month of the obligatory fasting.

p.332. 6. Birth of Alḥusain b. 'Alî according to all authorities except Alsalâmî.

7. Alma'mûn adopted the green colours.

10. Death of Khadîja.

17. The cursed 'Abd-alraḥmân b. Muljim Almurâdî struck 'Alî b. Abî-Ṭâlib on the head so as to injure the brain.

On the morning of the 17th the battle of Badr occurred; according to another report, it occurred on the 19th. But this is not correct, because there is an uninterrupted tradition saying that it occurred on a Monday in the second year of the flight. If we compute the 1st of Ramaḍân for this year, we find that it was a Saturday, and the Monday in question falls upon the 17th.

19. Mekka was conquered. The Prophet did not perform the pilgrimage, because the Arabian months were back behind real time in consequence of the *Nasî'* (postponement of certain months in the times of heathendom). Therefore he waited till the months returned to their proper places, and then he performed the farewell-pilgrimage, and forbade to use the *Nasî'*.

21. Death of the Prince of the Believers, 'Alî b. Abî-Ṭâlib; also death of 'Alî-Alriḍâ Ibn Mûsâ Alkâẓim b. Ja'far Alṣâdiḳ b. Muḥammad Albâḳir b. 'Alî Alsajjâd Zain-al'âbidîn b. Alḥusain, the protomartyr, son of the Prince of the Believers 'Alî b. Abî-Ṭâlib. According to others, his death (that of 'Alî-Alriḍâ) occurred on the 23rd Dhû-Alḳa'da.

22. Birth of 'Alî b. Abî-Ṭâlib, according to Alsalâmî.

25. 'Abû-Muslim 'Abd-alraḥmân b. Muslim first raised the standard of the 'Abbâsides in Khurâsân.

26. Revolt of Alburḳu'î in Baṣra; according to some, he was 'Alî b. Muḥammad b. 'Aḥmad b. 'Îsâ b. Zaid b. 'Alî b. Alḥusain b 'Alî b. Abî-Ṭâlib; according to others, he was 'Alî b. Muḥammad b. 'Abd-alraḥîm b. 'Abd-alḳais. There is a report saying that Alḥasan b. Zaid, the Prince of Ṭabaristân, wrote to him at the time when he came forward in Baṣra, asking for his genealogy, in order to learn the truth of the matter, whereupon he received this answer: "Do you mind my business as much as I mind yours (*i.e.* as little). My compliments." A wonderfully short and cutting answer, very much like that which Walî-aldaula 'Abû-'Aḥmad Khalaf b. 'Aḥmad, the Prince of Sijistân, gave, when Nûḥ b. Manṣûr, the Prince of Khurâsân, had written to him threatening him with various things. He answered: "O Nûḥ, you have quarrelled with us a great deal. Now carry out that which you threaten us, if you are a true-speaking man."

27. The night of this day is called *Lailat-alkadr* (Night of Fate), of which God says (Sûra xcvii. 3) that it is better than a thousand months.

The date of this night rests on universal agreement, because its real date is not known. People say: "See, this night is the night of the 17th or the 19th, for it was between these two nights that the battle of Badr occurred, the conquest of Mekka, the descending of the angels as a help, marked with certain badges (Sûra iii. 121)." This may be correct, for God says (Sûra xcvii. 4): "The angels descend and the Spirit. There is freedom from everything in that night by the permission of their Lord."

People say that on the following days the holy books were communicated to the Prophet—

on the 1st of Ramaḍân, the *leaves* to Abraham, p.333.
the 6th the Thora to Moses,
the 12th the Psalms to David,
the 18th the Gospel to Jesus, and
the 24th the *Furḳân* to Muḥammad.

As regards the Coran, God says (Sûra ii. 181): "The month of Ramaḍân in which the Coran was sent down." Thereby we learn that it was revealed in this month. Some people quote besides the passage (Sûra viii. 42): "And that which we have sent down upon our servant on the day of the *decision* (Alfurḳân), on the day when the two hosts met," inferring from this passage that the Coran was revealed on the 17th of Ramaḍân, because on this day the two *hosts* (that of Muḥammad and his opponents) met at Badr. But God knows best!

Regarding the Thora, we have already mentioned that it was revealed on the 6th of Sîwan, on the feast of *congregation* ('Azereth). If, at that time, Ramaḍân coincided with Sîwan, the matter is so as has been said. But there is no possibility of settling this question, because the year in which the Thora was revealed is not known; if it were known, we should inquire into the subject by chronological computations. The report regarding the Gospel is the saying of a man who does not know its character, nor arrangement, nor composition, and the revelation of the other books is altogether unknown and cannot be found out. God knows best!

Shawwâl.

1. Feast of fast-breaking, also called the *day of mercy*. God selected Gabriel as the bearer of His revelation. He inspired the bees and taught them how to make honey (Sûra xvi. 70).

People maintain that on this day God created Paradise. But why do they mention in their report such a thing with all that it may be supposed to indicate and that may be inferred therefrom? They go even so far as to attribute to Him an ugly anthropomorphism—as to say that on this day He planted the tree Ṭûbâ with His own hand. And this they have not tried to explain in any way; on the contrary, they believe it just as it stands, from sheer ignorance.

2. Beginning of a voluntary fasting of six consecutive days.

4. Muḥammad and the Christians of Najrân argued with each other. Muḥammad installed Ḥasan and Ḥusain in the right of sons of his, and Fâṭima in the right of his wives, and 'Alî b. 'Abî-Ṭâlib he made his intimate friend, complying with the order of God in the *verse of the cursing*.

17. Battle of 'Uḥud; according to others, it occurred in the middle of the month. In this battle Ḥamza was killed, and Muḥammad lamented over his loss.

19. Death of Abû-Ṭâlib.

28. On this day, they say, Yonas was devoured by the fish.

Dhû-Alka'da.

5. The Ka'ba was sent down. God took compassion on Adam. Abraham and Ishmael raised the bases of the temple of Mekka.

14. Jonas, they say, came forth from the belly of the fish. According to this view he must have stayed there twenty-two days, whilst according to the Christians he stayed only three days, as is mentioned in the Gospel.

29. On this day, they say, the tree *Yaḳṭîn* grew over Yonas.

Dhû-Alḥijja.

1. The Prophet of God married his daughter Fâṭima to his cousin, p.334. 'Alî b. Abî-Ṭâlib. The first 10 days of this month are also called *Dies noti* and *Dies sacri*. According to some, they are the time by which God completed the time which He had promised to Moses, saying (Sûra vii. 138): "And we have promised Moses thirty nights—which are the nights of Dhû-alka'da—and we have completed their number by ten"—which are the *Dies sacri*.

8. This day is called *Altarwiya*, because the pilgrim's-well in the holy mosque of Mekka used to be full of water about this season in the time of both heathendom and Islâm, and the pilgrims drank from it so much as *to quench their thirst*. According to another view, it was called so because they used to carry the water from Mekka on *Rawâyâ*, *i.e.* camels which are used to draw water from a well. According to a third opinion, because God made spring forth for Ishmael the well Zamzam, from which he drank so much as to quench his thirst. According to a fourth opinion, because on this day God revealed Himself to the mountain, as has been mentioned in the history of Moses.

9. This day is called *'Arafa*, the day of the great pilgrimage on *'Arafât*. It is so called because on that day people recognise each other at the time when they assemble for the performance of the rites of

pilgrimage, or, because Adam and Eve recognised each other after they had been driven out of Paradise in the place where people assembled, *i.e.* in *'Arafât*.

On this day God selected Abraham as a friend (*Khalîl*). It is also called *the day of forgiving*.

10. It is called *the day of the victims*, also *Dies mactationis*, because on this day the animals, that had been brought to Mekka to be sacrificed, were slaughtered. It is the last day of the days of the pilgrimage. On this day Isaak was ransomed with the ram. On this day, too, *the Road (via strata) to the Last Judgment* is said to have been created.

11. *The day of sojourning*, because on this day people sojourn in Minâ.

12. *The day of going away*, because on this day people go away from the holy district hurrying.

11, 12, 13. *The days of Tashrîḳ*, so called because on these days the meat of the sacrificed animals was cut to pieces and exposed to the sun for drying. The name is also derived from the saying, "*'Ashriḳ thabîr kaimâ nughîr*" (*i.e.* Shine forth, O mountain Thabîr, that we may break up). According to Ibn-Al'a'râbî they were so called because the victims (*hostiæ*) were not killed before the sun had *risen*.

These are the days which God means in His words (Sûra ii. 199): "And ye shall remember God on certain counted days."

In the time immediately before and after these days people say *Allâh akbar* after every prayer. Among the theologians there are differences regarding the beginning, the end, and the limits of the prayer of *Takbîr* (*i.e. Allâh akbar*), differences peculiar to their science.

17. 'Uthmân b. 'Affân the Khalif was killed.

18. It is called *Ghadîr Khumm*, which was the name of a station on the road-side where Muḥammad alighted when returning from the farewell pilgrimage. He gave orders to collect the saddles and all the riding-instruments into one heap; this he ascended, supported by the arm of 'Alî b. 'Abî-Ṭâlib, and said: "O men, am I not nearer to you than you yourselves?" They answered, "Yes." Then he said: "To every man whose friend I am, also 'Alî is a friend. O God, befriend him who befriends 'Alî, and oppose him who opposes 'Alî, help him who helps 'Alî, and desert him who deserts 'Alî. Let truth go about with him wherever he goes." Then he is said to have raised his head towards heaven and said

(*Lacuna.*)

24. 'Alî gave away his seal-ring as alms, in praying. p.335.

25. 'Umar b. Alkhaṭṭâb was killed, and the Sûra *Hal 'Atâ* (Sûra lxxvi.) was revealed.

26. David was inspired to ask for pardon (Sûra xxxviii. 23).

29. Battle of *Alḥarra*, in which the Banû-'Umayya killed the people of Medîna, when the honour of the *Muhâjirûn* (companions of the flight of Muḥammad) and of the *'Anṣâr* (his partisans in Medîna) was stained and their wives were given up to the enemies. Therefore may God curse all those whom His Prophet cursed, of those who rebelled in Medîna against the law of God, and may He let us belong to those who do not like wickedness on earth. God is the best Helper, and infinite thanks are His due!

CHAPTER XXI. p.336.

ON THE LUNAR STATIONS, THEIR RISING AND SETTING, AND ON THEIR IMAGES.

It is now time for us to finish, after we have, as best we could, fulfilled our promise in explaining the science of that subject which our friends wanted to know, and in relating all we know regarding it. But above every knowing man there is God all-wise! To complete the representation of this science, only one more chapter is required, that of the rising of the Lunar Stations in the days of the solar year. For this science is practised on account of its general usefulness for the purpose of prognosticating all meteorological occurrences which revolve together with the Lunar Stations. Therefore we shall now proceed to explain this subject both at large and in detail, and we shall add some of the proverbial sayings relating to them, which we gather from the literature of this kind, *e.g.* from the book of Alkulthûmî, that of Ibrâhîm b. Alsarrî Alzajjâj, that of Yaḥyâ b. Kunâsa, of Abû-Ḥanîfa Aldînawarî on the *'Anwâ*, the book of 'Abû-Muḥammad Aljabalî on the science of the configurations of the stars, the book of Abû-Alḥusain on the fixed stars, and from other books.

The Hindûs divided the globe, in conformity with their 27 Lunar Stations, into 27 parts, each Station occupying nearly $13\frac{1}{4}$ degrees of the ecliptic. From the stars entering these Stations, which are called *Jufûr*, they derived their astrological dogmas as required for every subject and circumstance in particular. The description of these Astrologoumena would entail a long explication of things, foreign to our purpose, all of which may be found in— and learned from—the books on Astrologoumena.

The Arabs divided the celestial globe into 28 parts, so that each Station occupies nearly $12\frac{6}{7}$ degrees of the ecliptic, and each zodiacal sign contains $2\frac{1}{3}$ Stations. Some poet says:

"Their number is, if you want to count them,
Twenty stars, and a number 8 after them.
In each of the zodiacal signs there are
Two Stations and one complete third of a Station.
A peculiar system of computation belongs to them, and they have their heliacal risings and settings,
Which are the reason that winter and summer revolve."

The Arabs used the Lunar Stations in another way than the Hindûs, as it was their object to learn thereby all meteorological changes in the seasons of the year. But the Arabs, being illiterate people, could not recognize the Lunar Stations except by certain marks, visible to the
p.337. eye. Therefore they marked the Stations by those fixed stars which lie within them. And the rising of the fixed stars in the east early after the rise of dawn they considered as a sign of the sun's entering some one of the Stations, and so they could do, since the stars do not recede from their places except after the lapse of long spaces of time, and, besides, the Arabs were not educated enough to notice such a variation. Further, they composed verses and rhymed poetry, so that these things could easily be remembered by illiterate people, and recorded therein the annual physical influences which, according to their observation and experience, coincided with the rising of each particular Station. These sayings and verses they use to indicate certain circumstances of theirs, *e.g.*:

"When the moon joins (*i.e.* stands in conjunction with) the Pleiades,
In a third night (of a month), then the winter is gone."

For the Pleiades occupy the place from 10° of *Taurus* till about 15° of Taurus. When, therefore, the moon joins the Pleiades in the 3rd night of a month, the distance between sun and moon is about 40 degrees. Then the sun stands in the first part of *Aries*. Further:

"When full-moon is complete and stands with the Pleiades,
Then you get the beginning of the cold season, the winter."

For when the moon stands in opposition to the Pleiades, the sun stands in the middle of Scorpio, and that time is the beginning of the cold season. Further:

"When full-moon joins Aldabarân
In the 14th night of a month,
Then winter encircles the whole earth,
Being like riders who ride about, telling people to warm themselves,

And full moon rises in heaven high overhead, so that
The shadow of the tent-poles disappears,
When the night has reached its middle
And the air is free from dark clouds."

For at that time the sun stands in Scorpio close to *Alkalb* (the 18th Lunar Station); it is the time of cold and of morning frosts. The moon stands in some degree of northern declination, and frequently she stands in such a latitude from the ecliptic towards the direction of the declination, that she culminates (stands right) over the heads of the Arabs. In consequence, the shadows of all bodies disappear at the time when she reaches the middle of heaven, *i.e.* at the time of midnight. Further:

"When the new moon of a month first appears
To the eyes of people at the beginning of a night, standing in *Alnaʿâʾim*,
Then you get cold winds from every side,
And you find it agreeable a little before dawn to wrap a turban round the head."

For at that time the sun stands in the first part of Sagittarius. Further:

"The complete night, with all that belongs to it, has become cold,
And the sun stands in the Station of *Alʿawwâ*."

For the stars of Alʿawwâ (the 13th Lunar Station) lie around the p.338.
vernal equinox, as the table of the Lunar Stations will show.

However, if I were to communicate to the reader all the verses and sayings in rhymed prose which relate to the rising of each Lunar Station, I should also have to interpret their meanings, and to explain the rare words that occur in them. This, however, we may omit, since it has been sufficiently done by the authors of the books of *'Anwâ*, whom we mentioned above.

Since the Arabs attribute all meteorological changes to the influence of the rising and setting of the stars, in consequence of their ignorance of physical sciences, thinking that all changes of the kind depend upon the bodies of the stars and their rising, not upon certain parts of the celestial globe and the sun's marching therein, they believe a great many things similar to that which we have mentioned of the *Sirius Jemenicus*, during the rising of which Hippocrates in his time forbade taking hot drugs and phlebotomizing.

And this subject reminds me of an occurrence in my life which serves to confirm the verses of Aḥmad b. Fâris:

"A wise man of by-gone times has said:
'The importance of a man lies in his two smallest things.'

I on my part also speak like a wise man, saying:
'The importance of a man lies only in his two dirhams.'
If he has not his two dirhams with him,
His bride does not care for him.
In consequence of his poverty he is despised,
So that people's cats piss at him."

For when I was separated from the court of His Highness, and was bereft of the happiness of the royal service, I met a man in Rai (Rhagæ) who was counted among the learned astronomers. He had studied the conjunctions of the stars which form the Lunar Stations, and he had commenced to collect them in order to derive certain sentences (astrologoumena) from the Stations and their single parts, and thereby to prognosticate all changes of the air. Now, I told him that the truth is the very reverse of his theory, that the nature and peculiarities which are attributed to the first Station, and all that which the Hindûs relate of the connection of this Station with others, are peculiar to the first part of Aries, and never leave this place, although the star (or stars which form the Lunar Station) may leave it. In a similar way, all that is peculiar to Aries does not move away from the place of Aries, although the constellation of Aries does move away. But then the man became very haughty, and treated me slightingly, though he was inferior to me in all his knowledge. He told me my theory was a lie, and behaved very rudely to me, being very lengthy about the difference between us in wealth and poverty, which changes subjects for glory into subjects for blame. For at that time I was in a miserable condition, tried (troubled) on all sides; afterwards, however, when my troubles had subsided (ceased) to some extent, he chose to behave in a friendly way towards me.

It is evident that, if the science of meteorology were to depend upon the rising of the bodies of the stars, as observed by eye-sight, the times and seasons of the *Meteora* would differ in the same proportion as the stars change their places; besides, they would be different in different
p.339. countries, and we should require for them as well as for the appearing and disappearing of the planets various kinds of tiresome methods of calculations.

In reality the rising of the Lunar Stations means this, that the sun on entering one of them covers it and the preceding one too, whilst the third one, according to the inverted order of the zodiacal signs, rises between the rise of dawn and that of the sun, at that time which Ibn Alraḳḳâ' describes in the following verses:

"The observers saw Sirius distinctly,
As he turned away, when the morning prayer approached.
I recognize Sirius shining red, whilst the morning is becoming white.
The night, fading away, has risen and left him.

The night is not afraid to lose him, since he follows her,
But the night is not willing to acknowledge that he belongs to the night."

The rising of a Lunar Station they called its *Nau'*, *i.e.* rising. The influence of the rising they called *Bârih*, the influence of the setting they called again *Nau'*. The interval between the risings of two consecutive Lunar Stations is 13 days, except the interval between the rising of *Aljabha* (the 10th Station) and of the following Station, which is 14 days. So the following verses:

"All time, you must know, consists of fourths,
And each fourth consists of sevenths.
A complete seventh belongs to the rising of a star,
And to the influence (*Nau'*) of a star setting in the west.
Between the rising of each star
And that of the following star there are *four* nights
And *nine* nights more."

There is a difference of opinion regarding the *'Anwâ*. Some maintain that each influence (of a Lunar Station) is brought about between the risings of two consecutive Stations, that therefore the influence is attributed to the former of these two Stations. According to others, a certain space of time is peculiar to the rising and setting of each Lunar Station, and everything that occurs in this time is attributed to the Station in question; occurrences which fall after the end of this space of time are no longer attributed to it. The last view is the generally adopted one.

Besides, there are differences about the length of these spaces of time, which we shall afterwards describe.

When the influence of some Station has been found out and is known, and nothing happens at its time, people say: the star was *empty*; or: the Station was *empty*, *i.e.* the time of its *Nau'* has gone by without there being any rain, or heat, or cold, or wind.

(On the Winds.)—Regarding the directions of the winds, the planes over which they blow, and their number, there are different opinions. Some maintain that the directions of the wind are six, as Ibn Kunâsa relates, on the authority of 'Abû-Muḥammad Ja'far b. Sa'd b. Samura b. Jundub Alfazârî, whilst, according to most others, there are only four, as Khâlid b. Ṣafwân relates; the latter is the opinion of most nations, although they differ regarding the planes of the blowing of the winds. Both these opinions of the Arabs are comprised in the following two circles; the former view is represented in the inner circle, the latter in p.340.
the outer circle. There you also find the names of the winds and the directions of their planes. Here follows the circle.

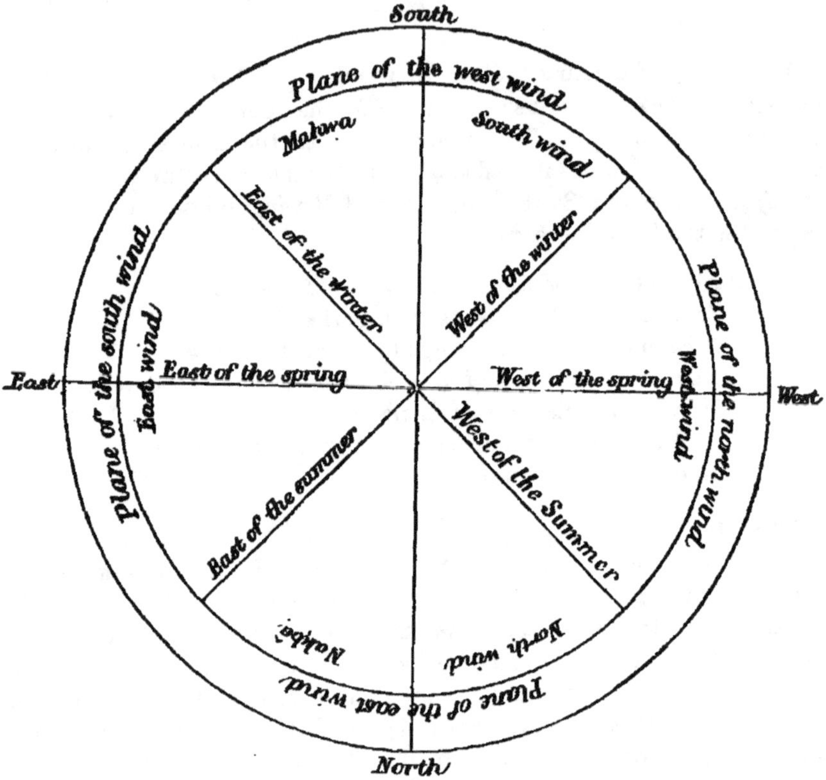

In the first theory the author (Ibn Kunâsa) places the wind *Maḥwa* near the south wind, whilst it is well known that *Maḥwa* is the north wind, because it *extinguishes* (destroys) the clouds when they are empty, after the south wind has driven them on, full of rain. In the same theory he assigns a separate plane to the wind *Nakbâ*, whilst it is well known that Nakbâ is every wind, the plane of which lies between the planes of any two other winds of the four cardinal winds. Dhû-alrumma mentions the winds, *Nakbâ* included, in this way:

"Heavy rain-showers of some *Anwâ* and the two *Haif* (south wind and west wind),
Which drove the sand-masses of the dusty-coloured mountains away over the house.
And a third wind, blowing from the side of Syria, a cold one,
Blowing with whirlwinds along its road over the sand.
And a fourth wind coming from the rising-place of the sun, driving
The fine dust of Almi'â and of Ḳurâḳir over the house.

> The side winds, carrying along the dust, excited it (the east wind) to still greater vehemence,
> So that it frequently roared like the she-camels in the tenth month of their pregnancy, when the throes are near."

The two *Haif* are the south wind and west wind; the wind blowing from Syria is the north wind; the wind coming from the rising-place of the sun is the east wind.

The planes of the winds with the Persians are the same as with the ancient Greeks, and all physical scholars; their centres correspond to the four directions. They are represented in the following circle:

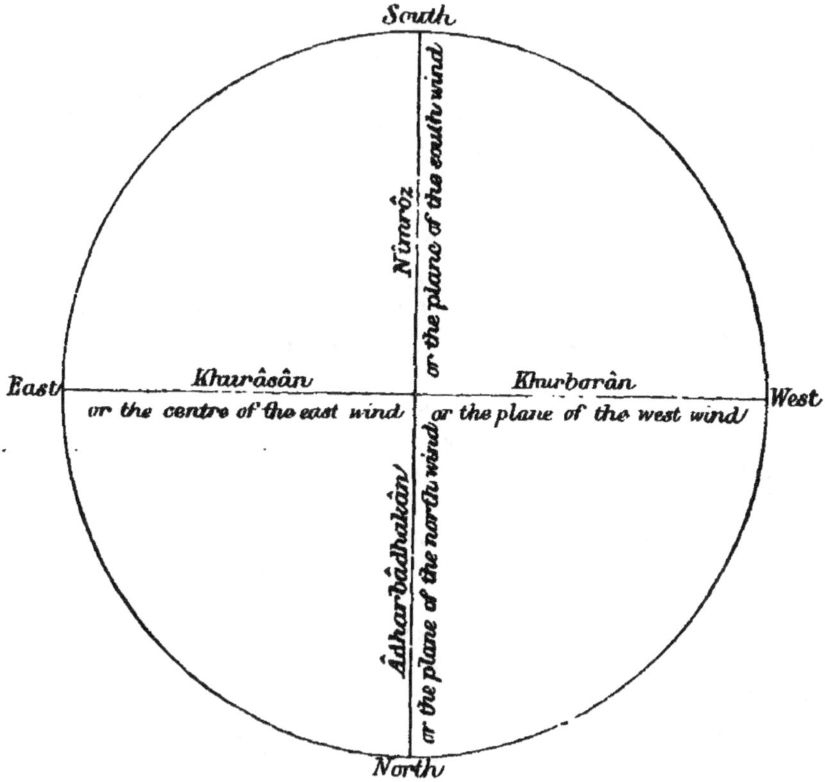

Any wind that lies between the centres of the planes of two other winds is referred to that centre which is the nearest (and receives its name therefrom). Other people refer an intermediate wind to the rising and setting places of the sun at the time of the solstices, and call it by a Greek name.

(Method for finding the time of the Nau' and Barih of a Lunar Station.)—The following is a good method to find the times of the influences (ἐπισημασία) of the rising and the setting of the Lunar Stations: Take the time from the 1st of Îlûl till that day the nature of which you want to find out, and divide the sum of days by 13. If there is no remainder, proceed in this way: If the moon stands opposite the sun or in one of her quadratures, you get rain, if it is the season for rain, or some change of the air in consequence of wind, or heat, or cold. For if there is no remainder (as in this case), it is the time of the rising of one Lunar Station and the setting of the opposite Station. On the 1st of Îlûl falls the *Bâriḥ* (influence of the rising) of *Alṣarfa* (the 12th Station) and the *Nau'* (influence of the setting) of *Sa'd-al'akhbiya* (the 25th Station). From this date you begin counting, for this special reason, that it is the first of a month and the beginning of autumn. If, besides, the moon happens to be in one of her *Foundations*, the influence (of the Lunar Station) will come out very strong.

Abû-Ma'shar says: "We have tried this method A. H. 279 in Shawwâl
p.341. at the time of full moon. We counted the days from the 1st Îlûl till this full moon. They were 130 days; dividing them by 13 you get no remainder, and the *Ascendens* of the full moon (or opposition) was *Amphora*. So we got rain on that day, and when the moon stood in her right quadrature, also on that day we had rain."

Further, he says: "We tried it also in the following year. We counted the days from the 1st Îlûl till Thursday the 13th of Kânûn I.; the sum of days we divided by 13, and there was no remainder; the distance between sun and moon was as much as half a zodiacal sign (*i.e.* 15 degrees), the moon had turned away from the hexagon of Mars and stood in conjunction with Venus. At that very time we got rain."

Now, this is a testimony of Abû-Ma'shar, showing that through this method you obtain correct results. If, besides, you take to help the mansions (the places of the Lunar Stations) of the Hindûs and their single parts, you are pretty sure in your calculation to come near the truth.

People relate that among the Arabs the Banû-Mâriya b. Kalb and the Banû-Murra b. Hammâm b. Shaibân had the most accurate knowledge of the configurations of the stars.

In enumerating the *Nujûm-al'akhdh*, *i.e.* the Lunar Stations, the Arabs commenced with *Alsharaṭân*, since in their time they stood in the first part of *Aries*. Other nations begin with the Pleiades. I do not know whether they do this because the Pleiades are more easily and clearly visible without any study or research than the other Stations, or because, as I have found in some books of Hermes, the vernal equinox coincides with the rising of the Pleiades. This statement must have been made about three thousand and more years before Alexander. God knows best what they intended!

We shall adopt the Arabian system in enumerating the Lunar Stations, and shall begin, as they do, with

1. *Alsharaṭân* (β, γ Arietis),

i.e. the two *signs*. They are called so for the same reason that the soldiers of the body-guard of a prince are called *Shuraṭ*, since they mark themselves by some *sign*, by the black colour, or something else. It consists of two stars belonging to Aries (β and γ). Sometimes, also, a third star near them is added, and then this Station is called *Al'ashrâṭ* (plural instead of the dual Sharaṭân). Between the two stars, when standing in the middle of heaven, there is an interval of two yards according to eye-sight; one of them belongs to the northern half, the other to the southern.

All measures of distances between the stars according to eye-sight are to be understood only for that time when they stand in the middle of heaven, for these distances appear greater near the horizon in consequence of the intense refraction of the ray of light in the watery vapours that surround the earth. This has been explained in the books on the geometrical configurations (of the stars). Further, the distance between two stars increases in the direction from north towards the south; frequently, too, when the stars march towards the horizon, it increases in the direction from east to west, or pretty nearly in the direction of one of the cycles of altitude. The reason of this is that the spheres decline from the perpendicular direction which they have on the equator.

The Station *Al'ashrâṭ* is also called *Alnaṭḥ* (*i.e.* horn), because the two *Sharaṭ* are placed on the root of the two *horns* of Aries. The meteorological influences of this Station are peculiar to the first (*i.e.* original) position of Aries, and in no way depend upon the stars from which the Station has got its name. These stars have migrated from their original place (in consequence of the precession of the equinoxes) and have in our time come to occupy a second position (different from the former). p.342.

2. *Albuṭain* (ϵ, δ, π Arietis).

It consists of three stars at the end of the womb of Aries, forming an isosceles triangle. The word is the diminutive of *Baṭn*, so as to mean the *little womb*, so called in comparison with *Baṭn-alḥût* (the womb of the fish), which is the 28th Station.

3. *Althurayyâ* (Pleiades)

consists of six stars close to each other, very similar to a cluster of grapes. According to the Arabs they form the *clunis* of Aries, but that is wrong, because they stand on the hump of Taurus.

The word is a diminutive of *Tharwâ*, which is originally identical with *Tharwa*, *i.e.* a collection and great number of something. Some people maintain they were called so because the rain, which is brought by their

Nau', produces *Tharwa*, *i.e.* abundance. They are also called *Alnajm* (i.e. *The Star*).

Ptolemy mentions only four stars of the Pleiades, since he had not observed more of them, because to eye-sight they seem to lie quite close together.

The forty days during which this Station disappears under the rays of the sun, are, according to the Arabs, the worst and most unhealthy of the whole year. Al'asadî says: "Althurayyâ never rises nor sets unless bringing some harm." And one of their medical men says: "Warrant me the time between the disappearing and the rising of Althurayyâ, and I shall warrant you all the remainder of the year." The Prophet is related to have said: "When the Star rises, all harm (mishap) rises from the earth;" and according to another tradition: "When the Star rises, all mishap is raised from every place."

4. *Aldabarân* (α Tauri),

a bright red star, so called because it *follows after* the Pleiades, standing over the southern eye of Taurus. It is also called *Alfanîḳ*, *i.e.* a great camel-stallion (not serving for riding), because they call the stars around it *Ḳilâṣ*, *i.e.* young she-camels (serving for riding). Other names of it are "*The follower of the Star*," because in rising and setting it follows immediately after the Pleiades, and *Almukhdij* (*i.e.* a she-camel giving birth to a young one of imperfect formation).

5. *Alhaḳ'a* (λ, φ′, φ″ Orionis)

consists of three small stars close to each other, looking like so many dots impressed upon the earth by the thumb, the fore-finger, and the middle-finger, the fingers being closely pressed together. They were so called because they were compared with a circle of hairs on the side of the horse at the joint of the foot; such a horse is called *Mahḳû'*. They are also called *Altaḥâ'i* (or *Altaḥâyi*). Ptolemy considers them as one cloudy star, and calls them the nebula on the head of Aljabbâr, *i.e.* Aljauzâ (Orion).

6. *Alhan'a* (γ, ξ Geminorum)

consists of two bright stars in the Milky Way between Orion and the head of Gemini, distant from each other as far as the length of a whip. The one is called *Zirr* (button), the other *Maisân* (walking along proudly); they stand on the foot of the second twin. According to Alzajjâj, *Han'a* is derived from the verb *Hana'a*, *i.e.* to wind and twine one thing round the other, as if each of them were winding and twining round the other. According to others, this name is to be understood of a third star, standing behind their middle, which gives them the appearance of an *inclined* neck. The Arabs consider Alhan'a and six other stars as the bow of Orion, with which he shoots at the p.343. Lion.

7. *Aldhirâ'* (α, β Geminorum)

consists of two stars, one yard distant from each other. The one is the *blear-eyed* Sirius or Sirius Syriacus, according to the Arabs, the outstretched arm of Leo; the other is Sirius *'Abûr* or Sirius Yemenicus, the arm of Leo which is not stretched out. According to the astronomers, the outstretched arm is the head of Gemini, and the other arm belongs to the stars of *Alkalb Almutakaddim* (Procyon). But people differ greatly regarding these stars and produce various futile traditions and stories in support of the names which they give them. The rising of *Ghumaiṣâ* (the blear-eyed Sirius) in the year 1300 of Alexander took place on the 10th Tammûz, and that of Sirius Yemenicus on the 23rd Tammûz.

8. *Alnathra* (Præsepe (ε) et duo Aselli (γ, δ) Cancri)

is the place between the mouth and the nostrils of the Lion. It is also called *Allahâ* (the uvula), and consists of two stars, between which there is a nebula, the whole belonging to the figure of Cancer.

9. *Alṭarf*,

the eye of Leo, two stars close to each other, one belonging to Leo, the other to the stars outside the figure of Cancer. In front of them there are stars called *Al'ashfâr*, *i.e.* the eyebrows of Leo.

10. *Aljabha* (ζ, γ, η, α Leonis),

the front of Leo, four stars, each star distant from the other by the length of a whip, lying athwart from north to south in a curve, not in a straight line. According to astronomers, they stand on the mane of Leo. The most southern star of them they call the *Heart of the Royal Lion*; it rises when Suhail rises in Alḥijâz. Suhail is the 44th star of Argo Navis, standing over its oar. Its latitude is 75 degrees in the southern half. Therefore it does not rise very high above the horizon, in consequence of which it has something unsteady for the eye. People say that a man, if his eye falls on this star, dies, as they also relate that on the island of Râmîn, belonging to Ceylon, there is an animal the sight of which kills a man within forty days afterwards. The most curious instance of the connection between animal life and its material influence is the fish called *Silurus Electricus*. For the hand of the fisherman who has caught it takes care not to touch it as long as it is in the net still living. If you take a reed and touch the living fish with one end and keep the other end in your hand, the hand becomes feeble and drops the reed.

Further, the worms in Raghad, one of the districts of eastern Jurjân. For there you find in certain places small worms; if a man carrying water treads upon them, the water becomes bad and foul; if he does not tread upon them, the water remains good and keeps its nice odour and sweet taste.

The death of a man bitten by a panther, when a field-mouse pisses at him——

[*Lacuna.*]

p.344.

11. *Alzubra* (δ, θ Leonis),

i.e. the shoulder of the Lion, the place where the neck begins. According to Alzajjâj, it is the place of the mane on his neck, because the mane *bristles up* when he is in wrath. According to Alnâ'ib Alâmulî, *Zubra* is a piece of iron by which the two shoulder-blades of a lion are imitated.

This station consists of two stars, distant from each other by the length of a whip. They are also called the *Two Khurt*, *i.e.* holes, as if each of them were penetrating into the interior of the Lion, but in reality they stand upon the shank of the Lion, one of them on the root of the tail. When they rise, Suhail is seen in Al'irâk.

12. *Alṣarfa* (β Leonis),

a bright star near to some very dim ones, called the *Claw of the Lion*. It stands on the end of the Lion's tail, and is called so because the heat *turns away* when it rises, and the cold *turns away* when it disappears.

13. *Al'awwâ* (β, η, γ, δ, ε Virginis)

consists of five stars in a line, the end of which is turned. And therefore the Station is called so because the verb *'Awâ* means *to turn*. Alzajjâj says: "I do not know of anybody else besides me who has explained the word in this way. Those who say that these stars are dogs running behind the Lion and *barking* are wrong." They stand on the breast and wing of Virgo.

14. *Alsimâk Al'a'zal* (Spica).

It is also called the *Calf of the Lion*, and *Alsimâk Alrâmiḥ* is his other calf.

This Simâk is called *'A'zal* (*i.e.* bare), because whilst the other Simâk Alrâmiḥ (the shooter) is accompanied by a star, said to be his lance, this one has no such accessory, and is therefore said to be *bare* of weapons.

According to Sîbawaihi, Simâk is called so on account of its rising high, or, according to others, because the moon does not enter this Station. But if that were the case, *Alsimâk Al'a'zal* would not deserve the name of a Lunar Station, for, of course, the moon enters it and frequently covers it (so as to make it disappear).

It is a brilliant star on the left palm of Virgo, which some people call *Sunbula* (the ear). But this is wrong, because the Ear (Spica) is *Alhulba* (*i.e.* hog's bristle), which Ptolemy calls *Alḍafîra*, i.e. *Crines plexi.* This is a number of small stars behind the tail of the Great Bear, very much like the leaf of *Lublâb*, *i.e.* helxine. The whole zodiacal sign is also called so (*i.e.* Spica).

According to the Arabs, *Alhulba* (the hog's bristle) stands on the end of the Lion's tail, being the small hairs on the end of the tail.

15. *Alghafr* (ι, κ, λ Virginis)

consists of three not very brilliant stars on the train and the left foot of Virgo. According to the Arabs, it is the best of the Lunar Stations, because it stands behind Leo and before Scorpio. The evil of the Lion lies in his teeth and claws, the evil of the Scorpion lies in its venom and the sting of its tail. A Rajaz poet says:

"The best night for ever
Lies between Alzubânâ and Al'asad (Leo)."

People say that the horoscopes of all the prophets lie in this Station; but this does not seem to be true except in the case of Messiah, the Prophet who keeps off all mishap. The birth of Moses—according to the report of the Jews—must have coincided with the rising of the tooth of Leo and the moon's entering the claws of Leo. p.345.

It is called *Ghafr*, because the light of its stars is imperfect, from the verb *Ghafara*, *i.e.* to cover a thing, or, because it rises above the claws of Scorpio and becomes to it like a *Mighfar* (*i.e.* coat of mail). According to Alzajjâj, the name is derived from *Ghafar*, *i.e.* the hair on the end of the Lion's tail.

16. *Alzubânâ* (α, β Libræ)

consists of two brilliant stars, separated from each other as far as five yards, and standing in a place where the two claws of *Scorpio* might be; they belong, however, to *Libra*. The word is also derived from *zabana* (i.e. *to push*), as if the one of them were being *pushed* away from the other, not united with it.

17. *Al'iklîl* (β, δ, π Scorpii)

is the head of Scorpio, consisting of three stars which form one line. Ibn-Alṣûfî declares this to be impossible, and maintains that it consists of the 8th star of *Libra* and the 6th one of the stars outside Libra, as also Ptolemy has it in his Almagest. According to Ibn-Alṣûfî, those who consider the three bright stars in one line as *Al'iklîl* are mistaken, for he says that the Crown, (*i.e.* Al'iklîl) could not be anywhere but upon the head. However, the general view of the Arabs—in opposition to that of Ibn-Alṣûfî—is this, that the three stars in one line are *Al'iklîl*. The Arabs have a proverb applicable to this subject, saying: "The two contending parties were content, but the judge declined to give a judgment."

[18. Alḳalb (α Scorpii)

is a red star behind Al'iklîl and between two stars called Alniyâṭ (præcordia).]

19. *Alshaula* (λ, ν Scorpii)

is the sting of *Scorpio*, so called because it is always *mushâla*, *i.e.* raised. It consists of two bright stars near each other on the top of the tail of Scorpio.

20. *Alna'â'im* (γ, δ, ε, η, σ, φ, τ, ζ Sagittarii)

consists of eight stars, four of them lying in the Milky Way in a square, which are the *Descending Ostriches*, descending to the water, which is the Milky Way; and four of them lying outside the *Milky Way*, also in a square, which are the *Ascending Ostriches*, ascending and returning from the water.

Alzajjâj reads the word *Alnu'â'im*, *i.e.* the beams placed above the mouth of a well, where the sheaves of the pulley and the buckets are fixed (attached).

The stars were compared to ostriches, as if four of them were descending, four ascending. The *Descending Ostriches* stand on the bow and arrow of Sagittarius, and the *Ascending Ostriches* stand on his shoulder and breast.

21. *Albalda*

is a desert district of heaven without any stars, at the side of the Horse, belonging to Sagittarius. According to Alzajjâj, this station was compared to the interstice between the two eyebrows, which are not connected with each other. You say of a man *'ablad*, which means that his eyebrows do not run into each other.

22. *Sa'd-Aldhâbiḥ* (α, β Capricorni)

consists of two stars, the one to the north, the other to the south, distant from each other about one yard. Close to the northern one there is a small star, considered as the sheep which he (Sa'd) slaughters. The two stars stand on the horn of *Capricorn*. p.346.

23. *Sa'd-Bula'* (μ, ν, ε Aquarii)

consists of two stars with a third and hardly visible one between them, which looks as if one of them had *devoured* it, so that it glided down from the throat to the breast. According to others, it was called so because Sa'd is considered as he who *devoured* the middle star, robbed it of its light and concealed it. According to Abû-Yaḥyâ b. Kunâsa, this *Station* was called so because it rose at the time when God said: "O earth, *devour* thy water" (Sûra xi. 46). This is a rather subtle derivation. These stars stand on the left hand of Aquarius or Amphora.

24. *Sa'd Alsu'ûd* (β, ξ Aquarii)

consists of three stars, one of which is more bright than the two others. It is called so because people consider its rising as a *lucky* omen, because it rises when the cold decreases, when the winter is past and the season of the continuous rains sets in. Two of these stars stand on the left shoulder of Aquarius; the third one stands on the tail of Capricorn.

25. *Sa'd-Al'akhbiya* (γ, ζ, π, η Aquarii)

consists of four stars, three forming an acute-angled trigone, and one standing in the middle, as it were the centre of a circumscribed circle. The central star is *Sa'd*, and the three surrounding stars are his *tents*. According to others, this Station was called so because at the time when it rises all reptiles that had been *hidden* in the earth come forth. These stars stand on the right hand of Aquarius. God is all-wise!

26. *Alfargh Al'awwal* (α, β Pegasi),

also called the *Upper Handle* (of the bucket), and the *First Two who move the Bucket in the Well* (in order to fill it). It consists of two bright stars, separated from each other, standing on the spine and shoulders of Pegasus.

27. *Alfargh Althânî* (γ Pegasi and α Andromedæ),

also called the *Lower Handle* (of the bucket), and the *Later Two who move the Bucket in the Well* (in order to fill it). It consists of two stars similar to *Alfargh Al'awwal.* According to the Arabs *Amphora* consists of these four stars.

28. *Baṭn-Alḥût* (β Andromedæ),

also called *Ḳalb-Alḥût,* is a bright star in the one half of the womb of a fish (a star) called *Ribbon,* which must not be confounded with the *Two Fishes,* one (the 12th) of the zodiacal signs. These stars stand above Libra and belong to Andromeda (*lit.* the chained wife who had not seen a husband).

The preceding notes we have condensed and have added thereto other notes relating to Lunar Stations; this we have arranged in the form of a table, showing the nature of the Lunar Stations according to the different theories. We have also noted the rising of the stars of the Stations for the year 1300 of Alexander according to mean calculation; this we have also deposited in a table of the conditions of the stars of the Lunar Stations. If you look into these two tables you will find that the superscriptions at the top of each column render it superfluous to consult anybody beforehand as to their use. Here follow the two tables.

pp. 347, 348.

Table of the Lunar Stations.

	The seasons as determined by the setting of the Lunar Stations.	Names of the Lunar Stations.	Distances between the beginning of Aries and that of each Station, according to calculation, no regard being had of the real places of the stars. Zodiacal Signs.	Degrees.	Minutes.	Seconds.	Whether they portend luck or misfortune.	Whether they indicate dry or wet weather.	On what days of the Syrian months they rise.		How many days each Bârih (influence of the rising of a Station) lasts—according to most Anwâ-books.	On what days of the Syrian months they set.		How many days each Naw' (influence of the setting of a Station) lasts—according to most Anwâ-books.	How many days each Naw' lasts, according to Yahyâ b. Kunâsa.	How many days each Naw' lasts, according to Abû Hanîfa Aldînawarî.
1	Stations of Autumn.	Al-sharaṭân	0	0	0	0	Middling	Intermediate	Nîsân	10	1	Tishrîn I.	10	1	3	3
2		Al-buṭain	0	12	51	26	Lucky	Dry		23	1		23	3	3	3
3		Al-thurayya	0	25	42	52	Lucky	Moderate	'Iyyâr	6	4	Tishrîn II.	5	7	5	7
4		Al-dabarân	1	8	34	18	Unlucky	Moist		19	1		18	1	3	3
5		Al-haḳ'a	1	21	25	44	Unlucky	Moist	Ḥazîrân	1	1	Kânûn I.	1	1	6	6
6	Stations of Winter.	Al-han'a	2	4	17	10	Lucky	Intermediate		14	3		14	3	3	3
7		Al-dhirâ'	2	17	8	35	Lucky	Moist		27	1		27	5	3	5
8		Al-nathra	3	0	0	0	Mixed	Intermediate	Tammûz	10	1	Kânûn II.	9	1	7	7
9		Al-ṭarf	3	12	51	26	Unlucky	Dry		23	1		22	6	6	6
10		Al-jabha	3	25	42	52	Lucky	Moist	Âb	5	1	Shubâṭ	4	7	7	7
11		Al-zubra	4	8	34	18	Lucky	Intermediate		19	3		17	3	4	4
12		Al-ṣarfa	4	21	25	44	Unlucky	Moist	Îlûl	1	3	Adhâr	2	3	7	3
13	Stations of Spring.	Al-'awwâ	5	4	17	10	Lucky	Intermediate		14	1		15	1	1	1
14		Al-simâk	5	17	8	35	Middling	Moist, Intermediate		27	1		28	4	4	4
15		Al-ghafr	6	0	0	0	Unlucky	Moist	Tishrîn I.	10	3	Nîsân	10	3	3	1
16		Al-zubânâ	6	12	51	26	Unlucky	Moist, Moderate		23	1		23	3	3	3
17		Al-iklîl	6	25	42	52	Unlucky	Moist	Tishrîn II.	5	4	'Iyyâr	6	4	4	4
18		Al-ḳalb	7	8	34	18	Lucky	Dry		18	1		19	1	1	1
19		Al-shaula	7	21	25	44	Unlucky	Moist	Kânûn I.	1	1	Ḥazîrân	1	3	3	3
20	Stations of Summer.	Al-na'â'im	8	4	17	10	Unlucky	Moist		14	1		14	1	1	1
21		Al-balda	8	17	8	35	Mixed	Moist		27	1		27	1	3	3
22		Sa'd-aldhâbiḥ	9	0	0	0	Lucky	Moist	Kânûn II.	9	1	Tammûz	10	1	1	1
23		Sa'd-Bula'	9	12	51	26	Middling	A little moist		22	1		23	1	1	1
24		Sa'd-alsu'ûd	9	25	42	52	Unlucky	Intermediate	Shubâṭ	4	1	Âb	5	1	1	1
25		Sa'd-al'akhbiya	10	8	34	18	Mixed	Dry, Moderate		17	1		19	1	1	1
26		Alfargh Almuḳaddam	10	21	25	44	Lucky	Dry	Adhâr	2	1	Îlûl	1	3	3	3
27	Stations of Autumn.	Alfargh Almu'akhkhar	11	4	17	10	Mixed	Moist		15	1		14	4	4	4
28		Baṭn-alḥût	11	17	8	35	Lucky	Moist		28	1		27	1	1	1

TABLE OF THE STARS OF THE LUNAR STATIONS.

The names of the Lunar Stations.	The number of their stars.	On what days of the Syrian months of A. Alex. 1300 they rise.		On what days of the Syrian months of A. Alex. 1300 they set.		The different constellations, represented by the stars of the Lunar Stations according to the Astronomers.	The different constellations, represented by the stars of the Lunar Stations according to the Arabs.
1. Al-sharaṭân	2	Nîsân	22	Tishrîn I.	22	The two horns of Aries	The two horns of Aries
2. Al-buṭain	3	'Iyyâr	5	Tishrîn II.	4	Clunis Arietis	The belly of Aries
3. Al-thurayyâ	6		18		17	The bunch of Taurus	Clunis Arietis
4. Al-dabarân	1		31		30	The eye of Taurus	The eye of Taurus
5. Al-haḳ'a	3	Ḥazîrân	13	Kânûn I.	13	The head of Orion	The head of Orion
6. Al-han'a	2		26		26	The two feet of the second twin	The bow of Orion
7. Al-dhirâ'	2	Tammûz	9	Kânûn II.	8	The head of the twins	The out-stretched fore-leg of Leo
8. Al-nathra	3		22		21	Cancer	The nose of Leo
9. Al-ṭarf	2	Âb	4	Shubâṭ	3	The neck of the Lion	The two eyes of Leo
10. Al-jabha	4		17		16	The mane and heart of the Lion	The front of Leo
11. Al-zubra	2		31	Adhâr	1	The root of the tail of Leo	The withers of Leo
12. Al-ṣarfa	1	Îlûl	13		14	The end of the flank of Leo	The tail of Leo
13. Al-'awwâ	4		26		27	The breast of Virgo	The hip of Leo
14. Al-simâk	1	Tishrîn I.	9	Nîsân	9	The hand of Virgo	The calf of Leo
15. Al-ghafr	3		22		22	The train of Virgo	The coat of mail of Scorpio
16. Al-zubânâ	2	Tishrîn II.	4	'Iyyâr	5	The scale of the Balance	The two stings of Scorpio
17. Al-'iklîl	3		17		18	The front of Scorpio	The head of Scorpio
18. Al-ḳalb	1		30		31	The heart of Scorpio	The heart of Scorpio
19. Al-shaula	2	Kânûn I.	13	Ḥazîrân	13	The tongue of Scorpio	The tongue of Scorpio
20. Al-na'â'im	8		26		26	The bow of Sagittarius	The Ostriches
21. Al-balda	0	Kânûn II.	8	Tammûz	9	The body of Sagittarius	A district without stars
22. Sa'd-aldhâbiḥ	2		21		22	The horn of Capricorn	Not belonging to any constellation
23. Sa'd-bula'	2	Shubâṭ	3	Âb	4	The left hand of Aquarius	The same
24. Sa'd-alsu'ûd	3		16		17	The left shoulder of Aquarius	The same
25. Sa'd-al'akhbiya	4	Adhâr	1		31	The right arm of Aquarius	The same
26. Alfargh almuḳaddam	2		14	Îlûl	13	The shoulder and right fore-leg of Pegasus	The upper cross-bar of Amphora
27. Alfargh almu'akhkhar	2		27		26	The wing and navel of Pegasus	The lower cross-bar of Amphora
28. Baṭn-alḥût	1	Nîsân	9	Tishrîn	9	The flank of Andromeda	The womb of Pisces.

(On the interstices between the Lunar Stations.)—The moon's standing in conjunction with a star or with stars which give the name to a Lunar Station and belong to it, is called her *Mukâlaḥa*; it is disliked as foreboding evil. p.351.

If the moon, accelerating her course, passes by (beyond) a Station, or if her course is slackened and she has not yet reached the Station, so that she is seen standing, as it were, in an interstice between two Lunar Stations, this is called the moon's *'Udûl*; and this phase is liked as foreboding something good.

Some of these interstices are called by special names, *e.g.* the interstice between the Pleiades and Aldabarân is called *Alḍaiḳa*. This interstice they consider as a bad omen, foreboding evil. It is called *Ḍaiḳa*, because it sets very rapidly, for between the degree of the setting of the Pleiades and the degree of the setting of Aldabarân there are *six* degrees on the ecliptic, and nearly *seven* degrees on the equator. According to some authors of 'Anwâ-books *Ḍaiḳa* consists of the 21st and 22nd stars of Taurus, which the Arabs call the *Dog of Aldabarân*, but this is not correct.

Sometimes the moon, not reaching *Alhan'a* stands in *Al-taḥâyi*, *i.e.* the 24th, 25th, and 26th of the stars of Gemini. According to others *Altaḥâyi* is identical with *Alhaḳ'a*; whilst others again maintain that it is neither the one nor the other.

Sometimes the moon, not reaching Alsimâk (Spica), stands in her throne of Alsimâk, which some Arabs call the *Backside of the Lion*, *i.e.* the 3rd, 4th, 5th, 7th of the stars of *Alghurâb* (Corvus).

Sometimes, not reaching *Alshaula* (Aculeus Scorpii), the moon stands among the *Kharazât*, i.e. *the vertebræ* of the tail of Scorpius.

Further, not reaching *Albalda*, the moon stands in *Alḳilâda* (Monile), also called *Al'udḥiyy* (Nidus Struthiocameli) *i.e.* the 9th, 10th, 11th, 12th, 13th, 14th, of the stars of Sagittarius. Some people take these stars to be the bow, but they are the head of Sagittarius and his two locks.

Sometimes, not reaching *Sa'd-alsu'ûd*, the moon stands in *Sa'd-Nâshira*, *i.e.* the 23rd and 24th of the stars of Capricornus.

Sometimes, not reaching Alfargh Althânî, the moon stands in *Alkarab*, which means the place where the two cross-woods of the bucket meet, where the string is fastened, *i.e.* the 5th and 7th of the stars of the Great Horse (Pegasus). Or (not reaching Alfargh Althânî), the moon stands in the *Balda of the Fox*, *i.e.* an empty starless region between Alfargh Althânî and Alsamaka (Pisces).

Some one of the authors of 'Anwâ-books thinks that *Al'anîsân*, *i.e.* the 1st and 2nd of the stars of the *Triangulum*, stand between *Baṭn-ulḥût* and *Alsharaṭân*, where he saw them setting *after* Alsharaṭân; therefore he maintains that the moon, not reaching Alsharaṭân, stands in Al'anîsân. But this is wrong, for Al'anîsân stands in Aries more westward (*lit.* at more degrees) than Alsharaṭân. However, the retardation of the setting of Al'anîsân (that they set after Alsharaṭân) was caused by their northern latitude. For it is peculiar to the stars that those which have much northern latitude rise earlier than those that have less, that in consequence the former set earlier than the latter, and *vice versâ* in the south.

Because, now, the fixed stars which give the forms and names to the p.352. Lunar Stations move on in one and the same slow motion, you must add one day to the days of their rising and setting in every 66 solar years, since in such a period they move on one degree. We have represented in a table the places of the stars of the Lunar Stations for A. Alex. 1300, along with the names given to them by the astronomers, with their longitudes and latitudes, and the six degrees of magnitude to which each star belongs. Now, if the reader wants to know the reality about the Lunar Stations, he must correct their places for his time according to the progression we have mentioned, *i.e.* adding one degree for every 66 years. Further, as to their disappearing in the rays (of the sun) and their coming out of the rays, he uses the rules mentioned in the *Canons*. The demonstration of these things is found in the Almagest. The eastward and westward motions of the Lunar Stations differ at the same rate as the latitudes of the countries, further according to the six classes of magnitude to which the stars belong, and according to their distances from the ecliptic. In so doing he will arrive at certain astonishing facts when he has to do with high degrees of latitude north of the ecliptic; *e.g.* when Venus stands in conjunction with the sun in the sign of Pisces, the time of its being concealed under the rays is one day or nearly two days, whilst it is nearly 16 days when she stands in conjunction with the sun in the sign of Virgo.

Mercury is observed in the sign of Scorpius in the mornings as progressing towards the sun, whilst the interstice between them is as much as $\frac{4}{5}$ths of a sign (*i.e.* 24°), and receding from the sun, whilst he is not at all seen in the evenings. The reverse of this takes place when he moves in the sign of Taurus, for then he is observed in the evenings progressing towards the sun and receding from him, whilst he is not seen in the mornings. All these particulars are explained and accounted for in Ptolemy's Almagest.

Here follows the table of the places of the stars of the Lunar Stations.

pp. 353, 354.

Names of the Lunar Stations.	Which places in the 48 constellations the single stars of the Lunar Stations occupy.	The numbers of the stars of the single constellations.	Longitude. Zodiacal Signs.	Longitude. Degrees.	Longitude. Minutes.	Latitude. Degrees.	Latitude. Minutes.	Whether northern or southern latitude.	The sizes according to Abû-alḥusain Alṣûfî.
Alsharaṭân	The foremost of the two stars which are the horns of Aries	1	0	18	13	7	30	north	3
	The following one	2	0	19	13	8	20	north	3
Albuṭain	The star on the root of the backside of Aries	7	1	2	53	4	50	north	5
	The foremost of the three stars in the backside of Aries	8	1	5	23	1	40	north	4
	The star on the second thigh of Aries	11	1	1	18	1	10	north	5
Althurayyâ	The northern end of the foremost rib of Althurayyâ in Taurus	29	1	13	43	4	30	north	5
	The southern end of the foremost rib	30	1	14	3	3	40	north	5
	The second end of Althurayyâ, the narrowest place in the constellation	31	1	15	13	3	20	north	5
	The small star outside Althurayyâ to the north	32	1	15	13	5	0	north	4
	Two stars not mentioned by Ptolemy, nor by any astronomer before him or after him		1	not determined by observation				north	4
			1					north	4
Aldabarân	The star on the southern eye of Taurus	14	1	24	13	5	10	south	1
Alhaḳ'a	The nebula on the head of Orion. Ptolemy counts the middle of the Triangle as one star	1	2	3	33	13	50	south	nebula
Alhan'a	The star on the left foot of the second Twin	17	2	23	33	7	30	north	3
	The star on the right foot of the second Twin	18	2	26	13	10	30	north	4
Aldhirâ'	The star on the head of the foremost Twin	1	3	4	53	9	40	north	2
	The star on the head of the second Twin	2	3	8	13	6	15	north	2
Alnathra	The centre of the nebular cluster in the breast of Cancer	1	3	21	33	0	20	north	nebula
	The most northern of the following two stars	4	3	21	33	2	40	north	4
	The most southern of them	5	3	22	33	0	10	north	4
Alṭarf	One of the four stars outside Cancer, following after the end of the southern sting	2	4	3	13	5	40	south	4
	The star on the cheek of Leo	2	4	2	43	7	30	north	4
Aljabha	The most northern of the three stars on the neck of Leo	5	4	11	43	11	0	north	3
	The following star, the middle one of the three	6	4	13	48	8	30	north	2
	The most southern one of them	7	4	12	13	4	30	north	3
	The star on the heart of Leo, also called *Regius (Regulus)*	8	4	14	3	0	10	north	1
Alzubra	Of the two stars on the belly of Leo the following one	20	4	25	43	13	40	north	2
	The northern star of the two stars on the upper parts of the haunches of Leo	22	4	27	53	9	40	north	3
Alṣarfa	The star on the end of the tail of Leo	27	4	6	3	11	50	north	1
Al'awwâ	The star on the end of the left wing of Virgo	5	5	10	33	0	10	north	3
	The following star	6	5	19	18	1	10	north	3
	The next following star	7	5	24	43	2	50	north	3
	The most northern of the three stars in the right wing, the *Provindemiator*	13	5	23	43	15	10	north	3
Alsimâk	The star on the left palm of Virgo	14	5	8	13	2	0	south	1
Alghafr	The middle one of the three stars on the train of Virgo	22	5	18	13	7	30	north	4
	The southern one of them	23	5	18	33	2	40	north	4
	The star on the left, the southern foot of Virgo	25	5	21	33	0	33	north	4

[*Continued on next page.*]

[*Continued from previous page.*]

Names of the Lunar Stations.	Which places in the 48 constellations the single stars of the Lunar Stations occupy.	The numbers of the stars of the single constellations.	Longitude.			Latitude.		Whether northern or southern latitude.	The sizes according to Abû-alḥusain Alṣûfî.
			Zodiacal Signs.	Degrees.	Minutes.	Degrees.	Minutes.		
Alzubânâ	The most brilliant of the two stars on the end of the southern sting, *i.e.* on the southern scale of Libra	1	6	29	33	0	40	north	3
	The most brilliant of the two stars on the end of the northern sting, *i.e.* on the northern scale	3	6	3	43	8	50	north	3
Al'iklîl	The most northern of the three brilliant stars on the forehead of Scorpius	1	7	17	53	1	20	north	3
	The middle one of them	2	7	17	13	1	20	south	3
	The most southern of them	3	7	17	13	5	0	south	3
Alḳalb	The heart of Scorpius	8	7	24	13	4	0	south	2
Alshaula	Of the two stars in the sting of Scorpius, the following one	20	7	9	3	13	20	south	3
	The preceding one	21	7	8	33	13	30	south	3
Alna'âm Alwârid	The star on the top of the arrow of Sagittarius	1	8	16	3	6	20	south	3
	The star in the handle of the left hand	2	8	19	13	6	30	south	3
	The star on the southern side of the bow	3	8	19	33	10	50	south	3
	The star on the foremost right ankle	25	8	18	13	13	0	south	3
Alna'âm Alṣâdir	The star on the left shoulder of Sagittarius	6	8	26	53	3	10	south	3
	The preceding star, on the arrow	7	8	24	33	3	50	south	4
	The star on the neck, the middle one of the three stars on the back	21	8	29	13	4	30	south	4
	That one of these three stars which stands under the armhole	22	8	27	53	6	45	south	4
Albalda	A space without stars, lying in the south near the 11th and 12th stars of Sagittarius.								
Sa'd Aldhâbiḥ	The most northern of the three stars in the second horn of Capricorn	1	9	18	53	7	20	north	3
	The most southern of them	3	9	18	53	5	0	north	3
Sa'd Bula'	The middle one of the three stars on the left hand of Aquarius	7	10	27	43	8	0	north	5
	The preceding one of them	8	10	26	13	8	40	north	4
Sa'd Alsu'ûd	The star on the left shoulder of Aquarius	4	10	6	3	8	50	north	3
	The star under it on the back, standing as it were under the armhole	5	10	8	53	6	15	north	5
	The star on the end of the tail of Capricorn	28	10	10	13	4	20	north	5
Sa'd Al'akhbiya	The star in the right arm of Aquarius	9	10	21	3	8	45	north	3
	The most northern of the three stars in the right palm of Aquarius	10	10	23	13	10	45	north	4
	The foremost of the two other stars	11	10	23	33	9	0	north	4
	The next following one to the south	12	10	24	53	8	30	north	4
Alfargh Almuḳaddam	The star on the right shoulder of Pegasus and on the root of his foot	3	11	13	43	31	0	north	2
	The star on the back of Pegasus between the two shoulder-blades	4	11	8	13	19	40	north	2
Alfargh Almu'akhkhar	The star on the navel of Pegasus, belonging in common to Pegasus and to the head of Andromeda	1	11	29	23	26	0	north	2
	The star on the back and the end of the wing	2	11	23	43	12	30	north	2
Baṭn-alḥût	The most southern of the three stars above the drawers of Andromeda	12	11	15	23	26	20	north	2

(On projection, and the construction of star-maps.)—I have p.357. followed in this book a method which the student of this science will not disapprove, treating in each chapter the subject as fully as possible, and not referring the reader to other books until I had myself nearly exhausted the subject. Now, I must add to the book another chapter on the representation of the Lunar Stations and of other constellations *on even planes*, for the human mind, once knowing at what different times the different stars rise, forms an idea as to the positions which they occupy in the ecliptic. Our remarks in the preceding pages will enable the student to recognise the stars of the Lunar Stations by eye-sight, and to point them out. However, not everyone who requires these things knows the positions of the ecliptic. Besides, the representation of the Lunar Stations as well as the other stars comprehended by the 48 constellations (on an even plane), offers many conveniences in common to all classes of scholars. The same applies to the representation of countries, cities, and what else there is on earth, on an even plane. Therefore, not knowing any special treatise on this subject, I shall treat it myself, mentioning whatever occurs to my mind. The reader, I hope, will excuse!

The projection of great and small circles and points on globes may be done in this way, that you make one of the two poles the top of cones, the envelopes of which pass through them (the circles and points), and cut a certain plane which is assumed. For the parts (lines or points) which are common to this plane and the envelopes of these cones if they pass through circles, *or* common to this plane and the lines (of this cone) if they pass through points, are their projections on this even plane.

This is the method of the astrolabe (stereographic polar projection), for in the north the southern pole is made the top of the cones, and in the south the northern pole is made the top of the cones, and the plane which we want to find (the plane of projection) is one of the planes parallel with the plane of the equator. Then they (*i.e.* the cones) represent themselves as circles and straight lines.

'Abû-Ḥâmid Alṣaghânî has transferred the tops of the cones from the two poles, and has placed them inside or outside the globe in a straight line with the axis. In consequence the cones represent themselves as straight lines and circles, as ellipses, parabolas, and hyperbolas, as he (Abû-Ḥâmid) wants to have them. However, people have not been in a hurry to adopt such a curious plane. (This is the central projection, or the general perspective projection.)

Another kind of projection is what I have called *the cylindrical projection* (orthographic projection), which I do not find mentioned by any former mathematician. It is carried out in this way: You draw through the circles and lines of the globe lines and planes parallel to the axis. So you get in the day-plane straight lines, circles, and ellipses (no parabolas and hyperbolas). All this is explained in my book, which gives a complete representation of all possible methods of the construction of the astrolabe.

However, lines, circles, and points do not represent themselves in the same way on a plane as on a globe; for the distances which are equal on a globe differ greatly in a plane, especially if some of them are near to the one pole and others to the other pole. But it is not the purpose of the astrolabe to represent them (the lines, circles, points) as agreeing with eye-sight, but to let some of them revolve whilst the
p.358. others are at rest, so that the result of this process agrees with the appearances in heaven, including the difference of time. On the other hand, the purpose of the representation of the stars and countries (on even planes) is this, to make *them* correspond with their position in heaven and earth, so that in looking at them you may form an idea of their situation, always keeping in mind that the straight lines are not like the revolving (circular) lines, and that the spherical planes have no likeness to the even planes that are equal among each other.

We must give an illustration to make the reader familiar with these methods. One way serving for this purpose is the construction of the flat astrolabe.

Draw a circle as you like it, the greater the better. Divide it into four parts by two diameters which cut each other at right angles. Divide one of the radii into 90 equal parts. Then we make the centre of the circle a new centre, and describe round it circles with the distances of each of the 90 parts. These circles will be parallel to each other, and will be at equal distances from each other. Divide the circumference of the greatest circle into the (360) parts of a circle, and connect each part of them and the centre by straight lines.

In doing this we imagine the periphery of this first circle to be the ecliptic, and its centre to be one of the poles of the ecliptic. On the ecliptic we mark a point as the beginning of Aries. Then we fix the places of the stars according to Almagest, or to the Canon of Muḥammad b. Jâbir Albattânî, or to the *Book of Fixed Stars*, by 'Abû-Alḥusain Alṣûfî, taking into account the a[illegible] of precession up to our time, and changing accordingly the places of the stars as determined by our predecessors. Take one of the stars of that half (of heaven) for which you have constructed this circle, and count from this assumed point (the beginning of Aries), proceeding from right to left, as many degrees as the star is distant from Aries. That place where you arrive is the degree of this star *in longitude*.

Further count, from the same point in a straight line which extends to the centre, the corresponding number of the star's latitude in the 90 circles. Then the place you arrive at is the place of the body of the stars (*i.e.* the point determined by both the degrees of longitude and latitude). There you make a dot of yellow or white colour, according to the class of magnitude and brilliancy to which in the six classes the star may belong.

The same process you repeat with every star the latitude of which lies in the same direction, till you have finished all the stars of this direction. The same you continue to do with the stars of another direction, until you have fixed the stars of the whole sphere in two circles. We mark these circles with the blue of lapis lazuli, in order to distinguish them from the stars, and we draw round the stars of each constellation the image which the stars are believed to represent, after having fixed all the stars in their proper positions. In this way the object we had in view is realized.

This method, however, we do not like, because the figures on the ecliptic cannot completely be represented, since some parts of them fall into this half, some into the other half. If you drew round the circle p.359.
of the ecliptic, outside of it, 90 circles, parallel to and distant from each other as far as in the former construction, in the same way as is done with the flat astrolabe, the matter would evidently proceed in the same order. Further, we do not like this method, as the places of the stars in heaven and those in the design (drawing) greatly differ from each other. For the more southern the stars are, the distances between them which appear equal to the eye are the greater and wider in the figure, if its centre be the north pole, till at last they assume quite intolerable dimensions. The same applies to the method of him who wants to represent the stars in the plane of a circle which passes through the two poles of the ecliptic, in those points where the straight lines of their heights touch the plane (*i.e.* the foot-points of the verticals), which method is similar to the astrolabic projection; for then the figures of the stars are in an undue manner compressed towards the periphery, and they become too large about the centre.

We shall now try to find another method, which is free from the inconveniences of the process just mentioned.

We draw a circle, divide it into four parts (by two diameters cutting each other at right angles), and upon the points of the four parts (*i.e.* where the diameters touch the periphery) we write the names of the directions (*i.e.* north, west, south, east).

We continue the two diameters that divide the circle into fourths, straight on in their directions in infinitum.

Each radius we divide into 90 equal parts, and the periphery into 360 parts.

Next we try to find on the line of the east and west the centres of circles, each of which passes through one of the parts (degrees) of the diameter and through both the north and south poles.

When these centres have been fixed, and we draw all the possible circles round them within that first (and largest) circle, we get 180 arcs, which divide the diameter into equal parts, and which cut each other at each of the two points, the north and south points.

These circles are the circles of longitude.

Then we return to that line which proceeds from the north point as the straight continuation of the diameter. On this line we try to find the centre of a circle, passing through those points of the periphery which are distant from the east and west points 1 degree, 2 degrees, etc. until 90 degrees, and through those points of the diameter which are distant from the centre 1 degree, 2 degrees, etc. until 90 degrees.

The same we do in the southern half, on the line which proceeds from the south point as the straight continuation of the diameter.

The circles we get in this way are the circles of latitude, 180 in number, which divide each of the circles of longitude into 180 parts.

Further, we assume the west point to be the beginning of Aries, and the line from east to west to be the ecliptic. From the beginning of Aries we count the distance of each star; so we find its degree (of longitude).

Then we count the latitude of the star in the proper direction on the circle of longitude. Thereby we find the place of the star.

We make another figure similar to the first one, where we assume the west point to be the beginning of Libra In this way we can give a complete map of all the stars in the two figures. Lastly, in representing the single star-groups or constellations, we draw those images which we have heretofore described.

If we want to make a map of the earth, we construct a similar figure as described in the preceding. We count the assumed longitude of a place from the west point, and then we count the degrees of latitude of
p. 360. the place on the circle of longitude. So we find the position of the place. The same we continue to do with other places.

This is the technical (graphic) method for the solution of this problem.

As some people have a predilection for calculations, and like to arrange them in tables, and prefer them to technical (graphic) methods, we shall also have to show how we may find, by calculation, the diameters of the circles of longitude and latitude, and the distances of their centres from the centre of the (great) circle. And with that we shall finish our work.

We draw the circle ABCD round the centre H, and divide it into four parts by means of the two diameters AHC and BHD.

A is to be the west.
B " south.
C " east.
D " north.

The radii we divide into 90 parts, and the whole circle into 360 parts.

Now, *e.g.* we want to find the radius of the circle BZD, which is one of the circles of longitude, and the distance of its centre (from the centre H).

Now, it is evident that HZ is known, being determined by the degrees, of which the radius HC as well as the radii BH and HD hold 90.

The multiplication of HZ, which is known, by the unknown sum of HK+KZ, which is the diameter we want to find, *minus* ZH,

is equal to

the multiplication of HB by HD, *i.e.* the square of one of them.

We take the square of HB, *i.e.* 8,100, and divide it by ZH, which is known. Thereby we get the sum of HK+KZ. To this we add ZH, and take the half of the whole sum. That is ZK, the radius of that circle to which BZD belongs.

Now, after having found out so much, we open the compasses to such an extent as the new-found radius is long; one of the legs of the compasses we place on the point Z, which is known, and the other leg we place on the continuation of the line HA, to whatever point it reaches. The latter point is the centre of the circle, *i.e.* K.

In this way we can dispense with the knowledge of the distance between the two centres.

(In the following the text is corrupt.)

This is the solution of the problem by means of calculation.

If you want to find the distance of the passage, *i.e.* that point on the periphery of the circle where the line which connects the two points B and K cuts the periphery, viz. the arc AT, draw the line BK which cuts the periphery in T, draw the vertical (*i.e. Loth-Linie,* the line which represents the height of a trapeza or cone) TS upon BD, and draw the line TD.

Because, now, in the triangle BHK the sides are known according to the parts, of which the radius counts 90, we change each side into that measure, according to which the radius counts 60 degrees, *i.e.* we multiply it by 60 and divide it by 90. So it is changed into the sexage- p.361.
simal system.

The triangles BHK and BTD and BST are similar to each other. Therefore we multiply KH by BD, and the product we divide by KB. So we get DT as the quotient.

Next we multiply DT by HK, and divide the product by KB. So we get DS as quotient (changed into that measure, according to which DT holds 60 parts).

If we take the corresponding arc in the table of sines and subtract the arc from 90, we get AT as remainder.

If we want to find the passage (T) by an easier method, we change the triangle BHK, the sides of which are known, into that measure according to which the radius of the circle ABCD contains 60 parts. Then the angle TDB in the first figure, and the angle TBD in the second figure, is that which determines the whole distance of the passage (from A), as a chord determines the distance between the two ends of an arc.

If we want to change each side of this triangle into the measure according to which BK holds 60 parts, we multiply it by 60 and divide the product by BK, according to the measure of which the radius holds 60 parts. So we find what we wanted to find.

If we, then, know the side HK according to this measure, we take the corresponding arc from the table of sines. So we get the arc DT.

So, by whatever method we solve the problem, the end we aim at is the same, and also the results.

Here follow Figures I. and II.

I.

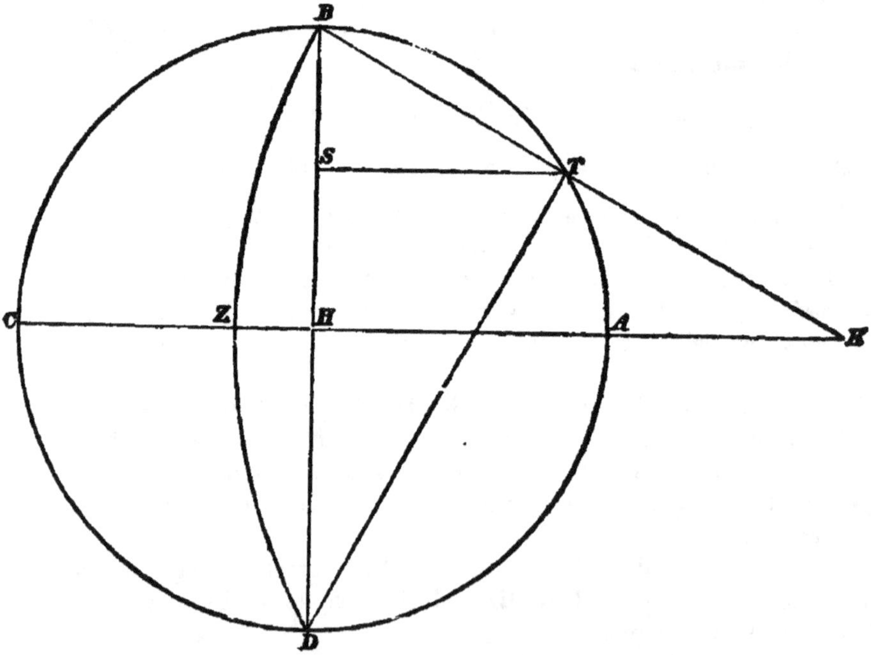

II.

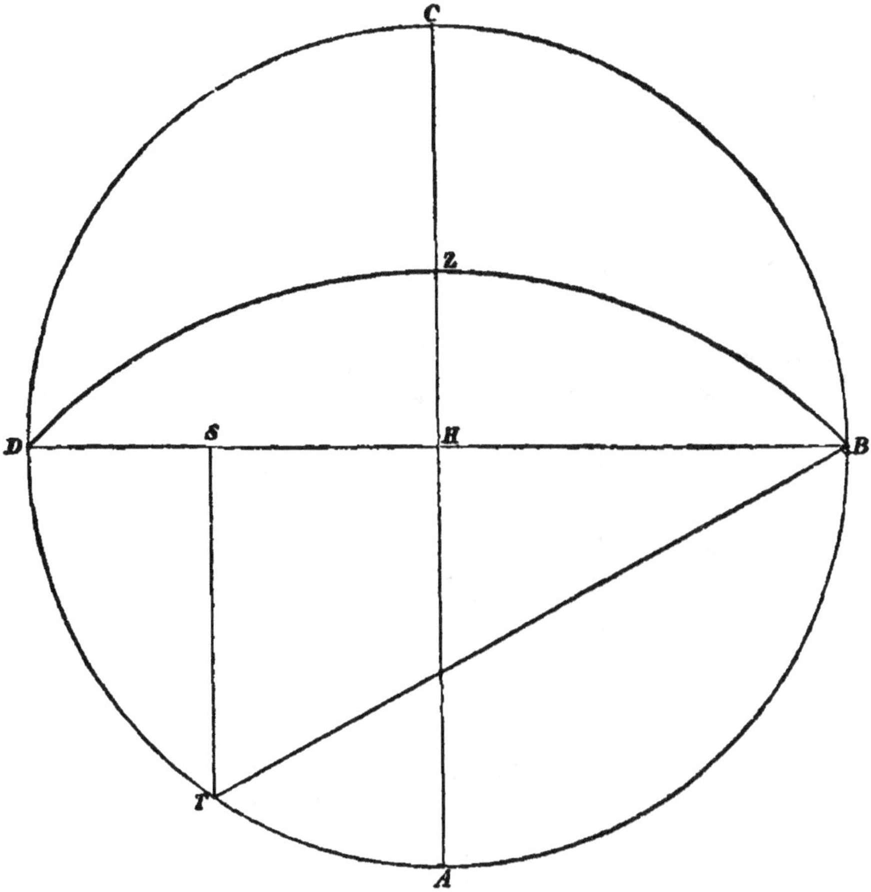

Let us again construct the same figure to show the same thing for the circles of latitude.

The circle of which we want to find the radius is that one to which MQL belongs. AM as well as HQ and CL are to correspond in number (*i.e.* to hold the same number of parts).

We draw the vertical (Loth-Linie) MX, which is the sine of DM, which is known, and we draw HX, *i.e.* the sine of AM, which is also known.

Then we subtract HQ from HX, after we have changed it from the nonagesimal system into the sexagesimal system. The remainder we get is QX.

By this we divide the square of MX, and add QX to the product. Of this sum we take the half, which is QK, the radius of the circle to which MQL belongs, according to the measure of which the radius of the circle ABCD holds 60 parts.

If we want to find the distance of the passage (T) from A, we draw the line AK, which cuts the periphery of the circle in T.

Further, we draw the line TC and draw the vertical (Loth-Linie) TS upon AC.

Then we multiply AC by HK and divide the product by AK. Thereby we get TC.

If we multiply this divisor by HK, and divide the product by AK, we get SC.

Multiply it by AS, and the root of the product is TS, which is the sine of the arc of the passage (*i.e.* of the arc AT).

Likewise, if we change AH into the measure according to which AK holds 120 parts, and we take the arc from the tables of whole chords, we get the arc AT, *i.e.* the distance of the passage (from A).

This method applies in the same way to the direction of C as to the direction of A, to that of B as to that of D, without the slightest difference.

And here ends my work. Here follows Figure III.

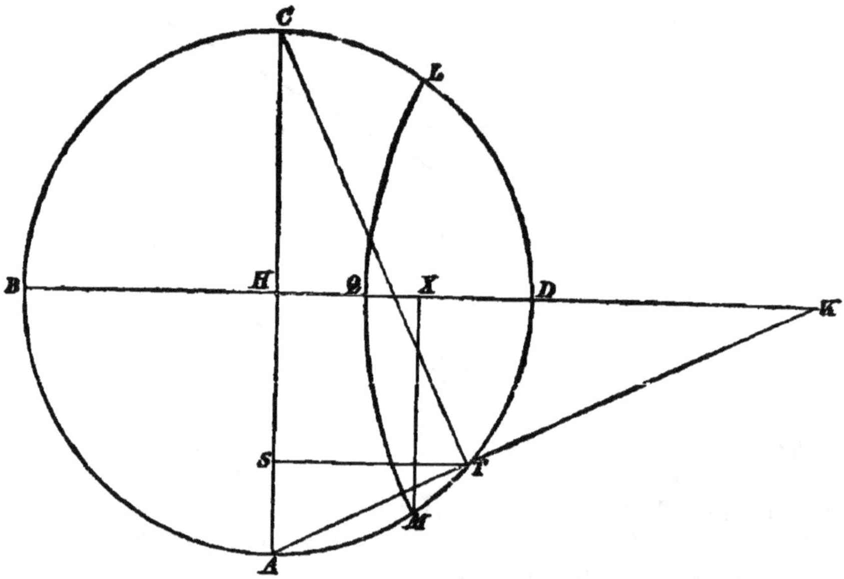

p.362. **(Conclusion.)**—Now I have fulfilled my promise, and I have comprehended in my exposition all the parts of this science, agreeably to the wishes of my friends, exerting myself to the best of my capability. Every man acts according to his fashion, and the value of a man lies in that which he understands. I hope that the elements which I have laid down are sufficient to train the mind of the student, and to lead him to a correct consideration of the *origines* of mankind, sufficient to lay open all that is doubtful in the eras of prophets and kings, and to give a correct idea of their own system to those of the Jews and Christians who are led astray.

If the reader be like me (in knowledge), he will thank me for the task I have carried out; if he be superior to me (in knowledge), he will be so kind as to correct my errors and to pardon whatever mistakes I may have made. If he be inferior to me in knowledge, he will not do me any harm, because he will either acquiesce in being led by me for the purpose of his instruction, or, in case he opposes me, he will offer opposition to things which he has not the power of mind to handle successfully.

But why should I mind—or be afraid of—the enmity of any adversary, since my badge is, wherever I am, the power of our lord, the noble prince, the glorious and victorious, the benefactor, *Shams-alma'âli*—may God give long duration to his power! Its firm column is my trust; from the fact that it spreads secretly and openly I derive strength; its brilliant light is the guide of my path; his undisturbed happiness is my trust and my hope. May God teach me and all Muslims to be truly thankful for his benefits by fulfilling all the duties of obedience as prescribed by the law, and by continually praying to God that He may reward him according to His mercy and grace.

Let us finish our book with the praise of God, who afforded me help and guidance, and who taught me to distinguish the path of truth from the path of blindness. "Let those who want to perish (as infidels, idolaters,) perish, after a clear proof (of the *true* religion) has been presented to them, and on the strength of it, and let those who want to live (the life of the *true* religion) live, after a clear proof (of the *true* religion) has been presented to them, and on the strength of it." (Coran, S. viii. 44).

The mercy and blessing of God be in all eternity upon the Prophet who was sent to the best of nations, and upon his holy family!

ANNOTATIONS.

p. 1, l. 25. *Shams-alma'âlî.* This prince, Kâbûs ben Washmgîr ben Mardâwîj, who had received from the Khalif the honorary name of *Shams-alma'âlî*, i.e. *Sun of the Heights*, belonged to the family of the Banû-Ziyâd, who ruled over Jurjân (Hyrcania), Ṭabaristân, and other countries south of the Caspian Sea during 155 years, viz. A.H. 315–470. Kâbûs, after having reigned A.H. 366–371, was driven away and fled into the dominions of the Sâmânian dynasty, where he lived as an exile, whilst his country was occupied by a prince of the family of Buwaihi A.H. 371–388. To this period the author alludes, p. 94, l. 19 ("at the time of the Ṣâḥib (*i.e.* Ṣâḥib Ibn-'Abbâd the Vazîr of the Buyide prince) and when the family of Buwaihi held the country under their sway").

Kâbûs returned to the throne A.H. 388, and was killed 403. This book was dedicated to him according to all probability A.H. 390 or 391 =A.D. 1000. The history of this prince is found in Sehir-eddin's "Geschichte von Tabaristan," etc., ed. Dorn, pp. 185–198.

p. 2, l. 27. *Wading.* Although all three MSS. have تدوّخ, I think it would have been more idiomatic to read تدويخ. The second form occurs ٣٦, 17; ٣٧, 7; ٣٠, 19.

p. 3, l. 14. *Which have come down from them.* The words احوال عنهم are very bad Arabic, next to impossible. It seems rather likely that between them a word has fallen out, *e.g.* صادرة, or some synonym of باقية.

p. 5, l. 15. *Nychthemeron.* To meet the inconvenience that the word *day* means the totality of day and night as well as the light half alone, I have ventured to adopt for the former meaning the word *Nychthemeron*, for which I beg the reader's pardon.

p. 6, l. 37. *Canon of Shahriyârân Alshâh.* The word *Canon* (so I translate the word *Zîj*) means a collection or handbook of astronomical tables of various kinds. They were always the depositories of the latest discoveries of Eastern astronomy. For more information on these *Canons* I refer to the excellent work of L. A. Sêdillot, " Prolégomènes des Tables Astronomiques d'Olough-Beg," Paris, 1847, p. viii. ff. (*Table vérifiée*, etc.)

A *Canon of Shahriyârân Alshâh* is not known to me. However, there is a *Zîj-i-Shahryâr* and a *Zîj-Alshâh*, with either of which this Canon may be identical.

The former was of Persian origin, and was translated into Arabic by Altamîmî, *vide* " Hamzæ Ispahanensis Annalium," libri x., p. ٣, and "Kitâb-alfihrist," ed. Flügel, p. 241, 244, and notes.

The second, with several other canons, was composed by the famous mathematician and astronomer, a native of Marw, Ḥabash ('Aḥmad ben 'Abd-Allâh), who lived in Bagdâd, end of the second and beginning of the third century. Cpr. Sédillot, "Prolégomènes," p. x., note 3, and the " Kitâb-Alfihrist," ed. by G. Flügel, p. ٢٧٥, and the annotations.

p. 6, l. 42. *A variation which during the eclipses*, etc. I have not been able to ascertain what relationship between the eclipses and the different length of the days is meant by the author in this passage.

p. 7, l. 39. *Except one Muslim lawyer.* The following discussion is of more interest to a Muḥammadan theologian than to us. The canonical time of fast was from the rise of dawn till sunset, and some too pious people mistook this *day of fast* for the astronomical day.

p. 8, line 22. *Likewise it had been forbidden*, etc. This passage refers to a custom which existed among the Muslims in the time before the verse Sûra II., 183, was pronounced by Muḥammad. We learn from the commentaries of Alzamakhsharî and Albaiḍâwî that the time of fast extended over the whole Nychthemeron, except the time from sunset till the second or last night-prayer, *i.e.* about midnight, or till a man fell asleep, during which eating and drinking and women were allowed. Omar once had intercourse with one of his wives after he had said the last night-prayer; this breach of the custom made him feel penitent, and he apologized to the Prophet. Thereupon the Prophet abolished the old custom, and pronounced the verse in question, allowing his people to eat and drink and have intercourse with their women not only from sunset till the second night-prayer, but also farther till the rise of dawn, *i.e.* nearly the whole night. For the traditions concerning this subject *vide* " Bokhârî," ed. Krehl, I., ٤٧٧, ٤٧٨.

In my translation, p. 8, l. 23, read " *after the last night-prayer* " instead of " *after night-prayer.*"

p. 9, l. 14. *Tradition which relates*, etc. It is not quite clear to what tradition the author refers in this passage. Prof. L. Krehl kindly directed my attention to a tradition which occurs several times in Albokhârî ("*Recueil des Traditions Mahométanes*," publié par L. Krehl), *e.g.* i. 149, ii. 50, etc. Here Muḥammad compares Jews, Christians, and Muslims to workmen. The Jews work from sunrise to noon, and receive one Ḳîrâṭ as wages. The Christians work from noon till the afternoon-prayer, صلوة العصر and receive one Ḳîrâṭ. The Muslims work from the afternoon-prayer till sunset, and receive two Ḳîrâṭ. So the Muslims receive twice the wages of the Jews, whilst they only work half the time.

To a similar tradition the author seems to refer. He calls the Muslims "*those who hasten to the mosque on a Friday*," thereby distinguishing them from Jews and Christians.

In any case this tradition must have proved, according to the author, that Muḥammad represented the day, whether it be short or long, as divided into twelve equal parts, the so-called ὧραι καιρικαί.

p. 9, l. 36. *What is he*, etc. The reading of the MSS. واين, I have changed into اين. It seems preferable, however, to read فاين, as Prof. Fleischer suggests.

p. 9, l. 46. *The prayer of the day is silent* (or, rather, *mute*).

The two prayers of the night (صلوة المغرب and صلوة العشاء) are جَهْرِيَّتَان i.e. these prayers are spoken with a clear audible voice. The two prayers of the day (صلوة الظهر and صلوة العصر) are سِرِّيَّتَان *i.e.* these prayers are not spoken with an audible voice or whisper; the lips move, but no sound is produced. Therefore they are called العَجْمَاوَان *i.e. the two silent or mute ones*, which name is derived from the tradition quoted by the author.

This division does not comprehend the morning-prayer صلوة الصبح. But it can be proved from tradition that this prayer is to be spoken with an audible voice. In the *Muwaṭṭa'* of Mâlik ben 'Anas (published with the commentary of Alzurḳânî at Bûlâḳ, A.H. 1279, tom. i. p. ١٥٥), Alfurâfiṣa ben 'Umair Alḥanafî relates that he learned the Sûrat-Yûsuf, *i.e.* Sûra xii. by hearing it from Omar, who recited it repeatedly as his *morning prayer*.

The author wants to prove that the صلوة الصبح is not a day-prayer, because it is spoken at the rise of dawn, *i.e.* before the beginning of the day. According to holy tradition *the prayer of the day is silent* (or *mute*), whilst the morning prayer is not.

p. 10, l. 1. *The "first" prayer*, etc. The five canonical prayers of the Nychthemeron are these:

صلوة العشاء الاولى or صلوة المغرب
صلوة العشاء الآخرة or صلوة العشاء } Prayers of the night.
. . . صلوة الصبح

صلوة الظهر
صلوة العصر } Prayers of the day.

The author's argument is this, that صلوة الصبح or morning-prayer is not a prayer of the day, because the صلوة الظهر is called the first, *i.e.* the first of the two day-prayers, and because the صلوة العصر is called the *middle* prayer, *i.e.* the middle between the first day-prayer and the first night-prayer. If the صلوة الصبح belonged to the day-prayers, the صلوة العصر would not be the exact *middle* in the way we have described, for in that case the صلوة الظهر as well as the صلوة العصر would be in the midst between the first day-prayer and the first night-prayer.

p. 11, l. 12. *Sindhind*. An astronomical hand-book of Indian origin, edited the first time by Alfazârî, A.H. 154, and a second time by Albêrûnî's famous countryman Muḥammad ben Mûsâ Alkhwârizmî. Albêrûnî wrote a book (commentary?) on the Sindhind with the title جوامع الموجود لخواطر الهنود فى حساب التنجيم. For the literature on this subject I refer to Flügel's "Kitâb-alfihrist," p. 274, notes.

The word *Sindhind* is supposed to be the Sanskrit *Siddhânta*. I must, however, observe that Albêrûnî writes this word in the more correct form of سدهاند, *e.g.* in the title of his book, "On the mathematical methods of the *Brâhmasiddhânta*," ترجمة ما فى براهم سدهاند من طرق الحساب.

p. 11, l. 15. *The four seasons*. Read الاربع instead of الاربعة.

p. 12, l. 38. The quotation of Theon refers to the introduction of his Πρόχειροι κανόνες, where he speaks of the Julian year of the people of Alexandria, of the Egyptian year of 365 days, and of the Sothis period of 1460 years; *vide* "Commentaire de Théon d'Alexandrie sur les tables manuelles astronomiques de Ptolemée," par M. l'Abbé Halma; Paris, 1822, p. 30. On the Sothis period, *vide* R. Lepsius, "Chronologie der Aegypter," Berlin, 1849, p. 165 ff.

p. 12, l. 40. On the year of the Persians, cf. a short treatise of A. v. Gutschmid, *Ueber das iranische Jahr*, "Sitzungsberichte der Kgl. Sächsischen Gesellschaft der Wissenschaften," 1862, 1 July.

p. 13, l. 22. *The Hebrews, Jews, and all the Israelites*. It is difficult to explain what differences the author meant to express by these three words, which to us mean all the same. Perhaps he meant by *Hebrews* the ancient Jews, Samaritans, and other kindred nations; by *Jews*, the monotheistic people in particular; and by *all the Bani-Israel*, the totality of the Jewish sects, Rabbanites, 'Anânites (Karäer), and others. *Vide* a similar expression on p. 62, ll. 16, 17.

p. 13, l. 34. *In a similar way the heathen Arabs*, etc. Cf. with Albêrûnî's theory, the description of ancient Arabic chronology by A. Sprenger, "Leben und Lehre des Mohammad," iii. p. 530 ff., and "Zeitschrift der Deutschen Morgenländischen Gesellschaft," tom. xiii. p. 134, and tom. xxxi. p. 552.

p. 13, l. 42. The genealogy of the Ḳalâmis occurs also, but with some differences, in *Ibn Hishâm*, "Life of Muḥammad," ed. Wüstenfeld, i. pp. 29, 30.

p. 14, l. 16. The text of this verse is incorrect, for, according to the context, it does not contain a description of Fuḳaim, but of 'Abû Thumâma; and, secondly, the metre is disturbed, for I do not think that the *licentia poetica* allowed a poet to distort the word القَلَمَّس into القَلَمَّسا.

p. 15, l. 13. *The leap-year they call Adhimâsa.* According to the double construction of the verb سمّى, the Indian word may either be بذماسة or ذماسة. It seems preferable to read باذماسة, and to explain اذماسة. as *Adhimâsa*, although it must be observed that this means "intercalary month," not "intercalary year," as the author maintains. Cf. Reinaud, "Mémoire sur l'Inde," p. 352.

p. 15, l. 15. *And their subdivisions.* Read جفورها (as depending on لاستعمالهم), *i.e.* the *Jufûr* of the Lunar Stations.

Our dictionaries do not explain the meaning which the word *Jafr*, pl. *Jufûr*, has in this passage. It is a term peculiar to the Indian system of astrology ('Uṭârid ben Muḥammad wrote a book *On the Indian Jafr*, كتاب الجفر الهندى, *vide* "Kitâb-alfihrist," p. 278), and it means something connected with the Lunar Stations, perhaps certain *subdivisions* (but this translation of mine is entirely conjectural).

The word occurs in four other places in this book:—p. 336, l. 9. The Indians derive their ἀστρολογούμενα from the fact of the stars entering the *Ribâṭât* (*i.e.* resting-places, road-side inns) of the *Lunar Stations.* These *Ribâṭât* are called *Jufûr*, and each of them is thought to refer to some special matter or event (life, death, travelling, victory, defeat, etc.). Their number must have been very great, because the author says that he will refrain from enumerating them, as this would detain him too long from the subject of his book. p. 338, l. 14: The author, speaking of the same subject, mentions the Ribâṭât and the *Jufûr* of the Lunar Stations side by side; also, p. 341, l. 7. On p. 347, both words occur in the superscription of column 5, but here the writing of the manuscript is such a bad scrawl that I do not feel sure of having made out a correct text.

As the subject-matter seems to be of Indian origin, one may presume that the word also is derived from the same source.

p. 15, l. 17. *Abû Muḥammad Alnâ'ib Alâmuli* (*i.e.* a native either of Âmul in Ṭabaristân or of Âmul or Amû on the Oxus), the author of a *Kitâb-alghurra* is mentioned four times, *vide* p. 53, l. 34; p. 235, l. 8; p. 344, l. 2. He is not known to me from other sources.

p. 16, l. 14. *Cannot be dispensed with.* Read غَنِى instead of غَنَى (Fleischer).

p. 17, l. 6. *A tradition for which*, etc. Instead of مشفوع به (١٣, 4), read مشفوع or مشفوع له (Fleischer).

p. 17, l. 8. Perhaps we shall facilitate the understanding of the following pages, if we state the order of the author's arguments.

A. Notions of the Persians regarding the Era of Creation, p. 17, l. 8.
B. Notions of the Jews on the same subject, p. 18, l. 5.
C. Notions of the Christians, p. 19, l. 10.
D. Refutation of the Jewish theory, p. 19, l. 41.
E. Refutation of the Christian theory, p. 21, l. 5, and Biblical prophecies relating to Muḥammad, p. 22, l. 17.
F. On the Thora of Jews, Christians, and Samaritans, p. 24, l. 1.
G. On the difference of the Gospels, p. 25, l. 36.
H. On sectarian Gospels, p. 27, l. 9.

p. 17, l. 9. *For the Persians*, etc. Cf. with the following traditions, chapter xxxiv. of *Bundehesch*, ed. F. Justi, 1868.

p. 18, l. 5. *The Jews and Christians differ*, etc. An extract of the following by Almaḳrîzî has been published by S. de Sacy, "Chrestomathie Arabe," tom. i. p. 284.

p. 18, l. 16. By Ḥisâb-aljummal the author understands the notation of the numerals by means of the letters of the Arabic alphabet, arranged according to the sequence of the Hebrew alphabet.

p. 18, ll. 19, 20. *Alrâ'i, Abû-'Îsâ Aliṣfahânî.* Of these two pseudo-Messiahs the latter is well known. For his history, *vide* H. Graetz, "Geschichte der Juden," 2nd edition, tom. v. p. 167 and p. 438. Of the former name there is no pseudo-Messiah known in Jewish history. However, Graetz reports of a pseudo-Messiah (*loc. cit.* p. 162), whom he calls *Serene.* The oldest Hebrew report concerning this man begins:

ושאלתם בשביל מטעה שעמד בגלותינו ושריע שמו והיה
אומר וגו׳

"What you have asked regarding the deceiver (*or* heretic) who has risen in our exile, and whose name is ריע," etc. Whether this name has any connection with the Arabic الراعى, whether the reading ריע is to be changed into ירע, students of Jewish history may decide. Certainly a later Latin chronicle calls him Serenus (Graetz, *loc. cit* p. 434 ff.) A pseudo-Prophet, *Rā'ī*, in Tiberias, is also mentioned by Al-Jaubarî in "Zeitschrift der Deutschen Morgenländischen Gesellschaft," xx. 490.

p. 18, l. 25. The author's transliteration of Hebrew words resembles very much the present pronunciation of the Jews of Galizia. Between the words מהם and ביום the Arabic has the signs وهاس, and the last word ההוא is written هاهوم for both of which variations I am unable to account.

p. 18, l. 35. *Since the time when.* The Arabic translation of this passage is not quite correct, and next to unintelligible. It betrays a certain likeness to the translation of the Syriac Bible (Peshîṭâ), where this passage is rendered by—

ܘܡܢ ܥܕܢܐ ܕܢܥܒܪ ܕܒܚܐ ܬܬܝܗܒ ܛܢܦܘܬܐ ܠܚܒܠܐ

"And from the time when the sacrifice passes away, impurity will be given to destruction."

Accordingly I read the Arabic text:

منذ الوقت الذى يجوز القربان تصيرُ النجاسةُ الى الفساد

although I am aware that this is bad and ungrammatical Arabic. On p. ١٥, l. 20, read يصير with PL instead of يصير.

p. 19, l. 12. *Sum of* 1335. Read وثلثين instead of وثلثين يوما on p. ١٦, l. 8.

p. 19, l. 22. *Urishlîm, i.e. Jerusalem,* etc. The author gives to this prophecy of Daniel a wrong date. It falls into the first year of Darius, *v.* Dan. ix. 1, not in the time *some years after the accession of Cyrus to the throne.* This latter date the author has taken from Dan. x, 1 ("in the third year of Cyrus," etc.), and Dan. x. 4.

Perhaps in the Arabic text (p. ١٦, l. 11) the word ثلاث has fallen out between the words مضى and سنين.

l. 19, p. 31. *And before this,* etc. This is a blunder of the author's. It ought to be, "*And after this,*" etc.

l. 20, p. 43. *Jerusalem.* Here, p. 17, l. 21, and in all other passages (p. 16, ll. 1, 13, etc.) the correct reading is بيت المقدس according to Yâḳût,

"Geographisches Wörterbuch," iv. 590, not بيت المقدس. In the following line, p. 18, l. 1, read الذي instead of الا (Fleischer).

p. 22, l. 17. *For Muḥammad*, etc. Read لمحمد instead of بمحمد on p. 19, l. 3.

p. 22, l. 40. *Almathnâ*. Read بالمثنى instead of بالمثنى; and in the following line read في فيه instead of من فيه (p. 19, l. 13).

p. 23, l. 16. *Legions of saints who*, etc. This passage is of Koranic origin, and formed upon the pattern of Sûra iii. 121. The idea of warriors wearing certain badges (as *e.g.* the cross of the Crusaders) occurs also in a tradition, *vide* Albaiḍâwî ad Sûra iii. 121, and Lane, Arabic Dict. *s.v.* تسوم.

p. 24, l. 5. *After Nebukadnezar had conquered*, etc. The last source of this tradition regarding the origin of the version of the Seventy is the letter of Aristeas, well known to Biblical scholars, and now generally admitted to be apochryphal, *vide* De Wette, "Lehrbuch der historisch-kritischen Einleitung," edited by E. Schrader, part i. p. 92.

p. 24, l. 28. *And each couple*, etc. Read "*and every one of them had got a servant to take care of him.*" And in the Arabic, ٢١, 2, read رجل instead of رجلين, بشأنه instead of بشأنهم, and l. 3, ترجمتها instead of ترجمته. On the same page, l. 6, read تقول instead of يقول.

p. 25, l. 2. *Allâmasâsiyya*. This name is derived from the expression لا مساس ("*do not touch*"), in Sûra xx. 97; *vide* S. de Sacy, "Chrestomathie Arabe," i. pp. 339, 342, 344. It is identical with 'Αθίγγανοι, the Greek name of a heretical sect, *vide* Du Cange, "Lexicon infimæ Græcitatis," and "Etymologicum Magnum," ed. Gaisford.

p. 25, l. 25. *Anianus*. The Arabic manuscripts give the name اثنيوس *i.e. Athenæus*; but the well-known Athenæus cannot be meant here. I prefer to read انيانوس, *Anianus*. This author, an Egyptian monk, contemporary of Panodorus, is known as a chronographer; he is quoted in the fragmentary chronology of Elias Nisibenus, cf. Forshall, "Catalogue of the Syriac MSS. of the British Museum," p. 86, col. 2, no. 5.

p. 25, l. 28. *Ibn-albazyâr*, from whose Kitâb-alḳirânât the author has taken the statement of Anianus, was a pupil of Ḥabash, and lived in the 9th century, *vide* "Kitâb-alfihrist, p. 276.

p. 26, l. 30. *But no male children*, etc. Read لا بنين instead of لا بنون, p. ٢٣, 20.

p. 27, l. 5. *Now Joseph and Mary*, etc. Read قاد instead of قادا, on p. ٢٣, 6, and طها instead of طهى, l. 15 (Fleischer).

p. 28, l. 11. *It is related that Tahmûrath*, etc. The last source of this report is the *Book on the Differences of the Canons* (astronomical handbooks), by Abû-Ma'shar, cf. "Kitâb-alfihrist," p. 240, and also "Hamzæ Ispahanensis Annalium," libri x. ed. Gottwaldt, p. 197. The word علوم in this report, p. 24, l. 9, means *scientific books*, as also in the "Kitâb-alfihrist," p. 240, l. 28: "And he ordered a great quantity of scientific books (علوما كثيرة) to be transported from his storehouses to that place."

p. 28, l. 16. *Least exposed*, etc. Read منها instead of منه, on p. 24, l. 10.

p. 28, l. 23. *That Gayômarth was not*, etc. The same tradition occurs in the chronicle of Ibn-Alathîr, ed. Tornberg, i. p. 34, l. 5.

p. 28, l. 34. Some genealogists make the *Lûd* of Genesis x. (in Arabic لاوذ) the father of the Persians, Hyrcanians, of Ṭasm and Amalek, etc. (Ibn-Alathîr, i. 56). The Arabs have mistaken the Hebrew אמים (*Emaei*, the original inhabitants of the country of Moab) for a singular, and for the name of a man ('Amîm ben Lûd, Ibn-Alathîr, i. 56).

p. 29, l. 4. *Abû-Ma'shar*, a native of Balkh, one of the fathers of astrology among the Arabs. He wrote numerous books on all branches of astrology, many of which are still extant in the libraries of Europe. He lived in Bagdad, was a contemporary of Alkindî, and died A.H. 272, at Wâsit. Cf. "Kitâb-alfihrist," p. ٢٧٧ and notes; Otto Loth, "Alkindî als Astrolog," p. 265. In the middle-ages he was well known also in Europe as *Albumaser*, and many of his works have been translated into Latin; whilst modern philology has hitherto scarcely taken any notice of him. Wherever Albêrûnî quotes him, he wages war against him, and, to judge by the quotations from his books which our author gives, it seems that the literary work of Abû-Ma'shar does not rest on scientific bases.

p. 29, l. 18. On the *star-cycles*, cf. J. Narrien, "Historical account of the Origin and Progress of Astronomy," London, 1833, p. 112.

p. 29, l. 28. *Days of Arjabhaz* and *days of Arkand*. According to Reinaud, "Mémoire sur l'Inde," p. 322, the correct form of the former name would be *Aryabhatta*, and the latter would be the Sanskrit *ahargana*. Albêrûnî made a new edition of the *Days of Arkand*, putting it into clearer words and more idiomatic Arabic, since the then existing translation was unintelligible, and followed too closely the Sanskrit original, *vide* my "Einleitung," p. xl., in the edition of the Arabic text.

p. 29, l. 31. Muḥammed ben Isḥâḳ ben Ustâdh Bundâdh Alsarakhsî, and Abû-alwafâ Muḥammad ben Muḥammad Albûzjânî.

The latter was born at Bûzjân in the district of Nîshâpûr, A.H. 328, he settled in 'Irâḳ, A.H. 348, and died 387. Cf. Sêdillot, "Prolégomènes," p. 58; "Kitâb-alfihrist," p. 283; Ibn-Al'athîr. ix. ٩٧, 3.

The former scholar is not known to me.

p. 31, l. 15. *In Hebrew*, "Nebukadnezar." In the Arabic, ٢٧, 3, read نبوخذنصّر instead of بوخذنصّر (De Goeje, Noeldeke).

p. 31, l. 35. *Callippus was one of the number*, etc. Behind the words ومن يدين او قومه there seems to lurk a gross blunder of the copyists.

p. 32, l. 16. *Zoroaster, who belonged to the sect*, etc. The passage, وهو نصف العربانيّة, ٢٨, 2, seems hopelessly corrupt. My translation is entirely conjectural (وهو من صنف).

p. 32, l. 22. *Philip the father of Alexander*. This is a mistake of the author's. He ought to have said: *Philip the brother of Alexander*. The source of this statement regarding the era of Philippus Arridæus is Theon Alexandrinus, *Πρόχειροι κανόνες*, ed. Halma, p. 26; cf. L. Ideler, "Handbuch der mathematischen und technischen Chronologie," ii. 630.

p. 33, l. 8. *Ḥabîb ben Bihrîz, metropolitan of Mosul*, is known as one of those scholars who translated Greek books into Arabic at the time of the Khalif Alma'mûn (A.H. 198–218). Cf. "Kitâb-alfihrist," p. 244, l. 7; p. 248, l. 27; p. 249, l. 4.

p. 33. l. 18. *'Aḥmad ben Sahl*. This man of Sasanian origin was a *Dihkân* (*i.e.* great landholder) in the district of Marw. He played a great *rôle* in the history of his time, and was commander-in-chief to several princes of the house of Sâmân. His history is related by Ibn-Alathîr, viii. 86; "Histoire des Samanides," par M. Defrémery, Paris, 1845, p. 134.

p. 33. l. 28. *It was Augustus who*, etc. On the origin of the *Æra Augusti*, cf. Theon Alexandrinus, *Πρόχειροι Κανόνες*, ed. Halma, p. 30, l. 32; Ideler, "Handbuch der mathematischen und technischen Chronologie," i. 153 ff.

p. 33, l. 34. *Ptolemy corrected*, etc. The source of this information is Ptolemy, *μαθηματικὴ σύνταξις*, book vii. ch. 4 (ed. Halma, tom. ii. p. 30).

p. 33, l. 44. *The prognostics*, المسائل, are questions relating to the decrees of the stars (فى احكام النجوم). The books on this subject contain the astrological answers to all sorts of questions, and the methods by which these answers are found.

p. 34, l. 7. *Maimûn ben Mihrân*, a dealer in cloths and stuffs of linen and cotton, was at the head of the administration of the taxes of Northern Mesopotamia (Aljazîra) under the Khalif Omar ben 'Abd-al'azîz, and died A.H. 117; *vide* Ibn-Ḳutaiba, "Kitâb-alma'ârif" ed. Wüstenfeld, p. 228.

p. 34, l. 26. *Alsha'bî, i.e.* 'Âmir ben Sharâḥîl ben 'Abd Alsha'bî, of South-Arabian origin, was born in the second year of the reign of 'Uthmân; he was secretary to several great men of his time, *e.g.* to 'Abdallâh ben Yazîd, the governor of Alkûfa, for the Khalif Ibn-alzubair, and died A.H. 105 or 104; *vide* Ibn-Ḳutaiba, "Kitâb-alma'ârif," p. 229.

p. 36, l. 10. *Reform of the calendar by the Khalif Almu'taḍid.* Cf. Ibn-Alathîr, vii. p. 325.

p. 36, l. 14. *Abû-Bakr Alṣûlî, i.e.* Muḥammad ben Yaḥyâ ben 'Abdallâh ben Al'abbâs, most famous as a chess-player in his time, the companion of several Khalifs, died A.H. 335 or 336, at Baṣra. In his *Kitâb-al'aurâḳ* he related the history of the Khalifs, and gave a collection of their poems and those of other princes and great men. Cf. Ibn-Khallikân, ed. Wüstenfeld, nr. 659, and "Kitâb-alfihrist," p. 150.

p. 36, l. 19. 'Ubaid-allâh ben Yaḥyâ ben Khâḳân was made the Vazîr of the Khalif Almutawakkil, A.H. 236 (Ibn-Alathîr, vii. 37), and died A.H. 263 (*loc. cit.* p. 215).

p. 36, l. 42. Khâlid ben 'Abdallâh Alḳasrî was made governor of Al'irâḳ by the Khalif Hishâm ben 'Abd-almalik A.H. 105 (Ibn-Alathîr, v. 93), and held this office during 15 years, till A.H. 120, (*loc. cit.* p. 167). Cf. Ibn-Ḳutaiba, "Kitâb-alma'ârif," p. 203.

p. 37, l. 6. The Barmak family were accused of adhering secretly to the religion of Zoroaster, cf. "Kitâb-alfihrist," p. 338, l. 14.

p. 37, l. 9. *Ibrâhîm ben Al'abbâs Alṣûlî*, an uncle of the father of Abû-Bakr Alṣûlî (on p. 36, l. 14), a most famous poet and high official of the Khalif in Surra-man-ra'â, died A.H. 243. The family of Ṣûlî, a family of poets, of eloquent and learned men, of whom several acquired a great fame, descended from a princely house of Hyrcania. According to our author, p. 109, l. 44, the princes of Dahistân were called Ṣûl. For the biography of Ibrâhîm and the history of his family, *vide* Ibn-Khallikân, nr. 10 (ed. Wüstenfeld).

p. 37, l. 19. These verses of Albuḥturî form part of a larger poem in the poet's dîwân which exists in the Imperial Court-Library at Vienna (Mixt. 125 f. 293, 294), *vide* Flügel's Catalogue, i. 436.

p. 38, l. 5. 'Alî ben Yaḥyâ was famous in his time as an astronomer and poet, and as a friend of several Khalifs. He died A.H. 275 at Surra-man-ra'â. Ibn-Khallikân, nr. 479. He was one of a whole family of distinguished poets and scholars who traced their origin back to Yazdagird, the last Sasanian king. Cf. "Kitâb-alfihrist," p. 143.

p. 39, l. 16. On these mythological traditions, cf. L. Krehl, "Die Religion der Vorislamischen Araber," Leipzig, 1863, p. 83; Almas'ûdî, "Prairies d'or," ed. B. de Meynard, iv. 46; Ibn-Alathîr, ed. Tornberg, ii. 30.

p. 39, l. 33. *Banû-Kuraish.* In the Arabic, p. 34, l. 12, read لقريشَ instead of لقريشي; and p. 34, line 13, read قريشَ instead of قريشٌ (Fleischer).

p. 39, l. 34. The following famous battle-days of the ancient Arabs are well known to Arab historians. For more detailed information I refer to the chronicle of Ibn-Alathîr, of which nearly one half of tom. i. (p. 320 ff.) is dedicated to this subject. Cf. also Ibn Ḳutaiba,' "Kitâb-alma'ârif," p. 293; "Arabum Proverbia," ed. Freytag, tom. iii. p. 553 ff.

The pronunciation of the word *Alfaḍâ* (p. 39, l. 44) seems doubtful. Yaḳût, iii. 804, mentions *Alghaḍâ*, a place in the district of the Banû-Kilâb, where once a battle took place. Therefore it would perhaps be preferable to read "*The day of Alghaḍâ.*"

p. 40, l. 26. On this war of Alfijâr, in which Muḥammad took part, cf. A. Sprenger, "Das Leben und die Lehre des Moḥammad," i. 351, 423.

p. 40, l. 35. *Notwithstanding, we have stated,* etc. This passage proves that there is a *lacuna* in the order of the chronological tables, such as exhibited by the manuscripts. According to the author, his work contained also the tables of the princes of South-Arabia and of Alḥîra, but no such tables are found in the manuscripts. Their proper place would have been between the Sasanians and the Khalifs (after p. 128), but the table of the Khalifs is lost, too.

I am inclined to believe that the author had scarcely any other information but that of Ḥamza Alisfahânî (transl. by Gottwaldt, pp. 73 and 96). The manuscript of the University Library of Leyden proves a considerable help for the emendation of Ḥamza's work, but more manuscripts will be wanted before a reliable and clear text can be made out.

p. 40, l. 39. For the following report on the antiquities of Chorasmia I refer to my treatise, *Zur Geschichte und Chronologie von Khwârism I.*, published in the "Sitzungsberichte der Kais. Academie de Wissenschaften in Wien," Philosophisch-historische Classe, 1873, p. 471 ff.

p. 41, l. 7. On the name of Âfrîgh, *vide* my treatise *Conjectur zu Vendidad*, i. 34, in "Zeitschrift der Deutschen Morgenländischen Gesellschaft," xxviii. p. 450.

p. 41, l. 25. *Piece by piece.* Read قطاعا instead of قطاعا (text, p. 35, l. 16 (Fleischer).

p. 41, ll. 33, 41. I prefer to read *Azkâkhwâr* instead of Azkajawâr; also p. 42, l. 26.

p. 43, l. 24. The chief source of all information of eastern authors regarding Alexander is the book of Pseudo-Callisthenes (edited by C. Müller, Paris, 1846, Didot). The book has been treated with the same liberty both in east and west, and it seems that the eastern translations have not less differed from each other than the various Greek manuscripts of the book. The passage p. 44, l. 30 ff. does not occur so in the Greek original, but something like it, cf. book ii. ch. 20, p. 77. The murderer of Darius, p. 44, l. 8, has a Sasanian name (Naujushanas), whilst in the original there are two murderers, Bessus and Artabarzanes (ii. 20). That Nebukadnezar is introduced into the tale, occurs also elsewhere—Mas'ûdî, "Prairies d'or," ii. 247; Ṭabarî (Zotenberg), i. 516. That Alexander was originally a son of Darius, is the tradition of the Shâhnâma of Firdausi, *vide* also Ṭabarî, i. 512; Ibn-Alathîr, i. 199, 1. For more information I refer to Fr. Spiegel, "Die Alexandersage bei den Orientalen," Leipzig, 1851.

p. 45, l. 3. *Ibn-'Abd-Alrazzâḳ Alṭûsî.* A man of this name, *i.e.* Ibn 'Abd-alrazzâk is mentioned in the history of the Buyide prince Ruknaldaula, by Ibn-al'athîr, viii. p. 396, among the events of A.H. 349.

p. 45, l. 5. Abû-Isḥâḳ Ibrâhîm ben Hilâl, the Sâbian, was the secretary of the Buyide prince 'Izz-aldaula Bakhtiyâr, famous as an eloquent writer in prose and verse. He died A.H. 384, or, according to another statement, before A.H. 380. Cf. "Kitâb-alfihrist," p. 134; Ibn-Khallikân, nr. 14; F. Wilken, Mirchond's "Geschichte der Sultane aus dem Geschlechte Bujeh," Berlin, 1835, p. 105.

The title of his book (p. 45, l. 6) read *Altâji* instead of *Altâj.*

On the pedigree of the family of Buwaihi, cf. Ibn-Ḳutaiba, "Kitâb-alma'ârif," p. 36; Ibn-Alathîr, viii. 197; F. Wüstenfeld, Genealogische Tabellen T. 10 and Register, p. 152. Most of the names which occur in this pedigree are also found in Sehir-eddin's "Geschichte von Tabaristan, Rujan und Mazandaran," ed. Dorn, p. 101, and the whole pedigree, *loc. cit.* p. 175.

p. 45, l. 9. Read فناخسرو instead of فناخسرو (text, p. 38, l. 3).

p. 45, l. 22. Abû-Muḥammad Alḥasan ben 'Alî ben Nânâ, mentioned as the author of a history of the Buyide princes, is not known to me.

p. 46, l. 8. Read باسل instead of باسل (text, p. 38, l. 10); and باسل instead of باسل (p. 38, l. 11). Read تسمّى instead of يسمّى (p. 38, l. 11).

p. 46, l. 12. The names of Lâhû and Layâhaj (p. 15) are unknown to me; perhaps they have some sort of relation with the word *Lâhijân*, لاهجان, which is the name of one of the two capitals of Ghîlân, cf. Dorn, "Sehir-eddin's Geschichte von Tabaristân," etc. Vorwort, p. 11, note 1.

p. 47, l. 23. *Ghîlân.* Read الجبل instead of الجبل (text, p. 39, l. 5).

p. 47, l. 25. *Asfâr ben Shîrawaihi.* Under the Khalifate of Almuḳtadir (A.H. 295–320) the party of the Alid. . tried to occupy the countries south of the Caspian Sea, Tabaristân, Dailam, Ghîlân and Jurjân, fighting against the troops of the Samanian princes of Khurâsân and those of the Khalif. The first Alide whose efforts were crowned with success was Ḥasan ben 'Alî, called *Alnâṣir Al'uṭrush*, about A.H. 302. Soon, however, the generals of the Alide princes, Lailâ ben Alnu'mân, Mâkân ben Kâkî, Asfâr ben Shîrawaihi, were more successful than they themselves. The latter, Asfâr, who abandoned the party of the Alides, succeeded, A.H. 315, in occupying Tabaristân, and in rendering himself an independent ruler. He did not long enjoy the fruits of his labours. After having made himself thoroughly unpopular, he was killed by his generals, at the head of whom was Mardâwîj, A.H. 316. Mardâwîj was now the ruler of Ṭabaristân and Jurjân, and tried to extend his sway over the neighbouring countries. He was the founder of a dynasty who held the supreme power in those countries during one hundred and fifty years. He abandoned the party of the Alides, and adopted the black colour of the Abbasides. To the Khalif he made himself so formidable that he was invested and proclaimed as the legitimate governor of all the provinces which his sword had conquered. Cf. Weil, "Geschichte der Khalifen," ii. 613–621. A history of this man and of his descendants is found in Sehir-eddin's "Geschichte von Ṭabaristân, Rujan und Mazandaran," ed. by Dorn, 1850, on pp. 171–201 and 322.

Mardâwîj was a جيلى, *i.e.* native of Ghîlân (not جبلى, native of Aljabal or Media). The name of his father is written زيار and زيان, and I have not been able to make out which form is the correct one. In Sehir-eddin's chronicle, the name is always written زيار.

In the text, ٣٩, 6, read الجبل instead of الجبل. Between the words وله and شهر (line 6), there seems to be a *lacuna* which I have no means of filling up. This *lacuna* is the reason why the following words do not offer a clear meaning. It is not clear who was the *son of Ward-inshâh* who instigated Mardâwîj to free the people from the tyranny of Asfâr.

p. 47, l. 29. *Khurâsân*. Read خراسان, ٣٩, 8, instead of طبرستان.

The name *Farkhwârjirshâh* may possibly be identical with that name which Anôshirwân is said to have had as the governor of Ṭabaristân in the lifetime of his father, Hamza Isfahânî, ed. Gottwaldt, ٥١, 3, 4. Cf. فرهواد كر and فرهواد كر شاه in Sehir-eddin's "Chronik von Tabaristan," ed. Dorn, pp. 19, 31, 42; P. de Lagarde, "Beiträge zur Baktrischen Lexikographie," p. 50 ff.

p. 47, l. 30. In the text on p. 39, read المَلَك instead of المَلَك, l. 9; كمثل instead of كمثل, l. 13; and العَلَويَّة instead of العَلَويّة, l. 19 (Fleischer).

p. 47, l. 32. The Ispahbad *Rustam*, the uncle of Shams-alma'âlî is also mentioned by Ibn-Al'athîr, viii. 506. To a son of this Rustam, Marzubân ben Rustam, the Ispahbad of Jîljîlân, our author has dedicated one of his books, *vide* my edition of the text, Einleitung, p. xl. nr. 7.

The history of the ancestors of the Ispahbad Rustam is related in Sehir-eddin's "Geschichte von Tabaristan," etc., ed. Dorn, pp. 201–210, 270, 322. They are called "*the family of Bâwand.*"

In p. 47, ll. 34 and 38, read هروين instead of هروين. Cf. Yâḳût, "Geographisches Wörterbuch," iii. 283.

p. 48, l. 5. The same pedigree of the house of Sâmân is also given by Ibn-al'athîr, vii. 192, and in the geography of Ibn-Hauḳal, pp. 344, 345.

p. 48, l. 16. *The Shâhs of Shirwân*. According to Ḳazwînî, "Âthâr-albilâd," p. 403, Shirwân was first colonized by Kisrâ Anôshirwân. The kings of the country were called اخستان. Anôshirwân is said to have installed the first governor and prince of Shirwân, a relative of his family. Cf. Dorn, "Versuch einer Geschichte der Shirwânshâhe," p. 12 and 25. Mas'ûdî, "Prairies d'or," ii. 4, makes the Shirwânshâh of his time descend from Bahrâm-Gûr.

p. 48, l. 24. *Ubaid-Allâh*, etc., founded the empire of the Fâṭimide dynasty in Ḳairawân and Egypt, A.H. 296. He pretended to be a descendant of 'Alî ben Abî-Ṭâlib. Cf. Ibn-Al'athîr, viii. 27; Ibn-Ḳutaiba, "Kitâb-alma'ârif," p. 57; Weil, "Geschichte der Chalifen," ii. 598.

That prince of this dynasty who ruled at the time of our author, Abû-'Alî ben Nizâr, etc. (p. 48, l. 31) was Khalif of Egypt, A.H. 386–411, and is better known under the name of Alḥâkim, cf. Ibn-Al'athîr, ix. 83, 2; 82, 14; 221, 14.

p. 48, l. 41. I feel inclined to suppose that in this pedigree there is a *lacuna* between يافث and سوعون, that يافث the son of Noah was originally the end of the first pedigree, and that the second commenced with Alexander ben Barka, etc. This opinion is supported by Mas'ûdî, "Prairies d'or," ii. 248, and Spiegel, "Alexandersage," p. 60. However, I must state that the pedigree—such as it is given by Albêrûnî—also occurs in Mas'ûdî, ii. 293, 294, and Ibn-Al'athîr, i. 200, 5–9. If, therefore, there is a *lacuna*, as I suppose, it is a blunder of older date, and must have occurred already in the source whence all, Mas'ûdî, Albêrûnî, and Ibn-al'athîr have drawn.

Some of the names of this pedigree exhibit rather suspicious forms.

II. مصريو, perhaps مصريم *Egyptus*? Cf. Ibn-Alathîr, i. p. 200, l. 6.

VI. لبطى. Read لطنى *Latinus*.

XV. الاصفر is a corruption of צְפוֹ. Genesis, xxxvi. 11, 15; *vide* Ascoli, "Zeitschrift der Deutschen Morgenländischen Gesellschaft," xv. 143.

p. 49, l. 22. The combination of Dhû-alkarnain with Almundhir ben Imru'alkais, *vide* Ḥamza Isfahânî, translated by Gottwaldt, p. 82.

p. 49, l. 26. *'Abdallâh ben Hilâl.* On this famous juggler, *vide* "Kitâb-alfihrist," p. 310, and note.

p. 49, l. 37. On the supposed South-Arabian origin of Dhû-alkarnain, *vide* Mas'ûdî, "Prairies d'or," ii. 244, 249; A. v. Kremer, "Südarabische Sage," pp. 70–75; Ḥamza, transl. p. 100.

p. 49, l. 41. The name *Ṣubaiḥ* occurs also in *Ibn-Hishâm.* The life of Muhammad, i. 486. It seems to be the diminutive of اصبح *'Aṣbaḥ* (Ibn-Duraid, "Kitâb-alishtiḳâḳ," p. 41), as *Nu'aim* نُعَيم is the diminutive of *'An'am* انعم according to Ibn-Duraid, *loc. cit.* p. 85, l. 14. Another name of the same root is صباح in Ḥamza, ed. Gottwaldt, p. 132.

The spelling of the name *Alhammâl*, l. 34, is uncertain.

For the spelling of the name *Tan'um*, *vide* Ibn-Duraid, *loc. cit.* p. 84, n[illegible]

p. 50, l. 4. *Fever-water.* Read *muddy water.* Read حمأ instead of حُمّى (text, p. 41, l. 3), and read الاحمر instead of الاحمر (text, p. 41, l. 5) (Fleischer).

p. 50, l. 7. The following reasoning occurs already in Ḥamza, transl. p. 100.

p. 50, l. 26. *Ibn-Khurdâdhbih* was postmaster in Media, and wrote about the middle of the third century of the Flight (between 240–260). His geographical work has been edited and translated by B. de Meynard, "Journal Asiatique," 1865.

p. 51, l. 1. *Sawâr.* This nation is mentioned by Byzantine authors under the name of *Σαβιροί.*

p. 52, l. 24. Abû-Sa'îd 'Aḥmad ben Muḥammad, a native of Sijistân, is not known to me from any other source.

p. 53, l. 31. The same names occur in Mas'ûdî, iii. 415. These days have also Arabic names (*loc. cit.* p. 416, and this book, p. 246, l. 16).

p. 53, l. 34. Read الناكب instead of الثابت (text, p. 43, l. 22).

p. 53, l. 37. *Zâdawaihi ben Shâhawaihi*, a native of Isfahân, is mentioned in "Kitâb-alfihrist," p. 245, as one of those who translated Persian works into Arabic. He is also mentioned on p. 202, l. 7, and p. 207, l. 11.

p. 54, l. 1. Abû-alfaraj Aḥmad ben Khalaf Alzanjânî; "Kitâb-alfihrist," p. 284, mentions an Aḥmad b. Khalaf among those who made astronomical and other instruments; also mentioned p. 118, l. 31.

p. 54, l. 4. Abû-alḥasan Adharkhûr ben Yazdankhasîs is not known to me from any other source; *vide* p. 107, l. 40 and p. 204, l. 14.
Read آذرخور بن instead of آذرخواى (text, p. 44, l. 6).

p. 54, l. 29. Read وإنفاقِ instead of وأنفاقَ (text, p. 44, l. 15).

p. 54, l. 39. The reason why the Persians did not like to increase the number of days of the year was, according to Mas'ûdî, iii. 416, that thereby the established sequence of lucky and unlucky days would have been disturbed.

p. 55, l. 3. The words l. 5–27 do not in the least harmonize with the preceding, which makes me believe that after the word *Âdhar-Mâh* there is a gap, although the manuscripts do not indicate it. The explanation which is commenced in ll. 3, 4, is continued in l. 28 ff.

p. 56, l. 7. *Yazdajird Alhizârî* is also mentioned by Yâḳût, "Geographisches Wörterbuch," iv. 970. Yâḳût may have drawn his information from this book.

p. 56, l. 22. As these names, the scanty remnants of a long-lost Eranian dialect, are of considerable philological interest, I shall add the readings of the Canon Masudicus of Albêrûnî according to two manuscripts, MS. Elliot (now the property of the British Museum, dated

Bagdad, A.H. 570, Rabî' I.), and MS. Berlin (the property of the Royal Library, acc. ms. 10, 311, or MSS. Orr. 8°. 275).

MS. Elliot, f. 14a.	MS. Berlin.
نومرد	نومرد
حرحن	حوجن
نسسن	نسس
نساك	پساك
اهناحندا	اهناحندا
مرنخندا	مرنخندا
معكان	نعكان
الابانج	بانج
فوغ	فوغ
مسافوغ	مسافوغ
ريمد	ريمذ
حسوم	خشوص

Whoever wants to explain these names will also have to consult the six manuscripts of the *Kitâb-altafhîm* of our author, and the most ancient copy of the Canon Masudicus in the Bodleian Library.

In this book Albêrûnî does not mention the months of the Armenians, but I have found them in a copy of the "Kitâb-altafhîm" (MS. of the Bodleian Library) in the following form (p. 165):

(!) زجانی اریك مهیکن اراتس (read کاغوتس) کاغوس دری مهمی هوری فالسرقی
هرودتس (MS. . . هرود) مارکاس (read مارکاتس) ماریری

Cf. E. Dulaurier, "Récherches sur la chronologie Armenienne," p. 2.

p. 57, l. 17. I am sorry to state that there are no tables of these Chorasmian names in the Canon Masudicus, nor in the "Kitâb-altafhîm."

The form اهمن (p. 57, l. 2) reminds one of the Cappadocian name Οσμαν, *vide* Benfey und Stern, "Ueber die Monatsnamen einiger alter Völker," Berlin, 1836, pp. 110–113.

The name دذو (name of the 8th, 15th, and 23rd days) is, like the Persian *Dai*, to be retraced to *Dadhvâo* (Benfey and Stern, *ib.* pp. 109, 110).

The corresponding Sogdian name (p. 56) is written دسست, which is, perhaps, a metathesis for دثس, which would be equal to *Dathushô*, the genitive of *Dadhvâo*, and would resemble the Cappadocian Δαθουσα (Benfey und Stern, *ib.* p. 79).

The reader will easily recognize the relationship between the Sogdian and Chorasmian names of the days of the month and the Persian names; this is more difficult in the case of some of the names of the months.

p. 58, l. 10. *And relied*, etc. Read يعولون instead of هولون (text, p. 48, l. 14).

p. 58, l. 16. *Dai*, *vide* note ad p. 57, l. 17.

p. 58, l. 33. It is not known that the Egyptians called the single days of the month by special names.

p. 59, l. 3. On the names of the Egyptian months, their forms and meanings, *vide* R. Lepsius, "Chronologie der Aegypter," pp. 134–142.

p. 59, l. 22. *The small month.* The Coptic name for the Epagomenæ is *p abot n kouji*, "the small month," cf. R. Lepsius, "Chronologie der Aegypter," p. 145; and this book, p. 137, l. 22. On the Egyptian names of the 5 Epagomenæ, cf. R. Lepsius, *loc. cit.* pp. 146, 147.

p. 59, l. 25. البسط. It seems, one must read this word البقطا, since the Coptic word for leap-year is ϯⲁⲡⲟⲕⲧⲓ, *i.e.* Ἐπακτή, as Mr. L. Stern kindly informed me. In that case the author was wrong in translating the word by علامة *i.e. signum.*

p. 59, l. 26. Abû-al'abbâs Alâmulî, the author of a book on the Ḳibla, is mentioned by Ḥâjî Khalîfa, iii. p. 236. His full name is Abû-al'abbâs 'Aḥmad b. Abî-'Aḥmad Alṭabarî Alâmulî, known as Ibn-alḳâṣṣ, and he died A.H. 335.

The months which this author ascribes to the *People of the West* are our names of months in forms which can hardly be traced back to a Latin source (ancient Spanish?). I suppose that by the *People of the West* he means the inhabitants of Spain.

p. 60, l. 21. *Kitâb-ma'khadh-almawâḳît.* This book is not known to me.

p. 61, l. 1. Twenty-four hours are = 86,400 seconds, which, divided by 729, give a quotient of $118\frac{378}{729}$.

p. 61, l. 13. Read المتقدّم instead of المتقدّم (text, p. 51, l. 17).

p. 61, l. 45. Thâbit ben Ḳurra was born A.H. 221 and died 288; *vide* "Kitâb-alfihrist," p. 272, and notes. On his astronomical theories, *vide* Delambre, "Histoire de l'astronomie du moyen age," p. 73.

On the family of the Banû-Mûsâ *vide* "Kitâb-alfihrist," p. 271. Muḥammad died A.H. 259.

p. 62, l. 16. *The Hebrews and all the Jews.* The word *'Ibrânî=Hebrew*, was a learned name, known only to scholars; it meant that people of antiquity who spoke the Hebrew tongue and who lived in Syria under the law of Moses. *Jew* is a popular name which means the descendants of that people, who no longer live in Syria, but are scattered all over the world, who no longer speak Hebrew, but who still live under the law of Moses.

p. 62, l. 18. The names of the months of the Jews occur also in Assyrian, cf. E. Norris, "Assyrian Dictionary," p. 50.

Part of the following chapter has been edited by S. de Sacy, "Chrestomathie Arabe," i. p. ٨٨ (taken from Almaḳrîzî).

p. 62, l. 40. "*Remember the day*," etc. This quotation is an extract from Exodus xiii. 3, 4 (Deut. xvi. 1). The words *in that month when the trees blossom* are the rendering of the Hebrew בחדש האביב. The month 'Âbîb has always been identified with Nîsân by the whole exegetical tradition of both Jews and Christians, but I do not see for what reason.

p. 63, l. 15. This view, that Adhâr II. is the leap-month, was held by the Karaeans, according to Eliah ben Mose in Selden, "Dissertatio de civili anno Judaico," cap. v. p. 166 (היו מעברין אותו אדר שני).

p. 63, l. 31. On the invention of the Octaeteris by Cleostratos of Tenedos (about 500 B.C.) *vide* Ideler, "Handbuch der mathematischen und technischen Chronologie," ii. 605.

The cycle of 19 years is the cycle of Meton, invented about 432 B.C., *vide* Ideler, *loc. cit.* i. 297 ff.

The cycle of 76 years is the improvement of the Metonian cycle by Callippus of Cyzicus (about 330 B.C.), Ideler, *loc. cit.* i. 299, 344.

The cycle of 95 years (5 × 19) has been used by Cyrillus for the computation of Easter, *vide* Ideler, *loc. cit.* p. 259.

The cycle of 532 (= 19 × 28) was invented by the Egyptian monk Anianus, *vide* Ideler, *loc. cit.* pp. 277, 451.

p. 63, l. 37. In the author's statement regarding the 4th cycle of 95 years there is a mistake: we must read 1,175 *months* instead of 1,176 *months*.

The synodical month or one lunation is = 29 d. 12h. 793 Ḥ.

1,176 lunations = 34,727 d. 23 h. 528 Ḥ = 900,149,208 Ḥ.

If we divide this sum by the length of the solar year, *i.e.* 365 d. $\frac{890}{1080}$ h. = 9,467,190 Ḥ, we get as quotient 95 (years), and a remainder of 29 d. 13 h. 438 Ḥ, *i.e.* 1 lunation plus 725 Ḥ., *i.e.* one lunation too much.

If we reckon 1,175 lunations, we get as the remainder 725 Ḥ., and this result is correct, because it is five times the remainder of the cycle of 19 years, of which this cycle is a five times multiplication.

95 years	=	5 × 19
1,175 lunations	=	5 × 235
35 leap-months	=	5 × 7
725 Ḥ. remainder	=	5 × 145

This remainder represents the difference between the rotations of the sun and the moon at the end of the cycle.

p. 64, l. 3. *Ḥalaḳ*, as I have written, according to the Arabic, is the Hebrew word חֵלֶק, which in the Canon Masudicus is sometimes rendered by حلق. Of the still smaller division of time, of the רְגָעִים (one Rêga' = 76 Ḥalaḳ), I have not found any trace in the works of Albêrûnî.

For the convenience of those who want to examine the following computations, I give a comparison between the Ḥalaḳs and the other measures of time:

I.	1 hour	= 1,080 Ḥ.
	1 minute	= 18 Ḥ.
	1 II.	= $\frac{3}{10}$ Ḥ.
	1 III.	= $\frac{1}{200}$ Ḥ.
	1 IV.	= $\frac{1}{12000}$ Ḥ.
	1 V.	= $\frac{1}{720000}$ Ḥ.
II.	1 Ḥalaḳ	= $\frac{1}{1080}$ hour.
	1 Ḥ.	= $\frac{1}{18}$ minute.
	1 Ḥ.	= $2\frac{1}{3}$ seconds.
	1 Ḥ.	= 200 III.
	1 Ḥ.	= 12,000 IV.
	1 Ḥ.	= 720,000 V.
III.	1,080 Ḥalaḳs	= 1 hour.
	1 Ḥalaḳ	= $\frac{1}{1080}$ h.
	1 Rêga'	= $\frac{1}{82080}$ h.

In Jewish chronology there occur two kinds of years, the Julian year (in the calculation of R. Samuel), and a scientific year derived from the researches of Hipparchus, which is the basis of the calculation of R. 'Addâ bar 'Ahabâ.

The year which Albêrûnî mentions, consisting of 365 d. $5\frac{3791}{4104}$ h., is the year of R. 'Addâ, equal to

365 d. 5 h. 997 Ḥ. 48 Rêg.

Cf. Lazarus Bendavid, "Zur Berechnung und Geschichte des Jüdischen Kalenders," Berlin, 1817, p. 32.

Regarding the origin of this year there cannot be any doubt. The

Jewish chronologists found it by dividing by 19 the Enneadecateris of Meton, which consists of 235 Hipparchical synodical months (*i.e.* 6,939 d. 16 h. 595 Ḥ.).

It will not be superfluous for the valuation of the following calculations to point out the difference between the ancient Greek astronomers and the Jewish Rabbis who constructed the Jewish calendar.

The elements for the comparison of the rotations of sun and moon are two measures: that of the length of the synodical month and that of the length of the solar year. When Meton and Callippus constructed their cycles, these two measures had not yet been defined with a great degree of accuracy. Hence the deficiencies of their cycles.

Centuries later, when the sagacity of Hipparchus had defined these two measures in such a way that modern astronomy has found very little to correct, comparisons between the rotations of sun and moon could be carried out with a much higher degree of accuracy. Thereby the Jewish chronologists were much better situated than Meton and Callippus, and the following calculations prove that they availed themselves of this advantage.

p. 64, l. 10. Computation of the Octaeteris and Enneadecateris.

I. *Octaeteris.*

The ancient Greeks counted the solar year as $365\frac{1}{4}$ days (*i.e.* too long), and the synodical month as $29\frac{1}{2}$ days (*i.e.* too short). The Jews counted—

the solar year as 365 d. $5\frac{3791}{4104}$ h.
and the synodical month as 29 d. 12 h. 783 Ḥ.

The 99 lunations of the Octaeteris, each lunation at 29 d. 12 h. 783 Ḥ., give the sum of—

2,923 d. 12 h. 747 Ḥ.

which is equal to the sum of—

75,777,867 Ḥ.

If we divide this sum by the length of the solar year, *i.e.* 365 d. $5\frac{3791}{4104}$ h.=9,467,190 Ḥ., we get as quotient 8 (years) and a remainder of—

1 d. 13 h. 387 Ḥ.

This would be the difference between the rotations of the sun and moon at the end of the first Octaeteris, *i.e.* the moon reached the end of her 99th rotation, when the sun had still to march during 1 d. 13 h. 387 Ḥ., till he reached the end of his 8th rotation.

According to the calculations of the ancient Greeks, this difference was less, viz. $1\frac{1}{2}$ days. Cf. L. Ideler, "Handbuch der mathematischen und technischen Chronologie," i. p. 294 ff.

As the author says, 387 Ḥalaḳs do not correspond to $\frac{11}{30}$ h. with mathematical accuracy (p. 64, ll. 24, 25). There is a difference of $\frac{1}{120}$ h., for

$$\frac{387}{1080}\text{ h.} = \frac{43}{120}\text{ h.}$$

whilst

$$\frac{11}{30}\text{ h.} = \frac{44}{120}\text{ h.}$$

II. *Enneadecateris.*

Meton discovered that 235 synodical months pretty nearly correspond to 19 solar years. In constructing his cycle of 19 years, he reckoned the solar year at $365\frac{5}{19}$ d., *i.e.* by $\frac{1}{76}$ d. longer than it had been reckoned in the Octaeteris (a mistake which afterwards Callippus strove to retrieve). More correct was the following Jewish calculation with Hipparchic measures:

235 lunations, each = 29 d. 12 h. 793 Ḥ., give the sum of—

$$6{,}939\text{ d. }16\tfrac{595}{1080}\text{ h.} = 179{,}876{,}755\text{ Ḥ.}$$

If we divide this sum of Ḥalaḳs by the length of the solar year of—

$$365\text{ d. }5\tfrac{3791}{4104}\text{ h.} = 9{,}467{,}190\text{ Ḥ.},$$

we get as quotient 19 (years), and a remainder of only 145 Ḥ.

According to this computation, the difference between the rotations of sun and moon at the end of the first Enneadecateris would not be more than 145 Ḥ., or $\frac{29}{216}$ h., *i.e.* a little more than $\frac{1}{7}$ h., or than $\frac{1}{168}$ d., whilst, according to Callippus, this difference was greater, viz. $\frac{19}{76}$ d. $=\frac{1}{4}$ d.

This reform of the Metonic Enneadecateris enabled the Jews to dispense with the 76 years cycle of Callippus, which he constructed of four-times the Enneadecateris with the omission of *one* day. The Jewish calculation is more correct than that of Callippus, who reckoned the solar year too long.

p. 64, l. 33. On the meaning of the word מחזור cf. an interesting chapter in the ספר העיבור of Abraham Bar Chyiah, edited by H. Filipowski, London, 1851, book ii. ch. iv. (בפירוש שם המחזור).

At the beginning of this exposition (p. 64, l. 31, text, p. 55, l. 8) there seems to be a *lacuna*. It is not likely that the author should introduce a technical foreign work (like Maḥzôr) without having previously explained what it means (and this is not the case).

p. 65. The difference of the *Ordines intercalationis* is caused and accounted for by the difference of the beginning of the Jewish *Æra Mundi*.

The world was created at the time of the vernal equinox, *i.e.* the Teḳûfat-Nîsân. But the year as reckoned by the Jewish chronologists does not commence at the time of the vernal equinox, but at that of the autumnal equinox, *i.e.* the Teḳûfat-Tishrî. Now, the question whence to begin the first year of the *Æra Mundi*, has been answered in various ways. Some commence with the vernal equinox preceding the creation of the world, others with the first vernal equinox following after the creation of the world. Some counted the year in the middle of which the creation took place as the first, others counted the following year as the *first* year of the *first* Enneadecateris. Cf. ספר העבור of Abraham bar Chyiah, iii. 7, p. 96. In conformity with this difference also the order of the leap-years within the Enneadecateris has been fixed differently.

The *Ordo intercalationis* גבטבב, which reckons the second (complete) year of the creation as the first year of the first Enneadecateris, occurs also in the valuable *Teshûbhâ* (Responsum) of R. Hâi Gâôn ben Sherîrâ, a contemporary of the author, *vide* Abraham bar Chyiah, p. 97, l. 36.

The *Ordo intercalationis* גוחאדזט which has become canonical since and through Maimonides, is not mentioned by Albêrûnî.

The three *Ordines intercalationis* which the author has united in the circular figure, are constructed upon this principle:

Of the seven intervals between each two leap-years, there are five intervals each of 2 years, and two intervals each of 1 year.

p. 66, l. 7. The solar cycle (מחזור לחמה) of 28 years consists of Julian years of $365\frac{1}{4}$ days. At the end of this cycle time returns to the same day of the week. Cf. L. Ideler, "Handbuch," etc., i. 72.

p. 66, l. 23. Of the five *Deḥiyyôth* of the Jewish calendar יח · אדו · בטותקפט · גטרד · יח אדו which are certain rules ordering a date, *e.g.* New-year's-day, to be transferred from one week-day to another, our author mentions only the first one, viz. אדו, *i.e.* the rule that New-year's-day can never be a Sunday or a Wednesday or a Friday.

The words *that Passover by which the beginning of Nisân is regulated* I understand in this way, that Passover, *i.e.* the 15th Nîsân, and the 1st Nîsân always fall on the same week-day.

The rule אדו is connected with the rule בדו *i.e.* that Passover shall never fall on a Monday, Wednesday, or Friday, in the following way:

Passover must be the 163rd day from the end of the year. The division of 163 by 7 gives the remainder of 2.

If New-year's-day were a Sunday, the last day of the preceding year would be a Saturday, and the 163rd day from the end would be a *Friday*.

If New-year's-day were a Wednesday, the 163rd day from the end would be a *Monday*.

If New-year's-day were a Friday, the 163rd day from the end would be a *Wednesday*. Cf. Lewisohn, "Geschichte und System des Jüdischen Kalenderwesens," Leipzig, 1856, § 92, § 127.

On the correspondence between the four days that *can* be New-year's-days (called אדבעה שערים) and Passover, cf. Abraham bar Chyiah, ii. ch. 9.

p. 67, ll. 28, 35. I should prefer to read الجبال instead of الجبل and فى حسابهم من instead of من حسابهم عن (text, p. 57, ll. 18 and 21).

p. 68, l. 4. On the calculation of the *arc of vision* קשת הראיה *i.e.* that part of the moon's rotation between conjunction and the moment of her becoming visible at some place, *vide* Selden, "Dissertatio de anno civili Judaico," cap. xiii.; Lazarus Bendavid, "Zur Geschichte und Berechnung des Jüdischen Kalenders," § 36.

The *mean* motion of the moon is called in Hebrew מהלך אמצעי, the *real* motion מהלך אמתי, *vide* Maimonides, קדוש החדש, vi. 1; xi. 15.

p. 68, l. 32. *Pârûaḥ* فروح is a Biblical name, *vide* 1 Kings iv. 17.

p. 68, l. 35. If the M'lâdîtes commenced the month with the moment of the conjunction, they differed from the Rabbanites in this, that the latter made the beginning of the month (*e.g.* the beginning of the first month or New-year's-day) depend *not alone* upon conjunction, but also upon certain other conditions, *e.g.* the condition יח (Lazarus Bendavid, § 36). The Rabbanites tried in everything to assimilate their calendar, based upon the astronomical determination of conjunction, to the more ancient calendar which had been based upon the observation of New Moon. The conservative tendency of this reform of the Jewish calendar is pointed out by A. Schwarz, "Der Jüdische Kalender," pp. 59–61. Cf. also Abraham bar Chyiah ספר העבור, p. 68, l. 6; p.69, l. 21.

p. 68, l. 36. Read القرّاء instead of القرّاه as plural of قارئ (text, p. 58, l. 17).

p. 69, l. 5. *'Ânân*, the founder of the great schism in the Jewish world, lived in Palestine in the second half of the 8th century. For his history, *vide* Graetz, "Geschichte der Juden," ii. ed., tom. v. p. 174; for 'Ânân's reform of the calendar, *ib.* p. 454.

The pedigree of 'Ânân has been the subject of much discussion, *vide* Graetz, *ib.* pp. 417, 418, and J. Triglandii, "Notitia Karaeorum," Hamburg, 1714, p. 46.

p. 69, l. 25. Read أَخَذَ instead of أُخِذَ (text, p. 59, l. 9).

p. 70, l. 16. Read لَا instead of لَ (text, p. 60, l. 4, after كانون حراى).

p. 72, l. 36. *Ismâ'îl ben 'Abbâd*, born A.H. 326, was Vazîr to the Buyide princes Mu'ayyid-aldaula, and afterwards to Fakhr-aldaula. He died A.H. 385. Cf. Ibn-Al'athîr, ix. p. 77. The same man is quoted by Albêrûnî as *the Ṣâḥib*, p. 94, l. 19. On this title, *vide* Hammer, "Länderverwaltung unter dem Khalifat," pp. 34, 35; "Abulfedæ Annales Moslemici," ii. p. 586.

p. 74, l. 7. The *farewell pilgrimage* is described by A. Sprenger, "Leben und Lehre des Moḥammad," iii. p. 515 ff. On Muḥammad's prohibiting intercalation, etc., *ib.* p. 534 ff.

Read حِجّة instead of حَجّة (text, p. 63, ll. 1, 3). (Fleischer.)

p. 74, l. 15. Ibn-Duraid, a famous philologist of the school of Baṣra, died A.H. 321, in Bagdad. Cf. G. Flügel, "Grammatische Schulen der Araber," p. 101.

p. 74, l. 25. Abû-Sahl 'Îsâ ben Yaḥyâ Almasîḥî, a Christian physician, was a contemporary of Albêrûnî, who lived at the court of 'Alî ben Ma'mûn and Ma'mûn ben Ma'mûn, princes of Khwârizm. The year of his death is not known; probably he died between A.H. 400–403. Cpr. Wüstenfeld, "Geschichte der Arabischen Aerzte und Naturforscher," p. 59, nr. 118.

p. 75, l. 26. Read معها instead of فيها (text, p. 64, l. 6, لتبرّز الشمس معها).

p. 76, l. 36. Abû-'Abdallâh Ja'far ben Muḥammad Alṣâdiḳ is one of the twelve Imâms of the Shî'a. He was born A.H. 80, and died A.H. 146. On the sect who derived their name from him, *vide* Shahristânî, ed. Cureton, p. 124. Cf. also Wüstenfeld, "Geschichte der Arabischen Aerzte und Naturforscher," nr. 24.

p. 77, l. 4. This tradition occurs in Bukhârî, "Recueil des traditions Mahométanes," ed. L. Krehl, i. p. 474. The other traditions to which the author refers in the course of his discussion (p. 78) are also mentioned by Bukhârî, i. 476 ff. Cf. the *Muwaṭṭa'* of Mâlik ben 'Anas, ed. Bûlâḳ, ii. chap. 84.

p. 77, l. 22. Read والكديرة, instead of والكدرة, (text, p. 65, l. 14), and اذ instead of اذا (p. 65, l. 15). (Fleischer.)

p. 80, l. 4. Read وتبطل, instead of ويبطل, (text, p. 67, l. 17).

p. 80, l. 5. The same fact is related by Ibn-Al'athîr, vi. p. 3. In consequence of his killing 'Abd-alkarîm, the governor of Kûfa, Muḥammad was removed from his office A.H. 155 (or 153). The story shows that the falsification of tradition has at certain times been practised wholesale in the Muslim world. Ibn-'abî-al'aujâ, also mentioned in "Kitâb-alfihrist," p. 338, l. 9.

p. 80, l. 27. Read *and its origin* instead of *and of its original*, etc. Read وأمَلَه instead of وأمِلَه (text, p. 68, l. 4). (Fleischer.)

p. 80, l. 34. Read موجَبات instead of موجِبات (text, p. 68, l. 6), and والممتعنى – فتخمة – تمير (text, p. 68, ll. 9, 10), as in the manuscripts.

p. 82, col. 1. Ḳubâ was the second largest town of Farghâna, not far from Shâsh. It is described by Ibn-Ḥauḳal, p. 394; Yâḳût, iv. 24.

The word بخارقلك (بحارلك) I have not been able to explain hitherto. Perhaps the word bears some relation to بخارا *i.e.* Bukhârâ.

p. 82, coll. 1, 8. The names of col. 1 are in use among the eastern Turks (of Kashghar and Yarkand), *vide* R. B. Shaw, "A Sketch of the Turki Language as spoken in Eastern Turkistan," Lahore, 1875, p. 77; J. Grave, "Epochæ celebriores," London, 1650, p. 5.

The names of col. 8 seem to be in disorder; they mean: The Great Month, the Small Month, the First Month, the Second Month, the Sixth Month, the Fifth Month, the Eighth Month, the Ninth Month, the Tenth Month, the Fourth Month, the Third Month, the Seventh Month. Cf. Shaw, "Sketch," etc., p. 75.

Both columns are of particular interest in so far as they exhibit the most ancient specimen of the Turkish language.

p. 82, col. 5. *Octombrius.* Perhaps it would be better to read *Octembrius,* in conformity with *Octembre,* which occurs in Provençal beside *Octobre, vide* Reynouard, "Lexique Roman ou dictionnaire de la langue des troubadours," tom vi. p. 390.

p. 86, l. 13. The 210 years for the stay of the Jews in Egypt are found in this way:

Interval between the birth of Abraham and that of Moses - - - - - - -	420 years.
Moses was 80 years of age when he left Egypt -	80 ,,
Interval between the birth of Abraham and the Exodus - - - - - -	500 ,,
Further:	
Abraham was 100 years of age when Isaak was born	100 ,,
Isaak was 60 years of age when Jacob was born -	60 ,,
Jacob entered Egypt when he was 130 years of age	130 ,,
Interval between the birth of Abraham and Jacob's entering Egypt - - -	290 ,,

Now, the difference between the two numbers (500—290), *i.e.* 210 years, represents the time during which the Jews stayed in Egypt.

p. 87, l. 11. Read ثانية instead of ثاوية (text, p. 75, l. 1). (Fleischer.)

p. 87, l. 13. The *Sêder-'Ôlâm, i.e. Ordo Mundi,* is a well known Hebrew book on the Chronology of Jewish history, carrying it down as far as 22 years after the destruction of the Temple by Titus. It is the סדר עולם רבא to which our author refers, not the סדר עולם זוטא. Cf. "Chronicon Hebræorum Majus et Minus," ed. Joh. Meyer, Amstelodami, 1699. I am, however, bound to state that some of the numbers which Albêrûnî quotes on the authority of this book are not found in—or do not agree with—the text as given in the edition of Meyer.

pp. 88, 89. In these tables there are three blunders in the addition. The last three numbers in the addition of the years of the Sêder-'Ôlâm ought to be 460, 500, 503 (on p. 88); and in the same column on p. 89 the eleven last numbers of the addition ought to be: 781, 810, 865, 867, 898, 909, 920, 990, 1080, 1563, 2163.

p. 90, l. 18. On Kûshân, *vide* Judges, iii. 8, 10.

p. 90, l. 35. *Ḥashwiyya and Dahriyya.* The Ḥashwiyya or Ḥashawiyya are a heterodox sect of Muslim philosophers who adhere to an exoteric interpretation of the divine revelation, and consider God as a bodily being, *vide* "Dictionary of Technical Terms," i. p. 396.

The word *Dahr* seems nearly to correspond to the *Zrvânem akerenem* ("*endless time*") of the Avastâ. The *Dahriyya* are a heathenish school of philosophers who believe the *Dahr* (time) to be eternal, and who trace everything to the Dahr as last cause, *vide* "Dictionary of Technical Terms," i. p. 480.

p. 90, l. 44. In the following the author attacks 'Abû-Ma'shar, the author of the book *De nativitatibus* (p. 92, l. 2; p. 91, l. 31; p. 94, l. 44; p. 96, l. 1). Cf. note ad p. 29, l. 4.

The subject of the discussion is the *Dona astrorum* (*vide* Delambre, "Histoire de l'astronomie ancienne," ii. 546), *i.e.* the question how long a man may live, if at the moment of his birth the planets occupy such places and stand in such relations to each other as are considered the most favourable.

For a detailed explication of the astrological terms which occur in the following, and all of which are of Greek origin, I refer to the *Dictionary of the Technical Terms used in the Sciences of the Musalmans*, Calcutta, 1862.

The *Materfamilias* (كدبانو) is the *indicium corporis*, the *Paterfamilias*, the *indicium animæ* (p. 90, l. 45).

The house of the Sun is *Leo*, his *altitudo* is the 19th degree of *Aries*.
Cardines are four points of the ecliptic:

I. *Cardo horoscopi*, or *Cardo primus*, that point which rises in the east at the moment of the birth.

II. *Cardo occasûs* or *Cardo septimus*, that point which at the same moment sets in the west.

III. *Cardocoeli* or *Cardo decimus*, the point between the preceding two points, but above the earth.

IV. *Cardo terræ* or *Cardo quartus*, the point between the points I. and II., but under the earth. Cf. "Dictionary of Technical Terms," i. 465.

In a concordant masculine quarter. By *quarter* I understand the division of the signs of the Zodiac into four trigones, the *trigonum igneum*, *trigonum terreum*, etc., which are either masculine or feminine. Cf. M. Uhlemann, "Grundzüge der Astronomie und Astrologie der Alten," pp. 66, 67.

The term *concordant* is applied to any two places of the ecliptic which lie at equal distances from one of the two equinoctial points so as to form with each other the constellations called *Tasdîs* or *Tathlîth* or *Mukâbala*. Cf. "Dictionary of Technical Terms," ii. 1392, *s. v.* تناظر.

p. 91, l. 10. *Have no aspect*. The word سقط is the contrary of نظر. There are five aspects:

Tasdîs, *i.e.* the planets are distant from each other by 60 degrees.
Tarbî', *i.e.* the distance between them is 90 degrees.
Tathlîth, *i.e.* the distance between them is 120 degrees.
Mukâbala, *i.e.* the distance between them is 180 degrees.
Istikbâl, is the Mukâbala of Sun and Moon. Any other relation between two planets is called *Sukût* (*i.e.* falling out).
Cf. "Dictionary of Technical Terms," ii. 1385, *s.v.* نظر.

p. 91, l. 13. The *Caput Draconis* is that point of the ecliptic which a planet cuts when moving northward. If sun and moon meet at this point in the same zodiacal sign and degree, they are said to stand within the ὅροι ἐκλειπτικοί (Ptolemy, "Almagest," vi. cap. 5; *limites ecliptiques*, *vide* Delambre, "Histoire de l'astronomie ancienne," ii. 226), and an eclipse takes place. Every eclipse is considered as unlucky.

p. 91, l. 16. The elements of this sum (215 years) are not quite clear. If the Sun gives 120 and 30 years, Moon, Venus, and Jupiter, 25, 8 and 12 years, we get the sum of 195 years. Whence the astrologers derive the missing 20 years is not stated. They are hardly to be considered as a gift of Saturn or Mars, since they are *unlucky* stars; perhaps they are traced to the influence of Mercury. One may suppose that there is somewhere a *lacuna* in the text.

p. 91, l. 31. Read اختلافهم فى وقت العمارة الاولى instead of فى وقت العمارة الاولى فى اختلافهم (text, p. 78, l. 19).

p. 91, l. 34. The *middle conjunction* of Saturn and Jupiter is 240 years, the *minor* conjunction 20 years, the *major* c(junction 960 years. Cf. O. Loth, "Al-Kindî als Astrolog in Morgenländische Forschungen," Leipzig, 1875, p. 268.

p. 93, l. 10. *Tûzûn* was 'Amîr-al'umarâ in Baghdâd A.H. 331–334, at the time of the Khalif Almuttaḳî, whose eyes he put out. He was of Turkish origin, and commander of the Turkish troops who held Baghdâd and some other parts of central Mesopotamia.

p. 93, l. 15. *Ghurûr-aldaula.* Read *'Izz-aldaula.* Mu'izz-aldaula died A.H. 356, and 'Izz-aldaula died A.H. 367, both princes of the family of Buwaihi.

p. 93, l. 23. *Nâṣir-aldaula,* prince of Moṣul and the north of Mesopotamia, of the family of Ḥamdân, died A.H. 358.

p. 93, l. 42. Read كانت instead of كان (text, p. 81, l. 7), أنْ تَمَقَّ فيُوجَدْ (p. 81, l. 9), and والإتمام instead of والاتمام (p. 81, l. 12). (Fleischer.)

p. 94, l. 19. *Ṣâḥib.* The author means 'Ismâ'îl ben 'Abbâd, Vazîr of the Buyide prince Fakhr-aldaula. *Vide* note ad p. 72, l. 36. The time during which Fakhr-aldaula held the country of Jurjân under his sway was A.H. 372–388.

p. 94, l. 40. *'Abû Sa'îd Shâdhân* is not known to me from other sources. A man called Shâdhân is mentioned by Yâḳût, i. p. 204, l. 20, and Ḥâjî Khalîfa, v. p. 102.

p. 95, l. 2. According to "Dictionary of Technical Terms," i. p. 568, *retrograde motion* is any motion which does not, like that of the planets, proceed conformably with the order of the zodiacal signs.

The ecliptic is divided into twelve equal parts, called *houses.* The 12th, 2nd, 6th, and 8th *houses* are called *Domus cadentes.*

p. 95, l. 22. *'Abû-'Iṣma.* A man of this name was general to the Khalif Alhâdî, and was killed by Hârûn Alrashîd A.H. 170. Cf. Ibn-Al'athîr, vi. p. 74. The epithet *Ṣâḥib alṣaffâr* I cannot explain.

p. 96, l. 12. Read تَحَمَّلَ instead of تَحَمُّلِ (text, p. 82, l. 21), and ومتحجّبين instead of متحيّنين (p. 83, l. 1).

p. 96, l. 25. *Jamâlabadhra,* a town in India, is not known to me; the word can be read in various ways.

p. 96, l. 43. Abû-'Abdallâh Alḥusain, etc. Alnâtilî, a native of Nâtila, a town in Ṭabaristân, is sometimes mentioned as the teacher of 'Abû-'Alî ben Sînâ. He lived in Bukhârâ, and afterwards at the court of the prince Ma'mûn ben Muḥammad of Khwârizm. Cf. my edition of the text, "Einleitung," p. xxxiv.

p. 98, l. 13. *Covering*. Read طَبَّق instead of طَبَق (text, p. 84, l. 10). (Fleischer.)

p. 98, l. 22. *In some book*. Albêrûnî does not mention the author of the work whence he took the chronological tables of the kings of Assyria; in any case it must have been derived from the "Chronicon" of Eusebius. Cf. A. Schoene, "Eusebii Chronicorum libri duo, Berolini, 1866 and 1875"; vol. i. p. 63, and vol. ii. p. 11 ff.

p. 100, l. 24. Another table of successors of Nimrod is given by Mas'ûdî, "Prairies d'or," pp. 96–100.

A similar table is also found in Albêrûnî's *Canon Mas'ûdicus* (MS. Elliot, fol. 28a).

	Years of reign.	Anni Adami.
Nimrod - - - -	59	2951
Interval after the confusion of languages and the destruction of the tower - - -	43	2994
ههسروسى - - - -	85	3079
سسروس - - - -	72	3151
حسرونوس - - - -	42	3193
ارها - - - -	18	3211
Interval - - - -	7	3218

Then follow the Assyrian kings, Belos, Ninos, etc.

p. 101. *Table of the Kings of the Chaldeans*. It is the table of Ptolemy. Cf. "Chronologie de Ptolemée," par l'abbé Halma, Paris, 1819, 2de Partie, pp. 3, 4, and "Georgius Syncellus," ed. Dindorf, Bonn, 1839, p. 390 ff.

p. 102. This table of kings of Egypt begins with the 20th dynasty of Manetho. Cf. "Eusebii Chronicorum libri duo," vol. i. p. 145; vol. ii. p. 62.

p. 103, l. 13. This table of Ptolemæans is based upon that of Ptolemy. "Chronologie de Ptolemée," par Halma, 2de Partie, p. 4. In l. 32 read: *Cleopatra, till the time when Gajus Julius obtained supreme power in Rome*. Read برومية instead of الرومية (text, p. 92, l. 15).

p. 104. The last source of this table of the Roman emperors seems to be the "Chronicon of Eusebius." Cf. also "Hamzæ Ispahanensis annalium," libri x., translation, pp. 51, 54. In the addition of the years there is a mistake; the last sum is 313, not 303.

p. 105. Part of this table of Byzantine emperors seems to have been taken from Ḥamza Isfahânî, translation, p. 52 and 55 ff. In this table the sum of the years is 526, not 528. In the text (p. 96, l. 12), read مشلمة instead of مسلمة (De Goeje).

p. 106. The tradition of the judge *Alwakî'*, see in Ḥamza Isfahâni, translation, p. 57–59. Alwakî' seems to have lived in the first half of the 4th century of the Flight, *vide* "Kitâb-alfihrist," p. 114.

The addition of the years of this table is in great confusion, and Albêrûnî has not made an attempt at correcting it.

In the text (p. 98, l. 10), read سطا instead of سطا.

p. 107, l. 1. The following chapter on Persian chronology bears a close resemblance to that of Ḥamza Isfahânî, translation, p. 6 ff.

The explication of the word Gayômarth, l. 5, see in Ḥamza, p. 48.

p. 107, l. 43. Abû-'Alî Muḥammad ben 'Aḥmad Albalkhî, mentioned only in this place, is not known to me from other sources. Ḥâjî Khalîfa, iv. p. 13, quotes from Albêrûnî.

p. 108, l. 3. The following sources of ancient Persian history are also quoted by Ḥamza, p. 7.

'Abdallâh b. Almuḳaffa' was killed in Albaṣra, probably A.H. 145. Cf. "Kitâb-alfihrist," p. 118; Ibn-Khallikân, nr. 186.

Muḥammad b. Aljahm, of the family of Barmak, lived under the Khalif Almu'taṣim (A.H. 218–227). Cf. "Kitâb-alfihrist," pp. 81, 245, 277, and notes; Ibn-Khallikân, nr. 31, p. 40.

Hishâm b. Alḳâsim and Bahram b. Mardânshâh, Zoroastrian priest in Shâpûrstown, in Persis, are mentioned in the "Kitâb-alfihrist," p. 245, among those who translated Persian books into Arabic.

p. 108, l. 19. The manuscripts have *khzûra*. My reading, *khrûra*, is a conjecture. The word may be identical with *khrûra* of the Avastâ (*vide* Justi, "Handbuch der Zendsprache," p. 92), and also with هوريا mentioned by Mas'ûdî, "Prairies d'or," ii. 88, in a very curious chapter, where the author enumerates Ahriman and his son *Hûriyâ* in a table of kings of the Syrians.

p. 108, l. 34. *A young man.* In text (p. 100, l. 7) read الشبابُ instead of الشرابُ ; يَسْتَنْفِرُ (l. 11) instead of يستفز ; and المقصِد instead of المَقْصَد (l. 12). (Fleischer.)

p. 109, l. 14. Similar tables of the words for king, emperor, prince, etc. in various languages are given by several authors, *e.g.* by Ibn-Khurdâdhbih, "Journal Asiatique," 1865, p. 249–257.

Tadan. Perhaps we must read *Tudun,* and compare the following note of the "Etymologicum Magnum," ed. Gaisford, p. 763: Τούδουνοι: οἱ τοποτηρηταὶ παρὰ Τούρκοις.

On *Sûl, vide* note at p. 37, l. 9.

The word قبّار *Ḳabbâr* (p. 110, l. 1) is supposed by my learned friends P. Lerch, of St. Petersburgh, and W. Tomaschek, of Gratz, to be a misspelling for قناز, *i.e. Knaz, Knaez* (a derivation from the Teutonic *cuninga*), a conjecture which I recommend to the students of Slavonian antiquities.

The title *Bukhârâ-Khudâh* has been found by P. Lerch on the coins of the satraps of Bukhârâ under Sasanian rule and later (as far as the time of Almahdî). The coins offer an original writing of Semitic origin; the legend is without any doubt to be read *Bukhârâ Khuddât* (or *Khuddâh, Khuddâi*). A number of these coins are found in the coin-collection of the Royal Museum of Berlin.

p. 110, l. 26. The following verses are also found in Mas'ûdî, "Prairies d'or," ii. p. 116.

p. 111. On the pedigree and family relations of the Pêshdâdhians from Hôshang till Frêdûn, cf. Bundihish, chap. xxxii. On the chronology of the Pêshdâdhians and Kayânians, *ib.* chap. xxxiv.

In the text (p. 103, ll. 11, 15), read بصنعة instead of بضيعة.

p. 112. On the descendants of Kawi Kawâta or Kaiḳubâdh and their names, cf. Nöldeke, "Kayanier im Awestâ, Zeitschrift der Deutschen Morgenländischen Gesellschaft," tom. xxxii. p. 570.

p. 113. With this table compare that of Ḥamza, translation, pp. 9, 10.

p. 114, l. 4. With this table, compare Ḥamza, pp. 17, 18.

p. 115, l. 3. A similar table occurs also in the author's *Canon Masudicus* (MS. Elliot, fol. 29a).

After the kings of Assyria and Arbaces the Median follow *the kings of Babylonia and Media.*

	Years of reign.	Anni Mundi.
Pûl بول, a descendant of Sardanapal	35	4709
Tiglatpilesar	35	4744
Salmanassar (سلمعس) *i.e.* Bukhtanaṣṣar I.	14	4758
Sanherib Sargon سحاريب سرجون	9	4767
Ezarhaddon سرحدوم	3	4770
Merodakh Baladan ben Baladan, *i.e.* Mardokempad	48	4818
Sanherib Minor	31	4849
Kiniladan مسلدن	17	4866
Nabopolassar the Magian	21	4887
His son Nebukadnezar, *i.e.* Bukhtanassar II., who destroyed Jerusalem	43	4930
Evilmerodakh ben Nebukadnezar	2	4932
His brother Belteshaṣṣar	4	4936
Darius the Median	[17]	[4953]

Then follow the kings of the Persians:

Cyrus	9	4962
His son Cambyses	8	4970
Darius the son of Vishtasp	36	5006
Xerxes, *i.e.* Xerxes Kisrâ b. Darius	20	5026
Artaxerxes (ارطحسسطا رزوح), *i.e.* Ardashîr Longimanus	41	5067
Darius Nothos	18	5085
Artaxerxes ذو الدراوير	40	5125
Artaxerxes Ochus, *i.e.* the black	27	5152
Arses ben Ochus	4	5156
Darius ben Arsaḳ	6	5162

Then follow Alexander and the Ptolemæans. In a special column the author mentions some contemporary events of Jewish, Egyptian, Greek and Roman history.

p. 115, l. 45. In the text (p. 112, l. 4) read بآذيم instead of بآذمْ (Fleischer).

p. 116, l. 8. Sa'îd b. Muḥammad Aldhuhlî is perhaps the same Dhuhlî with whom Bukhârî (died A.H. 256) had a controversy, *vide* Ḥâjî Khalîfa, iii. 172.

p. 116, l. 34. Mâh is Media or *Aljibâl* or *Aljabal* in the later geographical terminology. Read الجبال instead of the misprint احبال.

They were one of the families, etc. is a literal translation of the reading of the manuscripts, but I do not believe that this reading is correct, nor that Arabic grammar allows such a construction.

My conjecture, احرى instead of احدى is not satisfactory, as it is not conformable to the usual construction of this word.

One might think of reading أَجْسَرَ ("They were the most daring and enterprising of the petty princes," etc.), but this, too, does not seem to settle the difficulty.

I am sorry to state that I have not been able to find the original upon which the term *Mulûk-altawâ'if*, "Petty princes," has been coined.

Cf. with this passage Ḥamza, p. 30; Ṭabarî, ed. Zotenberg, i. 523 ff.; Ibn-Alathîr, i. 208–210, 271, 272; Mas'ûdî, "Prairies d'or," ii. 136.

The pedigree of Ashk is carried back to a son of Siyâwush, whose name I do not know how to pronounce. Another son of Siyâwush is mentioned by Ibn-Alathîr, i. 173 (Fêrôzad فيروزاد) and Ṭabarî, ed. Zotenberg, i. 467 (Afroud).

For another pedigree of Ashk, *vide* B. Dorn, "Sehir-eddin's Geschichte von Tabaristan, Rujan und Masanderan," p. 152.

For the chronology of the Ashkanians, cf. Mühlau-Gutschmid in "Zeitschrift der Deutschen Morgenländischen Gesellschaft," tom. xv. p. 664; Blau, *ib.* tom. xviii. p. 680; Gobineau, *ib.* tom. xi. p. 700; Mujmil-altawârîkh in "Journal Asiatique," 1841, p. 164: H. Schneiderwirth, "Die Parther," Heiligenstadt, 1874.

p. 117, l. 9. On the surnames of the Ashkanians I offer a few conjectures:

Khôshdih, i.e. *well-born*, *de race pure* = setrîvahya, *vide* Gobineau, "Zeitschrift der Deutschen Morgenländischen Gesellschaft," tom. xi. p. 702.

Zarrîn, i.e. *golden*.

Khûrûn seems to be a mistake for جودرز *i.e.* Gotarzes.

Gêsûwar, i.e. *curled*, cf. the Persian word Gêsûdâr = a man of authority.

Barâdih = ده فراخ *happy-born*.

Balûd = ده بالا *high-born*; but see note at p. 118, l. 21.

p. 117, l. 30. See this table in Ḥamza, translation, p. 10.

p. 118, l. 5. See this table in Ḥamza, p. 18.

p. 118, l. 21. Besides the name *Malâdhân* there occurs a Parthian name *Milâd*, in Mujmil-altawârîkh, "Journal Asiatique," 1843, pp. 393, 415, 416. Perhaps there is some connection between ملاد ميلاد and the surname of Fêrôz ben Bahrâm, mentioned p. 117, l. 17 (بلاد).

p. 119, l. 19. Abû-Manṣûr 'Abd-alrazzâḳ is not known to me from other sources.

p. 119, l. 37. In the text (p. 117, l. 13) read ما يظهر من instead of من ما يظهر فى

p. 120, l. 22. In the text (p. 118, l. 3) read ونقصِد, instead of ونقصُد,

p. 121, l. 6. *Shâbûrḳân.* Of this work of Mânî's very little is known, *vide* G. Flügel, "Mânî, seine Lehre und seine Schriften," Leipzig, 1862, pp. 365–367.

p. 121, l. 36. In the text (p. 119, l. 5) read خلف with the MSS., instead of خلاف

p. 121, l. 40. The following calculation is known in astrology by the name of Tasyîr تسيير (*Directio*). The calculation is this:

$$407 \times 93\tfrac{1}{4} = 37{,}925\tfrac{3}{4}.$$

If you divide this product by 360, you get a remainder of $152\frac{3}{4}$ degrees. The meaning of the $93\frac{1}{4}$ degrees, the nature of the solar cycle here mentioned, and the further details of the calculation, I do not understand, and cannot, therefore, guarantee the correctness of the text.

p. 122, l. 14. Mûsâ ben 'Îsâ Alkisrawî is also mentioned in the "Kitâb-alfihrist," p. 128. His chronological theory is stated by Ḥamza, translation, pp. 11–16.

p. 122, l. 32. For the pedigree of Ardashîr ben Bâbak, cf. B. Dorn, "Sehir-eddin's Geschichte von Tabaristan, Rujan und Masanderan," pp. 146, 151.

p. 123. With this table, cf. the history of the Sasanians according to Mirchond, translated by S. de Sacy in *Mémoires sur diverses antiquités de la Perse*, p. 273 ff.

Instead of بردة read تيردة *Tiridates*, surname of Shâpûr I.

The word ساهدة is explained by Mirchond as نيكوكار *bienfaisant* (Sacy, p. 296).

Instead of هاذدوست Mirchond has سپاذدوست

Instead of كرمان ماه Mirchond has كران مايه

Read ترك زاد instead of بول زاد with "Mujmil-altawârîkh" (*see* "Journal Asiatique," 1841, p. 265; 1843, p. 403).

I have to add in this place that opposite the name of Ardashîr ben Bâbak the MSS. have the following note:

ويلقب بالجامع لجمعه ملك الفرس وفى زمانه وضع الرد

I have not been able to make out the meaning of the last word.

In the note which is written opposite the name of Shâpûr ben Ardashîr, the MSS. have the reading استخرجت which I have altered into استخرة as the word عرد is in the masculine.

The surname of Shahrbarâz حرمان is perhaps to be read خرّمان or to be considered as a corruption of خوهان He is also called *Farkhân*.

In the text (p. 122, l. 7) read وخنق instead of وحنق.

p. 124. With this table compare Ḥamza, translation, pp. 10, 11.

p. 125. With this table, cf. Ḥamza, pp. 18, 19.

p. 126, l. 27. Jushanasptadha or Jushanastadha is the correct reading of the signs حسسسسه G. Hoffmann read first the beginning of the word as *Jushanas* or *Jushanasp* (كشن اسپ, Armenian form *Veshnasp, vide* Langlois, "Collection des historiens," etc., ii. p. 345). The second part of the compound I read *Tada* or *Tadha* (a word of unknown etymology), and found the whole name in the Armenian form of *Vishnasptad* (*vide* Langlois, "Collection des historiens de l'Arménie," tom. ii. p. 387). G. Hoffmann added a further support of this identification by pointing out the Greek form of the name, viz. Γουσαναστάδης (cf. P. de Lagarde, "Gesammelte Abhandlungen," p. 185).

p. 127, l. 23. In the text (p. 129, l. 9), read يَشْغَلَهم instead of يشغلهم; يَنْقُسّ (l. 11) instead of ينقس; and حِين (l. 14) instead of حتّى

p. 128. With this table, cf. Ḥamza, pp. 14, 15.

p. 129, l. 16. 'Aḥmad b. Alṭayyib Alsarakhsî, a pupil of Alkindî and companion of the Khalif Almu'taḍid, was killed A.H. 286. Cf. "Kitâb-alfihrist," pp. 261, 300, and Wüstenfeld, "Geschichte der Arabischen Aerzte und Naturforscher," nr. 80.

p. 129, l. 19. On the Indian astrologer Ḳanaka, *vide* "Kitâb-alfihrist," p. 270, and note.

p. 129, l. 24. In the text (p. 132, l. 10), read المُوالِي والمُعادِى instead of المَوالى والمُعادى, يَنْجَحْ حاجتهُ (l. 12) instead of يَنجِح حاجته and بالأدْلاءِ (l. 13) instead of أدِلّاء (Fleischer).

p. 130. This table contains a number of mostly well-known princes, statesmen, and generals:

No. 1 was Vazîr to the Khalif Almu'taḍid, and died A.H. 291. Cf. Weil, "Geschichte der Chalifen," iii. pp. 514, 539.

His son, 'Amîd-aldaula, is not known to me.

No. 3–5 are princes of the house of Ḥamdân in Syria (Mosul).

No. 6–11, 13, 14, 17–21, 23, are princes of the house of Buwaihi or Bûya, *vide* the pedigree of this family in F. Wilken, "Mirchond's Geschichte der Sultane aus dem Hause Bujeh," p. 12; the Turkish chronicle of Munajjim Bashy, ii. pp. 484, 488, 495, 501.

No. 12, 15, are two princes of the family of the Banû-Ziyâd of Jurjân.

No. 16 is not known to me.

No. 22, 28, 29, are the two founders of the famous Ghaznawî dynasty.

No. 24, 27, 32, belong to the family of Sîmjûr, governor of Khurâsân under the Samanide dynasty. Cf. Defrémery, "Histoire des Samanides," pp. 261, 169, 188, 201, 203.

No. 25. Abû-al'abbâs Tâsh was governor of Nîshâpûr under Samanide rule, and died A.H. 379. Cf. Defrémery, *ib.* p. 168.

No. 26. Abû-alḥasan Alfâ'iḳ, a general of the last Samanide princes, disappears before A.H. 389. Cf. Defrémery, *ib.* p. 196.

No. 31. Abû-alfawâris Begtûzûn was governor of Khurâsân and Vazîr to the last Samanide princes; he seems to have died before A.H. 389.

No. 33. Abû-Manṣûr Alp-Arslan Albâlawî was Vazîr to the last Samanide prince Muntaṣir, and was still alive when this book was composed. Cf. Defrémery, *ib.* p. 202.

p. 131, l. 18. On Bughrâkhân, prince of Kâshghar, the conqueror of Transoxiana, *vide* Weil, "Geschichte der Chalifen," iii. Anhang 1.

p. 131, l. 23. Here the author speaks of the prince of Jurjân, Ḳâbûs ben Washmgîr, to whom he has dedicated his book, *vide* note at p. 1, l. 25.

p. 131, l. 41. In the text (p. 135, l. 6) read حِدّا instead of احدا (Fleischer).

p. 132, l. 3. *Ṭailasân.* (Cf. p. 152, l. 34.) By the term *twofold* (or redoubled) *Ṭailasân*, the author means an oblong quadrangular field, divided into two equal parts by a diagonal. Ṭailasân is the name of a piece of dress, *vide* Dozy, "Dictionnaire des noms des vêtements chez les Arabes," p. 278, and Lane, "Arabic Dictionary" under this word.

p. 132, l. 7. The Greek name of the sexagesimal system is ἑξηκοστά, *vide* Delambre, "Histoire de l'astronomie ancienne," ii. pp. 577, 608 (Hexécostades). There is a chapter on the sexagesimal system of calculation in Barlaam's λογιστικὴ ἀστρονομική (Delambre, *ib.* i. 320).

p. 133. A similar table of intervals between the epochs of the various eras is also given by Delambre, "Histoire de l'astronomie du moyen âge," p. 96, on the authority of Ibn-Yûnus. In the text of this table I had to correct some mistakes:

At notes a, c. PL have the correct reading, 101 4933, guaranteed by جهطلاى. The corresponding sexagesimal numbers 54, 7, 43, 4, are wrong in all manuscripts, for they represent the erroneous number 101, 9274.

I have printed instead of them the sexagesimal numbers which represent the number 101, 4933, *i.e.*

33, 55, 41, 4.

At notes b, c, d. The reading of the manuscripts 123,8523 is wrong, for the addition of the constituent numbers gives the sum of 123,8516. Accordingly also حصصحبا must be changed into اهصصبا,

The sexagesimal numbers have also been derived from the wrong number, for 3 (not 43), 2, 44, 5, represent the number 123,8523, whilst we must read

56, 1, 44, 5

as representing the number 123,8516.

At d, read ه صد ا نو instead of ه صد ب نو

p. 134. The chapter on the chess problem I have separately edited and explained in the "Zeitschrift der Deutschen Morgenländischen Gesellschaft," tom. xxix. pp. 148–156.

Regarding the English terminology of this chapter, I must say a word to justify the use of the word *check.* If I had used the common expression for a field on the chess-board, *i.e. square,* my translation would have become very ambiguous, as frequently in *one* sentence I should have had to speak of a *square* (in the mathematical meaning) and a *square* (a field on a chess-board). The *square* (former meaning) *of the number of a square* (latter meaning) would have been intolerable. To avoid this ambiguity I have adopted the word *check* in the common meaning of *square,* as *check* seems to be the next synonymous term, meaning a quadrangular field in a piece of Scotch cloth or tartan plaid.

p. 136, l. 7. The days of the epochs of the various eras according to Ibn-Yûnus have been communicated by Delambre, "Histoire de l'astronomie du moyen âge," p. 96.

Albâṭinî's rules for the comparison of eras between each other, see *ib.* p. 41.

p. 136, l. 20. The epochal day of the *Æra Diluvii* is a Friday, *vide* Ideler, "Handbuch der mathematischen und technischen Chronologie," ii. p. 627.

p. 136, l. 26. The epochal day of the *Æra Nabonassari* is a Wednesday, that of the *Æra Philippi* is a Sunday; Ideler, *ib.* ii. pp. 627, 628. The correspondence between the I. Tôt and the I. Daimâh, is also stated by Alfarghânî, "Elementa astronomica," ed. Golius, p. 5.

p. 136, l. 30. The epochal day of the *Æra Alexandri* is a Monday; Farghânî, p. 6; Ideler, ii. 628.

p. 137, l. 9. The Syrian year commences with the 1 Oct., the Greek year with the 1 January. The interval between these two New Year's Days is 92 days.

p. 137, p. 17. The epochal day of the *Æra Augusti* is a Thursday; Ideler, ii. p. 628.

p. 137, l. 37. The epochal day of the *Æra Diocletiani* is a Wednesday, *see* Ideler, ii. 628.

p. 138, l. 9. The epochal day of the *Era of the Flight* is a Thursday; Ideler, ii. 629.

p. 138, l. 30. The epochal day of the *Æra Yazdagirdi* is a Tuesday, *see* Farghânî, p. 6, and Ideler, ii. 629.

p. 139, l. 7. Read *Alnairîzî* instead of *Altibrîzî* (also in the text, p. 142, l. 22). In the text, p. 142, l. 21, read ابو العبّاس instead of ابا العبّاس

p. 141, l. 29. The following lines (till p. 142, l. 2), are a *torso* of which I do not know a proper restoration. It seems the author gave an exposition of the length of the Jewish, the Christian, and the astronomical years, and pointed out some incongruity between Jewish and scientific astronomy. Both Jewish Years, that of R. Samuel (the Julian year), of 365 d. 6 h. and that of R. 'Addâ of 365 d. 5 h. 997 Ḥ. 48 Reg. are too long, *vide* Dr. A. Schwarz, "Der Jüdische Kalender," pp. 65, 120. In the present state of the text I am not able to say what the 165 days (p. 142, l. 2) mean.

p. 142, l. 12. The subtraction of two years in this calculation is necessitated by the Babylonian *Ordo intercalationis*, גבטבג, which the author uses in this place. Cf. p. 65, l. 6.

p. 142, l. 20. The *Assaying Circle* is based on the assumption that the Enneadecateris corresponds to 19 solar years (whilst there is a difference between them of 145 Ḥalâḳîm, *vide* p. 64, l. 16), and that the mean Lunar year has 354 days in a common year and 384 days in a leap year. The former, if compared with the Julian year, is too short by 11 days; the latter is too long by 19 days.

In the squares of the thirteenth year of the cycle read *Îlûl* 7, instead of *Îlûl* 6 (also in the text).

No regard has been had of the intercalation of the Julian years.

p. 143, l. 28. In the text (note i, last line,) read وسبع عشرة instead of وستّ عشرة

p. 144, l. 5. By *the apparent motion* the author means that motion which at any time is found by astronomical observation, no equation or correction being used.

p. 144. l. 17. This space of time, *i.e.* 2 d. 16 h. 595 Ḥ is the so-called *Character* of the Enneadecateris.

p. 144, l. 26. The 4 d. 8 h. 876 Ḥ. are the *Character* of the *Common Year*, the 5 d. 21 h. 589 Ḥ. the *Character* of the *Leap-year*. Cf. Lazarus Bendavid, "Zur Geschichte und Berechnung des Jüdischen Kalenders," Berlin, 1817, § 32.

p. 144, l. 30. These 5 d. 14 h. are the Môlêd of the Creation (ו' יד) *i.e.* Friday morning, 8 o'clock. Cf. Dr. A. Schwarz, "Der Jüdische Kalender," p. 50, note 2.

p. 145, l. 15. With the 12th year of the *Æra Alexandri* begins a new Enneadecateris of the Jewish *Æra Adami*, the 182 d. one.

The *Basis*, *i.e.* the Môlêd of *Æ. Alex.* 12 (*i.e. Æ. Adami* 3460) has been omitted in the tables of all manuscripts. It is, however, easy to find by the help of the tables on pp. 145–147. 3460 years are:

		d.	h.	Ḥ.
6 Great Cycles - - -	=	3	20	600
14 Small Cycles - - -	=	2	15	770
2 single years - - -	=	3	6	385
		9	18	675

Therefore the Môlêd of the 12th year of Alexander is 2 d. 18 h. 675 Ḥ. (cf. the astronomical calculation of this Môlêd on p. 148, l. 19).

p. 145, l. 30. The numbers of days, hours, and Ḥalâḳîm of this table the reader may check by always adding the Character of the Enneadecateris, *i.e.* 2 d. 16 h. 595 Ḥ., and by subtracting 7, as soon as the addition of the days gives more than seven days.

p. 146, l. 20. The number of days, hours, and Ḥalâḳîm the reader may check by always adding for a common year 4 d. 8 h. 876 Ḥ., for a leap-year 5 d. 21 h. 589 Ḥ., and by subtracting 7, as soon as the addition of the days goes beyond this number.

p. 147, l. 1. The Character of the Great Cycle is 5 d. 7 h. 460 Ḥ., which you get by multiplying the Character of the Enneadecateris, *i.e.* 2 d. 16 h. 595 Ḥ. by 28, dividing the sum by 7, and taking the remainder.

p. 147, l. 42. *Times*. One *time* is equal to four minutes.

p. 148, l. 15. In the following tables these measures have been used:

Character of the Enneadecateris 2 d. 16 h. $28^{\mathrm{I.}}$ $57^{\mathrm{II.}}$ $57^{\mathrm{III.}}$ $53^{\mathrm{IV.}}$

Accordingly the length of the Enneadecateris according to the sons of Mûsâ ben Shâkir was

6939 d. 16 h. $28^{\mathrm{I.}}$ $57^{\mathrm{II.}}$ $57^{\mathrm{III.}}$ $53^{\mathrm{IV.}}$

The division of this sum by the number of the lunations of the Enneadecateris, *i.e.* 235, gives the length of the synodical month as assumed by the sons of Mûsâ ben Shâkir, *i.e.*

29 d. 12 h. $44^{\mathrm{I.}}$ $2^{\mathrm{II.}}$ $17^{\mathrm{III.}}$ $21^{\mathrm{IV.}}$ $10^{\mathrm{V.}}$

Cf. p. 143, l. 28, where the same measure is mentioned, with this difference, that there the number of fifths is stated to be 12. Cf. note at p. 158.

The Character of the Common Year is

4 d. 8 h. $48^{\mathrm{I.}}$ $27^{\mathrm{II.}}$ $28^{\mathrm{III.}}$ $14^{\mathrm{IV.}}$

The Character of the Leap-Year is

5 d. 21 h. $32^{\mathrm{I.}}$ $29^{\mathrm{II.}}$ $45^{\mathrm{III.}}$ $35^{\mathrm{IV.}}$

The Character of the Great Cycle is

5 d. 5 h. $31^{\mathrm{I.}}$ $3^{\mathrm{II.}}$ $0^{\mathrm{III.}}$ $44^{\mathrm{IV.}}$

p. 150, l. 22. The *Limits* within which the Môlêd of a year may fall are determined by the four Deḥiyyôth, *i.e.* גטרד, יח״אדו, אדו and בטותקפט. Cf. Lazarus Bendavid, "Zur Berechnung und Geschichte des Jüdischen Kalenders," §§ 35–39.

On the relations between New Year's Day and the character or nature of the year, cf. Lazarus Bendavid, §§ 46–48.

p. 152, l. 34. *Ṭailasân*, *vide* note at p. 132, l. 3. Perhaps it would be better to read حمرناها instead of احمرناها (text, p. 159, l. 1).

p. 153, l. 4. In this table as it occurs in the MSS. there is a mistake. ש and כ *can* follow each other, as the *Table of Ḳebi'ôth* (at p. 154) plainly shows. Therefore, read in the text, in the corresponding field, ممكن ان تتوالى instead of ممتنع ان تتوالى.

The three values of the table give the following six permutations:

ח+ח }
כ+כ } cannot follow each other.

ש+ש }
כ+ח }
ש+ח } can follow each other.
כ+ש }

p. 153, l. 14. The *Table of Equation*, *vide* pp. 280, 281.

p. 153, l. 18. The number of 6,940 days is a round number, for in reality the Enneadecateris has only 6,939 d. 16 h. 595 Ḥ. Regarding the preponderance of the *Perfect years* over the *Imperfect* ones in the Enneadecateris, cf. Lewisohn, "Geschichte und System des Jüdischen Kalenderwesens," Leipzig, 1856, § 90. 125 months of 30 days each, and 110 months of 29 days each, give the sum of 6,940 days.

Table at p. 154. There was a fatal mistake in the first square of this table. The MSS. have *the* 20*th Îlûl* instead of *the* 19*th Îlûl* (text, pp. 166, 167). Dr. Schramm, of Vienna, kindly settled the question for me by computing the date in question by means of the formula of Gauss.

The New-Year's Day of the Jewish *A. Adami* 4754 was the

19*th Îlûl A. Alex.* 1304.

However, according to Albêrûnî, the corresponding Greek year is *A. Alex.* 1305, not 1304. This difference is to be explained in the following way:

The Jewish year 4754 falls together or runs parallel with *A. Alex.* 1305, with one difference: The New-Year's Day (or 1st Tishrî) of A. Adami 4754 was the 19th Îlûl A. Alex. 1304. The Jewish New-Year's Day (1st of Tishrî) fell 11 days earlier than the Greek New-Year's Day (or 1st of Tishrîn Primus).

Therefore—to speak accurately—the beginning (*i.e.* the first 11 days) of the Jewish year 4754 falls into the Greek year 1304, but the whole remainder of the year corresponds with A. Alex. 1305.

This seems to be the reason why the author has in this table compared the Jewish year 4754 with the Greek year 1305.

The Jewish New-Year always precedes the Greek New-Year by a small number of days, *vide* the Assaying Circle on p. 142.

The table comprises the years of Alexander 1305–1818, *i.e.* 532 years, or one Great Cycle of 28 Small Cycles.

The *Ordo Intercalationis* in each Small Cycle is גבהטיזג *i.e.* the 3rd, 5th, 8th, 11th, 14th, 16th, 19th years of the cycle are leap-years.

It is a noteworthy fact that in every 247 years (*i.e.* 13 Small Cycles) nearly (not accurately) the same Ḳebî'ôth return, which the reader will find confirmed if he compares the years 1305 ff. with 1552 ff. and 1799 ff. On this subject, cf. Lazarus Bendavid, "Zur Berechnung und Geschichte des Jüdischen Kalenders," § 45.

p. 155, l. 26. On the two beginnings of a Jewish month, or the two Rôsh-Ḥôdesh, cf. Lazarus Bendavid, § 11.

In the following tables I have printed the real 1st of a month in Arabic numerals, and the fictitious first of a month, *i.e.* the last day of the preceding *complete* month in Latin numerals.

p. 157, l. 10. The computation of this table rests on the theory that between the Môlêd of one month and that of the following there is an interval of 29 d. 12 h. 793 Ḥ. The half of this (the Fortnight) is 14 d. 18 h. $396\frac{1}{2}$ Ḥ.

The Character of the month is 1 d. 12 h. 793 Ḥ. *i.e.* the Môlêd of a month falls by 1 d. 12 h. 793 Ḥ. later in the week than the Môlêd of the preceding month.

The Character of the Fortnight is 0 d. 18 h. $392\frac{1}{2}$ Ḥ. The table consists of additions of these two values.

p. 158. The checking of this table gave some difficulty, as in the column of the fourths the fractions have been omitted in all the manuscripts of the text, whilst in the computation they have not been disregarded.

This table shows that Albêrûnî reckoned the interval between two consecutive conjunctions (after the sons of Mûsâ b. Shâkir) at

$$29\text{ d. }12\text{ h. }44^{\text{I.}}\ 2^{\text{II.}}\ 17^{\text{III.}}\ 21^{\text{IV.}}\ 12^{\text{V.}}$$

The half of this is

$$14\text{ d. }18\text{ h. }22^{\text{I.}}\ 1^{\text{II.}}\ 8^{\text{III.}}\ 40^{\text{IV.}}\ 36^{\text{V.}}$$

or

$$14\text{ d. }18\text{ h. }22^{\text{I.}}\ 1^{\text{II.}}\ 8^{\text{III.}}\ 40\tfrac{3}{5}^{\text{IV.}}$$

With this measure, cf. my conjecture on p. 143, l. 28, and note at p. 148, l. 15.

Accordingly the Character of this synodical month is

$$1\text{ d. }12\text{ h. }44^{\text{I.}}\ 2^{\text{II.}}\ 17^{\text{III.}}\ 21^{\text{IV.}}\ 12^{\text{V.}}$$

i.e. the beginning of a month falls by so much later in the week than that of the preceding month. The half of this Character is

$$0\text{ d. }18\text{ h. }22^{\text{I.}}\ 1^{\text{II.}}\ 8^{\text{III.}}\ 40\tfrac{3}{5}^{\text{IV.}}$$

The table consists of additions of these two values.

p. 159, l. 11. The reason of the following calculation is this, that Passover always falls on the 163rd day from the end of the year. The division of 163 by 7 gives 2 as remainder.

If, therefore, you add 2 to the week-day of the Passover of a year, you get the week-day on which the New-Year's Day of the following year falls.

p. 159, l. 29. The *universal equations* refer to the various inequalities in the rotation of both sun and moon, and they serve the purpose of changing their *real* motion into *mean* motion.

p. 160, l. 6. In the text (p. 176, l. 21), read وَاقْطِفُوا instead of واقطعوا.

p. 160, l. 29. In the text (p. 177, l. 10), read آنْضَخُوا instead of آنضخوا

p. 161, l. 4. The number 350 is the multiplication of the 7 years of the Cycle of Shâoû' by the 50 years of the Cycle of Yôbêl. After this cycle of 350 years the single years of both cycles in question return again in the same order.

p. 163, l. 20. *With sediment.* Read بُدْرِيَّة instead of بدرية (text, p. 182, l. 12).

p. 163, l. 38. The solar year of $365\frac{1}{4}$ days, *i.e.* the Julian year, is called the year of Rabbi Samuel, whilst the year of 365 d. $5\frac{3791}{4104}$ h., the second of the two kinds of solar years which occur in Jewish chronology, is called the year of R. 'Addâ bar 'Ahabâ. Converting this latter space of time into Jewish measures we get

365 d. 5 h. 997 Ḥ. 48 Regâ'îm.

This length of the solar year has been found by dividing by 19 the Enneadecateris of Meton (6,939 d. 16 h. 595 Ḥ.), which comprehends 235 synodical months of Hipparchus, and which has been adopted by the Jewish chronologists. Cf. Dr. Ad. Schwarz, "Der Jüdische Kalender," p. 65 ff.

p. 164, l. 1. For an astronomical examination of the following chapter (as far as p. 167), I refer the reader to *Eine Berechnung der Entfernung des Sonnen-Apogaeum's von dem Frühlingspunkte bei Albêrûnî Mitgetheilt von Prof. Ed. Sachau und Dr. Joh. Holetschek* (p. 19 ff), in the "Sitzungsberichte der Kaiserlichen Akademie der Wissenschaften in Wien, Phil.-hist. Classe," 1876, February.

p. 167, l. 7. Abû-Naṣr Manṣûr, etc., a mathematician and astronomer, lived in Khwârizm and Ghazna and died, as it seems, in the latter place in the first quarter of the 5th century of the Flight. Cf. the text, "Einleitung," p. xxxiii.

p. 168. The Teḳûfôth are the chronological, not the astronomical year-points. Their calculation is based upon the Julian years of Rabbi Samuel.

The following are the elements of this calculation:

1. The year contains four quarters, each of 91 d. 7 h. 540 Ḥ. Dividing this by 7, you get the remainder of 7 h. 540 Ḥ., *i.e.* $7\frac{1}{2}$ hours.

2. The *Character* of the Teḳûfâ is $7\frac{1}{2}$ hours, which is the amount of the precession of each year-point within the week. This precession amounts for one complete year to 30 hours or $1\frac{1}{4}$ day.

3. If you multiply 30 hours by 28 and divide the product by 24, you get no remainder, which means that after a cycle of 28 such years the year-points fall again on the time within the week.

4. The question is: whence to begin with this calculation? with the Teḳûfâ of Tishrî or that of Nîsân?

The author fixes the Teḳûfâ of Tishrî on the 5th Tishrî, a Wednesday, 9 o'clock in the morning, *i.e.* 4 d. 15 h. after the Môlêd of Tishrî.

By subtracting herefrom the amount of the weekly precession of two Teḳûfôth (*i.e.* 15 hours), the author finds the first hour of the night of Wednesday (or, according to our method, Tuesday, 6 o'clock in the evening,) as the time of the Teḳûfâ of Nîsân, *i.e.* 4 d. 0 h. after the Môlêd of Nîsân.

5. In the *Table of Teḳûfôth* the author has assumed as the beginning of his calculation the time of noon (of Wednesday), *i.e.* 4 d. 18 h. instead of the sunset (of Tuesday) or 4 d. 0 h.

On this subject, cf. Dr. Ad. Schwarz, "Der Jüdische Kalender," pp. 65–69.

p. 169, l. 10. The names of the planets as given by the author are well known in later Hebrew. As a matter of interest for the history of Hebrew pronunciation, I mention the spelling of חַמָּה=حمّو and كبخو حمّو =כּוֹכַב חמה which reminds one of the pronunciation of the Jews of Galicia.

p. 172. Regarding Oriental names of the planets, I refer the reader to Chwolsohn, "Sabier und Sabismus," ii. pp. 156–175.

In the square of the Syriac names of Venus there occur two other names, which I have not been able to decipher. The one, دكو is, perhaps, a corruption for بيدوك

p. 174. The author shows that a year-point, as calculated by the Teḳûfôth of Rabbi Samuel (*i.e.* according to the Julian year), in no way agrees with reality, *i.e.* with a year-point as determined by astronomical observation, and that on the other hand the Teḳûfôth, as calculated by the system of Rabbi Addâ, come pretty near reality. The proof for this latter assertion has fallen out, as the chapter is *a torso*.

Here (l. 21) the author states that the first Teḳûfâ of Tishrî fell 5 d. 1 h. after the Môlêd of the year, whilst on p. 168, l. 19, he has said that fell 4 d. 15 h. after the same Môlêd. I cannot account for this divergency.

p. 174, l. 16. The constituent parts of this sum are the following:

		d.	h.	Ḥ.
8 Great Cycles - -	=	1,554,490	11	440
26 Small Cycles - -	=	180,431	22	350
9 years or 111 months -	=	3,277	21	543
Sum -		1,738,200	7	253
		−5	1	
Remainder -		1,738,195	6	253

This is the interval between the Teḳûfâ of Tishrî of the first year of the *Æra Adami* and the Môlêd of A. Alex. 1311.

The division of this sum by 364¼ days gives 4,758 Julian years, and a remainder of 335¾ d. 253 Ḥ. *i.e.* one year minus 29 d. 11 h. 827 Ḥ.

Sunday, 7 h. 253 Ḥ. of daytime is 0 d. 19 h. 253 Ḥ., which, added to 29 d. 11 h. 827 Ḥ., gives the sum of

30 d. 7 h.

If we count 30 d. 7 h. from the beginning of a Sunday (*i.e.* the preceding sunset), the 1st of Elûl, we come as far as a Tuesday night, 7 h., the first of Tishrîn Primus.

In l. 33, read 7 h. instead of 9 h., and in the text, p. 194, l. 15, read الى سبع ساعات instead of الى تسع ساعات

p. 175, l. 2. The year of Rabbi Addâ contains 365 d. $5\frac{3791}{4104}$ h. Of this kind of fractions (*i.e.* 4104th parts of an hour) one day contains 98,496.

The following is the conversion of

1,738,195 d. 6 h. 253 Ḥ

into these fractions:

1,738,195 d. =	171,205,254,720
6 h. =	24,624
253 Ḥ. =	$961\frac{2}{5}$
Sum -	$171,205,280,305\frac{2}{5}$.

Hence it is evident that in the number

$171,280,305\frac{2}{5}$ (line 5, for so it is to be read)

205 millions have fallen out.

If we divide this sum of 4104th parts of an hour by 35,975,351 (which is the solar year of R. Addâ, reduced into the same kind of fractions), we get as quotient 4,758 years, and a remainder of 350 d. $21\frac{463}{4104}$ h.

If we compare this remainder with that of the former calculation, *i.e.*

335 d. 18 h. 253 Ḥ., we get a difference of 15 d. 3 h $\frac{810}{4104}$ Ḥ., which means that, according to Rabbi Addâ, the Teḳûfâ of Tishrî of A. Alex. 1311 falls by 15 d. 3 h. $\frac{810}{4104}$ Ḥ. earlier than that of Rabbi Samuel.

This difference shows that the system of Rabbi Addâ comes pretty near astronomical truth, for, whilst *his* autumnal equinox fell 15 d. 3$\frac{810}{4104}$ h. earlier than that of R. Samuel, the astronomical equinox fell 14 days earlier, as the author says himself on p. 174, l. 35.

p. 174, l. 21. Read آقسس instead of آقمس in the text, p. 194, l. 9.

p. 175, l. 5. That part of the text which is missing in this place (*i.e.* between the words تسمتاها and علامات, in the text, p. 194, l. 21, not between the number and وخمسى) originally contained rules by which to find the week-days on which the years of the eras of the Deluge, Nabonassar, and Philippus commence. Of the chapter relating to the Era of Alexander only the end is extant.

The table on p. 175 contains a cycle of 28 Julian years, after which the single years begin again on the same week-days.

The reason why the beginning of Tishrîn I. in the first year of the cycle is fixed upon 2, *i.e.* Monday, is this, that Monday is the epochal day of this era. Cf. L. Ideler, "Handbuch," ii. p. 628.

p. 176, l. 1 ff. Similar rules for the derivation of the beginnings of the years of the different eras are also given by Delambre, "Histoire de l'astronomie du moyen âge," p. 41.

The epochal days of the single eras are also given by Delambre, p. 96.

p. 176, l. 27. This rule for the derivation of the *Signum Muḥarrami* of any year of the Flight is very intricate, and the author does not explain the principle upon which it is based.

Most people take Thursday, others Friday, as the epochal day of the era of the Flight.

The lunar year of this era is generally reckoned at

$$354\tfrac{11}{30}\text{ d.} = 354\text{ d. } 8\text{ h. } 48',$$

but in reality the mean lunar year is longer. Ideler ("Handbuch," ii. p. 479) reckons it as

$$354\text{ d. } 8\text{ h. } 48'\,36'',$$

and the author seems to reckon it as

$$354\text{ d. } 22'\,1'' = 354\text{ d. } 8\text{ h. } 48\text{ min. } 24\text{ sec.}$$

(It must be noticed that in the former number *minutes* are 60th parts of a *day*, *seconds* 60th parts of a 60th part of a *day*, whilst in the second number minutes are 60th parts of an *hour*, seconds 60th parts of a 60th part of an *hour*.)

The author does not explain why he adds 34 to the minutes. To add 5 days and 34 minutes is the same as if you add 6 days. In this case we must assume Friday (6) as the epochal day of the era, and the addition of 6 days brings us back to Sunday, the beginning of the week in which the Flight occurred (cf. p. 177, l. 43 ff., and p. 180, l. 19).

Further: why does the author count all minutes above 15 as *one hour*, whilst, according to the general practice, the minutes below 30 ought to be disregarded, and those above 30 to be counted as one hour?

The intricacies of this rule have not revealed to me the mystery of their mathematical ratio. As it seems, the author intended by some contrivance to meet the incorrectness of the common year (of $354\frac{11}{30}$ d.) being too short.

p. 176, l. 38. The sum of the days of two months is 59, which, divided by 7, gives the remainder of 3, *i.e.* the day on which a month begins, advances in two months as far as 3 days within the week.

p. 177, l. 2. Muḥammad ben Jâbir Albattânî, a famous astronomer of Ḥarrânian origin, died A.H. 317; *vide* "Kitâb-alfihrist," p. 279.

Ḥabash the mathematician, a native of Marw, author of famous astronomical tables, *vide* "Kitâb-alfihrist," p. 275, and notes, and Ḥâjî Khalîfa, v. p. 515.

p. 177, l. 44. Read تبيل instead of تنبل in the text, p. 197. l. 15.

p. 179. This table is the invention of the mathematician Ḥabash, indicating the *Signa Muḥarrami* for 210 years; but some sectarian has in every place added 5 to the number of days, and thereby changed it into a table indicative of the *Signa Ramaḍâni* for the cycle of 210 years. In this form the table is given by Albêrûnî.

The title of the table, as given by me, must be corrected: "Table showing on what week-days the Ramaḍâns of the single years of the cycle of 210 years commence." Accordingly also the superscription of col. B. is to be altered.

For the intercalary system of the lunar calendar I refer the reader to L. Ideler, ii. p. 479 ff. As the lunar year is reckoned as 354 days, 11 days must be intercalated within 30 years.

After the cycle of 30 years the New-Year days do not again fall on the same week days, as there is a remainder of 5 days. There is no remainder of days if this cycle is repeated seven times, *i.e.* after a cycle of 210 years the New-Year days fall again on the same week-day.

This is the reason why the table was constructed for a period of 210 years, cf. p. 180, l. 26.

The following is the *Ordo intercalationis* according to which Ḥabash has constructed his table.

	Cycle of 30 years.	Portio intercalanda.		Cycle of 30 years.	Portio intercalanda.
		d.			d.
	1	$\frac{11}{30}$	L.	16	$\frac{26}{30}$
L.	2	$\frac{22}{30}$		17	$\frac{7}{30}$
	3	$\frac{3}{30}$		18	$\frac{18}{30}$
	4	$\frac{14}{30}$	L.	19	$\frac{29}{30}$
L.	5	$\frac{25}{30}$		20	$\frac{10}{30}$
	6	$\frac{6}{30}$	L.	21	$\frac{21}{30}$
	7	$\frac{17}{30}$		22	$\frac{2}{30}$
L.	8	$\frac{28}{30}$		23	$\frac{13}{30}$
	9	$\frac{9}{30}$	L.	24	$\frac{24}{30}$
	10	$\frac{20}{30}$		25	$\frac{5}{30}$
L.	11	$\frac{31}{30}$		26	$\frac{16}{30}$
	12	$\frac{12}{30}$	L.	27	$\frac{27}{30}$
L.	13	$\frac{23}{30}$		28	$\frac{8}{30}$
	14	$\frac{4}{30}$		29	$\frac{19}{30}$
	15	$\frac{15}{30}$	L.	30	$\frac{30}{30}$

According to Ḥabash, the following years of the cycle of 30 years are leap-years:

2, 5, 8, 11, 13, 16, 19, 21, 24, 27, 30,

whilst, according to the common *Ordo intercalationis*, the following years are leap-years:

2, 5, 7, 10, 13, 16, 18, 21, 24, 26, 29,

or

2, 5, 7, 10, 13, 15, 18, 21, 24, 26, 29.

The principle of Ḥabash is obvious: He intercalates the *portio intercalanda* as one whole day, when the fraction has risen to more than $\frac{20}{30}$, *i.e.* $\frac{2}{3}$rds of a whole day.

Ḥabash has used Friday as the epochal day, because IV. (Signum Ramaḍânî) minus 5 gives VI. (Friday) as the Signum Muḥarrami.

In the text (*vide* the screw-figure, p. 198) there are four mistakes:

1. In the first square د (the Signum of the first Ramaḍân) has fallen out.
2. The Signum Ramaḍânî for the year 9 has fallen out, viz. د
3. The Signum of the year 25, read د instead of و
4. The Signum of the year 131, read د instead of و

p. 180, l. 7. *Tabula mediorum.* The word *Wasaṭ* or *Medium* means the *corrected* or *mean* motion of any celestial body.

p. 180, l. 27. Read اقل instead of اولى in the text, p. 198, l. 22 (Fleischer).

p. 181. The Corrected Table of the author contains the *Characters* of the single years of the cycle of 210 years, *i.e.* the remainders which you get if you divide the sum of the days of the years by 7.

The addition of 5 d. 34′ shows that the table is calculated for Friday as the epochal-day.

It must be kept in mind that in order to find the *Signum Muḥarrami* for any year, we must look out in the *Corrected Table* for the *Signum* of the preceding year; *e.g.* to find the *Signum Muḥarrami* of A.H. 100, we take the *Signum* of the year 99,

viz.
$$\begin{array}{r} 5\text{ d. }18' \\ +5\text{ d. }34' \\ \hline 10\text{ d. }52' \\ -7\text{ d.} \\ \hline 3\text{ d. }52' = \text{IV. or Wednesday.} \end{array}$$

The author does not explain what system of intercalation he follows.

p. 182, l. 1. The following passage and table are also found in the *Kosmographie* of Alḳazwînî, ed. Wüstenfeld, p. 74.

The Octaeteris of lunar years is the basis of the Turkish calendar, *vide* Ideler, ii. p. 564. It rests on the observation that the beginnings of consecutive cycles of eight years fall nearly on the same identical time of the week, but there is a difference of four minutes, *i.e.* the beginning of one Octaeteris, falls by four minutes later than that of the preceding one.

If we compute the beginnings of the first Octaeteris by the help of the corrected table, we get the following Signa:

Years of the Cycle.		Signa Muḥarrami.		
1	- -	3	=	$4\frac{11}{30}+6$
2	- -	7	=	$1\frac{22}{30}+6$
3	- -	5	=	$6\frac{3}{30}+6$
4	- -	2	=	$3\frac{14}{30}+6$
5	- -	6	=	$0\frac{25}{30}+6$
6	- -	4	=	$5\frac{6}{30}+6$
7	- -	1	=	$2\frac{17}{30}+6$
8	- -	6	=	$6\frac{28}{30}+6$

I do not see the reason why the author orders 4 to be added to the complete years of the *Æra Fugæ* (ll. 4, 5).

The first *Signum Muḥarrami* of the table belongs to the second year of the Flight. In l. 7, read *under* 8 instead of *under* 7.

'Aḥmad ben Muḥammad ben Shihâb is not known to me from other sources. The "Kitâb-alfihrist," p. 282, mentions a mathematician Aḥmad ben Muḥammad, a contemporary of Muḥammad ben Mûsâ, who died A.H. 259.

p. 183, l. 12. 'Abû-Ja'far Alkhâzin, a famous astronomer and mathematician, *vide* "Kitâb-alfihrist," p. 282. He was a contemporary of Abû-Zaid Albalkhî, who died A.H. 322. Cf. "Kitâb-alfihrist," p. 138, and G. Flügel, "Grammatische Schulen der Araber," p. 204.

p. 183, l. 13. The Characters of 30, 10, 5 years, and of 1 year, as given by the author, will be found to agree with the *Corrected Table*, if converted into the sexagesimal system.

Character of	30 years	=5 d.	=5 d. 0′
"	10 years	=1 d. 16 h.	=1 d. 40′
"	5 years	=0 d. 20 h.	=0 d. 50′
"	1 year	=4 d. $8\frac{4}{5}$ h.	=4 d. 22′

The remainder of the rule does not require an explanation.

p. 183, l. 34. The second rule of Alkhâzin is as correct as the first one, but it is unnecessarily complicated.

The *character* of the lunar year is $4\frac{11}{30}$ d. It is easy to multiply any number of years by 4 (or half the number by 8), but for the multiplication by $\frac{11}{30}$ Alkhâzin has sought for a simplified method.

$\frac{11}{30}$ of a number is nearly equal to $\frac{3}{8}$ of it, *i.e.* $\frac{3}{8}$ of a number are more than $\frac{11}{30}$ of it by $\frac{1}{120}$ of the number, or $\frac{1}{60}$ of half the number, *e.g.*:

$$\frac{3}{8} \text{ of } 60 = 22\frac{1}{2}$$
$$\frac{11}{30} \text{ of } 60 = 22$$

The difference between both numbers is $\frac{1}{2}$, *i.e.* $\frac{1}{120}$ of 60 (or $\frac{1}{60}$ of 30).

If, therefore, we multiply a number of years by $\frac{3}{8}$ (*i.e.* if we multiply half the number by 3 and divide the product by 4), we must subtract from the product $\frac{1}{120}$ of the number (or $\frac{1}{60}$ of half the number), in order to get $\frac{11}{30}$ of the same number of years.

Example: A.H. 90.

The number of complete years is 89, an odd number.

We subtract 1 year, and write down its *character*, *i.e.* 4 d. 22′,
Half of the remainder (88)=44.

I.) 44×8=352 d.
II.) 44×3=132 : 4=33 d.

352 d.
33 d.
6 d. (we add 6, taking Friday as epochal day, in agreement with Wüstenfeld's Tables, whilst Alkhâzin adds 5, taking Thursday as epochal day.)

sum 391 d.
−0 d. 44′ (60th parts to the amount of half the number.)

390 d. 16′
+4 d. 22′ (*character* of the one year.)

394 d. 38′ (these 38′ are counted as one day.)

Therefore 395 : 7=remainder 3=C.
i.e. A.H. 90 commenced on a Tuesday (cf. Wüstenfeld, "Vergleichungs-Tabellen").

p. 184, l. 24. The same rule for the *Æra Yazdagirdi*, *vide* in Delambre, "Histoire de l'astronomie du moyen âge," p. 41.

p. 184, l. 33. On the *Æra Magorum*, cf. Ed. Sachau, "Zur Geschichte und Chronologie von Khwârizm I." ("Sitzungsberichte der Kais. Wiener Akademie der Wissenschaften, phil.-hist. Classe," 1873, tom. 93, p. 485).

p. 184, l. 42. The author's report on the intercalation of the ancient Persians, *vide* on p. 38.

p. 186, l. 27. Ibn-Sankilâ (the son of Syncellus?) is not known to me from other sources.

p. 187, l. 13. 'Abdallâh b. 'Ismâ'îl is unknown to me, but 'Abd-almasîḥ Alkindî seems to be the famous philosopher of this name. As an authority on Sabians he is also quoted by the "Kitâb-alfihrist," p. 318, *vide* also Chwolsohn, "Sabier und Sabismus," ii. pp. 3 and 56.

p. 187, l. 37. *Ṭailasân*, name of a piece of dress, cf. note at p. 132, l. 8.

p. 188, l. 35. *Aljâmida* was a large village in the district of Wâsiṭ, between this town and Baṣra. Yâḳût, ii. p. 10.

A *Nahr-alṣila* in Wâsiṭ is mentioned by Yâḳût, iv. p. 841.

A place, *Al-ja'far*, I do not know, but Al-ja'farî was a castle in the neighbourhood of Samarrâ, built by the Khalif Almutawakkil, *vide* Yâķût, ii. p. 86.

p. 188, l. 44. On the Σαμαναῖοι, *vide* a note in "Kitâb-alfihrist," p. 345.

p. 189, l. 2. *Bahâr* is the Sanskrit word *vihâra*; of *Farkhâra* I do not know the origin. The author seems to think of the Buddhistic monuments of the Kâbul valley.

p. 189, l. 14. On the sects of Bardesanes and Marcion, cf. "Kitâb-alfihrist," i. pp. 338, 339, G. Flügel, Mânî, "Seine Lehre und seine Schriften," 1862, pp. 159, 161.

p. 189, l. 20. The *Termini* ὅροι are an astrological term, meaning the division of each zodiacal sign into five parts. These parts stand under the influence of the planets (except Sun and Moon). They are determined differently in different systems (that of the Egyptians and that of Ptolemy).

p. 189, l. 43. The chapter on Mânî, cf. with G. Flügel, Mânî, "Sein Leben und seine Schriften," Leipzig, 1862.

p. 190, l. 37. Yâķût (iv. p. 317) mentions the canal of Kûthâ, but he does not mention the name of *Mardînâ* (sic.).

p. 191, l. 1. Yaḥyâ b. Alnu'mân, the Christian, is not known to me from other sources.

p. 191, l. 19. Jibrâ'îl b. Nûḥ is not known to me. Yazdânbakht was a Manichæan chief in the time of the Khalif Ma'mûn, *vide* Flügel, Mânî, pp. 108 and 99, etc.

p. 191, l. 44. On this prince Marzubân b. Rustam, *vide* note at p. 47, l. 32.

p. 192, l. 6. The "Kitâb-alfihrist" mentions two *books on Mazdak*, one by Ibn-almuķaffa' (p. 118), and one by 'Abân b. 'Abd-alḥamîd (p. 163). In the same book, p. 342, a chapter on Mazdak.

p. 192, l. 26. This correspondence took place A.H. 10, cf. Ibn-Hishâm, pp. 965. 946; Ibn-al'athîr, ii. 227; Ibn-Ķutaiba, "Kitâb-alma'ârif," p. 206.

p. 193, l. 3. In a different form this verse is quoted by Ibn-Ķutaiba, "Ma'ârif," p. 206.

p. 193, l. 6. The story of the idol that was eaten by its worshippers is told by many Arab authors. Cf. "Muḥîṭ-almuḥîṭ," s.v. نبع.

p. 193, l. 16. On Bahâfirîd, cf. Shahristânî, ed. Cureton, p. 187; "Kitâb-alfihrist," p. 344.

The village or town of Sírâwand is also mentioned by Yâḳût, s.v. خواف ii. p. 486.

p. 194, l. 12. I do not know an 'Abdallâh b. Shu'ba in the early history of the Abbasides. In the "Kitâb-alfihrist," p. 344, l. 24, one of the two officers who were sent out in pursuit of Bahâfirîd, is called 'Abdallâh ben Sa'îd, and a man of this name is known as provincial governor of the Khalifs Hârûn and 'Amîn, *vide* Ibn-al'athîr, vi. pp. 156, 214.

p. 194, l. 22. Ibn-al'athîr, vi. pp. 25, 35, relates the coming forward of Almuḳanna' under A.H. 159, and his death under A.H. 161.

p. 194, l. 45. The history of Alḥallâj is told by Ibn-al'athîr, viii. pp. 57, 92.

p. 195, l. 33. On the literature of the *Kutub-almalâḥim*, cf. M. Steinschneider, Apokalypsen mit polemischer Tendenz ("Zeitschrift der Deutschen Morgenländischen Gesellschaft," xxviii. p. 627 ff.).

p. 195, l. 37. On Almukhtâr, who was killed A.H. 67, cf. Ibn-Ḳutaiba, "Ma'ârif," p. 204, and Ibn-al'athîr, iv. p. 220.

p. 195, l. 42. Raḍwâ is a mountain in Alḥijâz, between Yanbu' and Medina, also mentioned by Yâḳût (ii. p. 790, l. 20) as the residence of Almahdî. Cf. also Alkazwînî, "Kosmographie," ii. p. 160.

p. 196, l. 1. On the island of Barṭâ'îl, cf. Alḳazwînî, "Kosmographie," i. p. 53.

p. 196, l. 13. On the history of the Ḳarmatians, cf. De Goeje, "Memoire sur les Carmathes du Bahrain," Leyde, 1862.

p. 196, l. 21. Ṭamâm is mentioned by Yâḳût, iii. p. 547, as a town near Ḥaḍramaut.

p. 196, l. 44. Abû-'Abdallâh Al'âdî is not known to me from other sources. An Abû-'Abdallâh *Aldâ'î* (العادى a mistake for الداعى?) is known in history as a chief of the Shî'a in Dailam, *vide* Ibn-al'athîr, viii. p. 424, at A.H. 355, and viii. p. 443.

p. 197, l. 39. The Sûra called *Alzumar* is Sûra 39. These verses were also translated by de Goeje, "Mémoire sur les Carmathes," p. 51.

p. 198, l. 1. Read الْعَزَاقِر Al'azâḳir instead of الغراقر Alghrrâḳir, and *from Shalmaghân* من شلمغان instead of b. *Shalmaḳân* بن شلمقان (text, p. 214, l. 9). Cf. Yâḳût, iii. p. 314, and Ibn-al'athîr, viii. p. 216. Shalmaghân was the name of a district belonging to Wâsiṭ.

p. 199, l. 1. An extract from the author's chapter on the festivals of the Persians is given by Alkazwînî, "Kosmographie," ii. pp. 79–84; *see also* Alfarghânî, "Elementa astronomiæ," notæ, pp. 20–42.

p. 199, l. 20. I am unable to tell what the author means by the sphere of Fêrôz (not Fêrôza) and by the sphere of Afranjawî (or Ifranjawî) on p. 208, l. 15.

p. 199, l. 28. Sa'îd b. Alfaḍl. (*vide also* p. 208, l. 27) is not known to me from other sources.

Yâḳût, ii. p. 584, mentions a place, *Dummâ,* below Baghdad. Perhaps it would be preferable to read "On the mountain of Dummâ," etc.

p. 200, l. 3. Kalwâdhâ, a place not far south from Baghdad. Yâḳût, iv. 301.

The year in which 'Aḍud-aldaula entered Baghdad was A.H. 364.

On Abû-alfaraj Alzanjânî, cf. note at p. 54, l. 1.

p. 201, l. 4 ff. If popular use may more easily determine the solstices than the equinoxes, it is just the reverse for their scientific determination, as the author himself observes on p. 167, l. 2. Cf. Sachau and Holetschek, "Berechnung der Entfernung des Sonnenapogaeums von dem Frühlingspunkt," p. 25.

p. 202, l. 4. *Bûshanj.* It would be better to translate "on the mountain of Bûshanj," as Bûshanj is a village in the district of Herat, not far from the road to Nîshâpûr, *vide* Yâḳût, i. p. 758.

p. 203, l. 32. With this innovation of Shâpûr the Hero, cf. p. 209, l. 37.

p. 204, l. 8. *Afâhtar.* Apâkhtara means in the Avestâ *north,* not *south.*

p. 204, l. 51. *Manifest in the Avastâ* (*vide also* p. 205, l. 24) is the common mode of quotation in Parsee books, *vide* "Bundihish," ed. Justi, Glossary, *s.v.* پيعاك.

p. 204, l. 52. On the Gâhanbârs, cf. F. Spiegel, "Avesta," ii. p. c. and p. 4, note 1; on the etymologies of their names, *vide* A. Bezzenberger, "Einige Avestische Wörter und Formen in Göttinger Gelehrte Anzeigen," 1878, p. 251.

p. 205, l. 13. *Cashn-i-nîlûfar* means *the feast of the water-lily.*

p. 205, l. 25. *Arish.* The older form of this name is *Arshan*, *vide* Noeldeke, "Zeitschrift der Deutschen Morgenländischen Gesellschaft," xxxii. p. 570.

p. 206, l. 28. *Sâwa* was a town half-way between Rai and Hamadân, *vide* Yâḳût, iii. p. 24. A town of the name of *Andish* or *Mandish* is not known to me.

p. 207, l. 11. Here something seems to be missing, viz. that Âdhar-cashn fell on Mihr-Rôz, *i.e.* the 16th of the month.

p. 208, l. 21. Salmân Alfârisî, originally a slave of Persian descent, afterwards one of Muḥammad's companions. He died in the beginning of the reign of 'Uthmân.

p. 208, l. 25. *Alêrânshahrî* (also p. 211, l. 19) is not known to me from other sources.

p. 208, l. 28. *Shâhîn.* A mountain of this name is not known to me.

p. 208, l. 32. Alkisrawî, *vide* note at p. 122, l. 14.

p. 208, l. 33. Ḥâmîn. This place seems to be something like the *Hamêstagân*, an intermediate place between heaven and hell, *vide* West, "Mainyo-i-Khard," Glossary, p. 97.

p. 209, l. 20. *Kustîk* means girdle, still an essential part in the costume of a Zoroastrian.

p. 210, l. 15. Alnâṣir Al'uṭrûsh was a descendant of Ali, who ruled for some years in Dailam and Ṭabaristân, and was killed A.H. 304. As a missionary he had endeavoured to spread Islâm among the Zoroastrian people of these countries. Weil, "Geschichte der Khalifen," ii. pp. 613–615.

In the text, p. 224, l. 9, read الناصر *Alnâṣir* instead of الناظر *Alnâẓir.*

p. 210, l. 18. For a description of the feast of Farwardagân, *vide* F. Spiegel, "Avesta," ii., "Einleitung," p. ci.

p. 211, l. 3–8. The lines between brackets have been taken from the *Canon Masudicus* of Albêrûnî, MS. Elliot (British Museum), fol. 50A.

p. 211, l. 15. Ṭâhir b. Ṭâhir, unknown to me.

p. 212, l. 28. The name of the feast of the 15th Âbân is also in the *Canon Masudicus* (ff. 49a, 50a) written سكان.

p. 212, l. 33. The second name of this feast is written in the *Canon Masudicus* (ff. 49a, 50a) كاركل which is certainly more correct than كاكل.

p. 213, ll. 11–13. On these festivals, cf. *Canon Masudicus*, fol. 50a, ll. 18–22: "On the day of Bahmanja they cook in caldrons all sorts of plants, kernels, blossoms, and all sorts of catable meat. They drink the white Bahman-root, mixed with the purest white milk, maintaining that this helps to preserve the body and to defend it against evil.

"*Barsadhaḳ* means *above sadhaḳ*, because it precedes Sadhaḳ by five days. It is also called *Nausadha, i.e.* the new Sadhaḳ.

"*Sadhaḳ*. They say that on this day the creation of a *hundred* souls of the family of Mêshâ and Mêshâna had become perfect, and that therefore the day was called *Saddak*, i.e. *Hundred-day*. According to others there is an interval of 100 days between this day and Naurôz, if you count days and nights separately, and therefore the day was called *Hundred-day* in the same way as Nuwad-Rôz" (*vide* p. 212, l. 12).

p. 213, l. 24. Karaj, a town midway between Hamadân and Ispahân, also a village near Rai, and another between Hamadân and Nahâwand. Yâḳût, iv. p. 251.

p. 214, l. 28. Abû-'Uthmân Aljâḥiẓ is the well-known zoologist, author of a *Liber animalium, Kitâb-alḥayawân*, who died at Baṣra A.H. 255.

p. 214, l. 28. 'Ukbarâ, a village in the district of Dujail, near Ṣarîfîn and 'Awânâ. Yâḳût, iii. p. 705.

p. 214, l. 33. Aljaihânî, a famous polyhistor, vezir to the dynasty of the Samanides, beginning of the 4th century of the Flight, *vide* Reinaud, "Géographie d'Aboulfêda," i. p. lxiii.

p. 215, l. 15. Adharkhûrâ is not mentioned by Yâḳût, nor does he know anything about Kâm-Fêrôz (l. 39). Dârâ (l. 38) is not known as a place in Persis; perhaps Dârâbjird was meant.

p. 216, l. 22. In the text, p. 229, l. 16, read رز instead of رم as in the manuscripts.

p. 216, l. 31. Dînâr-Râzî. The spelling of this name is not quite certain. It is mentioned by Ibn-Ḥauḳal, ed. de Goeje, p. 275, l. pen. It is a place on the road from Jurjân to Khurâsân.

p. 216, l. 40. Zanjân, a town in Media towards Adharbaijân, not far from Ḳazwîn. Yâḳût, ii. p. 948.

p. 217, l. 25. On the *Feast of Kardfanndkhusra*, and the city of this name, cf. Yâḳût, iv. p. 258.

p. 221, l. 6. Râmush is mentioned by Yâḳût, ii. 737, as a village in the district of Bukhârâ.

p. 221, l. 21. The festivals called Mâkhîraj fell, according to the author's "Kitâb-altafhîm" (MS. of the Royal Library in Berlin, Peterm. 67, fol. 62b.), always on the 13th.

p. 221, l. 22. Instead of نكع اغام the "Kitâb-altafhîm" has بيكنج اغام *i.e.* the Âghâm or feast of Baikand, cf. Râmush-Âghâm, l. 15.

p. 221, l. 32. Al-ṭawâwîs, a town in the district of Bukhârâ, between this place and Samarḳand, *vide* Yâḳût, iii. 555. كميجكث is not known to me.

p. 222, l. 5. Shargh ("Kitâb-altafhîm," fol. 62b), called چرغ by Ibn-Ḥauḳal, p. 360, ll. 5, 6, was a large village near Bukhârâ, *vide* Yâḳût, iii. p. 276.

p. 224, l. 4. According to the "Kitâb-altafhîm" (fol. 63a) the festival Ajghâr fell on the 16th of Cîrî.

p. 224, l. 12. *Faghrubah* is *baga*=*God*, and some derivative from the root *srp* (رفتن).

p. 224, l. 18. Read *Azdâ* instead of *Azâd*, and cf. *Âzûiti*=*fat* in the Avestâ.

p. 224, l. 22. Instead of Ćîrî-Rôj the "Kitâb-altafhîm" has Cîr-rôz چهر روز.

p. 224, l. 28. Instead of *Akhîb* اخيب one may think of reading *Ikhshab* اخشب=شب *the night* (as اخشريوري=شهريور).

p. 225, l. 14. Yaḥyâ Grammaticus, a Jacobite bishop in Egypt who translated from the Greek and wrote philosophical and polemical books, lived in the first half of the 7th century of our era.

p. 225, l. 22. *Khêzh* I hold to be a derivative from the same root whence خاستن has sprung (cf. وستاخير and *uzgaêzanuha* in the Avestâ).

p. 225, l. 30. The following Chorasmian names of the Gâhanbârs are dialect-varieties of the names of the Avestâ in the following order:

Paitis-hahya (26–30 Shahrêwar).

Maidhyô-shema (11–15 Tîr).

Maidhyâirya (16–20 Bahman).

Maidhyô-zaremaya (11–15 Ardîbahisht).

Hamaçpathmaêdaya, which is omitted in this place, the five intercalary days at the end of Spendârmat.

Ayâthrema (26–30 Mihr).

p. 225, l. 33. What the author means by القرعى I do not know. قرعى means *young camels*, and قرعى means *relating to gourds* (قرع).

p. 226, l. 10. Whatever the true Chorasmian form may be, Akhar, Akhkhar, or Akhtar, it is certainly identical with the Persian اختر *Akhtar* = star.

p. 226, l. 14. The author's criticisms on the constellations of the single zodiacal signs as represented by Arabians and Chorasmians, may be compared with the book of L. Ideler, "Untersuchungen über den Ursprung und die Bedeutung der Sternnamen," Berlin, 1809.

p. 226, l. 21. *Adhûpaćkarîk* is a Bahuvrîhi compound of two words corresponding to the Persian *dû* دو = two, and *paikar* پیکر = figure.

p. 226, l. 37. Abû-Muḥammad ʻAbdallâh b. Muslim b. Ḳutaiba Aldînawarî is in Europe known as Ibn-Ḳutaiba. He was a native of Kûfa, and lived as judge in Dînawar. He died A.H. 270.

According to "Kitâb-alfihrist," p. 77, his books were highly esteemed, especially in Aljabal, *i.e.* Media, and to his Jabalî or Median character Albêrûnî seems to have certain objections (p. 227, l. 16). That one of his books which our author quotes is perhaps identical with that mentioned by "Kitâb-alfihrist," p. 78, l. 3 (كتاب التسوية بين العرب والعجم).

p. 227, l. 23. In Sûra ix. v. 98, Muḥammad blames the *'Aʻrâb*, *i.e.* the Arab Bedouins, in the strongest terms: "The 'Aʻrâb are the worst infidels and hypocrites, they do not deserve to learn the laws which God reveals unto His prophet, but God is all-knowing, all-wise," etc.

p. 227, l. 32. Three of the Sogdian names resemble the corresponding Sanskrit names:

Proshṭhapadâ = فرهمس باشت (No. 24).
Revatî = ریوات (No. 26).
Maghâ = مغ (No. 8.)

Cf. E. Burgess, Sûrya-Siddhânta ("Journal of the American Oriental Society," vol. vi. p. 327 ff.), and A. Weber, "Jenaer Literatur-Zeitung," 1877 (7 April), p. 211.

The name ستويس in No. 7 is the *çatavaêça* of the Avestâ; the name ونند, in No. 20, cf. with the *vanañt* of the Avestâ.

p. 229, l. 3. Aḥmad was the last prince but one of the ancient house of the Shâhs of Khwârizm, who undertook a reform of the calendar A. Alex 1270=A.D. 959, *i.e.* 13 years before the author was born. Cf. Sachau, Zur Geschichte und Chronologie von Khwârizm I. ("Sitzungsberichte der Wiener Akademie," phil. hist. Classe, 1873, p. 503). A short report of this reform is also found in the "Kitâb-altafhîm," fol. 63.

p. 229, l. 13. Alkharâjî and Alḥamdakî are not known to me from other sources.

p. 230, l. 8. According to the "Kitâb-altafhîm" the 1st Nâusârjî was fixed so as to fall on the 2nd Nîsân (فى اليوم الثانى من نيسان).

p. 230, l. 26. According to p. 258, l. 18, the Nile begins to rise on the 16th Ḥazîrân, *i.e.* the 16th Payni.

p. 233, l. 5. Sinân b. Thâbit died at Baghdad A.H. 331, and his father, Thâbit b. Ḳurra, A.H. 288. They were both famous as philosophers, mathematicians, and physicians, both Ḥarrânians, the last representatives of ancient Greek learning, through whom Greek sciences were communicated to the illiterate Arabs. Cf. "Kitâb-alfihrist," pp. 302, 272.

Sinân had made a collection of meteorological observations, called *Kitâb-al'anwâ*, compiled from ancient sources, and enriched by the observations of his father and his own. The work of Sinân has been incorporated by Albêrûnî into his chronology, and thereby he has preserved to us the most complete *Parapegma* of the ancient Greek world.

With the works of Sinân other works of a similar character may be compared:

Geminus, Ἐισαγωγὴ εἰς τὰ φαινόμενα, the 16th chapter, edited by Halma in "Chronologie de Ptolemée," Paris, 1819, pp. 79–87. (Cf. Boeckh, "Ueber die vierjährigen Sonnenkreise," p. 22 ff.)

Ptolemæus, φάσεις ἀπλανῶν ἀστέρων καὶ συναγωγὴ ἐπισημασιῶν edited by Halma, "Chronologie de Ptolemée."

Johannes Lydus ("Corpus scriptorum Historiæ Byzantinæ," Bonn, 1837), *De mensibus*, cap. iv., and *De ostentis*, in the same volume, pp. 357–382.

For calendaria of more recent times, *vide*

J. Selden, "De synedriis et præfecturis juridicis veterum Ebræorum 1734 (contains three calendaria).

Lobstein, "Nachrichten und Auszüge aus den Handschriften der Kgl. Bibliothek in Paris," i. pp. 415–424.

Vide Hammer, "Geschichte der Osmanischen Dichtkunst," i. pp. 76–81.

Fleischer, "Abulfedæ Historia anteislamica," p. 163 ff.

Ḳazwînî, "Kosmographie," ii. p. 75 ff. (extract from Albêrûnî).

A calendarium of Spanish-Arabic origin has been edited by R. Dozy, "Le Calendrier de Cordoue," Leyde, 1873.

Regarding the authorities quoted by Sinân, as Euctemon, Eudoxus, Philippus, Metrodorus, Dositheus, Conon, Cæsar, etc., I refer the reader to the excellent work of A. Boeck, "Ueber die vierjährigen Sonnenkriese der Alten, vorzüglich den Eudoxischen," Berlin, 1863.

p. 233, l. 9. By *Episemasia* ἐπισημασία I have translated the word *Nau'* نوء.

According to Albêrûnî, p. 339, l. 4, *Nau'* means the rising of a Lunar Station. The meteorological influence of this rising is called *Bâriḥ*; the influence of the sinking of a Lunar Station is called *Nau'*. Albêrûnî uses the word *Nau'* in either of these two meanings.

Comparing the conflicting opinions of the Arab philologists on this word (*vide* W. Lane, "Zeitschrift der Deutschen Morgenländischen Gesellschaft," iii. p. 97 ff.). I am led to believe that *Nau'* is an ancient Arabic word, probably much used in ante-Muḥammadan times, the meaning of which was no longer fully and distinctly understood by the Muslim Arabs. Afterwards when the Greek calendars were to be translated into Arabic, the word *Nau'* was used to render the Greek ἐπισημαίνει, as the comparison of Sinân's compilation with Gemînus, Ptolemy, and Johannes Lydus shows. The single days of these calendars do not correspond with each other, but the technical terms are everywhere the same.

p. 234, l. 40. Ibn-Khurdâdhbih, *vide* note at p. 50, l. 26.

p. 235, l. 19. Ispahbadhân, a town in Tabaristân, two miles distant from the Caspian Sea. Further inland in the mountains the castle *Tâḳ*, with caves and wells in the neighbourhood. Cf. Yâḳût, iii. pp. 490, 491; Ḳazwînî, ii. p. 270, l. 10 *ab inf.*

p. 238, l. 24. 'Alî b. Aljahm was a famous poet at the time of the Khalif Almutawakkil, who died A.H. 249. As he had made satirical verses on the Khalif, he fled, and was hunted about. At last, after having been wounded in a fight with his pursuers, in the agonies of expiring he is said to have recited this verse. Cf. Ibn-Khallikân, ed Wüstenfeld, No. 473.

p. 238, l. 39. Yaḥyâ b. 'Alî is not known to me from other sources.

p. 243, l. 31. Abû-Bakr Ḥusain Altammâr, a contemporary of Râzî, who died A.H. 320, is also mentioned by Wüstenfeld, "Geschichte der Arabischen Aerzte und Naturforscher," p. 46, l. 3.

p. 245, l. 29. 'Abdallâh b. 'Alî, a mathematician of Bukhârâ, is not known to me.

p. 247, l. 2. On the fire as a spherical body within the lunar sphere, cf. also Ḳazwînî, "Kosmographie," ii. p. 90; translated by Dr. Ethé, p. 185.

p. 247, l. 37. On the correspondence of Albêrûnî with Ibn-Sînâ, cf. my edition of the text, "Einleitung," p. xxxv.

p. 248, ll. 17, 31, 34. In the text, p. 257, ll. 16, 23, and p. 258, l. 2, read اورنيثيا instead of اوريسا The word اودرساوس if a genuine Greek word, might be read in various ways, but I hold it to be a mistake for اودكسوس *Eudoxus*.

p. 250, l. 15. Muḥammad b. Miṭyâr (also p. 258, l. 26) is not known to me from other sources.

p. 251, l. 33. Abû-Yaḥyâ b. Kunâsa, the author of a famous Kitâb-al'anwâ, was born at Kûfa A.H. 123, and died at Bagdad A.H. 207; *vide* "Kitâb-alfihrist," p. 70.

p. 252, l. 6 ff. The following discussion on the circumstances under which water rises, is of a technical nature, the due appreciation of which I must leave to physical scholars.

p. 255, l. 23. The word *Daḳj* does not occur in any Arabic dictionary. If the writing is correct, it is probably a word of foreign origin.

p. 255, l. 34. Kîmâk, or Kaimâk, a province of the Chinese empire, inhabited by Turkish nomades, *vide* Ḳazwînî, "Kosmographie," ii. p. 395, and Ibn-Khurdâdbih, in "Journal Asiatique," 1865, pp. 267–268.

p. 256, l. 5. *Alḳarya Alḥaditha* is not known from other sources.

p. 256, l. 10. Mihrjân was the ancient name of Isfarâ'în, a village between Jurjân and Nîshâpûr, also the name of a village in the district of Isfarâ'în and of another village between Ispahân and Ṭabs. Cf. Yâḳût, i. p. 246; iv. p. 699.

p. 258, l. 29. Hayawâniyya-sect, not known to me.

p. 261, l. 39. Abû-Nu'âs, the famous poet at the time of the Khalif Hârûn, died A.H. 199.

p. 261, l. 42. 'Alî b. 'Alî is not known to me from other sources.

p. 262, l. 37. *Naubakht.* If the text is correct, and we must not rather read *Ibn-Naubakht*, this man may have been the father of Abû-Sahl Alfaḍl b. Naubakht, librarian to the Khalif Hârûn, and a great astrologer. Cf. "Kitâb-alfihrist," p. 274.

p. 263, l. 21. Salamiyya, a village in the district of Ḥims, Yâḳût, iii. p. 123, l. 18; p. 124, l. 1.

p. 266, l. 40. *Thu'âliba.* This reading may seem doubtful, as no place of this name is mentioned anywhere. The nearest approach is *Thu'âlibât*, in Yâḳût, i. p. 925.

p. 268, l. 1. As useful material for the explication of the festal calendar of the Jews the following works have been used:

Canon Masudicus, MS. Elliot (British Museum, ff. 37b–38a).
Abulfedæ, "Historia anteislamica," ed. Fleischer, p. 156 ff.
Bartolocci, "Bibliotheca Rabbinica," ii. 553 ff.
A. G. Waehner, "Antiquitates Ebraeorum," Göttingen, 1742, sect. v.
T. C. G. Bodenschatz, "Kirchliche Verfassung der heutigen, sonderlich der teutschen Juden," Erlangen, 1748, vol. ii. pp. 87, 105.
M. Brück, "Rabbinische Caeremonialgebräuche," Breslau, 1837.
מגלת תענית ed. Jo. Mayer, Amstelodami, 1724 (cap. xii.).
מַסכת תענית *id est:* "Codex Talmudicus de Jejunio, ex Hebræo Sermone in Latinum versus commentariisque illustratus a Daniele Lundio Succo." Trajecti ad Rhenum, 1694.

p. 268, l. 20. Read דָּחִי instead of דָּחֵי. It is a Hebrew form فعيل meaning *protrusus*, i.e. *advanced* or *postponed*; feminine: דְּחִיָּה.

p. 269, l. 1. For the fasting of Gedalyâ, cf. 2 Kings xxv. 25, and Jerem. xli. 2.

p. 269, l. 20. The following story in a Hebrew garb is found in J. Zedner's "Auswahl historischer Stücke aus hebraeischen Schriftstellern," Berlin, 1840, pp. 6–11, as was pointed out to me by Prof. H. Strack.

p. 269, l. 27. Read "*quietly*" instead of "*following the course of the river.*"

p. 270, l. 29. *You shall celebrate a feast*, etc. The words وحجّوا حجّا (text, p. 277, l. 11), can only be explained as a too literal translation of וְחַגֹּתֶם אֹתוֹ חַג in Levit. xxiii. 41.

p. 270, l. 33. Abû-'Îsâ Alwarrâḳ (also p. 278, l. 22) is mentioned in "Kitâb-alfihrist," p. 338, as one of those who in public professed Islam, but in reality were heretics.

p. 270, l. 39. *On the same day* (*i.e.* the 21st) *is the Feast of Congregation*. This is a mistake. *Congregation*, or עֲצֶרֶת falls on the 22nd Tishrî, *i.e.* the following day. *Canon Masudicus* gives the following series:

21st Tishrî=ערבה
22nd „ =עצרת
23rd „ =عيد التبريك

Cf. Bodenschatz, ii. p. 235.

The word *Hârhâra* (l. 40) I cannot explain. The court of the temple is called הָעֲזָרָה and the place where the willows are gathered is called מוצא (Mishna) or ממציא (Talmud Palæstinensis).

Canon Masudicus (fol. 38a) says:

وعيد عرابا حجّ لهم حول المذبح

i.e. "The feast of Arâbhâ consists of a procession round the altar." The اورن of the text is the Hebrew אָרוֹן.

p. 270, l. 43. The *feast of benediction* is called שִׂמְחַת תּוֹרָה, *vide* Waehner, v. p. 111, and Bodenschatz, ii. p. 245.

p. 271, l. 14. *Between the 8th and 13th of this month*. As possibly between the 8th and 13th there is no Monday at all, it may happen that this fast-day does not occur in some year (if the rule is correct).

p. 271, l. 21. This fast-day falls according to Megillath-Ta'anîth on the 7th Kislêw, according to Waehner and Bodenschatz on the 28th Kislêw. Cf. on the origin of this fasting, Jerem. xxxvi. 27–32.

p. 271, l. 35. The following story occurs also in Abulfedæ, "Hist. anteislamica," pp. 160–162. Instead of the انطيخوس of the manuscripts, *Canon Masudicus* (fol. 38a) has اطناخوس *i.e.* Antiochus, and so I have translated.

In the text (p. 278, l. 11) the words الى ان يكون فى الثامنة are a rather short and incorrect expression for الى ان يكون عدد السرج ثمانية فى الليلة الثامنة.

p. 272, l. 17. Ptolemy is here called *Talmâ*, as if the initial *p* were the Coptic article. In the Megillath-Ta'anîth, which fixes this fast-day on the 8th Têbeth, he is called תלמי המלך (cap. 12).

p. 272, l. 30. According to Megillath-Ta'anîth, cap. 12, this fast-day falls on the 8th Shubâṭ.

p. 273, l. 17. In the text, p. 279, l. 20, there is a *lacuna*, which I have filled up with the help of *Canon Masudicus*, fol. 37b. and Waehner, v. *p.* 112.

p. 275, l. 8. In the "Kitâb-altafhîm" (Cod. Berolin. fol. 58b), the author says: ويعرف هذا اليوم بالكس; and in the *Canon Masudicus* (fol. 38v) he says: ويسمّى هذا اليوم الكس وهو العمل بالسريانى. In the same work, fol. 37v, the feast is called عيد الكس. Accordingly the reading كس seems to be preferable to that of مكس.

This word *Kas* seems in some way to be connected with the Syriac ܟܣܐ which means *middle of the month*, indicating the 15th, on which this feast falls.

p. 275, l. 12. *Thirty men*, a mistake for *thirty thousand*. The *Canon Masudicus* (f. 38b) has correctly ثلثون الف رجل. Cf. 1 Sam. iv. 10.

p. 275, l. 15. *Canon Masudicus* (f. 37b) mentions a feast also for the 15th Iyâr:

عيد الفسح الصغير وهو ايضا وفاة اشمويل

"The feast of the Small Passover, also the day of fast in commemoration of the death of Samuel."

p. 275, l. 19. The plural حجوة (text, p. 281, l. 16) seems to be a form coined by Albêrûnî from a singular حج *i.e.* the Hebrew חַג, for the pure Arabic word حجّ has the plural حجج

In the *Canon Masudicus* the author gives two days to the *Feast of Congregation*, the second of which is called صوم البا كورة *i.e. Fasting of the First-fruit.*

p. 275, l. 26. With this biblical quotation, cf. Exod. xxiii. 14–17; Exod. xxxiv. 22, 23; Deut. xvi. 16.

The reading of the MS. (text, p. 281, l. 20), حجابكم is unintelligible to me. My conjecture حجاجكم must be explained as the infinitive of a verb حجّ a denominative formation from حجّ=חַג.

p. 276, l. 12. The Megillath-Ta'anîth, cap. xii., has the following note on this fast-day:

נשרף רבי חנניא בן תרדיון וספר תורה עמו

p. 276, l. 18. There are two unlucky days in the Jewish calendar, the 17th Tammuz and the 9th Âbh. A short review of the disasters that have happened on these two days is given in the *Massekheth-Ta'anîth*, p. 55.

The text (p. 282, l. 11) is not quite correct in the manuscripts. With the text as given by me, cf. *Canon Masudicus* (ff. 37b 38b):

صوم ابتداء حصر اورشليم فى الانهدام

p. 276, l. 37. With the story of the lamp, cf. 2 Chron. xxix. 7, and *Megillath-Ta'anîth*, cap. xii. pp. 113, 122.

The name of the prophet as given in the manuscript, ايحون etc., seems to be corrupt. There was at that time a prophet '*Ôdêd* (2 Chron. xxviii. 9), and Isaia (Isaia vii. and viii.), but no prophet of such a name.

In the *Canon Masudicus* (f. 39a), the author relates that it was the king Ahaz احاز ملكهم who extinguished the lamp. Therefore I have changed ايحون into احوز and "*Ahaz the prophet*" seems to be a mistake for "*Ahaz the king*."

p. 277, l. 4. There is no *lacuna* as I have indicated in the text, p. 283, l. 3.

This fast-day is fixed by some on the 7th Elûl, by others on the 17th (*vide* Bodenschatz, Waehner, and Megillath-Ta'anîth). If, therefore, the author says that some people place this fast-day within the last week of the month, I know nothing by which to test this assertion.

In later times there was a fast-day on the last of Elûl as an atonement for the sins of the past year, but this is an institute of modern times. Cf. Bodenschatz, p. 88, § 2, l. 1.

Also in the *Canon Masudicus* the *Fasting of the Spies* (on the 7th Elûl) is the last of the feast- and fast-days of the Jewish year. Therefore the words (text, p. 283, l. 3) "*Lücke*," etc. are to be cancelled.

p. 277, l. 10. To this table of Deḥiyyôth may be added that 'Arâbhâ can never be ז *i.e.* Saturday.

The reason why the feast-day cannot fall on certain days of the week is this, that they wanted to prevent two non-working days from immediately following each other, as this might interfere with the practical welfare of the people. Besides, certain feasts cannot fall on a Sabbath, because they require a certain amount of work (*e.g.* the burning of Haman, etc.).

p. 277, l. 29. The words ذكر القرية (text, p. 283, l. 11) are the rendering of the bible-words זִכְרוֹן תְּרוּעָה מִקְרָא־קֹדֶשׁ. Cf. P. de Lagarde, "Materialien zur Kritik und Geschichte des Pentateuchs," ii. p. 134. The Peshîṭtâ has translated thus: ܥܡ ܥܗܕܢܐ ܥܗܕܢܐ ܘܫܒܬܐ ܢܗܘܐ ܠܟܘܢ ܢܝܚܐ ܘܙܘܥܐ ܕܫܝܦܘܪܐ ܩܪܝܐ ܘܩܕܝܫܐ ܢܗܘܐ ܠܟܘܢ ܗܘܦ

p. 280. In the text, pp. 286, 287, I read the ح in the fourth columns, as محال *i.e.* impossible.

It stands always with ا which the MSS. write with black ink. It ought, however, to have been written with red ink, since a year beginning with ا is *impossible*. Therefore, in order to indicate what elsewhere is indicated by the red ink, the letter ح=محال has been added.

Impossible means that a year beginning on such a day is a calendarian impossibility.

Necessary means that in a year beginning on such a day there is no possibility of a דְּחִיָּה, *i.e.* of postponing or advancing.

Possible means that a year beginning on such a day is possible, if the year be ח (*Imperfect*) and a common year, whilst it is impossible, if it be ח and a leap-year, and *vice versâ*.

The single numbers of the table may easily be checked in this way:

1. The intervals between New-Year's Day and Kippûr, *i.e.* the 10th Tishrî and 'Arâbhâ, *i.e.* the 21st Tishrî, are the same in every kind of year.

2. The intervals between New-Year's Day and the other three festivals, *Pûrîm*, *Pêsaḥ*, and *'Asêreth*, are different in different years.

In a *common year*—

Pûrîm is in ח the 161st, in כ the 162nd, in ש the 163rd day of the year.

Pêsaḥ, in ח the 191st, in כ the 192nd, in ש the 193rd day.

'Asêreth, in ח the 241st, in כ the 242nd, in ש the 243rd day.

In a *leap-year*—

Pûrîm is in ח the 191st, in כ the 192nd, in ש the 193rd day of the year.

Pêsaḥ, in ח the 221st, in כ the 222nd, in ש the 223rd day.

'Aṣêreth, in ח the 271st, in כ the 272nd, in ש the 273rd day.

These sums of days are to be divided by seven, and the remainders represent the distances from New-Year's Day.

To this table the author has referred the reader already on p. 153, l. 15. It shows why two *intermediate* years, *i.e.* כ cannot follow each other, in this way:

Of the seven years כ only those two are possible that begin with III. and V.

1. If, now, after a common year כ beginning with III., another year כ

were to follow, it would begin with a VII., *Saturday*, and that is impossible, as the table shows.

If after a leap-year כ beginning with III., another year כ were to follow, it would begin with II., which is again impossible, as the table shows.

2. If after a year כ beginning with V. another year כ were to follow, it would begin with II. in a common year, with IV. in a leap-year; and both cases are impossible, as the table shows.

p. 283, l. 14. For the emendation of the names of saints in the following chapter, I have used the *Menologium Græcorum*, jussu Basilii Imperatoris olim editum Græce et Latine. Studio et opera Albani. Urbini, 1727.

p. 283, l. 35. Regarding the degrees of the clergy of the Oriental churches, cf. Assemani, "Bibliotheca Orientalis," iii. pp. 788–790; also Ami Boué, "La Turquie d'Europe," iii. p. 421; Maurer, "Das Griechische Volk," Heidelberg, 1835, i. pp. 389, 403, 410.

p. 284, l. 16. Abû-alḥusain 'Aḥmad b. Alḥusain Al'ahwâzî. An author of this name is mentioned by Hâjî Khalîfa, iv. p. 81.

p. 284, l. 20. There are certain Greek names which I have not been able to decipher, Χρνσχς (l. 22), ακσιοτς (l. 39), and some others. The answer to these questions I must leave to those who are intimately acquainted with the archæology of the Byzantine empire.

The word خرسنخس might be a corruption for خرسقبس χωρεπίσκοπος, but in that case the explication which Abû-alḥusain gives is not correct. According to the explication, one would expect the word 'Αρχιμανδρίτης.

On παρακοιμώμενος, *vide* Du Cange, "Lexicon infimæ græcitatis," where it is explained as ἀρχιευνοῦχος, πραιπόσιτος.

A word ῥογάτωρ (p. 285, l. 4), I do not know. It seems to be a derivation from ῥόγα *present* and *stipend*, *vide* Du Cange (rogator).

On the μαγλαβίτης, cf. Reiske, commentary to "Constantini Porphyrogeniti, De cerimoniis aulæ," ii. pp. 53–55.

On τεσσαρακοντάριος, *a soldier who received 40 aurei as stipend*, etc. *vide* Du Cange.

p. 285, l. 23. Muḥammad b. Mûsâ b. Shâkir, the eldest of three brothers, all great scholars in mathematical and technical sciences *for* whom Greek books were translated into Arabic. Muḥammad died A.H. 259, *vide* "Kitâb-alfihrist," p. 271.

p. 285, l. 26. 'Alî b. Yaḥyâ, cf. note at p. 38, l. 5.

p. 286, l. 31. The reading *Cornutus* is not beyond all doubt. A saint of this name is mentioned in the "Menologium Græcorum," at 12th Sept.

p. 287, l. 3. Johannes the Father seems to be identical with Johannes Scholasticus, who died A.D. 578, as Patriarch of Constantinople. He had made a new arrangement of the *canones* of the church. Cf. K. Hase, "Kirchengeschichte" (8th edition), p. 149.

p. 287, l. 14. On this Modestus, *vide* Le Quien, "Oriens Christianus," iii. 102 ff. p. 258. He was Patriarch after A.D. 614.

p. 287, l. 15. The word *Sîsîn* is perhaps to be changed into *Sîs*, *vide* p. 289, l. 40.

p. 287, l. 17. The epithet البولوطاكى seems to have been derived from the Syriac Bible, where Joseph is called ܒܘܠܘܛܐ Luke xxiii. 51.

p. 287, l. 19. Alma'mûn b. Aḥmad Alsulamî Alharawî (also p. 297, l. 25) is unknown to me.

p. 287, l. 39. Abû-Rûḥ is not known to me from other sources.

p. 288, l. 34. *Every child is born*, etc. On this well-known tradition, cf. L. Krehl, "Ueber die koranische Lehre von der Prædestination," p. 99 (in "Berichte der Kgl. Sächsischen Gesellschaft der Wissenschaften," hist.-phil. Classe, 1870, the 1st July).

p. 289, l. 39. The reading *Belesys* is entirely conjectural. The word might also be read *Blasius* (*vide* Calendar of Armenian saints, Assemani "Bibliotheca Orientalis," iii. 1, p. 645, at 10th Shubât), but *Blasius Episcopus* was killed by the Romans, not by the Persians.

p. 292, l. 12. *Jûrî-roses*, the most famous of the east, so called from Jûr, a town in Persis, *vide* Yâḳût, ii. p. 147; Ḳazwînî, "Kosmographie," ii. 121.

p. 293, l. 10. The author seems to mean that two straight lines, cutting each other in the middle, and connected at the bottom by another straight line give the Cufic form of the word *no*, *i.e.* لا, as it frequently occurs in monumental writing.

p. 293, l. 30. On the wood *Pæonia*, cf. Ḳazwînî, "Kosmographie," i. p. 260.

p. 295, l. 15. Dâdhîshû', author of a commentary on the Gospel, is not known to me from other sources.

p. 295, l. 25. A *Cyriacus anachoreta* is mentioned in the "Menologium Græcorum," at 29th Sept.

p. 296, l. 18. *Dometius* is mentioned in the "Menologium Græcorum," at 7th Aug., and *Thuthael* (l. 26) *ib.* at 5th Sept.

p. 297, l. 29. With his tale regarding the blood of John, cf. Ibn-al'athîr, i. pp. 214–216, also iv. p. 140.

p. 297, l. 36. Cf. with this Ṭabarî, traduit par Zotenberg, i. p. 569; Ibn-al'athîr, i. pp. 208, 215.

p. 298, l. 13. *Church of the Sweepings* كنيسة القمامة is a corruption of كنيسة القيامة *Church of the Resurrection*, invented by Muslim malice. Cf. Farghani, "Elementa astronomiæ," ed. Golius, p. 138.

p. 298, l. 15. Among 40 martyrs who were killed by the Persians, Maruthas (Assemani "Bibl. Orient." i. pp. 192, 193) mentions Paul, Sabinus (not Sabinianus) and ܛܛܐ. Besides Tattâ also the form ܛܛܘܢ *Tattûn* occurs, *ib.* i. p. 190, col. 1.

p. 299. The following chapter contains the computation of Easter, as Albêrûnî had learned it from the Christians of his time.

The Easter which he means is the πάσχα σταυρωσιμόν, not the πάσχα ἀναστασιμόν; cf. Augusti, "Christliche Archaeologie," ii. p. 30.

The chief elements of this computation are the following:

1. Easter depends upon the Jewish Passover, *i.e.* the full moon of the Jewish month Nîsân.

2. Counting $1\frac{1}{2}$ lunation, *i.e.* 44 d. 7 h. 10′ backward from the full moon of Nîsân, you find the new moon of the preceding month.

3. The Monday nearest to this new moon, if it does not fall earlier than the 2nd Shubâṭ nor later than the 8th Adhâr, is the beginning of Lent.

4. Lent ends on the 49th day from the beginning. It begins on Monday and ends on a Sunday. Easter lies between this Sunday and the preceding Sunday, *i.e. Palmarum.*

To use modern language: The πάσχα σταυρωσιμόν was celebrated on some day between Palm Sunday and Easter Sunday, but on what particular day it was celebrated the author does not explain.

p. 299, l. 27. It may seem doubtful whether I have correctly interpreted the word ابديقطيونا (text, p. 302, l. 13). *Indictio* is certainly not identical with the Great Cycle of 28 years.

p. 300, l. 4. The Jews count 3,448 years between Adam and Alexander. If you divide this sum by 19, you get 9 as a remainder, *i.e.* the first year of the *Æra Alexandri* is the 10th year of the cycle.

The division of 5180 by 19 gives a remainder of 12, *i.e.* the first year of the *Æra Alexandri* is, according to the Christians, the 13th year of the cycle.

p. 300, l. 8. The whole passage, from "it is also well known," etc. (line 8), till "you get as remainder 5180 years (as the interval between Adam and Alexander)," in the Arabic text, p. 302, l. 17 (وهو المشهور الخ) till p. 303, l. 2 (هى خمسة آلاف ومائة وثمانون), seems to be a later interpolation.

p. 300, l. 9. Khâlid b. Yazîd is considered as the father of alchemy among the Arabs, *vide* Ḥâjî Khalîfa, v. p. 280.

p. 300, l. 30. The Easter-limits extend over 28 days, *i.e.*
from 21st Adhâr to 18th Nîsân.
The limits of Lent extend over 48 days, *i.e.*
from the 2nd Shubâṭ to 22nd Adhâr,
or
from the 8th Adhâr to 25th Nîsân.

The smallest interval between the beginning of Lent and Easter is 42 days; the greatest, 49 days.

p. 301, l. 23. By 44 d. 7 h. 6 min. the author means $1\frac{1}{2}$ synodical month. One synodical month is reckoned at
29 d. 12 h. 44′.

p. 301, l. 36. If full moon falls on a Sabbath, the 21st Adhâr, count 44 d. 7 h. 6′, *i.e.* 45 days backward, and you find the new moon of the preceding month, viz. the 4th Shubâṭ, a Wednesday in a common year, a Thursday in a leap-year.

The next Monday is the preceding one, the 1st Shubâṭ in a leap-year, the 2nd Shubâṭ in a common year.

As, however, the 1st Shubâṭ lies *before* the *Terminus Jejunii*, the year in question must be a common year.

p. 302, l. 1. If the Jewish Passover fell into Nîsân, and the Jewish year was a leap-year, it might seem doubtful to the Christians whether they were to make the new moon of Adhâr I. (*i.e.* Shubâṭ, February) or that of Adhâr II. (*i.e.* Adhâr, March) the basis of their computation of the beginning of Lent. The author gives the computations for both cases.

I. Computation on the basis of the new moon of Adhâr II. (Adhâr) of a leap-year:

If full moon falls on the 18th of Nîsân, a Sunday, count 44 d. 7 h. 6′ backward, and you find the new moon of Adhâr II., or the Syrian Adhâr (March), viz. the 5th Adhâr, a Friday.

The next Monday is the following one, the 8th Adhâr, which is the latest day of the *Terminus Jejunii.*

II. Computation on the basis of the new moon of Adhâr I. (Shubâṭ) in a leap-year:

If full moon falls on the 18th Nîsân, a Sunday, count $2\frac{1}{2}$ lunations, *i.e.* 73 d. 19 h. 50′ backward, and you find the new moon of Adhâr Primus or Shubâṭ, viz. the 5th Shubâṭ, a Thursday, if the Christian year is a common year.

The next Monday is the preceding one, *i.e.* the 2nd Shubâṭ.

This calculation is impossible, for the reason which the author states on p. 302, ll. 14–16.

If the corresponding Christian year is a leap-year, we find the 4th Shubâṭ, a Thursday, to be the beginning of Lent.

The next Monday is the preceding one, *i.e.* the 1st Shubâṭ, and this date is impossible, as being outside the *Terminus Jejunii* (the 2nd Shubâṭ to 8th Adhâr).

p. 302, l. 30. In the genuine 20 *canones* of the Synod of Nicæa there is no mention of Easter. In the Arabic collection of 84 *canones* the 21st refers to Easter (*vide* Mansi, "Collectio nova," ii. p. 1048), but this collection is an invention of later times, *vide* Hefele, "Theologische Quartalschrift," Tübingen, 1851, p. 41.

That, however, the bishops of the Synod of Nicæa had handled the Easter-question, is evident from the letter of Constantine, *vide* "Eusebii vita Constantini," iii. p. 18.

To decide the question whether Albêrûnî is right in ascribing the authorship of this *Chronicon* to Eusebius and the Synod of Nicæa, I must leave to scholars in Church History.

p. 303. This table contains the beginnings of Lent for a period of 532 years (*i.e.* 19 × 28). It resembles the period of Victorius, cf. Ideler, "Handbuch der mathematischen und technischen Chronologie," ii. p. 278.

p. 304, l. 3. This fragment treats of Still-Friday in the Easter-week. Cf. Augusti, "Christliche Archaeologie," ii. p. 136.

p. 304, l. 5. *New Sunday*, or Dominica nova, Dominica in albis, also Dies neophytorum, cf. Augusti, ii. p. 302.

p. 304, l. 33. The *canon* to which the author refers is this: "Quoniam sunt quidam qui in die Dominico genuflectunt et ipsis diebus pentecostes, ut omnia similiter in omni parochia serventur, visum est Sanctæ Synodo *ut stantes Deo orationes effundant*." This is the last (the 20th) of the *canones* of the Synod of Nicæa, *vide* Mansi, "Collectio conciliorum nova," ii. p. 678, § 20. Hence it is evident that Albêrûnî used the ancient and genuine *canones* of this Synod, not the later spurious collection, *vide* Hefele, Acten des ersten Concils zu Nicaea, "Theologische Quartalschrift, Tübingen," 1851, p. 41.

p. 304, l. 38. On the ܥܘܕܪܐ ܕܡܫܝ̈ܚܐ and ܕܒܣܘܬܐ ܕܗܘܕܝܐ, cf. Assemani, "Bibliotheca orientalis," ii. p. 305, and iii. 2, p. 382; also Acts, iii. 2–8.

p. 305, l. 4. The *Table of Fasting* of seven columns, mentioned in this place, is not found in the manuscripts. It must have fallen out.

p. 306. With the chapter on the festivals of the Nestorians, cf. a similar chapter in Assemani, "Bibliotheca Orientalis," iii. p. 2, and "Abulfedæ Historia anteislamica," p. 162 ff.

p. 308, l. 3. Johannes Cascarensis and Phetion were killed in Ḥulwân between A.D. 430 and 465, *vide* Assemani, "Bibl. Orient." ii. p. 403, col. 1.

Phetion Martyr is mentioned by Assemani, iii. 2, p. 386, on the 25th October; *vide* also W. Wright, "Catalogue of the Syriac manuscripts of the British Museum," part iii. p. 1134, No. 66.

Yâḳût (ii. p. 683) mentions a "Monastery of Phetion" دير فثيون.

p. 308, l. 8. I have not been able to decipher the name قونيا.

p. 308, l. 16. The name Ḳûṭâ is known to me from Yâḳût, ii. p. 689, where a *Monastery of Ḳûṭâ* دير قوطا is mentioned.

p. 308, l. 19. The Syriac form of the name Solomonis (mother of the Maccabæans) is ܫܡܘܢܝ, *vide* W. Wright, "Catalogue," etc. iii. p. 1137, col. 1.

p. 308, l. 25. I do not know a saint of the name of برصيا or مرصيا.

p. 308, l. 37. The word احادر seems to be a corruption. In the *Canon Masudicus* (fol. 47a) a *Friday of Eliezer* اليعازر is mentioned, but this Friday falls 40 days after the beginning of Lent.

p. 308, l. 40. The two saints mentioned in this paragraph are not known to me. ܒܪܫܝܐ may be ܒܪܟܝܐ Bereshyâ, *i.e.* *Berekhyâ*, and a martyr child, the son of Cyrus, is mentionod by W. Wright, "Catalogue," etc., iii. p. 1136, col. 2.

p. 309, l. 1. Two Armenian martyrs of the names of Ourenius and Surinus are mentioned under the 11th March by Assemani, "Bibl. Orientalis," iii. 2, p. 650. The name *Duranus* occurs, *ib.* p. 653, under the 2nd October.

p. 309, l. 26. The common year is here called ܫܢܬܐ ܡܫܬܘܝܬܐ (text, p. 311, l. 18). This term the author must have borrowed from the source whence he took the information of this chapter, for everywhere else he calls the common year ܫܢܬܐ ܦܫܝܛܬܐ.

p. 310. This table is based upon the beginning of Lent, which in a common year falls between the 2nd Shubâṭ—8th Adhâr; in a leap-year, between the 3rd Shubâṭ—8th Adhâr. The festivals keep certain invariable distances from this date. The Latin numerals at the top of the table denote the week-days on which the single festivals fall.

p. 311, l. 7. The reason why Christmas and the Commemoration of Our Lady Mary should not immediately follow each other seems to be this, that each two feasts must be separated not only by a night, but by one complete day. The idea seems to be this, that each feast requires a certain preparation on the preceding day.

p. 311, l. 11. The '*Ibâdites* are the Arab tribe to which the poets 'Adî b. Zaid and Zaid b. 'Adî belonged, *vide* Masudi, "Prairies d'or," iii. p. 205, and Caussin, "Essai sur l'histoire des Arabes avant l'Islamisme," ii. 148.

p. 311, l. 18. The expression حيث انتصرت العرب الخ (text, p. 314, l. 8) was coined upon the pattern of a word of the Prophet, *vide* Ibn-Al'athîr, i. p. 352, l. 16.

The name '*Anḳafîr* is mentioned by Freytag and Muḥîṭ-almuḥîṭ, not by the Turkish Ḳâmûs; it occurs also in Ḥamza Isfahânî, text, p. 112, l. 6; translation, p. 88.

p. 311, l. 22. On the Ninive-fast, cf. Assemani, "Bibliotheca Orient." iii. 2, p. 387.

p. 313. This table is based upon the cycle of 28 Julian years, after which every date falls again on the same week-day.

In the *Column of the number* I have marked the leap-years by a star.

As all the festivals of this table are attached to certain week-days, they wander about within the space of 7 days.

In common years each festival falls *one* day, in leap-years *two* days, later than in the preceding.

The double column of numbers under the head *Commemoration of Solomonis*, is to be explained in this way:

The first column represents the common computation of this day, whilst the second column represents the practice of the people of Bagdad, who made it fall a week later. So, according to *Canon Masudicus* (fol. 46b).

The numbers of the column *Feast of Dair-altha'âlib* are corrupt in the manuscript; I have computed them according to the rule given on p. 308, ll. 26–30. The numbers I. 29, etc., mean that in this year the feast is celebrated twice, on the 1st Tishrîn I. and on the 29th Îlûl, whilst the blank means that in this year the feast is not celebrated at all, which is the case if the last Sunday of the year (or the *Commemoration of Bar Safâ* in the preceding column) falls on the 24th Îlûl.

The Latin numerals at the top of the table denote the week-days of the single festivals.

p. 314. The following chapter, of which the text in many passages seems to be corrupt beyond hope, is to be compared with the researches of Prof. Chwolsohn ("Die Sabier und der Sabismus"). It would require a special commentary of its own, and whoever wants to undertake it must be thoroughly imbued with the knowledge of the last phases of Neo-Platonism and of the popular belief and superstition of the dying Greek heathendom.

The author distinguishes between the heathens of Ḥarrân and the Mandæans of the south of Babylonia. His festal calendar is that of the people of Ḥarrân.

This calendar, on pp. 315–318, the author has transferred from a book of *Alhâshimî* just as it was, with all the mis-spellings, faults, and lacunas, and since the time of the author the text has become worse and worse. He expresses the hope (on p. 318) that he will be able one day to correct this chapter, but his hope does not seem to have been fulfilled.

It would have been more cautious not to translate this chapter at all, but I hope that the reader will accept my translation with indulgence as a first essay at unravelling the mysteries of this enigmatic but nevertheless most valuable chapter.

p. 314, l. 7. On the Shamsiyya, *vide* Chwolsohn, i. pp. 292–295.

p. 315, l. 4. Chwolsohn, i. p. 140, gives for this event another date, viz. A.H. 215.

p. 315, l. 26. Muḥammad b. 'Abd-al'azîz Alhâshimî is not known to me from other sources. Perhaps he was a son of Abû-Muḥammad 'Abd-al'azîz b. Alwâthiḳ, mentioned in "Kitâb-alfihrist," p. 39.

p. 315, l. 36. Dhahbâna, a place near Ḥarrân, *vide* Chwolsohn, i. p. 306, note 6; ii. p. 630; and also Yâḳût ii. p. 725, *s.v.* الذهبانية In Syriac the place is called ܕܗܒܢܐ, *vide* Assemani, "Bibliotheca Orientalis," i. p. 278.

p. 317, l. 1. The text of this passage is very uncertain. If the moon stands on the 31st Adhâr in *Cancer*, it must on the 8th Adhâr have stood in Gemini or in the first degree of Cancer.

p. 317, ll. 21, 22. On Dair-Kâdhî and Dair-Sînî, cf. Chwolsohn, ii. pp. 24, 37, 40, 41, 630, and 808.

p. 317, l. 38. I do not know a word *Kurmûs*, but Ṭurmûs طرموس من دقيق means *panis in cinere coctus*, *vide* Chwolsohn, ii. p. 27 at 15, Tammûz.

p. 318, l. 4. Dailafatân, an old name of Venus, known as Δελέφατ from the Greek lexicographers, *vide* Hesychius, ed. M. Schmidt: Δελέφατ ὁ τῆς Ἀφροδίτης ἀστήρ ὑπο Χαλδαίων. The name occurs also in Assyrian.

p. 318, l. 32. From the rules relating to the computation of the Lent of the Ḥarrânians the author infers that their year was not a vague lunar year, running through all the seasons, but a kind of luni-solar year, like that of the Jews, which, though based upon lunar years, is made to agree with the course of the sun by means of the cycle of 19 years. Further, he infers that the Ḥarrânian Fast-breaking depends upon the vernal equinox and their New Year upon the autumnal equinox.

The *double-bodied signs* البروج ذوات جسدين are *Gemini*, *Virgo*, *Arcitenens*, *Pisces*.

The *inclining signs* البروج المنقلبة are *Aries*, *Cancer*, *Libra*, *Caper*.

The stable signs البروج الثوابت are *Taurus*, *Leo*, *Scorpio*, *Amphora*.

In the following we give a survey of what seem to have been the elements of the calendar of Ḥarrân:

1. The day begins with sun-rise.
2. The month begins on the second day after conjunction.
3. The year begins with Kânûn II.
   or with Tishrîn I. (with new moon, the next to the autumnal equinox),
   or with the winter solstice.
4. Lent begins the 8th Adhâr, when the sun stands in Pisces, and ends after 31 or 29 days, when the sun stands in Aries. The last quad-

rature of the moon before Passover (of the Jews) is the time of their fast-breaking.

Hence the luni-solar character of the calendar.

5. As *Terminus Paschalis* Albêrûnî adopts the time between the 16th Adhâr and 13th Nîsân, *i.e.* 28 days.

6. The difference between the lunar and solar years (11 d. 5 h. 45′, etc.) they insert in every fourth month as a leap-month, viz. Hilâl Adhâr I. (after Shubâṭ).

p. 320, l. 14. I am not able to explain this table in a satisfactory manner; however, I offer a few remarks to which an examination of the nature of this table has led me:

1. *On an average* the common year is counted at 354 days, the leap-year at 384 days. The one is 11 days shorter, the other 19 days longer, than the Julian year of 365 days.

2. *Once*, however, in the 19 years of the cycle, the common year has been counted at 353 days, so that between its beginning and that of the following year there is an interval of 12 *days*.

This subtraction of one day is perhaps to be explained in this way:

12 common years, each of 354 days	=4,248 days.
7 leap-years, each of 384 days	=2,688 days.
19 years - - - -	=6,936 days.

As, however, 19 solar years, each of 365 days, give only the sum of 6,935 days, this difference of one day was to be removed by the subtraction of one day in one of the 19 years of the cycle.

3. The following are the years which are counted at 353 days:

Year 5 in col. 3.
Year 4 in col. 4.
Year 2 in col. 6.
Year 2 in col. 8.
Year 2 in col. 10.

4. In the columns of the Sabian New-Year and Fast-breaking we find the *Ordo intercalationis* [illegible] *i.e.* the 2nd, 5th, 7th, 10th, 13th, 16th, 18th years are leap-years, whilst in the columns of the Corrected Passover, Mean Fasting of the Christians, and the 1st of Tishrîn I. of the following year, we find the *Ordo intercalationis* [illegible] *i.e.* the 3rd, 5th, 8th, 11th, 14th, 16th, 19th years are leap-years.

I cannot say on what principle this difference rests.

5. The New-Year in col. 3 and Fast-breaking in col. 4 are Ṣâbian.

The Corrected Passover in col. 6 is Jewish.

The Mean Fasting and the 1st of Tishrîn I. of the following year in col. 10 are Christian.

6. The computation of the Corrected Passover rests upon an astro-

nomical computation of the vernal equinox, special regard being had to the precession of the equinoxes which had been neglected.

The author does not communicate this astronomical computation of his.

7. The comparison between col. 3 and col. 10 shows that the Sabian New-Year falls always by one day (in some cases by two days) earlier than the Christian New-Year.

8. The author gives directions (p. 320, l. 7), to add 16 to the years of the *Æra Alexandri*, or to subtract 3 therefrom. This indicates that at the epoch of the era which he uses, already three years of an Enneadecateris had elapsed. Which era this is, I have not been able to find out.

p. 322, l. 28. Ibrâhîm b. Sinân, known as an astronomer, of Ḥarrânian origin, lived about the middle of the 10th century of our era. Cf. Fihrist, p. 272, and Wüstenfeld, "Geschichte der Arabischen Aerzte," p. 37.

p. 324, l. 4. Albaghâdî died A.H. 204, *vide* Ḥâjî Khalîfa, v. p. 411.

p. 324, l. 22. The reading *Alrâbiya* is uncertain, as I do not know a place of this name in Ḥaḍramaut. A place of this name is mentioned by Yâḳût, iv. p. 391, *s.v.* المارحين but it cannot be identical with that mentioned by Albêrûnî.

Some of these fairs are also enumerated by Ḳazwînî, "Kosmographie," ii. p. 56.

p. 325, l. 1. A festal calendar of the Muslims is also found in *Canon Masûdicus* (fol. 48) and in Ḳazwînî, "Kosmographie," i. p. 67 ff. (taken from Albêrûnî).

p. 326, l. 27. The same verses occur in Ibn-al'athîr, iv. p. 76.

p. 328, l. 21. Two of these verses occur also in Ḳazwînî, "Kosmographie," i. p. 68.

p. 328, l. 39. *On this day the pilgrimage of the forty men*, etc. This fact is not mentioned by *Canon Masûdicus* nor by Ḳazwînî. I have not been able to find out what is the historical basis of this statement.

p. 329, ll. 3–26. This passage is missing in the manuscripts. I have supplied it from the *Canon Masûdicus*.

p. 330, l. 4. Alsalâmî. A poet of this name of the 4th century of the Flight is mentioned in "Kitâb-alfihrist," p. 168. Another author of the same name, author of the "Kitâb-nutaf-alṭuraf" is mentioned by Yâḳût, iv. p. 203.

p. 330, l. 32. Ḥasan b. Zaid, the Alîde prince of Tabarîstân and Jurjân, died A.H. 270, *vide* Weil, "Geschichte der Chalifen," ii. p. 450.

p. 330, l. 37. Khalaf b. 'Aḥmad ruled over Sijistân at the end of the 4th century of the Hijra. It was Maḥmûd ben Sabuktegîn who put an end to his rule. Weil, "Geschichte der Khalifen," iii. p. 62.

p. 331, l. 39. The Arabic text of this passage (text, p. 333, l. 12, ولم يذكر فى قولهم آلج) seems to be corrupt.

p. 332, l. 7. The *verse of the cursing* is Sûra iii. 54. Muḥammad's negotiations with the Christians of Najrân are related in a special chapter of Ibn-Hishâm, ed. Wüstenfeld, p. 401 ff.; *vide* A. Sprenger, "Leben und Lehre des Mohammed," iii. p. 488 ff.

p. 332, l. 21. The tree Yaḳṭîn is mentioned in the Coran, Sûra xxxvii. 146.

p. 333, l. 17. Thabîr is a hill near Mekka, cf. Yâḳût, i. p. 917, where also this saying is mentioned (line 18). Muḥîṭ-almuḥîṭ, i. p. 1077, l. 8 ff.

p. 333, l. 19. Ibn-al'a'râbî, a famous philologist of the school of Kûfa, died A.H. 231, *vide* G. Flügel, "Grammatische Schulen der Araber," p. 148.

p. 334, l. 1. The battle of Alḥarra occurred A.D. 683, the 26th Aug. The troops of the Khalif Yazîd b. Mu'âwiya stormed Medîna under the command of Muslim b. 'Uḳba. Weil, "Geschichte der Chalifen," i. p. 331.

p. 335, l. 15. *Alkulthûmî* is not known to me from other sources.

Ibrahîm b. Alsarrî Alzajjâj, a famous philologist, died A.H. 310, *vide* "Kitâb-alfihrist," p. 61. Ibn-Khallikân, ed. Wüstenfeld, No. 12, mentions his "Kitâb-al'anwâ."

Abû-Yaḥyâ b. Kunâsa, *vide* note at p. 251, l. 33.

Abû-Ḥanîfa Aldînawarî, a grammarian and mathematician, died A.H. 150; cf. Ibn-Khallikân, ed. Wüstenfeld, No. 775, and "Kitâb-alfihrist," pp. 78, 88.

Abû-Muḥammad Aljabalî, better known under the name of Ibn-Ḳutaiba, *vide* note at p. 226, l. 37, and "Kitâb-alfihrist," p. 88.

Abû-alḥusain is 'Abd-alraḥmân b. 'Umar *Alṣûfî*, who died A.H. 376. His book of fixed stars has been translated by Dr. Sjellerup, St. Petersburg, 1874. Cf. "Kitâb-alfihrist," p. 284, and notes.

p. 335, l. 22. *Jufûr*, *vide* note at p. 15, l. 15.

p. 336, l. 20. *Verses and rhymed poetry*. The author means the *Kutub-al'anwâ* (*vide* p. 337, l. 29), frequently mentioned in the more ancient Arabic literature. As far as I know, there is no standard work of this kind in the libraries of Europe, but it is highly desirable to search for one and to publish it, since most likely many of the verses, to the interpretation of which these books are dedicated, may claim a much higher antiquity than the Kaṣîdas and Rajaz-poems of the earliest Arab poets.

p. 337, l. 40. *Aḥmad ben Fâris*, a native of Rai, a famous philologist and writer both in prose and verse, died at Rai, A.H. 390. Cf. Ibn-Khallikân, ed. Wüstenfeld, No. 48.

p. 338, l. 7. The author alludes in this passage to certain events in his own life, but unfortunately in such vague terms that we learn very little for his biography. Was he banished from the court of the prince? Of what kind were his troubles, mental or material? At present it is impossible to give an answer to these questions.

p. 338, l. 39. *Ibn-alraḳḳâ'* is not mentioned by Ibn-Khallikân, nor by the Fihrist, nor by Ḥâjî Khalîfa.

p. 339, l. 35. Abû-Muḥammad Ja'far Alfazârî is unknown to me. Perhaps he was a relative of the two brothers Muḥammad and Isḥâḳ, the sons of Ibrâhîm Alfazârî, *vide* "Kitâb-alfihrist," p. 164, and Reinaud, "Mémoire sur l'Inde," p. 310.

p. 339, l. 37. Khâlid ben Ṣafwân, a famous orator at the time of the first Abbâside Khalif *Alsaffâḥ*, *vide* Ibn-Khallikân, ed. Wüstenfeld, No. 808 (end), p. 315, and Ibn-Ḳutaiba, "Ma'ârif," p. 206.

p. 341 *med.* The same division is mentioned by Reinaud, "Géographie d'Aboulféda," i. p. 231. Instead of باختر *i.e. Apâkhtara*, the north is here called by the name of Âdharbaijân (Atropatene).

p. 342, l. 13. *Foundations.* I do not know the word تأسيسات as denoting some particular part of the path of the moon, but I suppose that the author means the four *Cardines*, *vide* note at p. 90, l. 44 (on p. 395).

p. 342, l. 31. *Banû-Mâriya ben Kalb and Banû-Murra ben Hammâm.* Assuming that the writing of the manuscripts is correct, I must state that these two clans are not known to me from any other source.

p. 343, l. 1. With the following description of the Lunar Stations, cf. Ideler, "Untersuchungen über den Ursprung und die Bedeutung der Sternnamen," Berlin, 1810, and Schier, "Globus coelestis cuficus," Dresden, 1866.

p. 345, l. 33. *Râmîn.* I do not know an island of this name. Perhaps the author meant the island *Râmanî*, described by Alkazwînî, "Kosmographie," i. p. 107. An island *Râmî*, in the Indian ocean, is mentioned by Yâkût, ii. p. 739.

p. 346, l. 1. *Raghad.* The reading of this name is conjectural, as I cannot prove it from other sources.

p. 351. The distance between two Stations is not, as the second number of the table, 12° 51′ 26″, would have us believe, but 12° 51′ $25\frac{5}{7}$″, as the reader will find if he examines the addition in the column of seconds. The author did not think it necessary to note the fractions in this column, but he did not disregard them in his calculation.

p. 357, l. 1. The following chapter on the projection of a globe on a plane (علم التسطيح) is purely mathematical. Dr. H. Bruns, Professor of Mathematics in the University of Berlin, has kindly undertaken the trouble of revising my translation of this chapter.

For purposes of comparison as regards both the subject-matter and the *termini technici* I refer the reader to—

L. Am. Sédillot, "Mémoire sur les instruments astronomiques des Arabes," Paris, 1844.

B. Dorn, "Drei in der Kaiserlichen öffentlichen Bibliothek zu St. Petersburg befindliche astronomische Instrumente" ("Mémoires de l'Académie," tom. ix. No. 1) 1865.

F. Woepcke, "Ueber ein in der Kgl. Bibliothek zu Berlin befindliches Astrolabium" ("Abhandlungen der Kgl. Akademie," etc.), 1858. On pp. 7–10 and 15, *vide* an explication of the stereographic projection and the graphic method, of other methods, on p. 17.

p. 357, l. 34. Abû-Ḥâmid 'Aḥmad ben Muḥammad Alṣâghânî, a famous constructor of astronomical instruments in Bagdad, died A.H. 379. *Vide* L. A. Sédillot, "Prolégomènes des tables astronomiques d'Olough-Beg," Paris, 1847, Introduction, p. 56, note 1.

p. 358, l. 1. *My book.* The author means his كتاب استيعاب الوجوه الممكنة في صنعة الاصطرلاب i.e. *the book in which all possible methods for the construction of the astrolabe are comprised*, of which there are several copies in European libraries, *e.g.* Royal Library of Berlin, Sprenger 1869; Bodleian Library, Marsh. 701.

INDEX.

A.

B.

C.

D.

E.

F.

G.

H.

I.

J.

K.

L.

M.

N.

O.

P.

R.

S.

T.

U.

V.

W.

Y.

Z.

PRINTED BY W. H. ALLEN AND CO., 13 WATERLOO PLACE, S.W.

Défauts constatés sur le document original

Contraste insuffisant ou différent, mauvaise qualité d'impression

Under-contrast or different, bad printing quality

www.ingramcontent.com/pod-product-compliance
Lightning Source LLC
Chambersburg PA
CBHW050253230426
43664CB00012B/1939